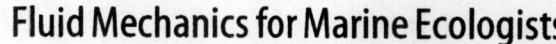

Fluid Mechanics for Marine Ecologists

Springer-Verlag Berlin Heidelberg GmbH

Stanisław R. Massel

Fluid Mechanics for Marine Ecologists

With 213 Figures and 21 Tables

 Springer

Prof. Dr. Stanisław Ryszard Massel
Australian Institute of Marine Science
Cape Ferguson, Queensland
PMB No. 3, Townsville MC, QLD 4810
Australia

The photo on the cover is a photo curtesy of Neville Coleman

Library of Congress Cataloging-in-Publication Data
Massel, Stanisław R.:
Fluid mechanics for marine ecologists / Stanisław R. Massel.
Includes bibliographical references and indexes.

Additional material to this book can be downloaded from http://extras.springer.com

ISBN 978-3-642-64305-7   ISBN 978-3-642-60209-2 (eBook)
DOI 10.1007/978-3-642-60209-2

1. Marine ecology. 2. Fluid mechanics. I. Title.
QH541.5.S3M287 1999   532'.0024577--dc21   99-34010

© Springer-Verlag Berlin Heidelberg 1999
Softcover reprint of the hardcover 1st edition 1999

Typesetting: Camera-ready by author
Cover design: E. Kirchner, Heidelberg
Production: ProduServ GmbH Verlagsservice, Berlin
SPIN:10629301   32/3020-5 4 3 2 1 0 - Printed on acid -free paper

To my wife Barbara

Preface

The future quality of life on Earth fundamentally depends on trends in world climate. The oceans play a dominant role in regulating global climate by helping to control greenhouse gases, global heat transfer and weather patterns. If we also consider the energy and other resources oceans offer, their essential importance to human existence become obvious. Ocean waters are in endless motions. Scales of these motions vary from micro–turbulence through to surface and internal waves, tides, and global currents of planetary dimensions. However, in each case the water motion satisfies the fundamental fluid mechanics and geophysical fluid mechanics principles.

Oceans are not empty, but support a large variety of organisms and plants. Many of the basic attributes of life of these creatures raise such questions as: why are the ocean's living resources distributed as they are?; how can cyclical changes overthrow the system?; why do organisms have a particular size and shape?; how do they move and reproduce?; how do they capture food and many others.

In the majority of situations, biological processes appear to be strongly influenced by the physics, while the physical processes are largely independent of the biology. However, when we consider a time scale of thousands years, the feedback from biology to physics is not so simple. The present composition of the Earth's atmosphere and the present average temperature at the surface of the Earth have been strongly influenced by living organisms.

Physical factors leading to fertile and infertile areas are very different on land and in the ocean. The nutrients required by terrestrial plants are generated nearby from decaying remains of previous generations, but decaying matter in the ocean tends to sink below the sunlit euphotic layer where plants grow. The nutrients supplied by the decay are thus unavailable for plant growth unless some physical mechanisms, such as currents and waves bring the nutrients back up to the surface.

Most biologists are relatively unfamiliar with how the mechanics of fluids (air and water) can help to explain the observed diversity of aquatic life. A successful explanation requires an approach in which, instead of putting the organisms at the centre of the picture and considering them in relationship with other organisms and the environment, it is necessary to work with marine ecosystems in which physical, chemical and biological components are equally important in defining the total system properties. At present there are many reasons for such an approach. Firstly, the fluid mechanical processes underlying some of the large-scale biological phenomena are now better understood. There have also been important advances in our ability to obtain continuous, fine-scale biological data with a coverage and resolution comparable with the best physical data. Also, the need to understand marine ecological processes influencing the greenhouse effect and other aspects of world climate, such as the flux of carbon dioxide from the atmosphere into surface waters and on down into the deep ocean, is becoming more urgent. Finally, the increased understanding of fundamental processes is beginning to influence the management of the ocean's living resources.

For all of these reasons, there is a need to bring the attention of marine ecologists, oceanographers and marine engineers to how mechanics of ocean waters interacts, influences and constrains the ocean's biological life. This proposed book is dedicated to such an effort. The main objective of this book is to make physical processes intelligible to biologists with a modicum of mathematics, and make the nature of life in the sea understandable to oceanographers and engineers. The book is directed to a diverse audience, such as marine biologists and ecologists who would like to know more about the mechanisms by which plants and animals live in the oceans, as well as physical oceanographers, engineers and fluid dynamicists who would like to see how their knowledge of mechanics can be used to study marine ecology. Above all, I hope the book will be interesting to readers who want to think about ocean water dynamics and its biological implications.

The book is designed for graduate students, PhD students, postdoctoral students and scientists – biologists, ecologists, physicists and engineers working on ocean environmental problems. Because of the large variety in the potential audience, the presentation of the physical and biological subjects will be given from slightly different viewpoints. The presentation of the marine physics will emphasize the recent development in this field, although given in a simple way. On the other hand, the biological part will be presented through many examples illustrating the interaction of ocean fluid mechanics with biological life.

In organizing the material to be included, I was faced with the dilemma to use either physics or biology as a framework. However, the physical phenomena, at least from my perspective, flow more easily in an orderly and useful sequence. Therefore the physics of oceanic flows was chosen provide the skeleton for the book.

The text is divided into three parts. In Part I (Chapters 1-2), the classical fluid mechanics phenomena such as laminar and turbulent flow, boundary layers, and forces induced by flow are examined. All these phenomena depend on the properties of the fluid itself and the character of fluid motion.

Part II (Chapters 3-9) of the book deals with large scale flows, such as waves, large ocean currents, or tides which are beyond the scope of classic fluid mechanics. However, all these phenomena play an important role for ecology of marine organisms.

Finally in Part III (Chapters 10-15), the link between hydrodynamics of ocean flows and marine ecology is demonstrated through examples of selected well-established phenomena and processes.

Only the final versions of mathematical formulas are given in the main text. More lengthy derivations of basic equations are given in the Appendices. In particular, Appendix A provides a list of symbols. The diversity of topics covered in this book leads to problems with nomenclature. There is a limited number of Roman and Greek letters, and certain symbols are used with different meanings in different fields. I have chosen, where possible, to use the symbols that are conventional for the particular field being discussed. As a result, several symbols are used in multiple ways, but each symbol is carefully redefined when it is given a new meaning.

In Appendix B the basic definition of the System International of Unit is given. The basic as well as derived units are discussed and relationships between various units of the SI system are given.

The mathematical fundamentals are collected in Appendix C. In particular, the derivation of the mass conservation equation as well as the equations of fluid motion are given. Also, the formulas of small amplitude wave theory which are used extensively in the text are developed.

Some figures in the book have been adapted from previously published figures, and this is indicated in the legends. My thanks are due to all those who gave permission to reproduce their original figures.

I apologize for the inevitable errors which may occur in this book, despite my efforts to eliminate them. Please bring these errors to my attention.

I would like to express my thanks to the Australian Institute of Marine Science for support during the preparation of this book. I have benefited enormously from the criticism of those who have read the book in manuscript and made helpful suggestions. In particular, I am grateful to Professor John Fenton from Monash University (Melbourne) for his precise commentary and very valuable suggestions. Thanks are also due to Richard Brinkman from the Australian Institute of Marine Science (Townsville) for his very accurate proof-reading and repeated correction of the text; their suggestions materially improved its quality. I also wish to acknowledge the stimulating discussions provided by many colleagues and members of the staff of the Institute.

Special thanks goes to my wife Barbara Massel for the enormous amount of word processing and correcting, and the final professional editing the text, and to my son Bartosz Massel for preparing all figures for this book. I thank them also for their support and encouragement during the writing.

Townsville, May, 1999

Contents

Part I

Basic Fluid Mechanics

1 Introducing Sea Water

1.1 Water on Earth

The general term **fluid** describes both **gases** and **liquids**. They both do not have a definite shape and vessels are needed to contain them. A **liquid** has a definite volume and a free surface, which is horizontal when the fluid is at rest. In contrast, **gases** do not have a definite volume and will expand to fill a container enclosing them. **Solids** are substances which are considered to have both a definite volume and shape.

The most important liquid on Earth is water. Water is a ubiquitous, life-sustaining substance covering 71% of the Earth surface. Of the total Earth's water content, some 97.2% is contained in the oceans, 2.15% is stored in ice sheets and glaciers, 0.62% is a ground water, and only 0.03% flows through rivers, streams and fresh-water lakes (Strahler and Strahler, 1992). For major portions of the Atlantic, Pacific and Indian oceans, the average depth is nearly 4 kilometres, but it is the surface water, together with the very small amount of ground water, that supports all life on the lands and in the oceans.

Depending on the temperature, water can exist as a liquid, a gas (water vapour) or a solid (ice). Although existing in all three phases, the total volume of water in the atmosphere, on the land, and in the oceans remains essentially constant over time. However, waters of the oceans, atmosphere, and lands move in a great and continuous cycle known as the **hydrologic cycle** (see Fig. 1.1).

Evaporation from the oceans, which are the basic reservoirs of free water, is approximately 419,000 km^3/year, while evaporation from soil, plants and water on the continents is only about 69,000 km^3/year. The total quantity of the evaporated water, 488,000 km^3/year, must be returned annually to a liquid or solid state through precipitation over the oceans and continents. Precipitation over continents is about 37,000 km^3/year larger than evaporation. This excess of quantity flows over or under the ground to reach the sea. Because of the cycling of water, freshwater is a 'renewable' resource in the sense that supplies of freshwater are constantly renewed by rainfall over the land and ocean.

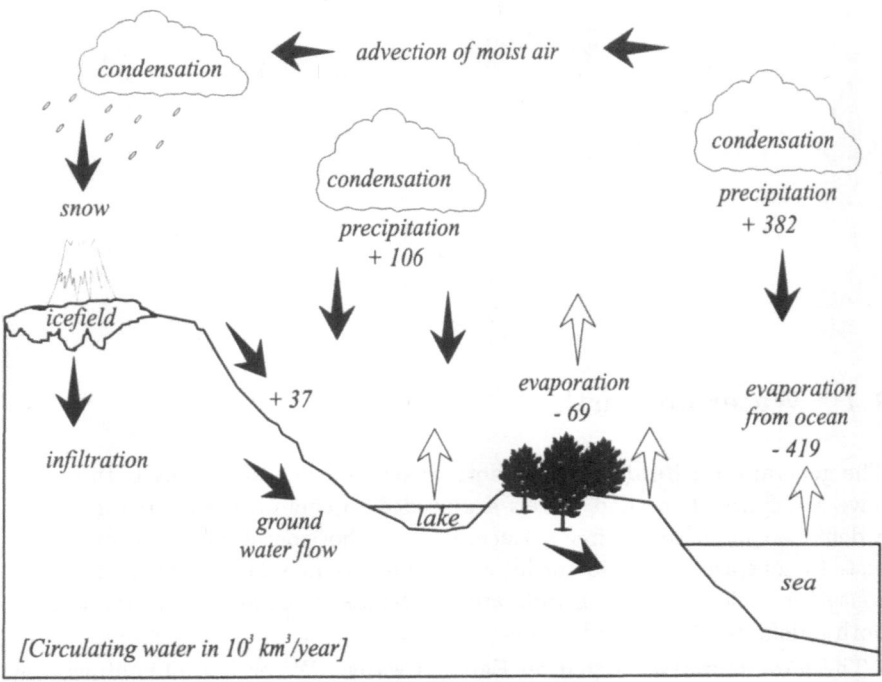

Fig. 1.1: The principal pathways of the water circuit (adapted from Strahler and Strahler, 1992)

The hydrologic cycle is a pronounced evidence of the strong link between the atmosphere and ocean. The input of energy from the atmosphere drives water motion in the form of waves and currents. Moreover, through the ocean surface, oxygen and carbon dioxide – gases vital for the growth of marine organisms – enter the ocean from the atmosphere. Contrary to the upper ocean layer, at great ocean depths water motion is extremely slow and water temperature remains uniform and low. Interaction between atmosphere and ocean is a topic discussed in some detail in Part II of this book.

1.2   Physical and Chemical Properties of Sea Water

1.2.1   Sea Water Density and Related Measures

Water has many unique chemical and physical properties. Unlike most substances which contract when frozen, water expands, allowing ice to float on the surface of water basins. It is well known that water forms a necessary constituent of the cells of all animals and plant tissues. Most of the water present in living organisms acts as an irrigant, distributing nutrients and removing waste products. Because of its unusual physical properties, a quite different set of environmental conditions is presented to amoebae and fish that both live in the same waters (Franks, 1972; Gordon *et al.*, 1992).

The principal sea water state parameter is **density**, or mass per unit volume. The density of water controls many fundamental processes in the ocean, e.g. the hydrostatic equilibrium or motion of water particles and propagation of internal waves, turbulence and mixing in water column, heat transfer, concentration of plankton and sediment transport, locomotion of marine species and many others. Even small alterations in the density of sea water result in great changes in water flow and its thermal and chemical status.

Density is normally symbolized by ρ and in the SI system it is expressed in kilograms per cubic meter (kg/m^3). For example, the density of pure water is approximately 1000 kg/m^3 being 770 to 890 times that of air at sea level. Detailed information on unit systems used in the physics of fluids is given in Appendix B.

Unlike measuring salinity, temperature and pressure, there are no practical methods of measuring the density of sea water *in situ*. The density of sea water is usually determined through its dependence on pressure, p, temperature, T, and salinity, S. Pressure, p, has an insignificant effect on the density of water for most applications, unless one is dealing with water at great depths within the ocean. For example, at a depth of 1 km the density of water is only 0.5% greater than at the surface, and at the deepest point in the earth's oceans, the Marianas Trench, water is about 6% denser than water at the surface, assuming that the temperature and salinity are constant (Denny, 1993). The density is much more dependent on temperature, generally decreasing as the temperature increases. However, the density of freshwater is not a monotonic function of temperature; water density reaches a maximum at temperature of 3.98°C under normal atmospheric pressure (see Fig. 1.2). Normal atmospheric pressure at sea level, called the *normal atmosphere*, is defined to be 1.01325 $\times 10^5$ Pa at 15° C, in which 1 Pa is equal to 1 N/m^2 (see Appendix B). This anomalous dependence of the density of freshwater on temperature is a result of the specific molecular structure of water (Franks, 1972; Harvey, 1985; Dera, 1992).

Salinity is another factor which influences water density. Away from coasts, the salinity of ocean water varies from 32 to 37 ppm (parts per million). The variations in salinity result from the differences in the relative rates of precipitation and evaporation from the surface of the ocean. When the temperature, T, and pressure, p, are constant, density of sea water increases with salinity and a relative change in the density of sea water is roughly proportional to a change of salinity. Ocean water does not demonstrate the anomalous thermal expansion of freshwater. At constant pressure and at salinity greater than $\sim$ 24.7ppm, the density of sea water increases monotonically with decreasing temperature right down to the freezing point, as shown in Fig. 1.2. At 0°C, freshwater turns into ice, and its density abruptly decreases from 999 kg/m^3

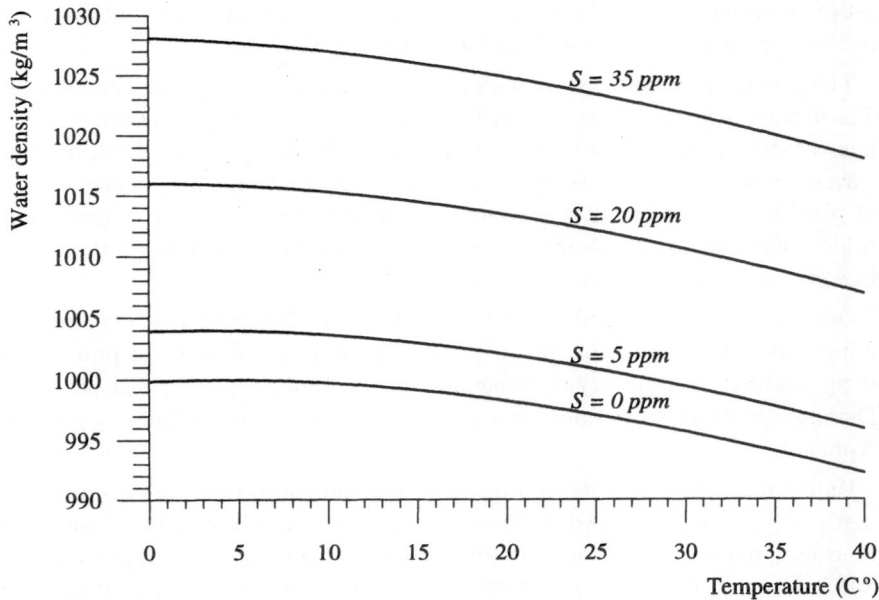

Fig. 1.2: The density of fresh and sea water as a function of temperature and salinity

to about 917 kg/m³. When sea water freezes, its salt is extruded and sea ice floats like freshwater ice at the surface.

A difference of 1 ppm in salinity has an effect on the density of sea water which is about five times greater than the change caused by a difference of 1°C in temperature. Large scale density structure of the ocean is dominated by variations in temperature while salinity differences have very appreciable effects on smaller scale motions. The densest waters are formed under the ice of the Weddell Sea in the Antarctic, in the North Atlantic off Greenland and in the Norwegian Sea.

The general dependence of the sea water density $\rho_w(S, T, p)$ on the salinity, S, temperature, T, and pressure, p, is called the **equation of state** for sea water. There have been many attempts to establish the relationship $\rho_w(S, T, p)$ in the past (for more details see Dera, 1992) and the modern algorithm of determination of density for given salinity, S, temperature, T, and pressure, p, was developed by the international group of experts on standards (UNESCO, 1987). The expression for $\rho_w(S, T, p)$ is rather complicated and it will not be given here.

For convenience, in Appendix D a computer program, D11, for calculating density of water for arbitrary values of temperature, T, and salinity, S, is given. Here, only some values of $\rho_w(S, T, p)$, frequently used in the book, are given:

$$\rho_w(0, 10°, p_{norm}) = 999.702 \text{ kg/m}^3;$$
$$\rho_w(35, 10°, p_{norm}) = 1026.952 \text{ kg/m}^3;$$
$$\rho_w(0, 20°, p_{norm}) = 998.206 \text{ kg/m}^3;$$
$$\rho_w(35, 20°, p_{norm}) = 1024.763 \text{ kg/m}^3,$$

in which p_{norm} denotes normal atmosphere.

In place of the water density, the specific weight of water will be sometimes used in this book. This quantity, usually symbolized by γ (gamma), is equal to the product of density and gravitational acceleration, g, so $\gamma = \rho_w g$. The specific weight is expressed in newtons per cubic metre (N/m^3); for details see Appendix B.

1.2.2 Sea Water Viscosity

Viscosity is a property which is a measure of a fluid's resistance to 'deformation' during motion. Within a fluid, momentum of rapidly moving particles is exchanged with the momentum of relatively slower particles. Those exchanges produce a shearing stress. For simplicity, let us imagine two horizontal layers of fluid slipping one over another (Fig. 1.3). The lower layer is moving with velocity u while the upper has slightly greater velocity $(u + \Delta u)$. As a result, the fluid between these layers shears with a variation in flow rate.

It is reasonable to assume that the unit force, F, or stress, τ (equal to F/surface area), needed to create the shear flow is proportional to the vertical shear of velocity $\Delta u/\Delta z$. If $\Delta u/\Delta z = 0$, horizontal velocity is uniform over the water column and shear stress $\tau = 0$. In general we can write for shear stress τ:

$$\tau = \mu \frac{\Delta u}{\Delta z}, \tag{1.1}$$

or more precisely when $\lim \Delta z \to 0$, i.e. $\lim_{\Delta z \to 0} \Delta u/\Delta z = du/dz$ and

$$\tau = \mu \frac{du}{dz}, \tag{1.2}$$

in which τ is shear stress acting on the horizontal (x, y) plane. The symbol du/dz is a differentiation operator expressing the rate of change with z. It was introduced by German philosopher and mathematician Gottfried Leibniz (1648–1716) and it should not be regarded as a ratio but as an operator. The proportionality coefficient, μ, between the shear stress, τ, and vertical shear of velocity, du/dz, is the **coefficient of dynamic molecular viscosity**. It has units of newton × second per square meter (N s/m^2).

A fluid which shows this direct proportionality between the applied shear stress and the resulting rate of deformation is called a 'Newtonian' fluid. Many biological materials, such as blood, synovial fluid, mucus of various consistencies, can not be treated as Newtonian fluids. Some of them have a memory of previous shape and elasticity. However, description of such fluids is a subject

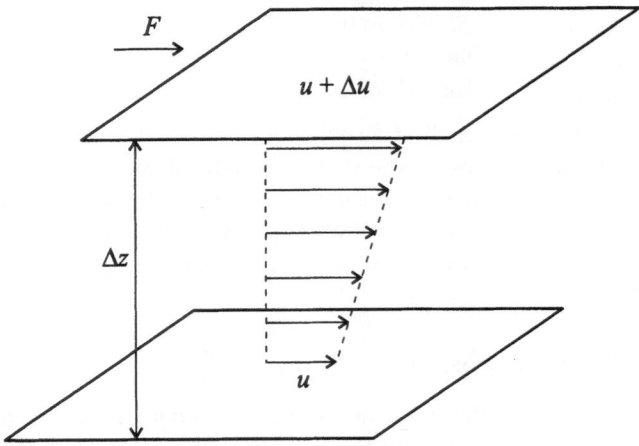

Fig. 1.3: Two layers of fluid slipping one over another in the shear flow

of another branch of fluid mechanics, so-called rheology, and is far beyond the scope of this book.

The dynamic molecular viscosity of sea water depends on temperature, T, and salinity, S, as follows from Table 1.1. Colder water is more 'resistant' to motion than warmer water. The dependence of μ on salinity is weaker than its dependence on temperature.

As force on an element of fluid varies like μ but the mass of that element varies like ρ, the acceleration and hence the velocity field is determined by the ratio μ/ρ, known as the **coefficient of kinematic viscosity**, ν:

$$\nu = \frac{\mu}{\rho}, \tag{1.3}$$

where ν has units of m^2/s. For example, the kinematic viscosity for sea water of salinity $S = 35$ ppm and of temperature $T = 20°$C is 1.064×10^{-6} m^2/s. It is interesting to note that there is almost no liquid with viscosity lower than that of water, and that $\nu_{\mathrm{air}} \approx 15 \times \nu_{\mathrm{water}}$. The viscosity of water is of particular significance in the regions close to solid boundaries and this subject will be discussed in detail in Sect. 2.5.

Both coefficients, μ and ν, are physical properties of fluid, independent of fluid motion. When a fluid is completely at rest or is moving very slowly, the rate of diffusion of momentum is essentially determined by molecular motion. Eddies are small and the velocity shear is great, and the internal resistance of the water due to molecular viscosity smooths out the gradients in velocity. This smoothing of the flow by viscosity is the way the energy in turbulence is finally converted to heat and dissipated.

Table 1.1: The coefficient of dynamic molecular viscosity, μ (x$10^{-3}\times$ N s m^{-2}), for selected temperatures, T, and salinity, S, (adapted from Dera, 1992)

Temperature	Salinity (ppm)			
[°C]	0	5	20	35
0	1.79	1.80	1.84	1.88
4	1.57	1.58	1.62	1.66
10	1.31	1.32	1.36	1.40
16	1.11	1.12	1.16	1.20
20	1.01	1.02	1.05	1.09
24	0.91	0.93	0.96	0.99
30	0.80	0.81	0.83	0.87

However, as soon as the fluid is stirred and the elements of fluid much greater than molecules mix, the coefficient μ increases considerably. In the oceans and atmosphere, eddies and turbulent motions in the flow can be so effective in moving particles among themselves that the effects of molecular diffusion are overwhelmed. For such situations, the coefficient of kinematic viscosity, ν, is replaced by the **coefficient of turbulent viscosity**, A, which is several hundred to many thousand times larger. By analogy to molecular exchange, the turbulent viscosity coefficients result from the hypothesis that the turbulent momentum flux is proportional to the averaged turbulent flow velocity. Here we only note that the estimates of the turbulent viscosity coefficient in the ocean vary enormously, from 10 m^2/s to 10^4 m^2/s in the horizontal plane, and from 10^{-4} m^2/s to 10^{-1} m^2/s in the vertical plane (see Sect. 7.3). The smaller values are obtained, for example, from the rate of spread of dye spots. The larger values can be found for the horizontal motion on an oceanic scale such as the diffusion of mass or momentum associated with the meandering flow of the Gulf Stream.

1.2.3   Surface Tension

Within a body of water, a water molecule is attracted by the molecules surrounding it on all sides, but molecules at the surface are only attracted by those beneath them. Therefore, there is net force downwards which creates tension on the water surface. Because the surface layer is under tension, any change

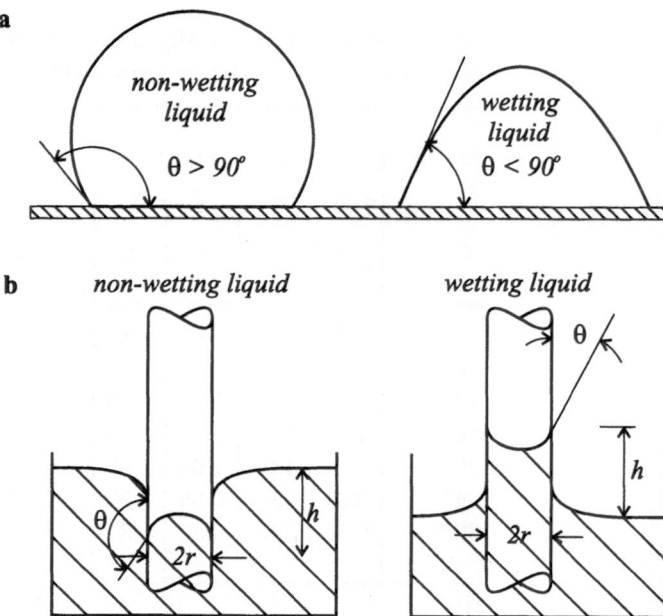

Fig. 1.4: Contact-angle effects at liquid-gas-solid interface and change of capillary rise for 'wetting' and 'non-wetting' liquid (adapted from White, 1994)

in shape which would add more area is resisted. In other words, the surface tension tries to minimize the area of surface films. This explains why water drops or submerged air bubbles are almost perfectly spherical as a sphere has less surface area per unit of volume than other shapes.

The surface tension, which lies tangential to the air-water interface, is symbolized by σ and has units of N/m, a force per unit length across which it acts. Existence of surface tension produces some pressure difference across the interface with a second liquid or gas, *i.e.* a droplet of water in air or air bubble under water, the pressure being higher on the concave side. This increase in pressure is given by (White, 1994):

$$\Delta p = \frac{\sigma}{r},\tag{1.4}$$

where Δp is the increase in pressure (in N/m^2) due to surface tension and r is the radius of droplet or bubble. The ability of surface tension to separate water from air is used by several bugs and beetles to breathe underwater. This and other biological consequences of the water properties are discussed in Parts II and III of the book.

When a liquid droplet touches a solid surface, the surface tension determines whether it forms beads or spreads out on the surface. The measure of this spreading is the contact angle θ (see Fig. 1.4a). If the contact angle is less than 90°, the liquid is 'wetting'. If the angle $\theta > 90°$, the liquid is 'non-wetting'. Water is extremely wetting to a clean glass surface. On the other hand, mercury is non-wetting to this surface with contact angle $\theta = 130°$.

Surface tension is the mechanism which causes water to rise in a capillary tube, and is of great biological importance. This phenomenon is called **capillarity**. The rise is positive (capillary rise) if liquids are wetting and negative (capillary depression) if liquids are non-wetting (see Fig. 1.4b). For wetting liquid the meniscus of the surface of liquid in the tube is concave, and convex for non-wetting liquid. The height of capillary rise can be found from the formula (White, 1994):

$$h = \frac{2\sigma \cos \theta}{\rho g r}, \tag{1.5}$$

in which r is the radius of the tube. The other symbols have been explained earlier in this section.

1.2.4   Inorganic Salts

Most of the water in the oceans contains dissolved salts and gases. Virtually every element is present in a dissolved state in sea water. However, there is a group of ions which make up the majority of the mass (of about 99%) of dissolved inorganic salts in sea water. They include: Na^+, K^+, Mg^{2+}, Cl^-, Br^-, SO_4^{2-}, Ca^{2+} and HCO_3 (see Table 1.2). Of the minor constituents of sea water which do not appear in Table 1.2, nutrient salts are vitally important for organic life in the ocean.

Table 1.2: The major ionic constituents dissolved in sea water (adapted from Horne, 1969, and Dera, 1992)

Sea salt	Percentage (%)
Chloride (Cl^-)	55.04
Sodium (Na^+)	30.61
Sulfate (SO_4^{2-})	7.68
Magnesium (Mg^{2+})	3.69
Calcium (Ca^{2+})	1.16
Potassium (K^+)	1.10
Others	0.72

1.2.5   Dissolved Gases

As well as salts, gases from the atmosphere are dissolved in sea water. There
are two gases, oxygen and carbon dioxide, which are of special interest due
to their involvement in biological and geochemical processes occurring in the
ocean. Oxygen enters water by diffusion at the atmosphere-ocean interface or
at the surface of air bubbles; it is also released into the ocean by the process
of photosynthesis (Harvey, 1985):

$$6\,CO_2 + 6\,H_2O \overset{\text{sunlight}}{\Longrightarrow} C_6H_{12}O_6 + 6\,O_2. \tag{1.6}$$

Oxygen is also utilized by organic respiration and decomposition in the break-
down of carbohydrates.

As temperature rises, the gas-absorbing capacity of water decreases rapidly
and reaches zero at 100°C. Under normal atmospheric pressure and a temperature of 20°C, water contains about 2% (by volume) dissolved air. The
dissolved air contains between 33% to 35% oxygen, O_2, depending on temperature. At any given temperature, the highest possible amount of dissolved
oxygen is called 'oxygen saturation'.

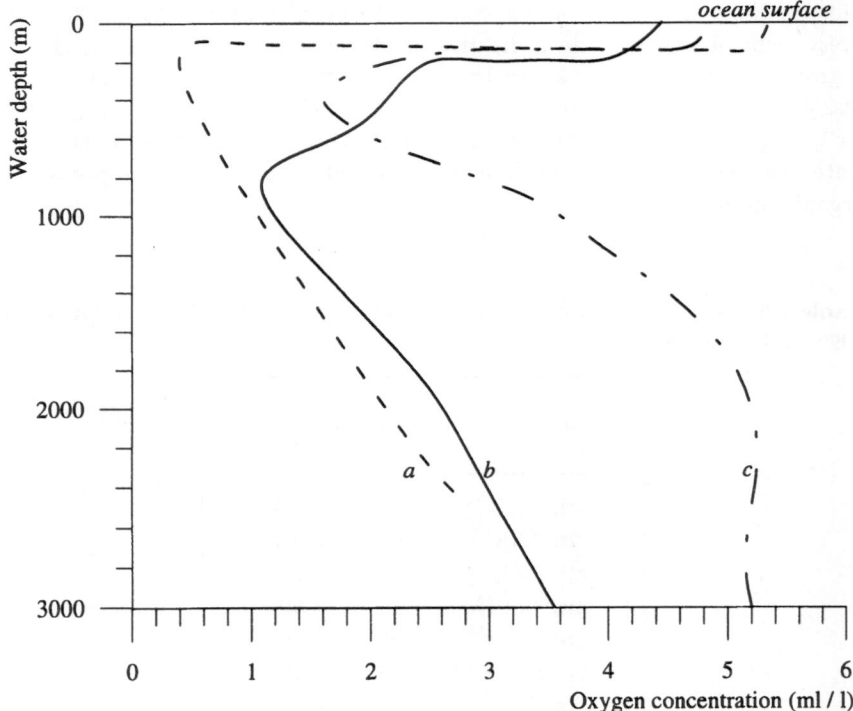

Fig. 1.5: Typical oxygen profiles: *a* Pacific Ocean, *b* Indian Ocean and *c* Atlantic
Ocean (adapted from Horne, 1969)

Usually, the oxygen concentration at the ocean surface, in the photic zone, is very close to its saturation level, and the concentration varies from about 4.5 ml per litre in tropical latitudes to more than 8 ml per litre in polar seas (Harvey, 1985). Oxygen concentration profiles with depth are very complex and dependent upon the location. Typical vertical distributions of dissolved oxygen in the oceans are given in Fig. 1.5.

Oceans provide a great reservoir for carbon dioxide. As follows from Eq. 1.6, carbon dioxide plays the reverse role to oxygen in the photosynthesis process and is utilized by respiration. The distribution of CO_2 in the ocean is controlled mainly by formation of carbonic acid through reaction with the water. This acid and the carbonate ions brought into the ocean by rivers (for example, flowing over limestone $CaCO_3$) react as follows (Harvey, 1985):

$$CO_2 + H_2O \rightleftharpoons H_2CO_3 \rightleftharpoons H^+ + HCO_3^- \rightleftharpoons 2H^+ + CO_3^{--}. \tag{1.7}$$

These reactions can produce supersaturation of the water with respect to the carbonate ion. This results in the removal of calcium carbonate by organisms whose remains eventually settle to the ocean floor. On the other hand, under-saturation will dissolve such sediments, reducing the calcium carbonate content of sediment, particularly at great water depths.

1.2.6 Concentration of Suspended Particles and Yellow Substances

Large varieties of suspended solid particles are present in the oceans and seas. This suspended matter comprises organic particles of biological origin and in-organic particles of mineralogical origin, and the concentration of suspended mass varies with locality as shown in Table 1.3.

Organic suspended material includes bacteria, fungi, phytoplankton and zoo-plankton. Bacteria and colloids are the smallest suspended particles in the sea, being less than 1 micron in size. The concentration of organic substances in sea water ranges from 0.05 to 0.5 mg/dm^3 for water in clean open oceans to about 5 mg/dm^3 for waters in enclosed seas and estuarine areas.

Many organic substances are carried into the sea by rivers, both in solution and in suspension. Heavy substances in suspension fall to the bottom close to shore and are subjected to further possible chemical reactions. As the sea and river waters have different ionic compositions, the substances dissolved in the river water are flocculated to a high degree on the contact with sea water to form colloids and fine suspensions. The coagulation process is a function of salinity and it affects the largest flocs (Gibbs and Konwar, 1986; Wolanski and Gibbs, 1995). Observations at the Amazon River estuary indicate that the mean diameter of suspended particles supplied by the river progressively increases seaward until a maximum is reached at about 10 ppm salinity. This results from the coagulation of small particles as they encounter ocean water. Seaward of 10 ppm salinity in the surface waters, the mean diameter of suspended particles decreases due to the settling of larger flocs. The maximum floc sizes are of the

Table 1.3: The concentration of suspended organic and inorganic mass in different parts of the ocean (adapted from Dera, 1992)

Water basin	Total dry mass of suspended matter in mg/dcm^3	% of organic mass
Littoral zone of Pacific	10.50	62
Pacific Ocean	3.80	29
Northeast Pacific	0.45 – 1.0	–
North Sea	6.00	27
Bering Sea	2.00 – 4.0	–
Baltic Sea	0.20 – 12.0	–
Average for all oceans	0.80 – 2.5	20 – 60

order of 100 μm. The flocs are structurally weak and easily destroyed by tidal turbulence (Wolanski and Gibbs, 1995).

There are many dissolved organic substances in the ocean which have a substantial effect on the absorption of light in water. Since they strongly absorb violet, these substances appear yellow in daylight and therefore they are called 'yellow substances' (Dera, 1992). The nature of yellow substances in sea water is very complex and it is not yet completely understood. Some of these substances are found in an environment where free hydrocarbons and amino acids are present (melanin type yellow substances).

Special types of organic, light-absorbing substances in sea water are the pigments contained in phytoplankton cells. They include chlorophyll a, b, c and d, carotenes, xanthophylls and other pigments essential to the photosynthesis of organic mass in sea water (Dera, 1992).

Inorganic suspended material is predominantly derived from the weathering and erosion of continental rocks. The products of weathering are carried to the ocean margins by rivers. The main dissolved products, found as constituents of river water, are calcium, sodium, potassium and magnesium cations and bicarbonate anions. Despite their immense significance, rivers are not the only means of transporting inorganic material to the ocean. Ice, wind, coastal erosion and volcanic eruptions also play a similar transportation role that varies in importance from one part of the world to another. The composition of the mean annual transfer of sedimentary materials to the ocean (in 10^9 tonnes per year) can be listed as follows: rivers 18.3, wind-blown dust 0.6, ice 2.0, coastal erosion 0.25, and volcanic eruption 0.15 (von Arx, 1977). Volcanic debris makes a small contribution to total amount of inorganic material transferred to the oceans, however, the input of volcanic material may be locally overwhelming.

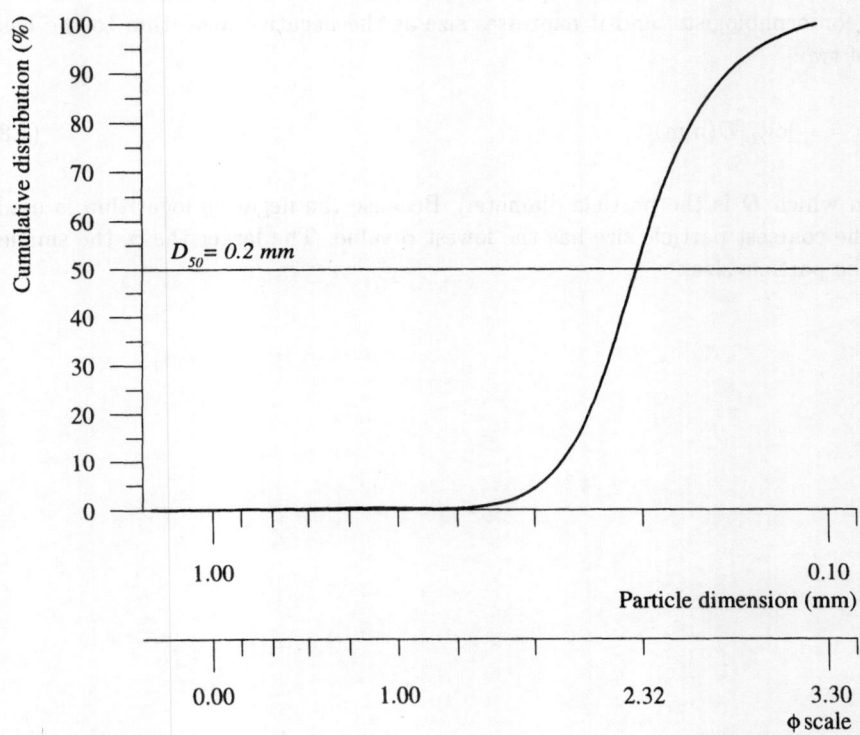

Fig. 1.6: Suspended particle distribution in bottom layer of coastal zone of Baltic Sea (based on data from Antsyferov and Kosyan, 1986)

For example, in 1815 the Indonesian volcano of Tambora erupted around 40×10^9 tonnes of volcanic material, much of which must have eventually been deposited in the oceans.

In general, in clean ocean waters, the number of suspended particles greater than 1 μm is of the order of 1000 per cm^3 of water. However, in polluted seas or rivers, the concentration can be as high as 10^7 particles per cm^3 of water (Dera, 1992).

When we examine particles under a microscope, only the projection of a randomly oriented particle is seen. To quantify the particle dimensions, the so-called 'Ferret diameter', 'Martin diameter' or 'Coulter diameter' techniques are used (Dera, 1992). However, the most objective technique to characterize the concentration and dimensions of suspended particles in the ocean is the use of the suspended particle distribution.

Fig. 1.6 presents the suspended particle size distribution in the bottom layer of the coastal zone (water depth 6 m) of the Baltic Sea. On the horizontal axis there are two scales for particle dimensions. Apart from the dimensions in (mm), the so called ϕ scale (phi scale) is used. This scale is commonly used by geomorphologists, and it expresses size as the negative logarithm to the base of two:

$$\phi = -\log_2\left[D(\mathrm{mm})\right].\tag{1.8}$$

in which D is the particle diameter. Because the negative logarithm is used, the coarsest particle size has the lowest ϕ value. The larger the ϕ, the smaller the particle size.

2 Water at Rest and in Motion

2.1 Introduction

This chapter has been divided into two main parts: hydrostatics and hydrodynamics. **Hydrostatics** is the study of water (hydro) at rest (static), and includes the investigation of distribution of pressure within the water body and buoyancy forces. Although both pressure and buoyancy are acting no matter whether water is at rest or in motion, it is useful to explain them first for the simpler case of water at rest. **Hydrodynamics** is the study of water in motion. As there are many types of water motion in the ocean, it is not possible to discuss them all in one chapter. However, some properties of the sea water motion are common and can be discussed independently of specific types of water motion. This is particularly true for such properties as steady and unsteady flow, rotational and irrotational motion, and laminar and turbulent flow. The basis of all these types of sea water motion is covered in this chapter. The following chapters are devoted to particular types of motions, such as surface waves, tides, currents and internal waves.

The significance of some physical properties of water and its motion reveal themselves at specific conditions of motion. However, prior to starting our discussion on the physics of water at rest and in motion, we must first establish a coordinate system through which objects can be located in space and water motion can be described.

2.1.1 Coordinates System

Throughout this book the Cartesian system of coordinates will be used (Fig. 2.1a). This system was invented by a French philosopher and mathematician René Descartes, as a result of a dream on the night of 10 November 1619 (Barber, 1969). For this coordinate system, the axes Ox, Oy, Oz are mutually perpendicular, and the x and y axes lie in a horizontal plane, leaving the z axis to point vertically. However, in some cases (for example, when discussing depths in the ocean) it will be convenient to point the z axis downward.

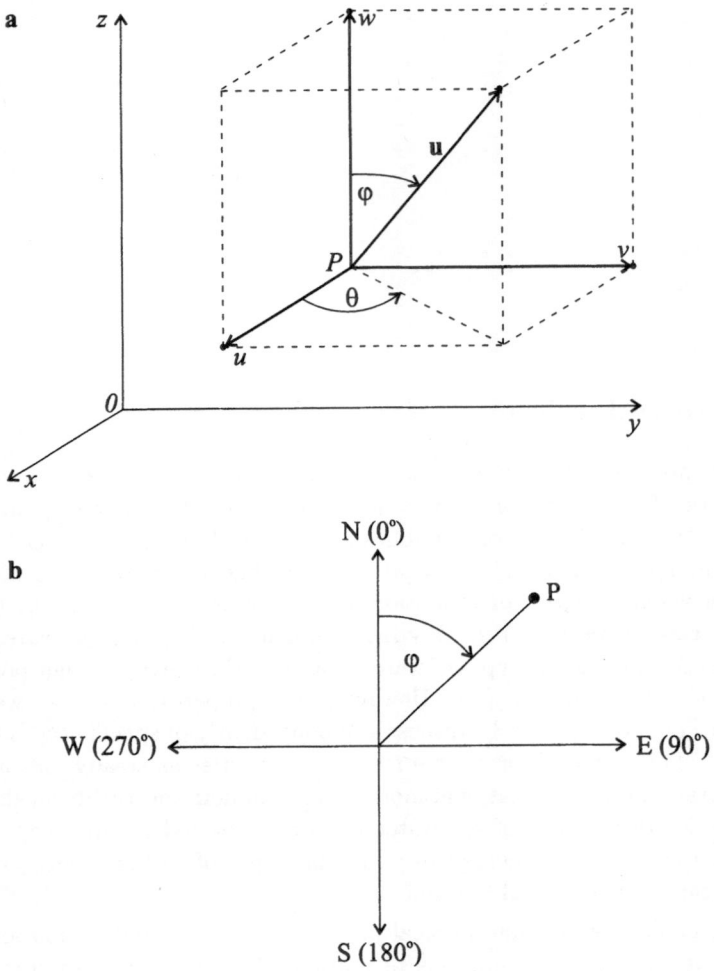

Fig. 2.1: Coordinates systems: **a** Cartesian, **b** geographic

The vector of velocity of fluid particles, $\boldsymbol{u}$, in Fig. 2.1 is inclined at an angle φ from the vertical axis and its projection on a plane O, x, y forms angle θ with the x axis. By the convention, positive angle θ increases in a counterclockwise direction, relative to the x axis, the reverse of the geographic convention. The components of vector $\boldsymbol{u}$ along x, y and z axes are u, v and w, respectively. Thus, $\boldsymbol{u} = \vec{u} = (u, v, w)$. A full definition of vector notations used in this book is given in Appendix C.1.

In instances where the Cartesian system is inappropriate, the conventional geographic system is used (Fig. 2.1b). Also for the specific applications it is useful to use the polar and spherical coordinate systems rather than the Cartesian one. An example of such application is given in Appendix C.

2.2 Water at Rest: Hydrostatics

2.2.1 Pressure Distribution in Water

The forces acting on any portion of fluid can be divided into volume forces due to external sources and surface forces exerted across the boundary by the surrounding matter. These forces must balance if the fluid is to remain at rest. Under the assumption of a uniform volume under only the influence of gravity, the necessary condition for equilibrium becomes (Batchelor, 1967):

$$\frac{dp}{dz} = -\rho_w(z)g, \tag{2.1}$$

in which dp/dz is the rate of change of pressure, and $\rho_w(z)$ is the water density which may be function of z. Thus, the pressure p is:

$$p(z) = p_a - \int_0^z \rho_w(z)g\,dz, \tag{2.2}$$

in which p_a is the surface atmospheric pressure.

In Sect. 1.2.1 it has been shown that water density increases only slightly with pressure (or water depth). Therefore, for simplicity of hydrostatic calculations we will assume that the density of water remains constant, $i.e.$ $\rho_w(z) = \rho_w$ and:

$$p(z) = p_a - \rho_w g z. \tag{2.3}$$

Water pressure is a scalar quantity acting at right angles to an object's surface, such as the sea bottom, a fish or a breakwater wall (Fig. 2.2a). In particular, pressure at a point P in Fig. 2.2b becomes:

$$p = p_a - \rho_w g z = p_a + \rho_w g h, \tag{2.4}$$

in which $z = -h$ is the submergence of point P. It is often convenient to express pressure by dividing by density and gravitational acceleration to give a quantity which has units of length and is the equivalent elevation, known as $head$. Equation (2.4) gives:

$$\frac{p}{\rho_w g} = \frac{p_a}{\rho_w g} + h. \tag{2.5}$$

The quantity $p/\rho_w g$ is known as the $pressure\ head$, which is seen to be equal to the atmospheric pressure head plus the submergence.

In hydrostatics the concept of $piezometric\ head$, H, is introduced, which is the sum of the pressure head plus the elevation:

$$H = \frac{p}{\rho_w g} + z. \tag{2.6}$$

a

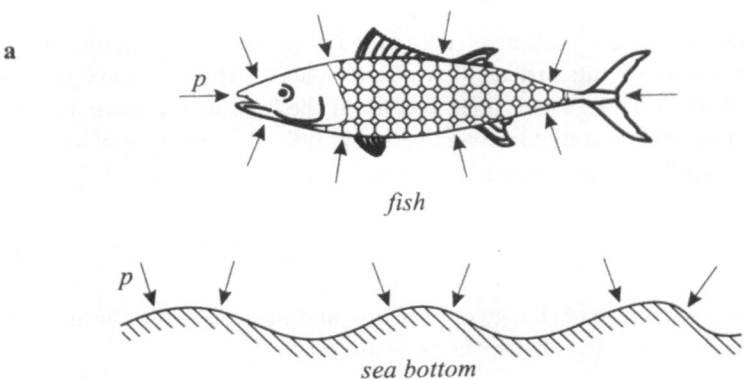

b

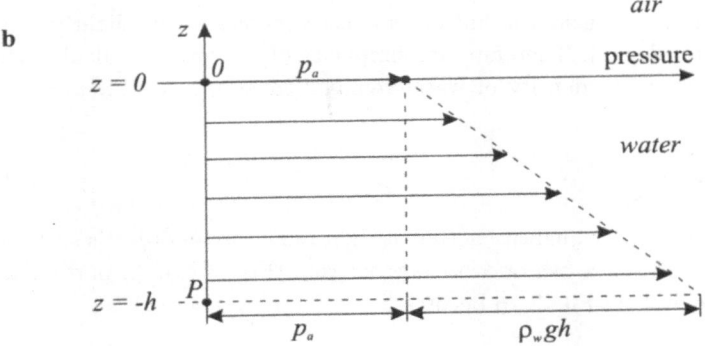

c

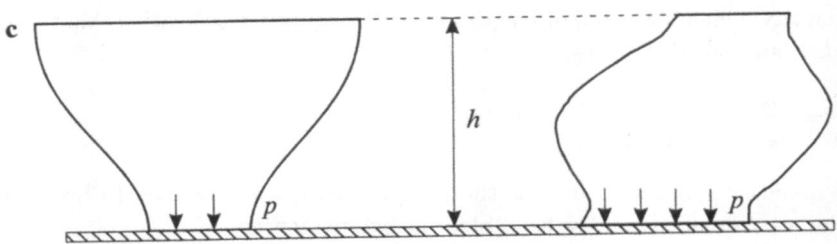

Fig. 2.2: Pressure distribution: **a** around body and sea bottom, **b** variation with water depth, **c** at the bottom of tanks of different shapes

Substituting Eq. (2.5) and using the fact that $z = -h$, we find that the piezo-metric head, H, is a constant throughout the fluid, $i.e.$:

$$H = \frac{p_a}{\rho_w g}. \tag{2.7}$$

If water is kept in containers being filled to the same height, the pressure on the base is the same and independent on shape of the container (Fig. 2.2c). The pressure at the container's base is controlled by the depth, h, not by the weight of water in container as given by Eq. (2.4). The resultant force acting on the container's base is a product of the corresponding pressure, p, and base surface A:

$$F = p\,A. \tag{2.8}$$

Sometimes the pressure at a given point in the water column is calculated relative to the atmospheric pressure. Therefore, pressure at point P simply is $p = \rho_w g h$.

From Eq. (2.4) it can be deduced that the pressure difference between two points of different submergence, h_1 and h_2 ($h_1 > h_2$), is:

$$p(h_1) - p(h_2) = \rho_w g(h_1 - h_2). \tag{2.9}$$

2.2.2   Buoyancy

The forces on submerged objects determine whether they sink, float or remain where they are. These three stages are usually categorized as negative, positive or neutral **buoyancy**. To examine the nature of the buoyancy force, let us consider a vertical submerged cylinder situated in a large volume of water (Fig. 2.3). Using Eq. (2.4), and assuming that $p_a = 0$, the vertical forces on cylinder top and bottom are respectively:

$$\left.\begin{array}{ll} \text{top}: & F_t = \pi\rho_w g h_t\, r^2 \\[2mm] \text{bottom}: & F_b = \pi\rho_w g h_b\, r^2 \end{array}\right\}. \tag{2.10}$$

When the cylinder weight is neglected, the net buoyancy force acting on the cylinder is obtained as:

$$F_{\text{buoy}} = F_b - F_t = \rho_w g(h_b - h_t)\,\pi r^2 = \pi\rho_w g l r^2, \tag{2.11}$$

where l is the cylinder height. This force acts in an upward direction and is exactly equal to the weight of water contained in the cylinder and does not depend on the depth of submergence. This conclusion can be extended to any completely submerged or floating body using the law of buoyancy discovered by Archimedes in the third century B.C.:

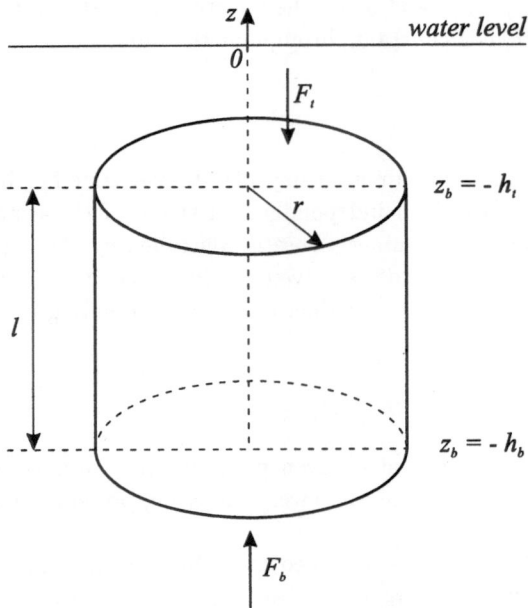

Fig. 2.3: Balance of forces on submerged cylinder

A body immersed in a fluid experiences a vertical buoyant force equal to the weight of the fluid it displaces.

If a body is only partly submerged, the buoyancy force is proportional to the volume of the submerged part of the body and it is in equilibrium with body weight. This is the essence of the second Archimedes law:

A floating body displaces its own weight in the fluid in which it floats.

Thus, if we assume that the density of the material from which the cylinder is made is ρ_c, we find that its weight is: $F_w = \pi \rho_c g l r^2$ and acts downward. The final net force is the difference between its weight and the buoyancy force, *i.e.*:

$$F_{\text{net}} = F_w - F_{\text{buoy}} = \pi(\rho_c - \rho_w)glr^2. \tag{2.12}$$

The term $(\rho_c - \rho_w)$ is called the *effective density*. If $F_{\text{net}} > 0$, the body sinks (negative buoyancy). When $F_{\text{net}} < 0$, the body floats (positive buoyancy); and when $F_{\text{net}} = 0$, it remains in equilibrium (neutral buoyancy). Small neutrally buoyant particles are frequently used in flow visualization, and neutrally buoyant bodies (floats) are used to track oceanographic currents. For floating bodies of arbitrary shapes, such as sitting birds or icebergs, only part of the object is submerged. Therefore, the buoyancy force results only from the volume of water displaced by the submerged part of the object. The discussion on how marine organisms can control buoyancy and the significance of other physical properties of water is given in Chap. 11.

2.2.3 Stability and Metacentric Height

In the example of the submerged cylinder, given in Sect. 2.2.2, it was assumed that the cylinder had uniform density ρ_c. Thus, the centre of gravity of the cylinder is located on its vertical axis. Also, because of symmetry, the buoyancy force acts along the vertical axis of the cylinder. As both forces act along the same line, the submerged cylinder either moves up, down, or remains stable, depending on the sign of the effective density (see Eq. 2.12), and does not rotate.

Let us now assume that a barge of uniform rectangular cross-section floats at the sea surface and the centre of mass, G, of the barge is exactly at the water line. Because of the barge's symmetry, the centre of buoyancy, B', is located at the line of symmetry, at the distance of $H/2$ above barge bottom (Fig. 2.4a).

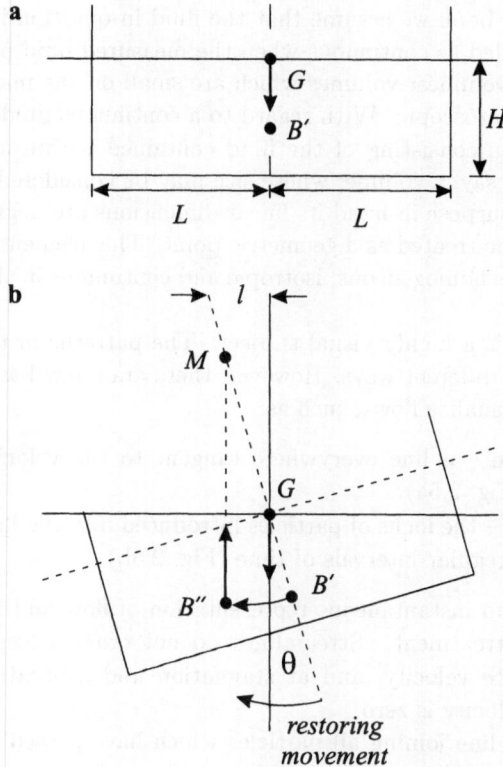

Fig. 2.4: Stability of rectangular barge: **a** neutral position, **b** tilted barge

Let the barge be tilted an angle θ (Fig. 2.4b). The centre of mass remains at point G, but this new centre of buoyancy, B'', is found at the centre of gravity of the water displaced by the barge. A vertical line drawn upward from B'' intersects the line of symmetry at a point M, called the metacentre. If point M is situated above G, that is, if the metacentric height $\overline{MG}$ is positive, a restoring moment is present and the original position of the barge is stable. If M is situated below G (negative $\overline{MG}$), the body is unstable and will overturn, if disturbed. The value of metacentric height gives an indication of the stability which increases with increasing $\overline{MG}$. For a body of varying cross-section and draft, such as a yacht, the computation of the metacentre is more complicated.

2.3 Water in Motion: Hydrodynamics

2.3.1 Methods of the Study

In a microscopic sense, fluids are aggregations of molecules and the distance between molecules is very large compared to the molecular diameter. However, the treatment of fluid mechanics from the molecular point of view, a subject dealt with in the kinetic theory of fluids, is beyond the scope of this book. For the purpose of this book we assume that the fluid in question is **continuous**. Fluid can be regarded as continuous when the measured fluid property is constant for sensitive volumes: volumes which are small on the macroscopic scale but large on the microscopic. With regard to a continuous fluid, we can define a **fluid particle** as consisting of the fluid contained within an infinitesimal volume. That is to say, a volume, whose size may be considered so small, that for the particular purpose in hand its linear dimensions are negligible. Thus, a fluid particle may be treated as a geometric point. This elementary fluid particle is assumed to be homogeneous, isotropic and continuous in the macroscopic sense.

Fluid mechanics is a highly visual subject. The patterns of the flow can be visualized in many different ways. However, there are a few basic types of line patterns used to visualize flows, such as:

- a **streamline** – a line everywhere tangent to the velocity vector at a given time (Fig. 2.5a),
- a **streakline** – the locus of particles introduced into the fluid at the same point at the regular intervals of time (Fig. 2.5b).

The streamline is an instantaneous representation of flow and it is convenient for mathematical treatment. Streamlines do not cross, except at points of theoretically infinite velocity, and at stagnation and separation points of a body where the velocity is zero.

A streakline is a line joining all particles which have passed through a particular point, such as a dye trace or a chimney plume, ignoring the finite width and diffusion (Fig. 2.5b). In addition to a streamline and a streakline, the path of a specific particle of fluid is defined by its position as a function of

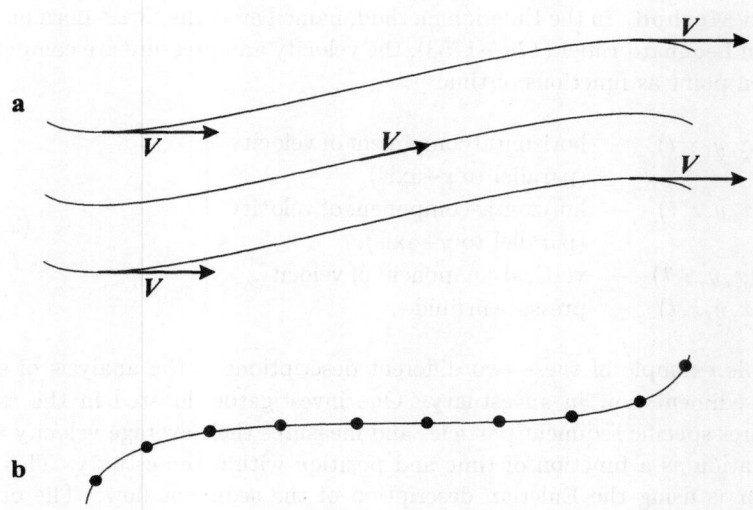

Fig. 2.5: Patterns of the flow: **a** streamlines, **b** streaklines

time. A pathline can be found by long exposures of a single marked particle moving with the flow.

As streamlines, pathlines and streaklines are dependent on time in various ways, their shapes are different for flow which is dependent on time (unsteady flow) and flow which is independent of time (steady flow). There are two basic methods for studying fluid motion: the Lagrangian method and the Eulerian method.

Lagrangian Method. This method is named after the Italian-French mathematician Joseph-Louis Lagrange (1736–1813). The approach consists of following fluid particles during the course of time and giving paths, velocities, and pressures in terms of the original position of the particles, and the time elapsed since the particles occupied their original positions.

If the initial position of a given particle at time t_0 is (x_0, y_0, z_0), the Lagrangian method gives the position (x, y, z) at the instant t as:

$$
\left.
\begin{aligned}
x &= F_1(x_0, y_0, z_0, t - t_0) \\
y &= F_2(x_0, y_0, z_0, t - t_0) \\
z &= F_3(x_0, y_0, z_0, t - t_0)
\end{aligned}
\right\}, \tag{2.13}
$$

where functions $F_1(\ )$, $F_2(\ )$ and $F_3(\ )$ provide the 'recipes' for changing the particle's x, y, z coordinates, respectively, during the elapsed time $(t - t_0)$. The Lagrangian method, seldom used in fluid mechanics, is more appropriate to solid mechanics.

Eulerian Method. In the Eulerian method, named after the Swiss-born mathematician Leonhard Euler (1707–1783), the velocity and pressure are computed at a given point as functions of time:

$$\left.\begin{array}{lll}
u = f_1(x,y,z,t) & - & \text{horizontal component of velocity} \\
& & \text{(parallel to } x-\text{axis),} \\
v = f_2(x,y,z,t) & - & \text{horizontal component of velocity} \\
& & \text{(parallel to } y-\text{axis),} \\
w = f_3(x,y,z,t) & - & \text{vertical component of velocity} \\
p = f_4(x,y,z,t) & - & \text{pressure in fluid.}
\end{array}\right\} . \qquad (2.14)$$

A simple example of these two different descriptions is the analysis of suspended sediment flow in an estuary. One investigator, located in the estuary, ignores specific sediment particles and measures their average velocity and concentration as a function of time and position within the estuary. This investigator is using the Eulerian description of the sediment flow. The other investigators may be interested in the path, speed and fate of specific sediments. They use the Lagrangian method.

2.3.2 Steady and Unsteady Flow

When flow is not time-dependent, the streamlines, pathlines and streaklines are identical. This is know as steady flow. However, when the flow changes with respect to time (unsteady flow) these lines are different. Figure 2.6 shows water velocity changes with time, recorded at a given point in the flow. In the first case, this velocity remains constant, while for the second case, the flow is unsteady and changes with time.

In some cases, when values fluctuate around some constant value, the mean motion of an unsteady flow with respect to time may be considered steady. An assumption of steady flow simplifies the solution of many hydrodynamic problems.

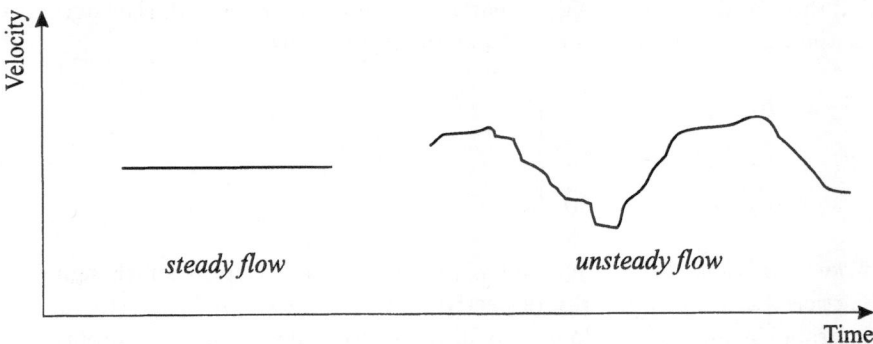

Fig. 2.6: Steady and unsteady flows

2.3.3 Rotational and Irrotational Flows

A classification of water flow as steady or unsteady, laminar or turbulent (see Sect. 2.4), is based on the physical properties of flow. However, there is another very important division in hydrodynamics which distinguishes rotational and irrotational flows. The irrotationality of water flow is an abstract, essentially mathematical concept which results in many simple and powerful methods used in the solution of hydrodynamic problems. As will be shown in Appendix C.3.1, for irrotational flow, unknown quantities such as velocity, pressure and acceleration can be expressed through one function called the **velocity potential**. A brief analysis of fluid rotationality (or irrotationality) is given in Appendixes C.3 and C.4, but more information can be found in many text books (for example, Batchelor, 1967; Le Méhauté, 1976; Pedlosky, 1979; Massel, 1989). It is difficult to establish simple practical rules for assessing the validity of the irrotationality assumption. Some of these rules, are given below, without proof;

- A fluid flow which initially is irrotational remains irrotational in absence of viscous and friction forces. A good example is that of still water onto which waves propagate. The initial flow is irrotational and after the waves arrive it will tend to remain irrotational, unless they move into a region where viscous forces are significant. Such forces are induced by jets, wakes, or solid boundaries. Near solid surfaces a boundary layer is created through which the stream velocity drops to zero.

- Flow becomes rotational when density gradients are caused by stratification, rather than by pressure gradients. This is a typical situation for large scale ocean circulation.

- Flow becomes rotational when there is significant forcing, apart from gravity. One example of such an effect, is the so called Coriolis acceleration due to the Earth's rotation (for more details, see Chaps. 5 and 7).

- Water motion is rotational in the neighbourhood of boundaries (sea bottom and sea surface), and motion may be considered irrotational only if the boundary layer is of little importance, *i.e.* relatively thin.

In this book, the property of irrotationality or rotationality will be used in many places.

2.3.4 Mass-Conservation Equation

In general, when studying the motion of ocean water, the following quantities must be known: three components of velocity (u, v, w), pressure p, sea surface elevation $\zeta(x, y, t)$ relative to the still water level, and water density ρ_w. As was shown in Chap. 1, water density, ρ_w, is dependent on water salinity, S, and temperature, T, and therefore can be considered as a known. The fundamental relationships for velocity and pressure result from the physical principles of conservation of mass and momentum.

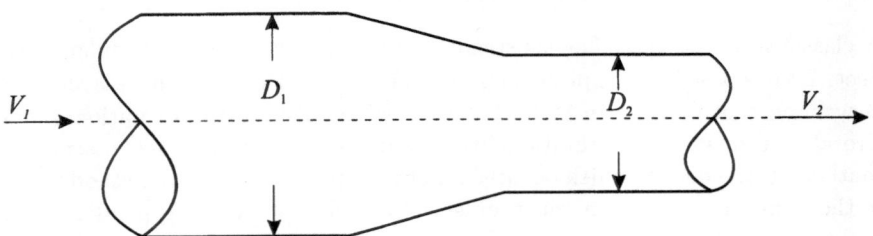

Fig. 2.7: Continuity of flow through cylindrical duct

The conservation of mass expresses the fact that the mass of water in any system must remain constant with time, *i.e.* fluid cannot be created or destroyed. It is interesting to note that this principle was first derived by Leonardo da Vinci in 1500 (White, 1994). Let us consider a simple example of steady flow through a circular pipe with a changing diameter (Fig. 2.7). The conservation of mass of the fluid requires that:

$$\text{Outflow} = \text{Inflow},\qquad\qquad\qquad\qquad\qquad\qquad (2.15)$$

which can be written as:

$$V_1 A_1 = V_2 A_2 = \text{Const},\qquad\qquad\qquad\qquad\qquad (2.16)$$

in which V_1 and V_2 are the inflow and outflow velocities, respectively; and A_1 and A_2 are the corresponding cross-sectional areas. From Eq. (2.16) and Fig. 2.7 it can be seen that when the area A_2 is decreased, the velocity V_2 increases.

The conservation of mass has to be satisfied in any volume of fluid, even for an infinitesimally small control volume. In this case, the principle is usually expressed in the differential form involving the partial derivatives of density and flow velocities:

$$\frac{\partial \rho_w}{\partial t} + \frac{\partial}{\partial x}\left(\rho_w u\right) + \frac{\partial}{\partial y}\left(\rho_w v\right) + \frac{\partial}{\partial z}\left(\rho_w w\right) = 0,\qquad\qquad (2.17)$$

which can be thought of as the rate of change of density of fluid at a point as being caused by the net mass influx, as expressed by the other three terms: $\partial(\rho_w u)/\partial x$, $\partial(\rho_w v)/\partial y$, and $\partial(\rho_w w)/\partial z$. For interested readers, the derivation of Eq. (2.17) is given in Appendix C.2.

Equation (2.17) is valid for any kind of flow under the assumption that there are no other sources or sinks in the volume under consideration. The first term $\partial \rho_w/\partial t$ in Eq. (2.17) vanishes for steady or incompressible flow. Further simplification arrives when we assume that water is homogeneous ($\rho_w = \text{Const}$). Then Eq. (2.17) becomes:

$$\frac{\partial u}{\partial x} + \frac{\partial v}{\partial y} + \frac{\partial w}{\partial z} = 0. \tag{2.18}$$

It should be noted that any homogeneous, incompressible fluid with a negligible coefficient of kinematic viscosity ($\nu \approx 0$) is called an inviscid, incompressible fluid or perfect fluid. For uniform, one-dimensional flow, Eq. (2.18) simplifies further as:

$$\frac{\partial u}{\partial x} = 0, \quad \text{or} \quad u = \text{Const.} \tag{2.19}$$

In professional literature, Eq. (2.18) is sometimes written in the more compact form:

$$\frac{\partial u}{\partial x} + \frac{\partial v}{\partial y} + \frac{\partial w}{\partial z} = \text{div}\,\boldsymbol{u} = 0, \tag{2.20}$$

where div denotes the divergence of a velocity vector $\boldsymbol{u} = (u, v, w) = u\,\boldsymbol{i} + v\,\boldsymbol{j} + w\,\boldsymbol{k}$. For a detailed explanation of the notation used in Eq. (2.20), see Appendix C.1.

2.3.5 The Momentum Principle and Bernoulli Equation

If the surroundings exert a net force, $\boldsymbol{F}$, on a unit volume of fluid, Newton's second law states that the mass begins to accelerate. In fluid mechanics Newton's law is called the conservation of momentum, which includes the fact that a fluid particle may be deformed. For steady, irrotational motion of a perfect fluid, conservation of momentum is identical to conservation of energy of an elementary particle of fluid, where the sum of kinetic and potential energy should be conserved. The momentum principle is obtained by equating the applied forces to the so called 'inertia force' which characterizes the natural resistance of matter to any change in its state of motion. The resulting equation for a no-viscous flow is called the equation of Euler:

$$\rho_w \left(\overbrace{\frac{\partial \boldsymbol{u}}{\partial t}}^{\text{Local acceleration}} + \overbrace{\boldsymbol{u}\nabla \cdot \boldsymbol{u}}^{\text{Convective acceleration}} \right) = \overbrace{\rho_w \boldsymbol{F} - \nabla p}^{\text{Applied forces}}, \tag{2.21}$$

in which $\boldsymbol{F}$ is the body force per unit mass, and ∇p is the surface force. For interested readers, the derivation of Eq. (2.21) is given in Appendix C.3. Extension of the Euler equation to a flow with viscosity is known as the Navier-Stokes equation, which is discussed in some detail in Appendix C.3. Here, as an example, the Navier-Stokes equation for motion along the Ox axis due to gravity is given by:

$$\rho_w \left(\frac{\partial u}{\partial t} + u \frac{\partial u}{\partial x} + v \frac{\partial u}{\partial y} + w \frac{\partial u}{\partial z} \right) = -\frac{\partial p}{\partial x} + \mu \left(\frac{\partial^2 u}{\partial x^2} + \frac{\partial^2 u}{\partial y^2} + \frac{\partial^2 u}{\partial z^2} \right). \tag{2.22}$$

Comparing Eq. (2.22) with the frictionless Euler's equation (C.30) shows an additional term on the right-hand side. The extra term can be written in a more compact way using vectorial notation (see Eq. C.14) as:

$$\mu \left(\frac{\partial^2 u}{\partial x^2} + \frac{\partial^2 u}{\partial y^2} + \frac{\partial^2 u}{\partial z^2} \right) = \mu \nabla^2 u. \tag{2.23}$$

In Appendix C it is shown that the Euler equation (2.21) provides a starting point for developing the very famous and widely used Bernoulli equation (see Eq. C.41):

$$\frac{|\boldsymbol{u}|^2}{2g} + \frac{p}{\rho_w g} + z = H = \text{Const.} \tag{2.24}$$

This is one of many forms of the Bernoulli equation for steady flow of a non-viscous fluid along a streamline. This particular form of the Bernoulli equation indicates that all terms represent vertical distances above some datum. By analogy with hydrostatics (see Sect. 2.2), the term $V^2/2g$ is the *velocity head*; $p/\rho_w g$ is the *pressure head*, and z is the *elevation head*. The sum $[(p/\rho_w g) + z]$ is called the *piezometric head* and the sum on left-hand side terms is called *total head*, and is constant for any streamline.

To provide a physical meaning of the Bernoulli equation, let us multiply both sides of Eq. (2.24) by a small mass, m, and by acceleration due to gravity, g. Thus, we obtain:

$$\frac{m|\boldsymbol{u}|^2}{2} + \frac{m\,p}{\rho_w} + mgz = \text{Const.} \tag{2.25}$$

Now we can see that Eq. (2.25) expresses the conservation of energy for an elementary mass m. In particular, the term $m|\boldsymbol{u}|^2/2$ is *kinetic energy*, while the term mgz provides expression for *potential energy* associated with vertical position of the elementary mass, m, against some reference datum. The third term represents the *pressure energy*, mp/ρ_w. This is different from kinetic and potential energy, as it represents the work that can be done because the elementary mass physically connects two areas of different elevation.

To illustrate the applicability of the Bernoulli equation (2.24), let us consider a simple tank filled with water which can escape through a nozzle (Fig. 2.8). We choose point 1 in the tank and point 2 at the nozzle cross-section. The tank

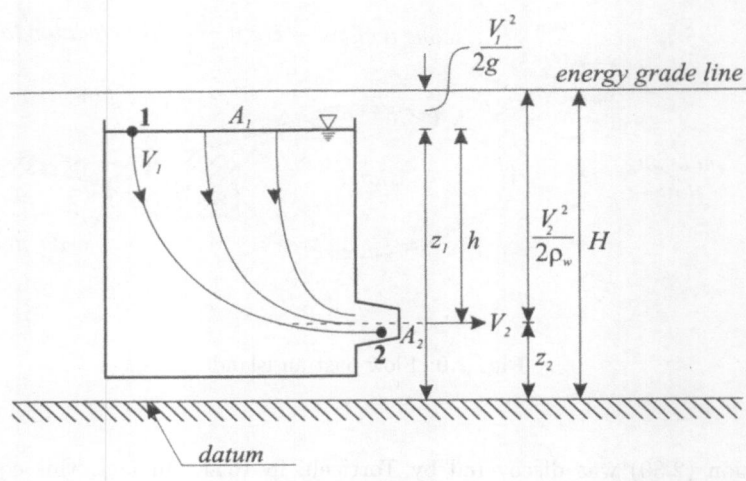

Fig. 2.8: Energy balance for water flow from tank

cross-section is A_1 and A_2 is the nozzle area. Then, by the mass conservation principle we have:

$$A_1 V_1 = A_2 V_2. \tag{2.26}$$

The application of the Bernoulli equation is facilitated by the fact that it does not require a control-volume analysis, only a selection of two points, 1 and 2, along a given streamline. The Bernoulli equation written for these two points yields:

$$\frac{V_1^2}{2g} + \frac{p_1}{\rho_w g} + z_1 = \frac{V_2^2}{2g} + \frac{p_2}{\rho_w g} + z_2 = H = \text{Const.} \tag{2.27}$$

As the points 1 and 2 are exposed to atmospheric pressure, $p_1 = p_2 = p_a$ and the pressure terms cancel on both sides of Eq. (2.27), which now becomes:

$$V_2^2 - V_1^2 = 2g(z_1 - z_2) = 2gh, \tag{2.28}$$

and using Eq. (2.26) gives:

$$V_2^2 = \frac{2gh}{1 - \left(\frac{A_2}{A_1}\right)^2}. \tag{2.29}$$

Usually, the nozzle area A_2 is much smaller than tank area, A_1, so $(A_2/A_1)^2 \to 0$ and finally, the outlet velocity is:

$$V_2 \approx \sqrt{2gh}. \tag{2.30}$$

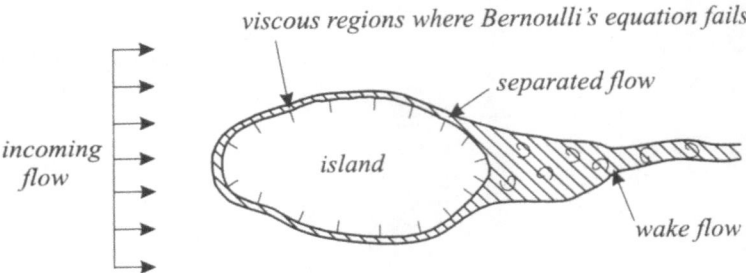

Fig. 2.9: Flow past an island

Equation (2.30) was discovered by Torricelli in 1644. In fact, this equation can also be derived from the conservation of energy principle. The potential energy of fluid in a tank (mgh) is entirely converted to kinetic energy $(mV_2^2/2)$. Assuming that motion is frictionless and that no net pressure work is done (the term $m\,p/\rho_w$ in Eq. (2.25) vanishes), equating both energies gives velocity V_2 as in Eq. (2.30).

Although the Bernoulli equation is widely used, there are, however, some assumptions under which this equation is valid. Here we stress them again:

- Flow is steady; for unsteady flow a 'Bernoulli constant', a function of time, and Eq. (C.39) should be used.
- Fluid is incompressible.
- It can be applied along any streamline, but not across different streamlines, as different streamlines may have different 'Bernoulli constant'.

Figure 2.9 schematically shows flow past an island in the ocean. The approach stream is irrotational but viscous stresses create a rotational shear layer beside and downstream of the island. The shear layer is laminar near the front of the island and turbulent toward the rear. A separated region occurs near the trailing edge, followed by an unsteady turbulent wake extending far downstream. In these regions Bernoulli's equation does not apply and more complete description of the water motion is needed which includes the existence of viscous stresses.

2.4 Laminar and Turbulent Flow

2.4.1 A Brief Overview

The basic principles of any fluid motion, namely the continuity and momentum principles, described in the sections above, are valid for particle movement along streamlines which are assumed to be smooth lines. For slow motion of highly viscous fluids, the 'layers' of fluid slide smoothly across one another with all particles moving in an orderly fashion. Such flow when all fluid particles move very nearly parallel to each other is termed **laminar flow**.

However, the motion of fluid is seldom orderly. Usually the flow is accompanied by tiny individual particles moving in a highly irregular manner, even if the fluid, as a whole, appears to travel smoothly in one direction. Intense small-scale motion in all directions is superimposed on the main large-scale flow. Wakes extending downstream of islands (Fig. 2.9) or breaking waves (Fig. 4.17) are examples. These small-scale chaotic motions of fluid are known as **turbulence**. The changeover between laminar and turbulent flow is called the *transition* to turbulence. Turbulence is essentially a statistical phenomenon. Descriptions of the overall motion in turbulent flows should not be presumed to describe the paths of individual particles.

In turbulent flow it is not only momentum that is transferred across the flow but similarly actual mass moves in directions other than that of the overall flow. The intensity of the momentum and mass exchange, or intensity of turbulence, is measured by the 'turbulent viscosity coefficient', which is analogous to the molecular viscosity, as was mentioned in Sect. 1.2.2. The process of exchange of momentum and energy in turbulent motion is a cascade from large structures (eddies) through increasingly smaller structures until energy is ultimately dissipated as heat by the action of viscosity. This pattern has been parodied in a piece of poetry by meteorologist Richardson (Perry and Walker, 1977):

> Big whirls have little whirls
> That feed on their velocity;
> And little whirls have lesser whirls,
> And so on to viscosity.

In the forties, the Russian mathematician Kolmogorov postulated that the dominant energy balance of the smallest eddies depends on the rate at which energy cascades down from large-scale motions and the viscosity of the fluid. The theory of the cascade exchange of turbulence energy and turbulence as a whole is beyond the scope of this book. The interested reader should consult Monin and Yaglom (1971) or Ozmidov (1986) for an in-depth discussion. However, some aspects of turbulence, for example, turbulent diffusion and its significance for life in the ocean, will be discussed in Chap. 8 and 13.

2.4.2 Reynolds Number

We are now back to the fundamental question of how to distinguish between laminar and turbulent regimes? Since turbulent flow is more prevalent than laminar flow, turbulence has been observed for centuries, but without any understanding of its nature. However, the abrupt character of the transition between the two regimes has been recognized for a long time. A basic experiment to understand the nature of the transition phenomenon was conducted by

Osborne Reynolds in 1883. He introduced a dye streak into a pipe flow and observed how it behaved as the current was altered (Fig. 2.10). A straight streak in Fig. 2.10a indicates a laminar flow with lower velocity, while in Fig. 2.10b the dispersal of the streak across the whole tube signals that the flow is turbulent.

The transition from laminar to turbulent flow is fairly sudden and can be induced by increasing flow speed, U, increasing the diameter of the pipe, D, increasing the density of the liquid, ρ_w, or by decreasing the liquid's viscosity, μ. Each of these factors, as well as a combination of all factors, induce a change. For practical use, a threshold combination which distinguishes laminar from turbulent flow is sought, and desirably dimensionless and independent of the system of units. Using a special technique called dimensional analysis (see Chap. 9 for further discussion), the set of variables U, D, ρ_w and μ can be reduced to a single non-dimensional parameter:

$$Re = \frac{\rho_w U D}{\mu} = \frac{U D}{\nu}, \tag{2.31}$$

where Re is known as the **Reynolds number**, named in honour of Osborne Reynolds. The accepted transition to turbulence for flow in a long circular pipe, with smooth walls, is $Re \approx 2300$. With roughened tubing, transition can happen at lower values, and for other geometries a particular value 2300 has no relevance at all.

For the case of an immersed, solid body, the characteristic length L is used instead of diameter D, in Reynolds number calculation, *i.e.*:

$$Re = \frac{U L}{\nu}. \tag{2.32}$$

The L value is typically taken as the greatest length of the body in the direction of flow.

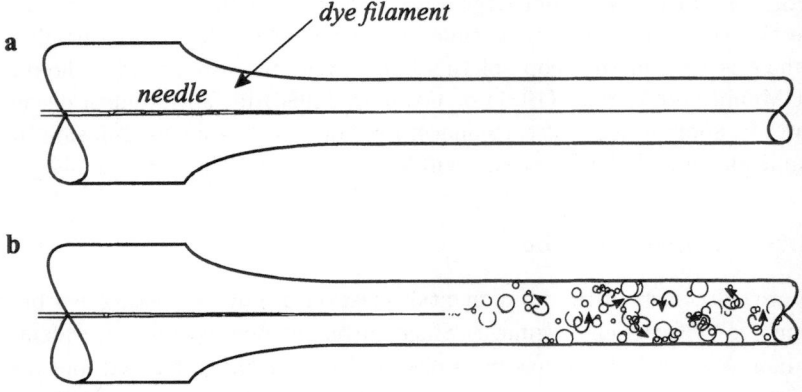

Fig. 2.10: Laminar and turbulent flows

How does the flow behave for Reynolds numbers lower and higher than the critical Reynolds number associated with transition from laminar to turbulent flow? In flow with a low Reynolds number, viscosity is sufficiently high to suppress the small perturbations in the flow. This is not the case for flow with a very high Reynolds number, where small perturbations, unsuppressed by viscosity, grow into turbulent eddies. The following approximate ranges of Reynolds number occur (White, 1994):

$$
\begin{array}{lll}
0 < & Re < 1 & : \quad \text{highly viscous laminar 'creeping' motion} \\
1 < & Re < 100 & : \quad \text{laminar, strong Reynolds number dependence} \\
100 < & Re < 10^3 & : \quad \text{laminar, usually in boundary layer} \\
10^3 < & Re < 10^4 & : \quad \text{transition to turbulence} \\
10^4 < & Re < 10^6 & : \quad \text{turbulent, moderate Reynolds number dependence} \\
10^6 < & Re < \infty & : \quad \text{turbulent, slight Reynolds number dependence.}
\end{array}
$$

These representative ranges are approximate only and can vary with flow geometry, surface roughness and the level of fluctuations in the inlet stream. As was shown in Chaps. 11 and 12, for marine ecologists who are dealing with systems that could span an enormous size range, the Reynolds number is a basic scaling parameter providing common basis for categorizing the flow induced by these various systems.

The Reynolds number is not only a quantity distinguishing laminar and turbulent flow regimes. It also has a well defined physical meaning which can briefly be explored as follows. For simplicity, a slow, irrotational flow, with constant viscosity μ, is assumed to be one-dimensional along the Ox axis. Therefore, the momentum equation (C.32) simplifies as:

$$
\rho_w \frac{\partial u}{\partial t} + \frac{\partial p}{\partial x} = \mu \frac{\partial^2 u}{\partial x^2}. \tag{2.33}
$$

Equation (2.33) contains the three basic dimensions: M (mass), L (length) and T (time). Let us non-dimensionalize this equation using density and two reference constants, *i.e.*: characteristic velocity, U, and characteristic length, L. For example, U may be the upstream velocity, and L characterizes the dimensions of a body immersed in the flow. Thus, we define the following dimensionless variables, denoting them by an asterisk:

$$
u^* = \frac{u}{U}, \quad x^* = \frac{x}{L}, \quad t^* = \frac{tU}{L}, \quad p^* = \frac{p}{\rho_w U^2}. \tag{2.34}
$$

Substitution of the new variables into Eq. (2.33) gives:

$$
\underbrace{\rho_w \frac{U^2}{L} \left(\frac{\partial u^*}{\partial t^*} + \frac{\partial p^*}{\partial x^*} \right)}_{\text{Inertia and pressure terms}} = \underbrace{\mu \frac{U}{L^2} \frac{\partial^2 u^*}{\partial x^{*2}}}_{\text{Viscous term}}, \tag{2.35}
$$

and dividing through we obtain:

$$\frac{\partial u^*}{\partial t^*} + \frac{\partial p^*}{\partial x^*} = \frac{\mu}{\rho_w U L} \frac{\partial^2 u^*}{\partial x^{*2}} = \frac{1}{Re} \frac{\partial^2 u^*}{\partial x^{*2}}, \tag{2.36}$$

so that the relative importance of viscous terms to the other terms is expressed by the ratio $1/Re$. The value of Re indicates the character of flow as being either under inertial or viscous forces. For low Reynolds numbers, flow is controlled by viscous forces, eddies are almost absent, strongly suppressed by viscosity. On the contrary, high Reynolds numbers favor turbulence with vortices of various dimensions.

Two other aspects of the Reynolds number are discussed later. In particular, in Sect. 2.6, the relationship between drag forces and Reynolds number is explained, while the importance of this number for laboratory modelling is discussed in Sect. 9.4. Equality of the Reynolds number for two different flows guarantees that the physical character of the flows is the same. Also, in the same section, other dimensionless parameters used in fluid mechanics, such as Euler number, Froude number, Weber number, and Mach number will be briefly discussed.

2.4.3   Mean and Fluctuating Components of Turbulent Flow

As was discussed above, at high Reynolds numbers, water motion becomes unpredictable and chaotic. Although the average flow may be quite regular and easily described, there are unsteady, random fluctuations (Fig. 2.11). The significance of turbulent motion for marine biology can not be underestimated. It is probably true to say that without turbulence, there would be no life on earth. Many examples of the link between biology and the physics of turbulent motion are described in Part III. Here, let us mention only the role of turbulent mixing for transporting planktonic larvae attempting to settle on the substratum from a position several hundred meters out to sea. The larva's feeble swimming abilities are not sufficient for this task and the process is facilitated by the turbulence present in the surf zone (Denny, 1988).

If turbulent motion is so chaotic and unpredictable, is there any possibility to describe the turbulence in a more quantitative way? Turbulence essentially is a statistical phenomenon, and its description should not be presumed to describe the paths of individual particles in a deterministic way. On the other hand, in oceanographic or biological practice, it is not always necessary to know the exact fine structure of the flow. Usually, only the average values and the overall and statistical effects of turbulent fluctuations are studied.

In turbulent motion, the instantaneous velocity, u, at a fixed point can be presented as the sum of the average velocity $\bar{u}$, with respect to time, and the fluctuation velocity, u', varying rapidly with time, $i.e.$:

$$u = \bar{u} + u'. \tag{2.37}$$

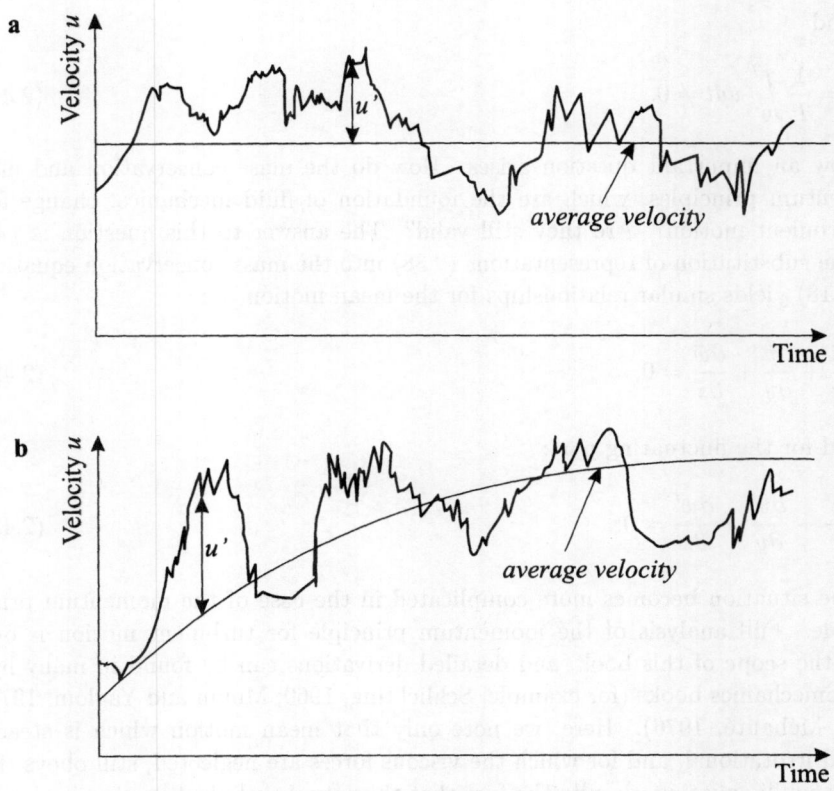

Fig. 2.11: Steady and unsteady random fluctuations

Mean velocity, $\bar{u}$, may be a function of time (see Fig. 2.11b). Even if velocity $\bar{u}$ is constant (see Fig. 2.11a), the real velocity u is a function of time. This means that the turbulence is always unsteady. A similar representation is valid for other velocity components and pressure. Thus, we have:

$$\left. \begin{array}{rcl} v &=& \bar{v} + v' \\ w &=& \bar{w} + w' \\ p &=& \bar{p} + p' \end{array} \right\} . \tag{2.38}$$

It should be noted that mean values of u', v', w' and p' are zero, *i.e.*:

$$\overline{u'} = \overline{v'} = \overline{w'} = \overline{p'} = 0. \tag{2.39}$$

A bar over each quantity denotes an averaging of the particular quantity over a time interval T. Thus, for $\overline{u'}$ and $\bar{u}$:

$$\overline{u'} = \frac{1}{T} \int_0^T u' dt = 0, \tag{2.40}$$

and

$$\bar{u} = \frac{1}{T} \int_0^T u\,dt \neq 0.$$
(2.41)

Now an important question arises. How do the mass conservation and momentum principles, which are the foundation of fluid mechanics, change for turbulent motion? Are they still valid? The answer to this question is *yes*. The substitution of representations (2.38) into the mass conservation equation (2.18) yields similar relationships for the mean motion, *i.e.*:

$$\frac{\partial \bar{u}}{\partial x} + \frac{\partial \bar{v}}{\partial y} + \frac{\partial \bar{w}}{\partial z} = 0,$$
(2.42)

and for the fluctuating part:

$$\frac{\partial u'}{\partial x} + \frac{\partial v'}{\partial y} + \frac{\partial w'}{\partial z} = 0.$$
(2.43)

The situation becomes more complicated in the case of the momentum principle. Full analysis of the momentum principle for turbulent motion is out of the scope of this book, and detailed derivations can be found in many hydromechanics books (for example, Schlichting, 1960; Monin and Yaglom, 1971; Le Méhauté, 1976). Here, we note only that mean motion which is steady and irrotational, and for which the viscous forces are neglected, still obeys the Bernoulli equation, despite the fact that the actual turbulent motion is always unsteady, rotational and dissipative.

The fluctuations of pressure, p', are usually very small when compared with the mean pressure $\bar{p}$, so the approximation $p \approx \bar{p}$ is used. Also, the term representing the viscous forces in Eq. (2.22) is generally small in comparison to the inertia terms on the left-hand side of Eq. (2.22) and is often neglected, except when investigating the boundary layer.

For example, the final form of the momentum equation for turbulent motion can be written as (only the momentum equation for the x direction is given):

$$\rho_w \left(\overbrace{\frac{\partial \bar{u}}{\partial t}}^{\text{Local acceleration}} + \overbrace{\bar{u}\frac{\partial \bar{u}}{\partial x} + \bar{v}\frac{\partial \bar{u}}{\partial y} + \bar{w}\frac{\partial \bar{z}}{\partial z}}^{\text{Convective acceleration}} \right) = \overbrace{-\frac{\partial p}{\partial x}}^{\text{Pressure forces}} +$$

$$+ \overbrace{\mu\left(\frac{\partial^2 u}{\partial x^2} + \frac{\partial^2 u}{\partial y^2} + \frac{\partial^2 u}{\partial z^2}\right)}^{\text{Viscous forces}} - \overbrace{\rho_w\left(\frac{\partial \overline{u'^2}}{\partial x} + \frac{\partial \overline{u'v'}}{\partial y} + \frac{\partial \overline{u'w'}}{\partial z}\right)}^{\text{Turbulent fluctuation forces}},$$
(2.44)

in which bars over the expressions denote time averaging.

This type of equation is called the Reynolds equation. It is very similar to the Navier-Stokes equation (2.22). Except that most quantities are time-averaged ones, the only real difference is the presence of the forces due to turbulent fluctuations. It should be noted that terms $\rho_w \overline{u'^2}$, $\rho_w \overline{u'v'}$ and $\rho_w \overline{u'w'}$ have the dimension of stresses, *i.e.* N/m^2, and are therefore called the **Reynolds stresses**.

To simplify the solution of turbulence problems, the concept of 'isotropic turbulence' is frequently introduced. In isotropic turbulence, the mean value of any function of the fluctuating velocity components and their space derivatives does not depend on the direction, *i.e.*:

$$\overline{u'^2} = \overline{v'^2} = \overline{w'^2} \quad \text{and} \quad \overline{u'v'} = \overline{v'w'} = \overline{v'w'} = 0. \tag{2.45}$$

The concepts of the Reynolds stresses and isotropic turbulence are used in the discussion of dispersion and mixing processes in Chaps. 8 and 13.

2.5 Boundary Layer Flows

2.5.1 Motivation

In the previous section we learnt how to study ocean waters depending on their physical characteristics and the character of the flow itself. Hence, we can distinguish between non-viscous and viscous fluids, steady and unsteady, and rotational and irrotational flows. However, the ocean is not empty, but full of various kind of living organisms and man-made structures, stationary, solid and moving. In this section we discuss the interaction mechanisms between fluids and bodies immersed in fluids. At the surface of a stationary body the velocity of fluid is zero. Therefore, a layer should exist in which the fluid velocity is reduced from its full, free stream value to zero at the body surface

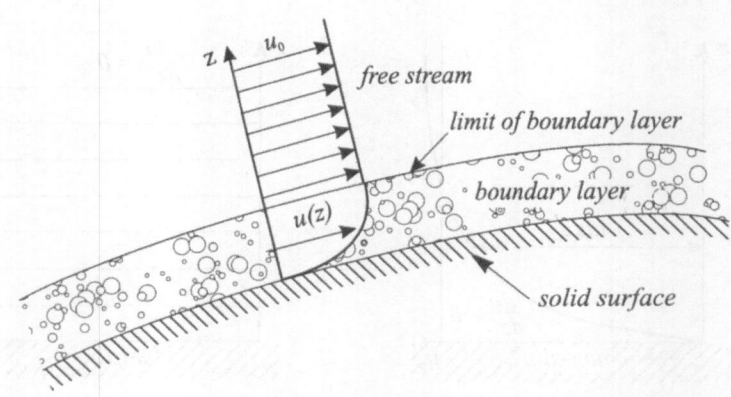

Fig. 2.12: Scheme of boundary layer

(see Fig. 2.12). This layer is called the **boundary layer**. Depending on the regime of motion, laminar and turbulent boundary layers are distinguished.

Any body of any shape, when immersed in a fluid stream, experiences forces and moments from the flow. The Sect. 2.6 provides the basis for determination of these forces for perfect as well as viscous fluids. In particular, it will be shown how a balance of forces acting on the body moving in the fluid, determines the 'terminal velocity' for biological and non biological particles. However, a key element for understanding the processes of interaction between a body and a fluid is the boundary layer, from which our analysis is started.

2.5.2   Non-Slip Condition and Boundary Layer Thickness

Everyday experience provides a confirmation that the fluid in direct contact with a solid surface does not slip in relation to that surface. This property of fluid is known as the **non-slip condition** (Fig. 2.13a). Slightly above the solid surface, fluid velocity increases gradually, eventually reaching its free-stream value at some distance from the surface. As was shown in Sect. 1.2.2, a velocity gradient is established and an accompanying shear stress, τ, is proportional to the velocity gradient. The proportionality coefficient between the shear stress, τ, and velocity gradient is the dynamic viscosity μ (see Eq. 1.2). On the other hand, if fluid slips by a solid surface (**slip condition**), no velocity gradient is established (Fig. 2.13b).

Velocity approaches zero almost linearly at the surface, and approaches the free-stream velocity asymptotically at some distance from the surface (see Fig. 2.13a). This creates some problem with defining the thickness of the boundary layer, δ. Mathematically it is quite correct to assume that the velocity gradient extends to infinity, asymptotically approaching zero, however, it is not helpful for practical purposes. From a physical point of view, the thickness of the boundary layer should be defined as the distance where the velocity gradient becomes so small that the effects of viscosity are negligible. The most

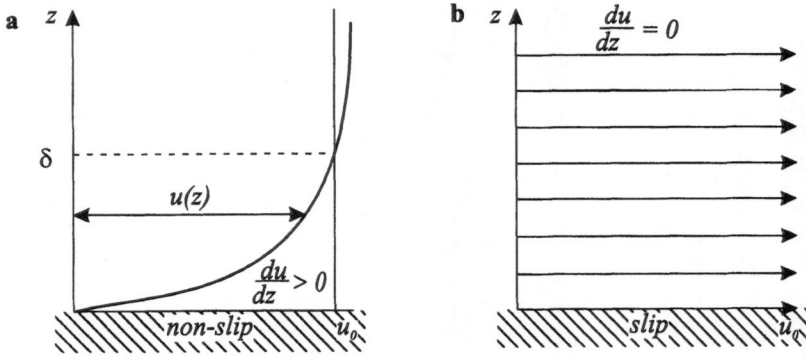

Fig. 2.13: Non-slip and slip bottom condition

commonly used boundary layer thickness is δ where the local velocity $u(z)$ has risen to 99% of the free-stream velocity, u_0, *i.e.*:

$$u(z) = 0.99u_0 \quad \text{for} \quad z = \delta. \tag{2.46}$$

The boundary layer thickness is a convenient, single number characterizing the boundary layer. However, thickness, δ, provides information about flow at one selected point only, and no information is given about the shape of the boundary layer along the solid surface.

The assumption of the slip condition, when $du/dz = 0$ and no force is imposed on the surface, implies that the fluid is non-viscous (perfect). However, when the no-slip condition is assumed, the velocity $u(z)$ depends on the regime of flow, laminar or turbulent, within the boundary layer. We now consider various forms of the boundary layer.

2.5.3   Laminar Boundary Layer

Consider a thin, flat plate oriented parallel to uniform flow. The plate retards the oncoming stream at the exact level of the plate. At points located further downstream from the leading edge of the plate, more space is available for boundary layer generation and the boundary layer becomes thicker (Fig. 2.14). The thickness of the boundary layer for a steady, uniform flow over a flat plate increases as a function of distance from plate edge according to the formula derived by von Kármán (Schlichting, 1960):

$$\delta(x) \approx 5.5x\sqrt{\frac{\nu}{u_0 x}} = \frac{5.5x}{\sqrt{Re_x}}, \tag{2.47}$$

in which $Re_x = u_0 x/\nu$ is called the local Reynolds number. In the von Kármán solution the velocity profile within the boundary layer has a parabolic shape:

$$u(x, z) \approx u_0 \left(\frac{2z}{\delta(x)} - \frac{z^2}{\delta^2(x)} \right), \quad 0 \leq z \leq \delta(x), \tag{2.48}$$

and for very small distance from the plate velocity, this profile becomes approximately linear, *i.e.*:

$$u(x, z) \approx \frac{2u_0 z}{\delta(x)}. \tag{2.49}$$

In general, a boundary layer is considered 'thin' when the ratio δ/x is less than about 0.1. Thus, in Eq. (2.47), when $\delta(x)/x = 5.5/\sqrt{Re_x} = 0.1$, local Reynolds number becomes ≈ 3000. For Re_x having value less than 3000, and $\delta(x)/x > 0.1$, the thick layer has a significant effect on the outer inviscid flow.

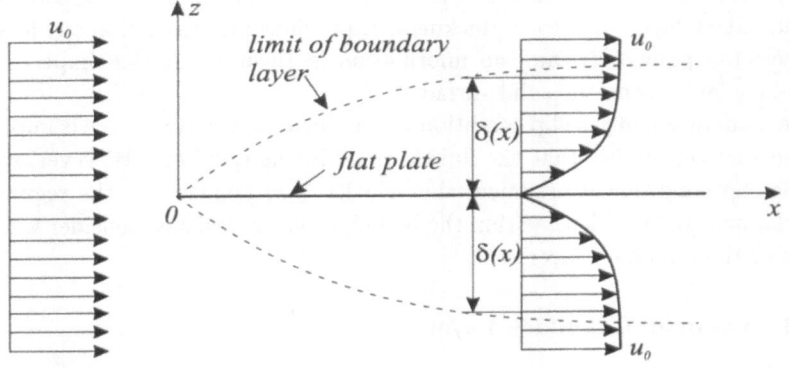

Fig. 2.14: Structure of boundary layer on plate

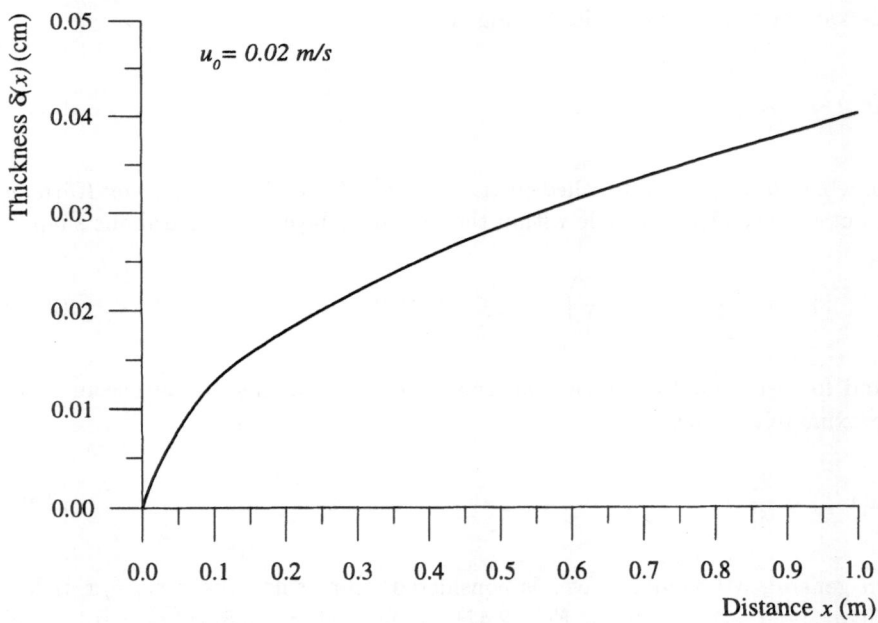

Fig. 2.15: Thickness of boundary layer versus distance from plate edge

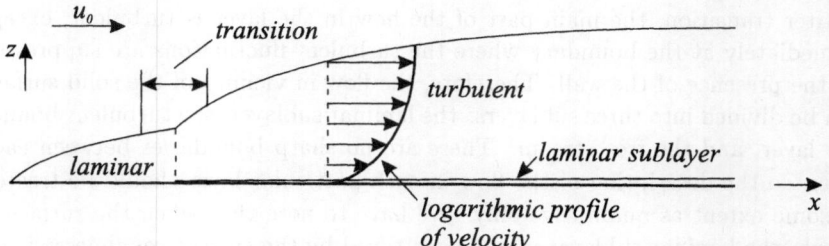

Fig. 2.16: Scheme of the transition of laminar boundary layer into turbulent one

In Fig. 2.15 the dependence of the boundary layer thickness, δ, on distance from the leading edge of the plate for mainstream velocity $u_0 = 0.02$ m/s, is shown. The figure suggests that a laminar boundary layer can grow endlessly. However, in real situations there are some limits for this growth. Firstly, boundary layers on objects are limited by size of vortices which separate from the surface and move downstream as wakes. The separation process is discussed in detail in Sect. 2.5.4. Secondly, when the viscous boundary layer extends into the water column, the flow in the velocity gradient region is likely to become turbulent and laminar boundary layer relationships are not valid. In practice, a boundary layer may by laminar near the leading edge and then turbulent somewhere downstream (Fig. 2.16) with the location of the transition depending on the local Reynolds number. The upper limit on Re_x for laminar flow is about 3×10^6 for experiments on a smooth plate (White, 1994).

So far the boundary layer induced by a uniform, stationary flow has been considered. However in shallow water, gravity waves produce oscillating water motion along the direction of wave propagation. Therefore, the mainstream velocity, u_0, is a periodic function of time (see Eq. 4.25), *i.e.*:

$$u_0 = \frac{gHT}{2L} \frac{1}{\cosh\left(\frac{2\pi h}{L}\right)} \cos\left(\frac{2\pi t}{T}\right). \tag{2.50}$$

The dependence of velocity, u_0, on time, t, implies that water in the boundary layer accelerates and decelerates in a similar manner. The solution of the boundary layer problem in an oscillating flow is of particular interest for studying boundary conditions for periodic gravity waves and wave damping by bottom friction. We will revisit these problems in Chap. 4.

2.5.4   Turbulent Boundary Layer

The value of the Reynolds number at which the boundary layer becomes turbulent depends on the turbulence level of the free incoming stream. In general, it ranges from 10^5 to 10^6. However, this figure may be considerably reduced if the surface is rough. For $Re_x < 10^5$, the laminar layer is stable; but at Re_x near 2×10^5, it is difficult to prevent transition.

After transition, the main part of the flow in the layer is turbulent, except immediately at the boundary where the turbulent fluctuations are suppressed by the presence of the wall. Therefore, the flow in vicinity of the solid surface can be divided into three sublayers: the laminar sublayer, the turbulent boundary layer, and the free stream. There are no sharp boundaries between each layer, so the description of the flow in each particular layer should be treated to some extent as qualitative only. We have to note that when the surface is rough, the laminar sublayer may be destroyed by the surface roughness.

Within a laminar sublayer (see Fig. 2.17), the velocity is controlled by the viscous stress, τ, as was given by Eq. (1.2), $i.e.$ $\tau = \mu \, (du/dz) = \rho_w \nu \, (du/dz)$. For a very thin laminar sublayer, stress, τ, is approximately constant and equal to its value at the wall τ_0. Then, an integration of Eq. (1.2) gives:

$$u(z) = \frac{u_*^2}{\nu} z, \tag{2.51}$$

in which:

$$u_* = \left(\frac{\tau_0}{\rho_w} \right)^{1/2}. \tag{2.52}$$

Equation (2.51) shows that in the laminar sublayer (a term 'viscous sublayer' is sometimes used), velocity u is a linear function of the distance z from the boundary. The quantity u_* is called the **friction velocity** because it has dimension (m/s), although it is not actually a flow velocity. Rather, it is a measure of the turbulent fluctuations in the velocity near the wall. The magnitude of u_* can only be estimated in an indirect way, and a more detailed discussion of velocity u_* will be given later. For now we only note that when, for example, wind blows over a rough sea surface, the ratio of the friction velocity to the mean wind velocity at a standard height of 10 m above sea level is of the order of 3 to 5% (Massel, 1996a). On the other hand, for water flow over the smooth bottom of a basin 10 m deep, the u_* velocity is about 3% of the mean flow velocity, while for a rougher bottom it varies from about 5% to 15% of the mean velocity (Vogel, 1994).

The laminar sublayer is at most only a few millimeters thick. However, its influence on the flow in the boundary layer is quite considerable. Whether the laminar sublayer exists or not depends on the free stream velocity and the bottom elements size (Fig. 2.17). When the bottom element diameter is less than one-third of the thickness of the laminar sublayer, the sublayer remains intact (Fig. 2.17a). The main turbulent flow above is not affected at all by the presence of the bottom elements. When bottom elements are greater than about one-third of the laminar sublayer thickness, the elements begin to disrupt the flow in the sublayer and once these elements reach about seven times the

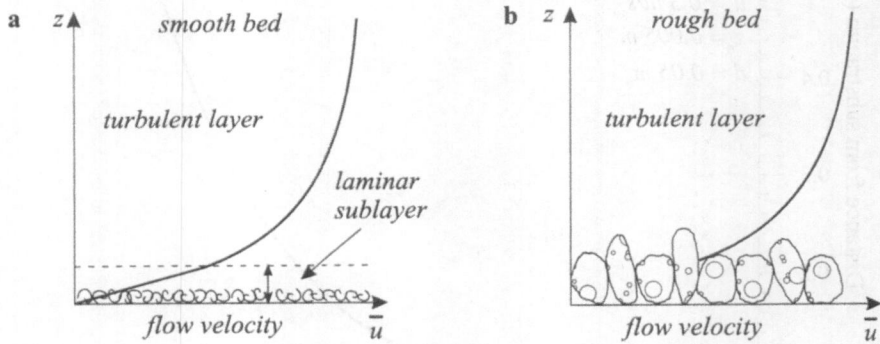

Fig. 2.17: Laminar sublayer and turbulent layer for smooth and rough substratum

thickness of the sublayer, they protrude far into the turbulent layer above. The laminar sublayer around the bottom elements breaks down and turbulent conditions extend down to the bed (see Fig. 2.17b).

As was shown in Fig. 2.16, a boundary layer may be laminar near the leading edge of a solid boundary and may then become turbulent from some location downstream, with a viscous sublayer remaining close to the boundary. For un-directional flow, the growth of the turbulent boundary layer in space can be estimated in a manner analogous to that of a laminar boundary layer, and the final result has the form (Schlichting, 1960):

$$\delta(x) = 0.37x\left(\frac{\bar{u}_0 x}{\nu}\right)^{-1/5} = \frac{0.37x}{(Re_x)^{1/5}}. \tag{2.53}$$

Comparison of Eqs. (2.53) and (2.47) indicates that growth of the boundary layer thickness is more rapid in the turbulent region, being proportional to the power $x^{4/5}$, whereas in laminar flow we have $\delta \approx x^{1/2}$.

The necessity of using an average velocity rather than an instantaneous value reflects the chaotic, unpredictable character of turbulent motion. For example, if we take many snapshots of the velocities near a flat sea bottom, each will show a different vertical velocity profile. Averaging these velocities as a function of distance from the bottom will yield an increase in the time-averaged velocity with distance from the bottom.

A turbulent layer may be treated in a similar manner as a laminar layer under the assumption that flow velocity is defined as the time-averaged velocity, $\bar{u}$. In particular, the thickness, $\delta(x)$, of a turbulent boundary layer given by Eq. (2.53) corresponds to the distance from a solid surface where the average velocity, $\bar{u}$, is 99% of the average mainstream velocity.

A dimensional analysis of the turbulent boundary layers in the atmosphere as well as in water suggests that near a bottom, the average velocity increases as the logarithm of the distance from the substratum (Schlichting, 1960;

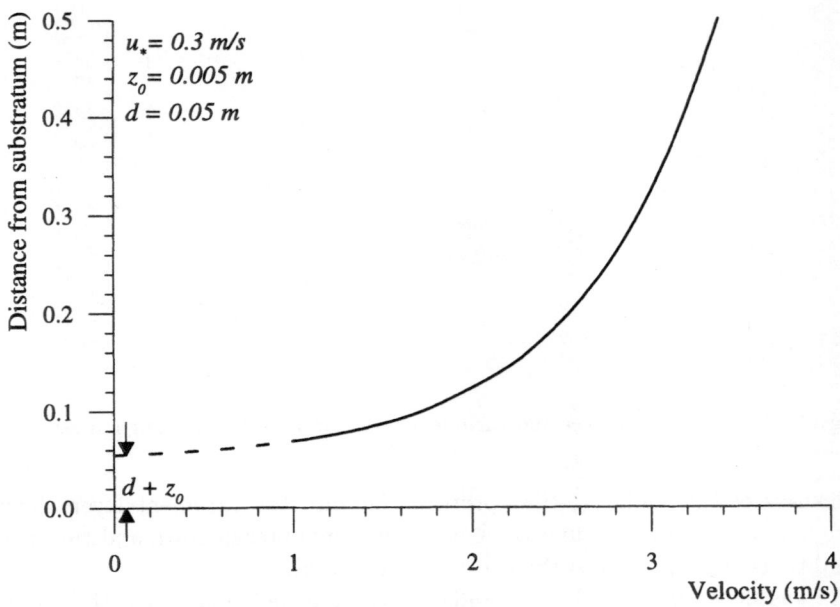

Fig. 2.18: Velocity profile within turbulent boundary layer

Townsend, 1976). Hence a typical equation for mean velocity distribution takes
the form (Fig. 2.18):

$$\bar{u}(z) = \frac{u_*}{\kappa} \ln\left(\frac{z-d}{z_0}\right),\tag{2.54}$$

in which u_* is the friction velocity given by Eq. (2.52) and κ is known as the von
Kármán's constant, having a value around 0.4; z_0 is the 'roughness parameter'
or 'roughness length', and d is the displacement thickness. From a physical
point of view, z_0 adjusts the steepness of the logarithmic velocity gradient,
as it is related to the size of the eddies generated at the surface. Thus, the
roughness length, z_0, depends on the flow around the roughness elements and
is a fraction of their height. The logarithmic profile can not be expected to
be accurate for z_0 smaller than the height of the roughness elements. The
velocity profile has to be corrected by inclusion of a displacement thickness, d,
which accounts for the fact that the logarithmic profile becomes zero somewhere
above the ground, especially on rippled, irregular or vegetated surfaces. In
many commonly encountered types of roughness, the relationship $d = 0.7\,h$,
where h is the roughness height, gives a good estimate. In particular, for sand
with a roughness of $h = 2$ mm, displacement $d = 0.72\,h$ and for grass with
$d = 0.02 - 0.20$ m, $d = 0.73\,h$ (Jackson, 1981). However, it is also clear that d
should strongly be dependent on the density of roughness elements.

The solid line in Fig. 2.18 shows an observed velocity profile which increases
with distance from the surface, while the dashed line denotes the velocity ac-

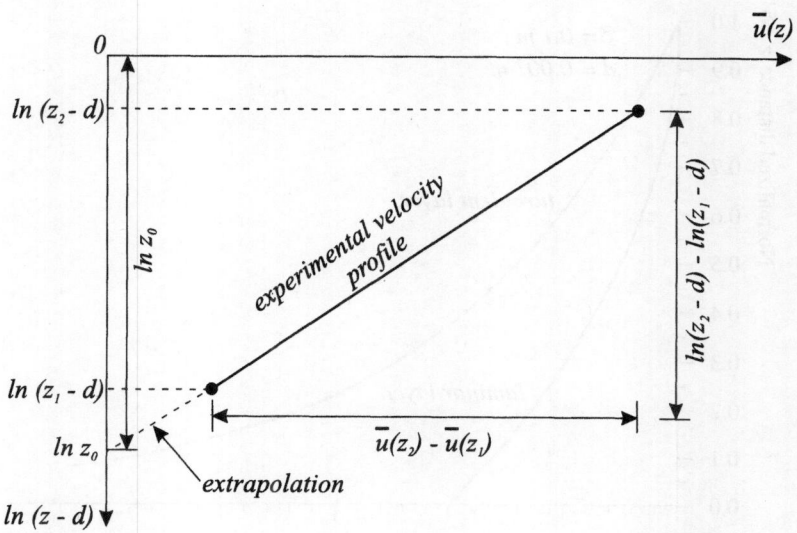

Fig. 2.19: Determination of friction velocity u_* and roughness length z_0

cording to Eq. (2.54); at height $(d + z_0)$ velocity is zero. To determine the unknown quantities of u_* and z_0, let us assume that the velocity profile $\bar{u}(z)$ is known in the range $z_1 < z < z_2$ (see Fig. 2.19). Now, Eq. (2.54) can be rearranged as follows:

$$\bar{u}(z) = \frac{u_*}{\kappa} \ln(z - d) - \frac{u_*}{\kappa} \ln z_0 = \frac{u_*}{\kappa} Z - A, \qquad (2.55)$$

in which $Z = \ln(z - d)$ and $A = (u_*/\kappa) \ln z_0$. Equation (2.55) is the equation of a straight line for variables $\bar{u}$ and Z with slope u_*/κ, as shown in Fig. 2.19. Value of height d should be taken from independent experiments or be chosen to give the best straight-line fit to the experimental data. Extrapolation of the experimental profile to the intersection with the $\ln(z - d)$ axis provides the $\ln z_0$ value (and consequently z_0 value), while the ratio $[\bar{u}(z_2) - \bar{u}(z_1)] / [\ln(z_2 - d) - \ln(z_1 - d)]$ gives a slope u_*/κ and velocity u_* value, as κ is a known constant.

We note that in order to avoid calculation of z_0 and u_* using logarithms, Prandtl pointed out that turbulent profiles can be approximated by a simpler one-seventh-power law (Schlichting, 1960; White, 1994):

$$\left(\frac{\bar{u}(z)}{\bar{u}_0}\right)_{\text{turb}} \approx \left(\frac{z}{\delta}\right)^{1/7}. \qquad (2.56)$$

At the end of this section, in Fig. 2.20 a comparison of the velocity distributions in laminar and turbulent layers is shown. To avoid determination of

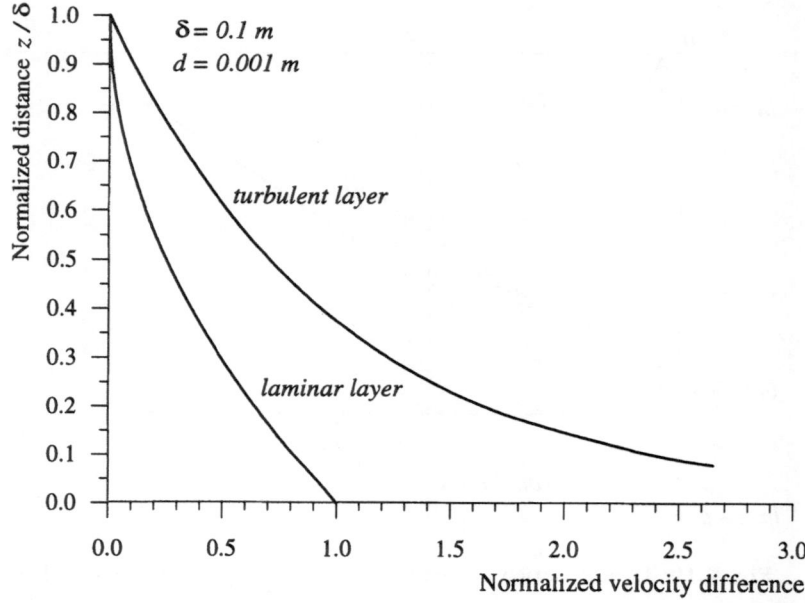

Fig. 2.20: Comparison of velocity distribution within laminar and turbulent layers

constants such as boundary layer thickness, δ, roughness length, z_0, and displacement thickness, d, the differences between the velocity at a particular level, z, and at the limit extent of the boundary layer, $z = \delta$, are determined.

Dimensionless velocity differences $(\bar{u}(\delta) - \bar{u}(z)/(u_*/\kappa)$ have been illustrated in Fig. 2.20, where an increased velocity gradient within the turbulent boundary layer is quite apparent.

Because a boundary layer is a region of reduced velocity, it can serve as a refuge from the mainstream flow for many organisms. Some organisms live partly or entirely within the boundary layer. How do they live and feed in this velocity gradient environment? How do they manage to disperse? This and other similar questions, which are of interest to biologists, are discussed in Part III of the book.

2.6   Forces Imposed by Fluid Flow

2.6.1   Introduction

So far, we have discussed flow in the vicinity of a flat surface and we described the boundary layer which controls the mechanisms of interaction of the fluid flow and the surface. Let us now pursue the problem of interaction somewhat further, discussing in particular the boundary layers and forces induced by flow on marine organisms and human-made structures of arbitrary shape. Generally

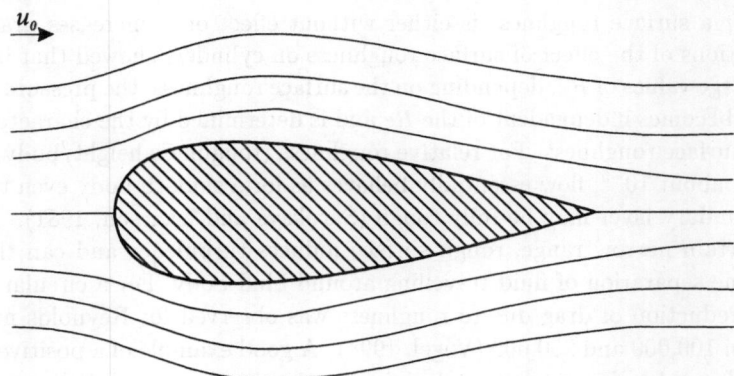

Fig. 2.21: Streamlined body

bodies can be characterized as either streamlined or bluff. An example of a streamlined body is shown in Fig. 2.21. In fact, this shape is a copy of many marine organisms; for example, it is the typical shape of fish. The streamlined shape is endowed with a long tail which induces a gradual deceleration of the fluid towards the rear and avoids the formation of a wake. Therefore, instead of being lost, the energy manifests itself as a forward-directed pressure that nearly counterbalances the dynamic pressure on the front, drastically reducing the force induced on the body.

On the other hand, a sphere or circular cylinder are bluff bodies, and experience high force from the flow. In general, there are a few types of forces imposed on body by fluid. The force on the body along the axis parallel to the free stream is called **drag**. Drag is essentially a rate of removal of momentum from a flowing fluid. Momentum is defined as the product of mass and velocity. Therefore, Newton's second law, which states that force equals mass multiplied by acceleration, can be regarded as equating force and the rate of change of momentum.

The drag on a streamlined body depends on the ratio of the length to the maximum diameter, which is known as the fineness ratio FR (see Eq. (2.70) and Sect. 2.6.2 for more details). The fineness ratio of about 4.5 provides the lowest drag. The fineness ratio of dolphins, large well streamlined marine animals with torpedo-shaped body, is about 5, which is very close to the ideal ratio of 4.5. Steven (1950) observed dolphins and seals swimming at night among plankton, which phosphoresced wherever the water was disturbed. While seals left a bright phosphorescent wake, the dolphins were far less conspicuous because they leave little disturbance in their wake.

Although drag on a streamlined body is very small, there is some disadvantage of streamlining from a biological point of view, namely a high sensitivity to the flow direction. A minor change in the direction can induce a much larger drag on a marine organism.

For the bluff bodies, a drag force is not so sensitive to the flow direction, however a surface roughness is either without effect or it increases drag. Investigations of the effect of surface roughness on cylinders showed that beyond some large values of Re, depending on the surface roughness, the pressure distribution becomes independent of the Re and is determined by the characteristics of the surface roughness. For relative roughness (roughness height/body diameter) of about 10^{-5}, flow essentially behaves as for a smooth body even though the boundary layer may be fully rough (Sarpkaya and Isaacson, 1981).

In certain narrow range, rough surface induces turbulence and can thereby postpone separation of fluid travelling around bluff body. For a circular cylinder, a reduction of drag due to roughness was observed for Reynolds number between 100,000 and 250,000 (Vogel, 1994). A good example of a positive effect of roughness for drag reduction during a motion in air is a golf ball.

A second important force acts perpendicular to drag and is called **lift**, because the most familiar example of a force of this kind is the upward force which acts on the wings of aircraft and keep them in the air. In spite its name, lift does not necessarily act upwards. Structures like aircraft wings, which are designed to produce lift to move in the air, are called aerofoils, while structures producing lift in water are known as hydrofoils.

To this point we have assumed that fluid flow is steady. When fluid is accelerating, an additional force is imposed on a stationary body, which is called **inertia force**. This force is usually accompanied by a drag force. The total force imposed on stationary or moving body is a vectorial sum of all forces involved. In the following sections each of these forces and their physical nature and methods of their determination are examined.

2.6.2   Drag Force

Let us start with a simple example of a circular cylinder with its axis normal to the uniform flow (Fig. 2.22a). When fluid is assumed to be non-viscid, ideal, streamlines can be determined and the total force imposed on the cylinder can be calculated. As is shown in Appendix C.4, the resulting force is nil! This is so called D'Alembert's paradox, which is valid for bodies of any shape. In the case of a circular cylinder, there are so called stagnation points S, at upstream and downstream extremities, where the fluid is locally stationary with respect to the cylinder (Fig. 2.22a). The fluid reaches maximal velocity $u = 2u_0$ at the sides of the cylinder, when $\theta = \pm 90°$, where the pressure drops to the value $p/\rho_w g = \text{Const} - (2u_0^2/g)$, as follows from the Bernoulli equation (see Eq. C.84); the Const is the total head. The whole pressure distribution is symmetrical about both the cross- and along-flow axes of the cylinder (Fig. 2.22b).

Pressure at a given point on the cylinder surface can generally be represented as a summation of ambient pressure p_0 and dynamic pressure induced by the interaction of the cylinder with the flow:

$$p(\theta) = p_0 + \frac{1}{2}\rho_w u_0^2 f(\theta), \tag{2.57}$$

in which the function $f(\theta)$ characterizes the pressure distribution along the cylinder circumference, and θ is an angle with respect to the x-axis. Rearranging Eq. (2.57):

$$\frac{p(\theta) - p_0}{\frac{1}{2}\rho_w u_0^2} = f(\theta), \tag{2.58}$$

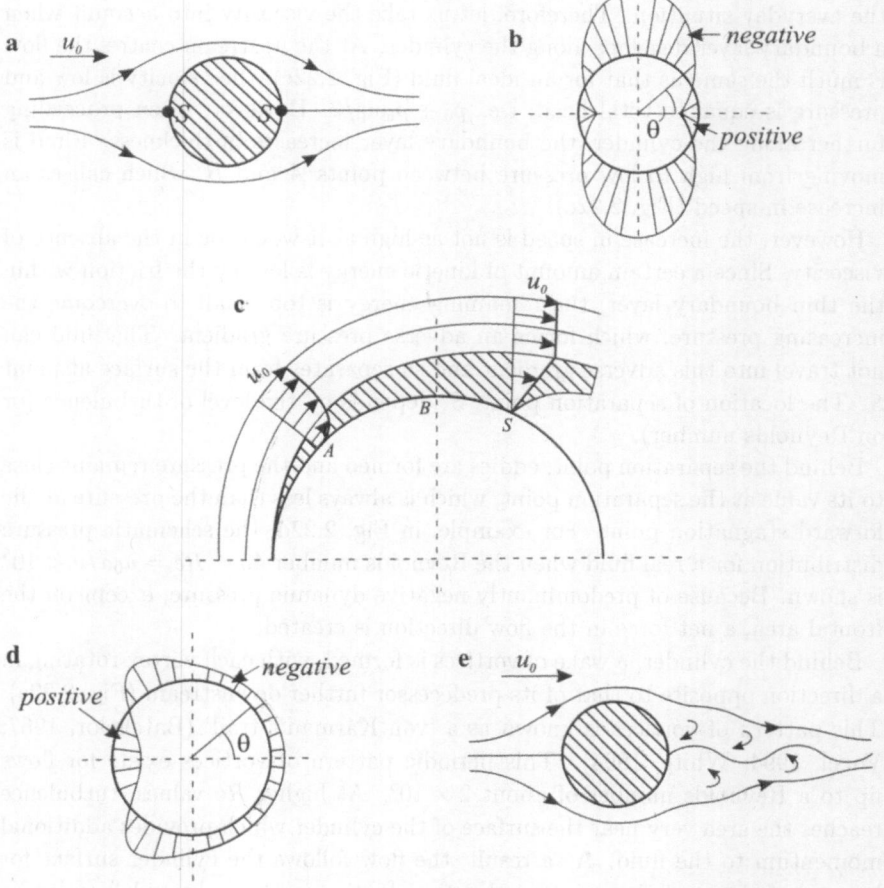

Fig. 2.22: Pressure distribution around circular cylinder for ideal and real fluids: **a** flow around cylinder and stagnation points position, **b** normalized pressure for ideal fluid, **c** developing of boundary layer for viscous flow, **d** normalized pressure for real fluid, **e** wake of vortices behind the cylinder

we can say that the function $f(\theta)$, represents an excess of pressure due to the cylinder presence, normalized by $(1/2)\rho_w u_0^2$. In the case of an ideal fluid, function (2.58) becomes (for derivation see Appendix C.4):

$$\frac{p(\theta) - p_0}{\frac{1}{2}\rho_w u_0^2} = f(\theta) = 1 - 4\sin^2\theta. \tag{2.59}$$

Relationship (2.59) is presented in Fig. 2.22b. Distribution of the normalized excess pressure is symmetrical with respect to both cross- and along-flow axis of cylinder. Along the flow axis, normalized pressure is equal to $+1$, while at lateral points it is equal to -3. The symmetrical distribution of pressure results in a zero value of force on cylinder.

It is clear that the assumption of an ideal, non-viscous fluid does not reflect the everyday situation. Therefore, let us take the viscosity into account when a boundary layer develops along the cylinder. At the upstream centre, the flow is much the same as that for an ideal fluid (Fig. 2.22c); the velocity is low and pressure is equal to total head, *i.e.* $p_0 + \rho_w u_0^2/2$. However, when proceeding further along the cylinder, the boundary layer increases in thickness. Fluid is moving from high to low pressure between points A and B, which causes an increase in speed (Fig. 2.22c).

However, the increase in speed is not as high as it would be in the absence of viscosity. Since a certain amount of kinetic energy is lost by the friction within the thin boundary layer, the remaining energy is too small to overcome the increasing pressure, which forms an adverse pressure gradient. The fluid can not travel into this adverse gradient and so separates from the surface at point S. The location of separation point, S, depends on the level of turbulence (or on Reynolds number).

Behind the separation point, eddies are formed and the pressure remains close to its value at the separation point, which is always less than the pressure at the forward stagnation point. For example, in Fig. 2.22d, the schematic pressure distribution for a real fluid when the Reynolds number $40 < Re = u_0 a/\nu < 10^5$ is shown. Because of predominantly negative dynamic pressure, except on the frontal area, a net force in the flow direction is created.

Behind the cylinder, a wake of vortices is formed, with each vortex rotating in a direction opposite to that of its predecessor further downstream (Fig. 2.22e). This pattern of vortices is known as a 'von Kármán's trail' (Batchelor, 1967; Vogel, 1994; White, 1994). This periodic pattern of vortices exists for flows up to a Reynolds number of about 2×10^5. At higher Re values, turbulence reaches the area very near the surface of the cylinder which provides additional momentum to the fluid. As a result, the flow follows the cylinder surface for longer before separation, and an abrupt reduction in the wake width and drag force appears. This reduction depends on the roughness of the cylinder and on the turbulence level in the free stream. A detailed description of the separation of boundary layer from a bluff body is given by Batchelor (1967).

What determines the drag force imposed on a stationary cylinder by real fluid flow? The starting point is to examine the pressure distribution around the

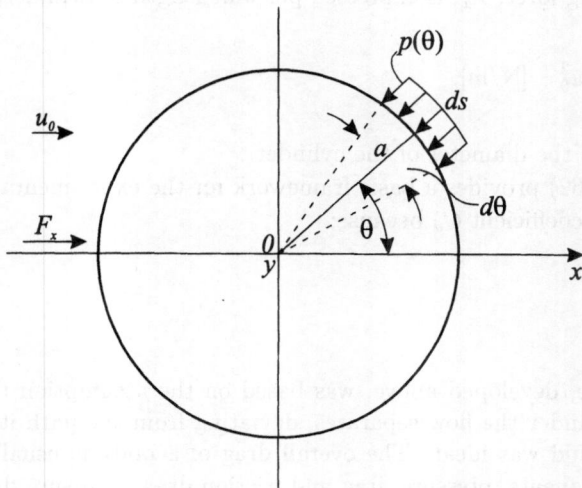

Fig. 2.23: Scheme for integration of pressure along cylinder circumference

cylinder, irrespective of the character of the fluid, perfect or real. Thus, assuming that this distribution is known (see for example Eq. 2.57), the resulting drag force in the x direction can be found by the integration of pressure along the cylinder's circumference (see Fig. 2.23):

$$F_x = -\int_0^{2\pi} p(\theta)ds \cos\theta \Delta y = -\Delta y \int_0^{2\pi} \frac{1}{2}\rho_w u_0^2 f(\theta) \cos\theta ds. \qquad (2.60)$$

The cylinder is elongated along the y axis, Δy is a unit length of cylinder, and the function cos is introduced to provide the x component of the drag force. In Eq. (2.60) the ambient pressure, p_0, does not contribute to the total force and is neglected. As a unit arc length $ds = ad\theta$, Eq. (2.60) becomes:

$$F_x = \frac{1}{2}\rho_w u_0^2 \underbrace{(2a\Delta y)}_{S} \underbrace{\int_0^{\pi} f(\theta) \cos\theta d\theta}_{C_d}. \qquad (2.61)$$

The product $2a\Delta y = S$ is the projected area of the cylinder onto the flow direction. The integral in Eq. (2.61) is an *a priori* unknown quantity for a general flow. It is usually denoted by C_d and is called the **drag coefficient**. Drag coefficient contains the influence of the body shape on flow as well as all of peculiarities and unknowns of flow around the body, as boundary layer theory can not accurately estimate the pressure distribution, especially in separated regions, as was discussed above. Thus a drag force, F_x, finally is:

$$F_x = \frac{1}{2}C_d\rho_w S u_0^2 \quad \text{[N]}. \qquad (2.62)$$

Sometimes drag force, F_x, is expressed per unit length of cylinder, *i.e.*:

$$F_x = \frac{1}{2} C_d \rho_w D u_0^2 \quad [\text{N/m}],$$ (2.63)

where $D = 2a$, the diameter of the cylinder.

Equation (2.62) provides a basic framework for the experimental determination of a drag coefficient C_d because:

$$C_d = \frac{F_x}{\frac{1}{2} \rho_w S u_0^2}.$$ (2.64)

The drag force, developed above, was based on the assumption that at some point on a cylinder the flow separates, deviating from the path it would have taken if the fluid was ideal. The overall drag of a body is usually separated into two components, pressure drag and friction drag. Pressure drag is a consequence of the difference between the high pressure of the front stagnation region, and the low pressure of the rear separated region. For streamlined bodies (for example, fish), the flow is less likely to separate and streamlined bodies cause very little disturbance to the flow. Therefore, it is reasonable to assume that almost total drag is due to shear stress in the boundary layer over the body. This type of drag is known as friction drag.

Let us examine friction drag for a thin plate. For simplicity, we will consider a two-dimensional plate of length L (in flow direction) and width b. Because of the non-slip condition (see Sect. 2.5.2), the fluid flowing very close to the plate is retarded to form a boundary layer and some force is needed to establish the velocity gradient. The drag due to friction is given by (Schlichting, 1960):

$$F_{x,\text{friction}} = \frac{1}{2} \rho_w C_{d,\text{friction}} S_w u_0^2,$$ (2.65)

where S_w is the total wetted area of the plate (equal to bL), u_0 is the mainstream velocity, and $C_{d,\text{friction}}$ is the frictional drag coefficient. The value of $C_{d,\text{friction}}$ depends on the boundary layer flow regime, and for the laminar boundary layer, with Reynolds number greater than 10^6, we have (Schlichting, 1960):

$$C_{d,\text{friction}} = \frac{1.33}{\sqrt{Re}},$$ (2.66)

and for turbulent flow:

$$C_{d,\text{friction}} = \frac{0.072}{Re^{0.2}}.$$ (2.67)

Table 2.1: Contribution of pressure and friction drags to the overall drag for a circular cylinder (in %)

Reynolds number Re	Pressure drag contribution	Friction drag contribution
10	57	43
100	71	29
1000	87	13
10000	97	3

The relative importance of friction and pressure drag can be determined by taking the ratio of Eq. (2.65) and Eq. (2.62), *i.e.*:

$$\frac{\text{friction drag}}{\text{pressure drag}} = 1.33 \frac{S_w}{S} \frac{1}{C_d \sqrt{Re}}, \qquad (2.68)$$

or:

$$\frac{\text{friction drag}}{\text{pressure drag}} = 0.072 \frac{S_w}{S} \frac{1}{C_d Re^{0.2}}, \qquad (2.69)$$

depending on whether flow in the boundary layer is laminar or turbulent, respectively. In Eqs. (2.68) and (2.69), S_w is the total wetted area, while S is the area that the object projects in the direction of flow. From these equations it follows that friction drag is substantial when Reynolds numbers are very small, or when the body is very small (wetted area S_w is very large compared to the projected area S).

Partition of the overall drag between pressure and friction components for a circular cylinder, at various Reynolds numbers, is listed in Table 2.1. The overall drag on a bluff body, such as a cylinder, is due primarily to pressure drag, and friction drag is only significant for viscous flow for low Reynolds numbers. However, even for high Reynolds numbers, this component of the drag force remains present.

From Eqs. (2.68) and (2.69) it follows that pressure drag becomes minimal when the ratio S/S_w becomes very small. This ratio is true for streamlined bodies (see Fig. 2.21). A streamlined profile is characterized by a slowly tapering tail and the ratio of maximum length, L, and maximum diameter, B, is between 2 and 6, *i.e.*:

$$FR = \frac{L}{B}, \qquad (2.70)$$

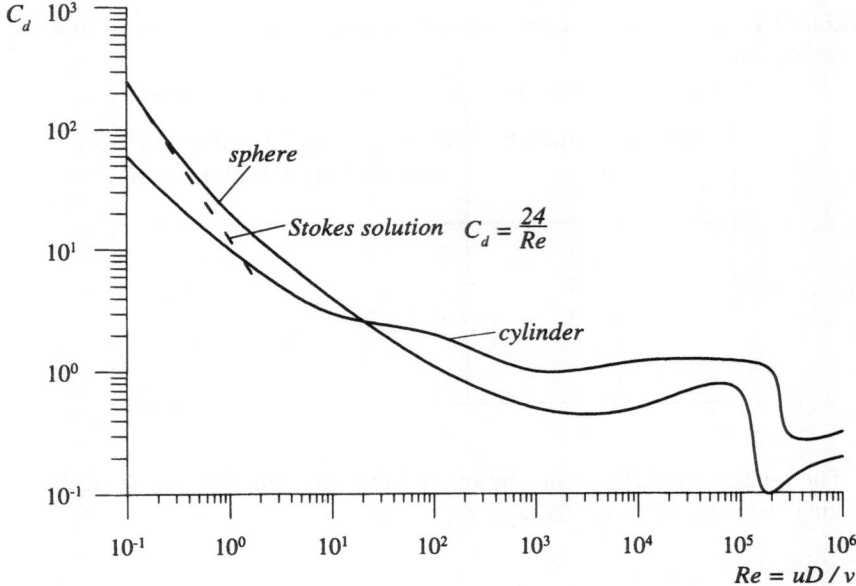

Fig. 2.24: Variation of the drag coefficient C_d with Reynolds number (adapted from Le Méhauté, 1976)

in which FR is known as the fineness ratio. In order to minimize the pressure drag, a streamlined body of a given volume should have an FR value of about 4.5 (Blake, 1983; White, 1994). Such shape has a drag coefficient of approximately 0.005 for the Re range of 10^4–10^6.

It should be noted that experimentally determined drag coefficients contain both pressure and friction contributions, and we will denote it by C_d. In particular, drag force for a sphere can be found in a similar way, as above, and the relationship is:

$$F_x = \frac{1}{2} C_d \rho_w \pi \left(\frac{D}{2}\right)^2 u_0^2, \tag{2.71}$$

in which D is the sphere diameter.

It is of a great practical convenience to summarize the physical phenomena of flow around bodies using a relationship between the drag coefficient C_d and the Reynolds number. Such relationships for a sphere and a circular cylinder with its axis normal to the direction of motion are shown in Fig. 2.24. Both curves are in fact very similar. However in some ranges of the Reynolds number, these bodies behave in different ways. For Reynolds numbers between 40 and 5000, asymmetrical vortex shedding induces a time-dependent circulation around the cylinder. Therefore a cylinder experiences a sideways push that reverses its

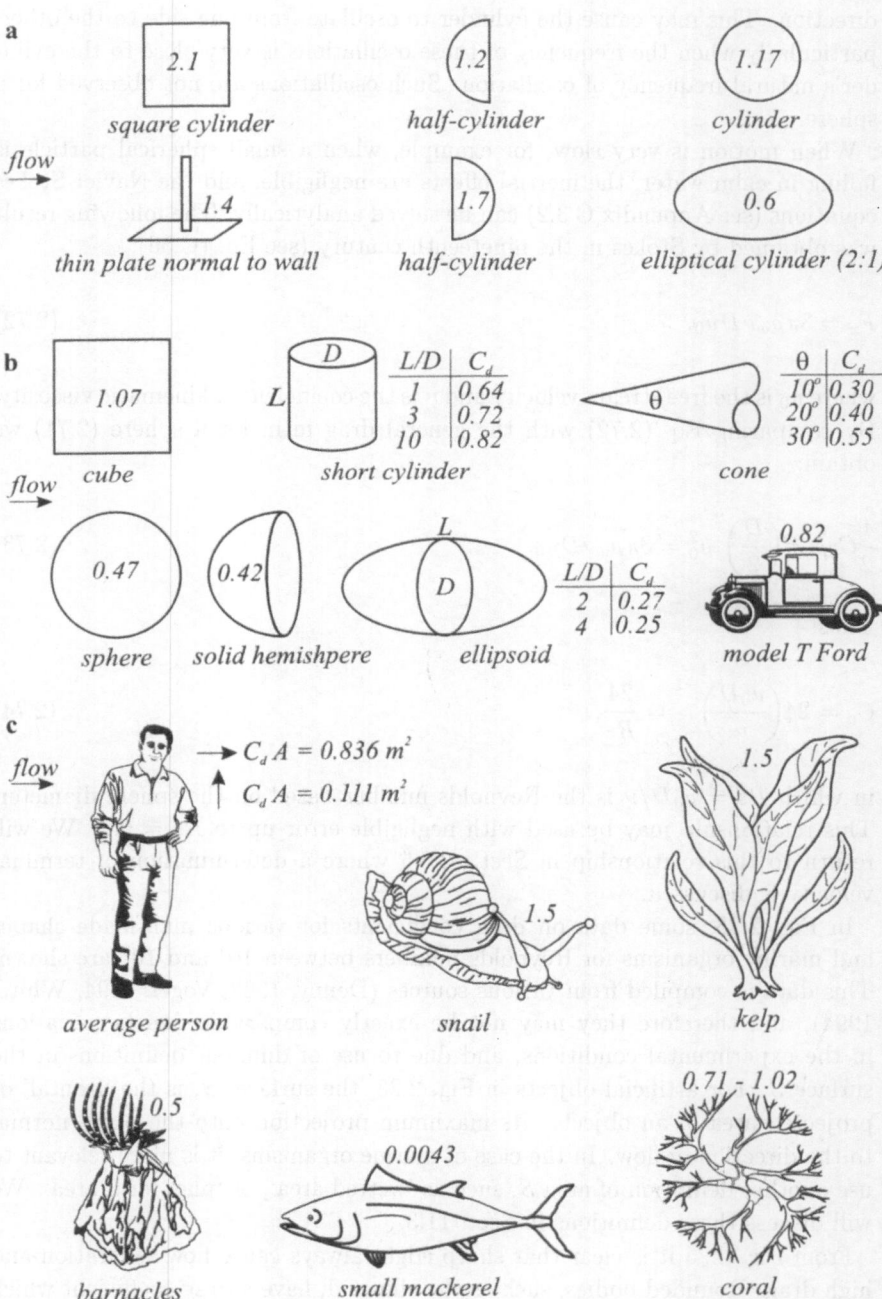

a

square cylinder 2.1

half-cylinder 1.2

cylinder 1.17

flow

thin plate normal to wall 1.4

half-cylinder 1.7

elliptical cylinder (2:1) 0.6

b

cube 1.07

short cylinder

L/D	C_d
1	0.64
3	0.72
10	0.82

cone

θ	C_d
10°	0.30
20°	0.40
30°	0.55

flow

sphere 0.47

solid hemishpere 0.42

ellipsoid

L/D	C_d
2	0.27
4	0.25

model T Ford 0.82

c

flow

$C_d A = 0.836 \ m^2$

$C_d A = 0.111 \ m^2$

average person

snail 1.5

kelp 1.5

barnacles 0.5

small mackerel 0.0043

coral 0.71 - 1.02

Fig. 2.25: Drag coefficient for various shapes: **a** two-dimensional, **b** three-dimensional, **c** living organisms; A denotes surface of human body (adapted from Denny, 1988; Vogel, 1994; White, 1994)

direction. This may cause the cylinder to oscillate from one side to the other, particularly when the frequency of these oscillations is very close to the cylinder's natural frequency of oscillation. Such oscillations are not observed for a sphere.

When motion is very slow, for example, when a small spherical particle is falling in calm water, the inertial effects are negligible, and the Navier-Stokes equations (see Appendix C.3.2) can be solved analytically. The following result was obtained by Stokes in the nineteenth century (see Eq. (C.56):

$$F_x = 3\pi \rho_w \nu D u_0, \tag{2.72}$$

where u_0 is the free stream velocity and ν is the coefficient of kinematic viscosity. By comparing Eq. (2.72) with the general drag form for a sphere (2.71) we obtain:

$$\frac{1}{2} C_d \rho_w \pi \left(\frac{D}{2}\right)^2 u_0^2 = 3\pi \rho_w \nu D u_0. \tag{2.73}$$

Thus:

$$C_d = 24 \left(\frac{u_0 D}{\nu}\right)^{-1} = \frac{24}{Re}, \tag{2.74}$$

in which $Re = u_0 D/\nu$ is the Reynolds number based on the sphere diameter. This relationship may be used with negligible error up to $Re = 0.2$. We will return to this relationship in Sect. 2.6.5 where a determination of terminal velocity is discussed.

In Fig. 2.25, some data on drag coefficients for various man-made shapes and marine organisms for Reynolds numbers between 10^4 and 10^6 are shown. This data is compiled from various sources (Denny, 1988; Vogel, 1994; White, 1994), and therefore they may not be exactly comparable due to variations in the experimental conditions, and due to use of different definitions of the surface S. For artificial objects in Fig. 2.25, the surface, S, is the 'frontal' or projected area of an object – its maximum projection onto the plane normal to the direction of flow. In the case of marine organisms, it is more relevant to use another definition of area S, such as 'wetted area', or 'plan form area'. We will discuss these definitions in Sect. 11.3.

From Fig. 2.25 it is clear that sharp edges always cause flow separation and high drag. Rounded bodies, such as the ellipsoid, have a drag coefficient which depends upon the point of separation; both the Reynolds number and the character of the boundary layer are important. A good example of this dependence are the drag coefficients of automobiles. The famous Model T Ford had a C_d of the order of 0.8–0.9, while modern cars have an average drag coefficient of about 0.30, with a constant trend to decrease (White, 1994).

2.6.3 Lift Force

As was shown in the previous section, the asymmetry of the pressure distribution is responsible for creating a drag force acting on a body immersed in a fluid, in the direction of relative motion. Apart from the drag force, the pressure asymmetry around some bodies creates a force component perpendicular to the drag, called lift. Due to this lift, birds, insects, gliders and airplanes can fly. The lift force is also important for life in the ocean and many marine organisms are subject to this lift. Limpets, flatfish, starfish, chitons and snails are but a few of the obvious examples.

To explore the nature of the lift force, let us again consider a horizontal cylinder, as in Fig. 2.22a. As the cylinder is symmetrical and the pressure distribution at its circumference is symmetrical with respect to horizontal and vertical axis, there is no net force in an ideal fluid. For a real fluid, however, there is some horizontal force, but still there is no force perpendicular to the flow.

What will happen if a body is located on the sea bed? For simplicity we will consider a half-elliptical cylinder sitting on the sea bottom (Fig. 2.26). Streamlines due to uniform flow become more dense at the top of the cylinder, thus by continuity, the fluid reaches maximum velocity there and from the Bernoulli equation it follows that the water there is at a lower than ambient pressure. The pressure difference imposes a lift force on the half cylinder. This force can be quite substantial; for example, for a limpet attached to the bottom, fluid flowing over the top of the shell creates a force even larger than drag (Denny, 1988).

As lift forces are caused by the same basic mechanism as drag forces, namely a pressure difference, it is reasonable to expect that the lift force is proportional to the dynamic pressure and to the area over which the pressure difference acts.

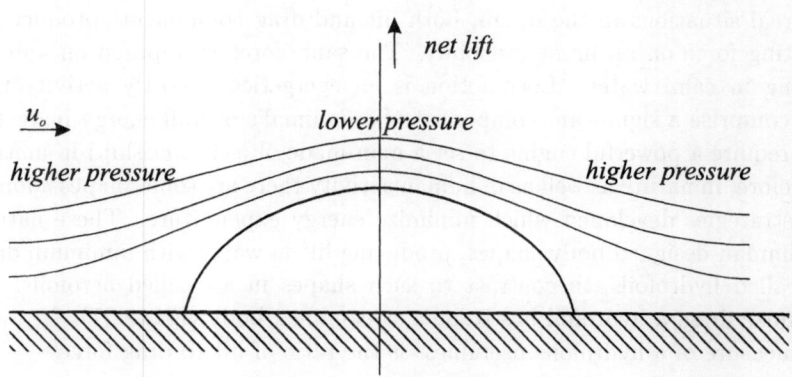

Fig. 2.26: Elliptical cylinder sitting at the sea bottom

Table 2.2: Measured lift coefficients for marine organisms (adapted from Denny, 1988)

Organism	Re	C_l
Limpets	$5(7) \times 10^4$	0.28–0.67
Barnacles	10^5	0.5
Snails	10^4	0.67
Coral *Acropora reticulata*	2×10^6	0.71–1.02
Kelp	7×10^4	1.5

In the case of lift forces, the relevant area is that projected perpendicular to the direction of flow, S_p, often called the platform area. Similarly to drag force, the dependence of the lift force on the shape of the object is included through an empirically determined **lift coefficient**, C_l. Thus we have:

$$F_l = \frac{1}{2} C_l \rho_w S_p u_0^2. \tag{2.75}$$

Lift force data are far less consistent than drag force data and the scatter (from 0 to 1.4) is attributed to various causes, such as the effect of the end gaps and free stream turbulence. Following Denny (1988), the values of lift coefficients for some marine organisms have been collected in Table 2.2. We also note here that Wilson and Reid (1963) showed that the C_l value for a pipeline seated on the sea floor reaches a value of 4.50.

In real situations in the ocean, both lift and drag components produce the resulting force on an immersed body. The same force is imposed on animals moving in calm water. Locomotion is an energetically costly activity that may comprise a significant component of an animal's overall energy budget or may require a powerful engine to set a man-made object (*i.e.* ship) in motion. Therefore, in nature as well as in human activity there are some shapes adopted and strategies developed which minimize energy expenditure. These natural and human-designed body shapes, producing lift in water with minimum drag, are called hydrofoils, in contrast to such shapes in air called aerofoils. To minimize drag, a hydrofoil has to be a streamlined body (Fig. 2.27).

A measure of a hydrofoils usefulness is the ratio of lift to drag forces:

$$\frac{\text{Lift}}{\text{Drag}} = \frac{\frac{1}{2} C_l \rho_w S_p u_0^2}{\frac{1}{2} C_d \rho_w S u_0^2} = \frac{C_l}{C_d} \frac{S_p}{S}. \tag{2.76}$$

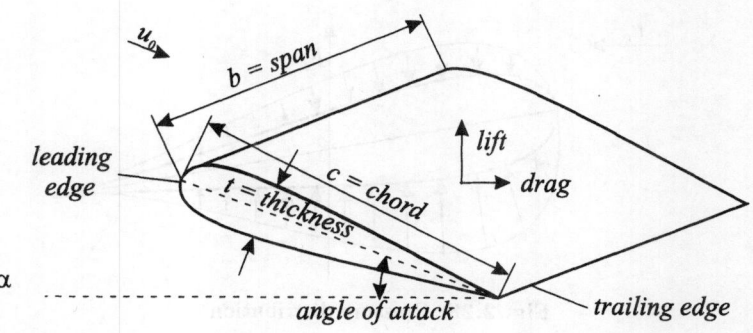

Fig. 2.27: A typical shape of hydrofoils

This ratio varies considerably for marine organisms and no general rule exists. We will discuss this problem in more detail in Chap. 11. Here, we only give some basic results and provide the terminology and some theoretical results.

For our purpose, a symmetric streamlined body is assumed. The angle between the free stream and the chord line is called the angle of attack. In some hydrofoils, as in aerofoils, the chord line between leading and trailing edges is not a line of symmetry, and such hydrofoils are known as 'cambered'.

How does a hydrofoil produce lift? The theory of aerofoils and hydrofoils was developed by Russian hydrodynamicist Zhukovskii (or Joukowsky) early in the present century. Rather than proceed with the derivation of the governing theoretical equations, we provide a physical explanation of lift creation by the hydrofoils. When a slightly inclined ($\alpha > 0$) hydrofoil starts to move it splits the flow into two streams: flow over the upper surface and flow under the lower surface. The velocities in these streams are not equal due to the inclination of the hydrofoil. When these two streams meet at the trailing edge, a separation appears and a starting vortex forms. This vortex is shed downstream and a smooth streamline develops over the hydrofoil and lift is fully developed. However, within the boundary layer the flow is viscous and due to the velocity gradient, vorticity exists (for definition of vorticity see Appendix C.1). Physically, the existence of vorticity means that the vortices in the fluid are generated in a boundary layer and their strength depends on the velocity gradient.

Because the velocity gradients on the top and bottom surfaces are of opposite sign, the vorticities also have opposite signs. When these vorticities are equal in strength, the resultant circulation around the hydrofoil is zero. This is a case of symmetrical hydrofoil with zero angle of attack. If the vorticity (also velocity gradients) over the top surface exceeds that over the bottom, the resultant circulation is a non-zero quantity. Velocities over the upper surface are higher than under the lower surface. According to the Bernoulli equation, such a

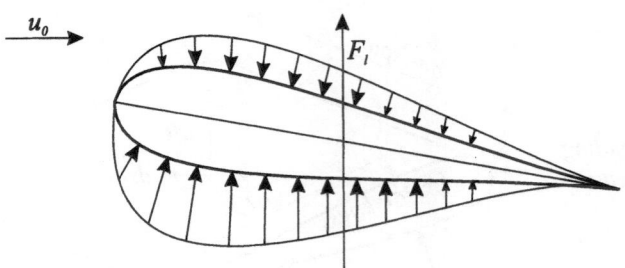

Fig. 2.28: Pressure distribution

velocity distribution induces higher pressure on the bottom of the surface and lower pressure on the top surface (see Fig. 2.28). Thus, the resulting vertical lift force is directed upward. The lift forces are strongly dependent on the angle of attack, α, by the relationship $C_l \approx \sin \alpha$. This means that for small angles of attack the lift is proportional to the angle α. For example, the lift coefficient, C_l, for a symmetric NACA 0009 aerofoil profile, with angle of attack $\alpha = 0$ is equal to 0, as should be expected, and increases linearly up to a value of $C_l = 1.15$ for $\alpha = 12°$ (White, 1994).

There is a simple relationship between circulation, Γ (see Eq. C.15), and lift force, F_l, known as the Kutta-Zhukovskii theorem:

$$F_l = \rho_w u_0 \Gamma, \tag{2.77}$$

in which:

$$\Gamma = \oint v_s ds, \tag{2.78}$$

where $\oint$ is the line integral of velocity, v_s, around the closed contour.

Until now we have dealt with the flow past an infinitely long hydrofoil. In real situations, the hydrofoils are of finite length (span) b (see Fig. 2.27) and flow is three-dimensional. Since there is end flow at the tips of the hydrofoil, the pressure difference between the top and bottom surfaces must decrease from a maximum at the middle section towards the tips where it is zero. Consequently, a finite span hydrofoil produces less lift and suffers more drag. The relationship between drag and lift coefficients for such hydrofoils takes the form:

$$C_d = \frac{C_l^2}{\pi(\text{aspect ratio})}, \tag{2.79}$$

in which aspect ratio = span/chord = b/c. A large aspect ratio minimizes the drag, as expected. For marine organisms, instead of the aspect ratio, another criteria should be used for designing shapes which minimize drag. This is a problem of minimizing drag per unit volume, as organisms require a volume in which to reside. This criterion, as well as the question of how marine organisms adapt to withstand frequently changing directions of flow will be examined in Chap. 11.

Let us give an example of the practical use of the lift coefficient and the lift force. The Boeing 727 commercial jet aircraft during takeoff has a gross weight of about 9.2×10^5N (about 94 tonnes); its wing area is 153 m^2 (Redding and Yenne, 1983). What is the minimum speed to provide a necessary lift force for takeoff? The lift force during takeoff must balance the aircraft's weight, thus:

$$W = \frac{1}{2}C_l \rho_a S_p U^2,$$ (2.80)

where W is the aircraft weight, and ρ_a is the air density, $\rho_a = 1.24$ kg/m^3. In the commercial aircraft, the lift performance of aerofoils is improved by adding flaps and slats. A combination of aerofoil and double-slatted flap results in the lift coefficient $C_l \approx 3.4$. Now we are in the position to calculate the minimum takeoff velocity U from Eq. (2.80) as:

$$U = \left(\frac{2W}{C_l \rho_a S_p}\right)^{1/2}.$$ (2.81)

Substituting all quantities into Eq. (2.81) yields the velocity U of about 190 km/hr., which is a realistic value. The above calculations are approximate only, as they are based on very few data for this aircraft. In many commercial aircraft, more complicated systems of flaps and leading-edge slots are used to increase the lift force.

2.6.4 Inertia Force

We have assumed that fluid moves past a body in a steady manner. Under steady conditions, the total force acting on a fixed body is nil for the case of a perfect fluid (D'Alembert's paradox) and for a real fluid, the force is non-zero, being a complex function of the Reynolds number, Re. What kind of additional force is created if the fluid is accelerating or a body accelerates in quiescent fluid? Such supposition is motivated by Newton's second law which states that an acceleration of mass is proportional to the applied force, $i.e.$ $\mathbf{F} = m(d\mathbf{u}/dt)$.

However, to apply Newton's second law, the effective mass of the body must be known. The visualization of the motion of a body in a fluid shows that the individual particles of fluid are pushed aside by the moving body. In addition to pushing the particle aside, the body also accelerates the neighboring fluid in the direction of its motion. This mass of fluid is called **added mass**, and is only evident if the body accelerates through the fluid.

In order to provide some insight into the added mass phenomenon, let us consider a fixed circular cylinder subjected to an unsteady ideal fluid flow perpendicular to the cylinder's axis. As usually, the total force per unit cylinder length imposed on the cylinder is given through integration of the pressure distribution, $p(\theta)$, along the cylinder circumference, $i.e.$:

$$F = - \int_0^{2\pi} p(\theta) a \cos \theta d\theta, \tag{2.82}$$

in which a is the radius of the cylinder.

The pressure distribution, $p(\theta)$, in the case of an unsteady motion can be deduced from Bernoulli equation (C.40) for a polar coordinate system as follows:

$$p(\theta) = f(t) - \rho_w \frac{\partial \phi}{\partial t} - \frac{\rho_w}{2} u_\theta^2, \tag{2.83}$$

in which u_θ is the tangential velocity at the cylinder's circumference (see Eq. C.83):

$$u_\theta = -2u_0(t) \sin \theta. \tag{2.84}$$

We note that normal velocity, u_n, is equal to 0 at the cylinder surface. Using the fact that the velocity potential, ϕ, is given by Eq. (C.80) and substituting Eq. (2.83) into (2.82) we obtain:

$$F = 2\rho_w \pi a^2 \frac{du_0}{dt}. \tag{2.85}$$

In general, the force induced by the accelerating fluid flow is usually represented as:

$$F = \rho_w C_m V \frac{du_0}{dt} = C_m \rho_w \pi a^2 \frac{du_0}{dt}, \tag{2.86}$$

in which V is the body volume, and C_m is the **inertia coefficient** which depends on the shape of the body. For the case of flow about a circular cylinder comparing Eq. (2.85) and Eq. (2.86), $C_m = 2$. It appears that in all cases, C_m is greater than unity; for example, 3/2 for a sphere.

To interpret the coefficient, C_m, from a physical point of view, let us rewrite Eq. (2.86) in a more general form:

$$F = \rho_w C_m V \frac{du_0}{dt} = \rho_w (1 + C_a) V \frac{du_0}{dt} = \rho_w V \frac{du_0}{dt} + \rho_w C_a V \frac{du_0}{dt}. \tag{2.87}$$

The quantity $\rho_w V$ is the mass of the object itself while the quantity $\rho_w C_a V$ is the added mass, with $C_a = C_m - 1$ termed the **added mass coefficient**. For the case of the cylinder, $C_a = 1$; thus the added mass exactly equals the cylinder mass. For the case of a sphere, the added mass is equal to half the cylinder mass, as $C_a = C_m - 1 = 1/2$.

As added mass is the mass of fluid 'carried along' by the object which remains stationary in accelerating fluid, an extra force $\rho_w C_a V (du_0/dt)$ is required to accelerate this added mass. The term $\rho_w V (du_0/dt)$ in Eq. (2.87) is a manifestation of the presence of the object in the fluid. The effect of this presence is the same as accelerating the volume of fluid equal to the object's volume, in the direction opposite to the fluid movement, in order to reflect the fact that the object is stationary.

It is instructive now to consider a case of a circular cylinder accelerating through a quiescent fluid. Is the force exerted on the cylinder by the fluid the same as in a case when the fluid accelerates past the cylinder? The cylinder induces a fluid motion, and this motion tends to zero when the distance from the cylinder tends to infinity. In each case, the exact rate of attenuation of the induced motion depends upon the shape of the body involved. Using a similar analysis to that above, we can find that the force induced by the fluid on the accelerating cylinder is:

$$F_a = \rho_w C_a V \frac{du_0}{dt}, \tag{2.88}$$

in which $C_a = 1$ for the cylinder, and $V = \pi a^2$ is the cylinder volume per unit length. Because the cylinder is moving in a fluid, it is obvious that a force is necessary to support its movement. The total force required to keep cylinder in motion through the fluid is:

$$F = F_a + \rho_c V \frac{du_0}{dt} = (\rho_c + C_a \rho_w) V \frac{du_0}{dt} = (\rho_c + C_a \rho_w) \pi a^2 \frac{du_0}{dt}, \tag{2.89}$$

in which ρ_c is the density of the cylinder, and $C_a = C_m - 1$.

Depending on whether the mass of the cylinder is greater or less than that of the displaced water, the total force required to accelerate cylinder in quiescent, ideal fluid could be greater or less than for the case of water accelerated past the cylinder. When dealing with various marine organisms either moving through the water column or residing on the sea bottom, we encounter both situations. There is quite extensive data of theoretically derived or empirically measured values of the inertia coefficient, C_m, for a variety of shapes. Some of them, mostly taken from Sarpkaya and Isaacson (1981), have been collected in Fig. 2.29.

In the general case of a real fluid, the inertia force is accompanied by a drag force induced by the fluid viscosity and its separation from the body. The following empirical formula is used to represent the combination of both forces acting on piles (Morison et $al.$, 1950):

$$F = \frac{1}{2} \rho_w C_d S u_0^2 + \rho_w C_m V \frac{du_0}{dt}. \tag{2.90}$$

It should be noted that in a real fluid both, C_d and C_m coefficients are not constants but complex functions of the Reynolds number and frequency of flow variation.

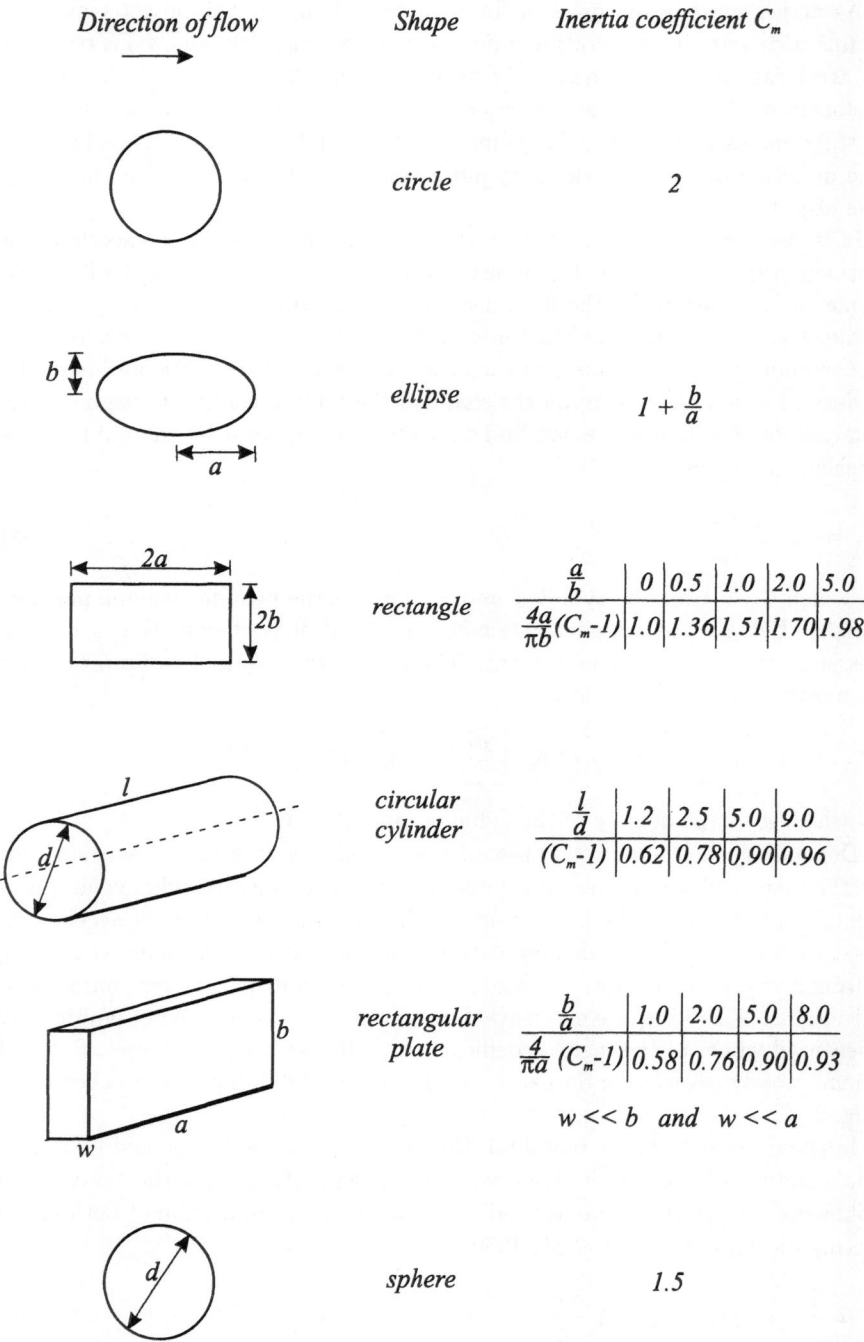

Direction of flow	Shape	Inertia coefficient C_m
	circle	2
	ellipse	$1 + \dfrac{b}{a}$

rectangle

$\dfrac{a}{b}$	0	0.5	1.0	2.0	5.0
$\dfrac{4a}{\pi b}(C_m-1)$	1.0	1.36	1.51	1.70	1.98

circular cylinder

$\dfrac{l}{d}$	1.2	2.5	5.0	9.0
(C_m-1)	0.62	0.78	0.90	0.96

rectangular plate

$\dfrac{b}{a}$	1.0	2.0	5.0	8.0
$\dfrac{4}{\pi a}(C_m-1)$	0.58	0.76	0.90	0.93

$w << b \quad and \quad w << a$

	sphere	1.5

Fig. 2.29: Inertia coefficient for various shapes (adapted from various sources)

Until now we have discussed the inertia force on a body far away from a solid surface, under the assumption of a perfect fluid. Experiments as well as theoretical analysis showed that the closer the body to a solid boundary, the higher the inertia coefficient. For example, Wilson and Reid (1963) reported that the C_m coefficient for a pipeline seated on the sea floor reached a value of 3.29. Massel and Done (1993) in their analysis of hard coral stability under cyclone induced waves theoretically showed that the inertia coefficient, C_m, for spherical coral on the substratum is $51/32 \approx 1.59$, which can be compared with the value of 1.5, corresponding to a sphere in infinite fluid.

2.6.5 Bodies Falling in Fluid

Organic and inorganic particles suspended in sea water are subjected to acceleration due to gravity and settle through the water column. Their vertical velocity increases, but there is some limit to the velocity growth due to increasing drag. The faster the particle moves, the greater the drag acting on the particle. At some velocity, the upward drag force becomes equal to the downward force of the particle's weight, reduced by the buoyancy according to Archimedes law (see Sect. 2.2). At this velocity there is no net force acting. Therefore according to Newton's first law, acceleration ceases and the body moves with a constant velocity. This highest velocity which a body reaches in free fall is called the **terminal velocity**.

To demonstrate the balance of forces controlling the terminal velocity, we will determine this velocity for a small sphere of diameter D (Fig. 2.30). At terminal velocity, the balance of forces can be written as follows:

$$\text{weight} - \text{buoyancy} = \text{drag}. \tag{2.91}$$

As the weight and buoyancy of sphere are $(4/3)\pi \rho_s g (D/2)^3$ and $(4/3)\pi \rho_w g (D/2)^3$, respectively, the left-hand side of Eq. (2.91) becomes:

$$\text{weight} - \text{buoyancy} = \frac{4}{3}\pi \rho_w g \left(\frac{\rho_s}{\rho_w} - 1\right) \left(\frac{D}{2}\right)^3, \tag{2.92}$$

in which ρ_s is the density of sphere. The drag force for a small sphere, moving slowly in calm water, is accurately predicted by Stokes' equation (C.56), i.e.:

$$\text{drag} = 3\pi \rho_w \nu D w, \tag{2.93}$$

where w is the terminal velocity. Equating drag and reducing weight gives the following expression for terminal velocity:

$$w = \left(\frac{\rho_s}{\rho_w} - 1\right) \frac{gD^2}{18\nu}, \tag{2.94}$$

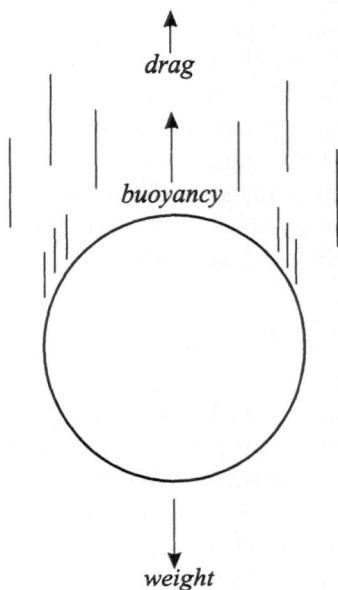

drag

buoyancy

weight

Fig. 2.30: Balance of weight and drag forces for sphere

which is valid for very small spherical particles and very small Reynolds numbers, $Re = wD/\nu < 0.2$. For example, for organic suspended particles with density close to water, say $\rho_s = 1.1 \times 10^3$ kg/m^3 (or 1.1 g/cm^3), a typical terminal velocity w is 1.23 m/year for particle diameter $D = 1\mu m$ and 1105 m/year for diameter $D = 30\mu m$ (Dera, 1992).

The terminal velocity for mineral particles which are not spherical combine the effects of grain size, shape, and composition of the fluid and viscosity. This velocity can only be determined experimentally. Numerous laboratory experiments suggest the following formulae for terminal velocities of commonly occurring fine, medium and coarse sands (Hallermeier, 1981):

$$\frac{wD}{\nu} = \begin{cases} \dfrac{A}{18} & \text{for} \quad A < 39 \\[2ex] \dfrac{A^{0.7}}{6} & \text{for} \quad 39 < A < 10^4 \\[2ex] 1.05\,A^{0.5} & \text{for} \quad A > 10^4, \end{cases} \tag{2.95}$$

in which $A = [(\rho_s/\rho_w) - 1]\,(gD^3/\nu^2)$ is the 'buoyancy index'. Normalized empirical terminal velocity (wD/ν) for sand particles and spheres, as a function of buoyancy index, is shown in Fig. 2.31. For very fine particles (or for buoyancy index $A < 39$), terminal velocity is equal to that given by the Stokes' formula

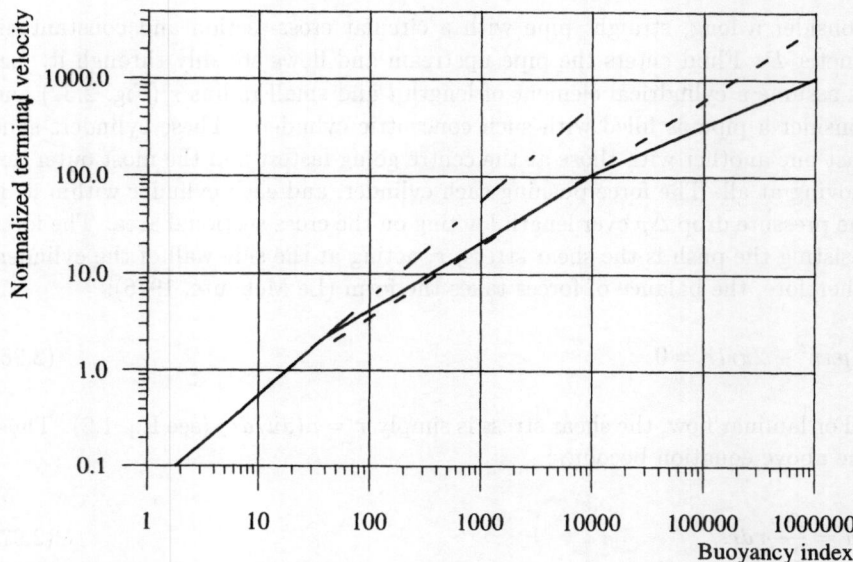

Fig. 2.31: Terminal velocity for natural sand and sphere: solid line – velocity for sand particles, dashed line – velocity for spheres, dash-dot line – velocity for very small spheres (accord. to the Stokes formula)

(2.94). When buoyancy index increases, terminal velocity for sand particles becomes smaller than the velocity for spheres, however Stokes' formula is not applicable for this buoyancy index range.

2.7 Laminar and Turbulent Flows in Ducts

2.7.1 Introduction

Previous sections have mostly been dedicated to flows outside a body or body movements in a fluid. However, everyday experience provides numerous examples of flow inside human-made installations as well as natural pipes and ducts. Marine organisms are filled with pipes and channels through which fluids flow. Internal flow becomes very complex because of the shape and varying cross-sectional areas of conduits. Moreover, most of the internal systems of animals involve flows of putative non-Newtonian fluids (see Chap. 12) in ducts.

In contrast to external flows, which are either laminar, transitional or turbulent, flow within pipes or other structures of biological interest is mostly laminar. Thus, we begin with laminar flows, with turbulent motion explored later.

2.7.2   Laminar Flow in Ducts

Consider a long, straight pipe with a circular cross-section and constant diameter D. Fluid enters the pipe upstream and flows steadily through it. Let us assume a cylindrical element of length l and small radius r (Fig. 2.32) and consider a pipe as filled with such concentric cylinders. These cylinders slide past one another with those at the centre going fastest and the most outer not moving at all. The force pushing each cylinder, and each cylinder within it, is the pressure drop Δp over length l acting on the cross-sectional area. The force resisting the push is the shear stress, τ, acting at the side wall of the cylinder. Therefore, the balance of forces takes the form (Le Méhauté, 1976):

$$\Delta p \pi r^2 - 2\pi r l \tau = 0. \tag{2.96}$$

For laminar flow, the shear stress is simply $\tau = \mu(du/dr)$ (see Eq. 1.2). Thus, the above equation becomes:

$$du = \frac{\Delta p}{2\mu l} r \, dr. \tag{2.97}$$

Integration with r as the variable gives:

$$u = \frac{\Delta p}{2\mu l} \frac{r^2}{2} + C. \tag{2.98}$$

The constant, C, is determined by the boundary condition at the pipe wall ($r = D/2$) where $u = 0$; thus, the velocity distribution in the pipe becomes:

$$u = \frac{\Delta p}{4\mu l} \left[\left(\frac{D}{2}\right)^2 - r^2 \right], \tag{2.99}$$

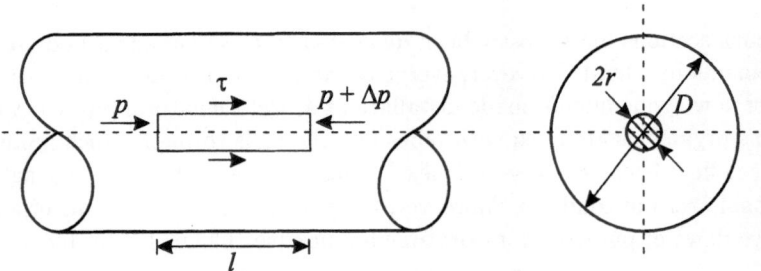

Fig. 2.32: Balance of forces within a pipe

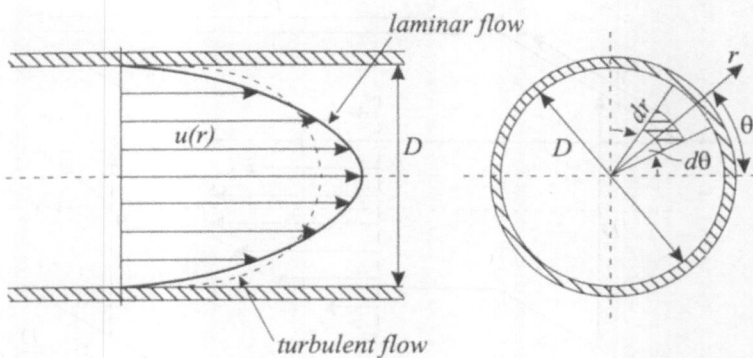

Fig. 2.33: Velocity distribution in a pipe for laminar and turbulent flow

showing that it has a parabolic shape (Fig. 2.33). The maximum value of $(\Delta p/4\mu l)(D/2)^2$ appears at the pipe centre $(r = 0)$. At the pipe wall $(r = D/2)$, velocity is zero, while the velocity gradient is linear, *i.e.*:

$$\frac{du}{dr} = \frac{-\Delta p}{2\mu l}r \quad \text{at} \quad r \to \frac{D}{2}. \tag{2.100}$$

A small area in the polar coordinate system (r, θ) is $ds = r\,dr\,d\theta$ (Fig. 2.33). Thus a small volume of flow through this area is:

$$dQ = u(r, \theta) \times dS = u(r, \theta)r\,dr\,d\theta. \tag{2.101}$$

The overall volume of flow becomes:

$$Q = \int_0^{2\pi} \int_0^{D/2} u(r, \theta)r\,dr\,d\theta. \tag{2.102}$$

As the velocity distribution is symmetrical along the pipe axis and is independent on angle θ, the integration versus θ simply is 2π. Hence, substituting velocity u into Eq. (2.102) and integrating with r finally gives a volume flow per unit time:

$$Q = \frac{\pi \Delta p D^4}{128\mu l}. \tag{2.103}$$

This is the 'Hagen-Poiseuille equation' for flow in pipes. Dividing the volume of flow per pipe cross-section we obtain the average velocity $\bar{u}$ of flow in the pipe as:

$$\bar{u} = \frac{\Delta p D^2}{32\mu l}. \tag{2.104}$$

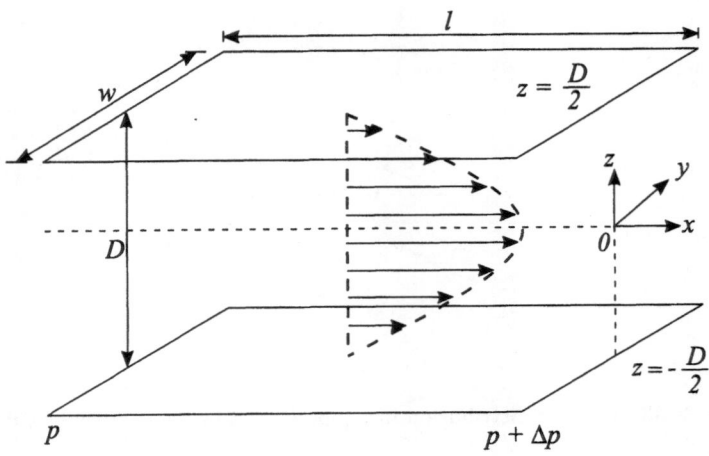

Fig. 2.34: Laminar flow between a pair of flat plates

It is easy to see from Eq. (2.99) that the maximum velocity is exactly twice the average velocity, *i.e.*:

$$u_{\text{max}} = 2\bar{u}. \tag{2.105}$$

The power, P, needed to push fluid through a pipe is equal to force multiplied by velocity. In particular, the total net force applied to the fluid is $\Delta p \pi (D/2)^2$. Using the average velocity to characterize the flow, for power, P, we obtain:

$$P = \frac{\pi \Delta p D^2}{4} \frac{\Delta p D^2}{32\mu l} = \frac{\pi D^4 (\Delta p)^2}{128\mu l}. \tag{2.106}$$

Without particular difficulty, the result for a circular pipe can be extended to the case of laminar flow within a channel between a pair of closely spaced flat plates (Fig. 2.34). Assuming that flow is uniform in the y direction, the vertical distribution of velocity becomes:

$$u(z) = \frac{\Delta p}{2\mu l} \left[\left(\frac{D}{2} \right)^2 - z^2 \right]. \tag{2.107}$$

The profile remains parabolic, however the parabola is slightly different from that for a circular pipe. An analogue of the Hagen-Poiseuille equation (2.103) for flat plates takes the form:

$$Q = \frac{\Delta p w D^3}{12\mu l}, \tag{2.108}$$

in which w is the width of the plates. Thus, average and maximum velocities are:

$$\bar{u} = \frac{\Delta p D^2}{12\mu l}, \qquad u_{\max} = \frac{\Delta p D^2}{8\mu l}. \tag{2.109}$$

Now, the maximum velocity is only one and a half times the average. There are quite a few biological situations which involve closely spaced and parallel flat plates such as, for example, flow through the gill of fish. This and other internal flows in marine organisms are discussed in more detail in Part III of this book.

At the end of this section we note that the above relationships are valid for laminar flow when the Reynolds number is below 2000 (see Sect. 2.4) or so, depending on the smoothness of the pipe or plate walls and entry upstream. The threshold Reynolds number $Re \approx 2000$ is about 0.4 m/s. Within organisms these limits are rarely exceeded and the internal flow is mostly laminar.

2.7.3   Turbulent Flow in Ducts

Although laminar flow dominates within marine organisms, there are many situations when organisms find themselves in a turbulent environment. As the flow becomes turbulent, the analytical determination of the velocity distribution is not possible. Within a pipe flow, momentum is transported across the pipe resulting in a velocity along the axis which is less than twice the average speed of laminar flow. In contrast to laminar flow, the roughness of the inside wall of a pipe plays a decisive role in the determination of the pipe's resistance.

To derive the velocity distribution in a pipe, we assume that velocity, u, at some distance from the pipe wall depends on the tangential stress, τ_0, at the wall, viscosity, μ, and density, ρ, of the fluid. Dimensional analysis as well as a detailed examination of experimental data and logical arguments given in Sect. 2.5.4 suggest that the velocity profile obeys the logarithmic law (see Fig. 2.33). An advantage of such laws, as compared with the $1/n$th-power law (see Eq. 2.56) consists of its being an asymptotic expression for very large Reynolds numbers. Experiments conducted by Nikuradse yields the following form of the velocity distribution for very large Reynolds numbers and for smooth pipes (Schlichting, 1960):

$$\frac{u(z)}{u^*} = 2.5 \ln\left(\frac{u_* z}{\nu}\right) + 5.5 \quad \text{for} \quad \frac{u_* z}{\nu} > 70, \tag{2.110}$$

or:

$$\frac{u(z)}{u^*} = 5.75 \log_{10}\left(\frac{u_* z}{\nu}\right) + 5.5 \quad \text{for} \quad \frac{u_* z}{\nu} > 70. \tag{2.111}$$

Note the different logarithms used in the above relationships. The equations above for turbulent flow are valid for situations where the laminar shearing stresses can be neglected in comparison with the turbulent stresses. Very close to the pipe wall, a laminar sublayer exists in which laminar stresses dominate and, as was shown for a flat plate (Eq. 2.51), velocity changes linearly with distance from the wall, *i.e.*:

$$\frac{u(z)}{u_*} = \frac{u_* z}{\nu} \quad \text{for} \quad \frac{u_* z}{\nu} < 5. \tag{2.112}$$

In an intermediate range $5 < u_* z/\nu < 70$, both contributions, laminar and turbulent, are of the same order of magnitude.

Assuming a velocity distribution as in Eq. (2.111), and integrating over a pipe cross-section, an average velocity, $\bar{u}$, results as follows:

$$\frac{\bar{u}}{u_*} = 5.75 \log_{10} \left(\frac{u_* D}{2\nu} \right) + 1.75. \tag{2.113}$$

Most pipes and ducts in marine organisms as well as those made by humans can not be regarded as smooth, at least at higher Reynolds numbers. Surface roughness induces resistance to flow and the effect of the roughness elements depends on the relationship between the height of elements and the laminar sublayer. When the sublayer is so thick that it covers the roughness elements, the surface can be considered as hydrodynamically smooth. However, if the size of the roughness elements are large compared with the laminar sublayer, the surface is completely rough and the effect of viscosity can be neglected.

To introduce the roughness effect into the formula for velocity distribution in a pipe, we present Eq. (2.111) in more general form:

$$\frac{u(z)}{u_*} = 5.75 \log_{10} \left(\frac{z}{k_s} \right) + B \left(\frac{u_* k_s}{\nu} \right), \tag{2.114}$$

where k_s is the roughness size; factor B depends on the so called 'shear Reynolds number', $u_* k_s/\nu$. A notation k_s was introduced by Nikuradse who used closely packed sand grain roughness elements in his study of flow resistance in pipes.

In general, it is necessary to consider three regions for the factor B (Schlichting, 1960):

1. hydraulically smooth regime when $0 \leq u_* k_s/\nu \leq 5$; the size of the roughness elements is very small, covered totally by the laminar sublayer (Fig. 2.17). The factor $B = 5.5 + 2.5 \ln (u_* k_s/\nu)$ (Fig. 2.35);

2. transition regime when $5 \leq u_* k_s/\nu \leq 70$. Some of the roughness elements extend outside the laminar sublayer and contribute some resistance. The value of B in the transition region is shown in Fig. 2.35;

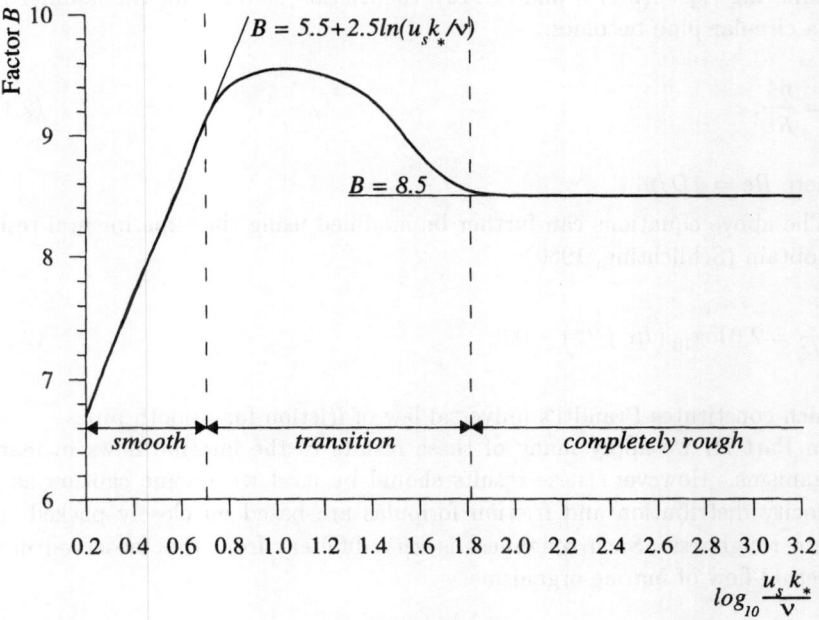

Fig. 2.35: Factor B (adapted from Schlichting, 1960)

3. fully rough regime when $u_*k_s/\nu > 70$. All the roughness elements reach outside the laminar sublayer and the predominant part of the resistance to flow is due to drag acting on them. Further increase of the Reynolds number brings no change of flow pattern, and consequently factor B remains constant, $B = 8.5$ (Fig. 2.35).

Finally, we return for a moment to an average velocity, $\bar{u}$, for laminar flow given by Eq. (2.104). After some rearrangements, this equation becomes:

$$\frac{\Delta p}{\rho_w g} = \frac{64\mu l}{\bar{u}\rho_w D^2}\frac{\bar{u}^2}{2g}. \tag{2.115}$$

This equation expresses a relationship between the piezometric head and velocity head (see Sect. 2.3.5) and is valid only for laminar flow in smooth pipes. However, for real fluid flow in non-smooth pipes, some pressure loss Δp or head loss ΔH can be expected due to drag at the pipe wall and shear in the turbulent motion itself. This is usually expressed by the Darcy-Weisbach equation as:

$$\Delta H = \frac{\Delta p}{\rho_w g} = f\frac{l}{D}\frac{\bar{u}^2}{2g}, \tag{2.116}$$

in which f is the friction factor. The head loss in Eq. (2.116) is proportional to the length of pipe and to the quadratic flow velocity, as is expected. Therefore, comparing Eqs. (2.115) and (2.116), the friction factor f for the laminar flow in a circular pipe becomes:

$$f = \frac{64}{Re},$$ (2.117)

where $Re = \bar{u}D/\nu$.

The above equations can further be modified using the experimental results to obtain (Schlichting, 1960):

$$\frac{1}{f^{1/2}} = 2.0 \log_{10}\left(Re\, f^{1/2}\right) - 0.8,$$ (2.118)

which constitutes Prandtl's universal law of friction for smooth pipes.

In Part III we apply many of these results to the internal flows in marine organisms. However, these results should be used with some caution as the velocity distribution and friction formulas are based on closely packed sand grain roughness. Such roughness is very different from that observed in the internal flow of marine organisms.

Part II

Oceanic Hydrodynamic Processes

Organic Thermodynamic Processes

3 An Introduction to Surface Waves

3.1 Introduction

In Part I of this book various phenomena which fall within the scope of classic fluid mechanics were described. However, in the world's ocean there are many large scale flows, such as waves, large ocean currents, or tides, which are beyond the scope of classic fluid mechanics and usually not treated in fluid mechanics textbooks. All of these flows are the subject of examination within a discipline known as **geophysical fluid dynamics**, which emerged in the late 1950s. Geophysical fluid dynamics deals with the large-scale motion observed in various atmospheric and oceanic systems, but nevertheless governed by similar dynamics.

In Part II of this book we will describe a number of phenomena which play an important role for ecology of marine organisms. Thanks to advances in geophysical fluid dynamics, reliable predictions of the behaviour of some elements of the ocean-atmospheric system are now available. These form the basis for subsequent prediction of the potential response of living oceanic resources.

In the next six chapters we will discuss small- and large-scale oceanic phenomena, such as waves, tides, internal waves, currents and water mass transport. The linkage of these phenomena with marine ecology will be the subject of discussion in Part III of this book.

3.2 Nature of Wave Motion

Waves are a common phenomenon in everyday life. Sound, the motion of plucked strings, ripples on a pond, oscillations of the ocean surface, signals from satellites, radio and television stations are all examples of various types of waves. However, in this book we will concentrate only on ocean waves. Before attempting to classify various types of water waves, let us consider some of the observable characteristics of wave motion:

1. A wave transfers a disturbance from one part of a material to another. For example, the disturbance caused by dropping a stone into a pond is transmitted across the pond in the form of ripples.

2. The disturbance is propagated through the material without any substantial overall motion of the material itself. A floating cork merely bobs up and down on the water surface ripples, but experiences very little overall movement in the direction of travel of the ripples. Closer observation shows, however, that the cork describes an almost circular path in a vertical plane parallel with the direction of wave motion.

Thus, as wave motion is characterized by the propagation of form while the material itself is not transported, the question arises as to what quantity is transported by waves? This quantity is 'energy'. In particular, waves on a water surface transport mechanical energy, electromagnetic waves transport electromagnetic energy and sound waves cause vibration of air particles. Therefore, we can generally define wave motion as a process whereby energy is transported across or through a material without any significant overall transport of the material itself. A wave is not a substance, such as water, air or rock; it is a pattern of behaviour.

3.3   Types of Waves

Waves which travel through materials are called 'body waves'. Seismic waves and sound waves are examples of body waves. However, in this book our main concern is with **surface waves**. The most familiar surface waves are those which occur at the interface between atmosphere and ocean, but surface waves can occur at the interface between any two bodies of fluid. For example, waves occur at the interface between two layers of ocean water of different densities. Because the interface is a surface, such waves are, strictly speaking, surface waves. However, in oceanography they are known as **internal waves**. They will be considered in more detail in Chap. 6.

All surface waves can be regarded as either **progressive waves**, where energy moves across the surface, or as **standing waves** which can be considered as the sum of two progressive waves of equal dimensions, but travelling in opposite directions. The motion of an elastic rope, horizontally oriented and fixed at one end but moving vertically up and down at the other, describes a standing wave.

Waves can also be classified depending on the nature of the restoring forces which maintain their propagation. There are two main kinds of restoring forces for surface waves, namely gravitational forces exerted by the Earth, Sun or Moon, and surface tension. The attraction of sea water by the Sun and Moon causes waves which have periods coinciding with the periodic motion of the Earth and Moon. Because of the special, global character of these waves commonly called **tides**, they are usually treated separately from other surface waves (see Chap. 5).

Most surface waves are generated by non-periodic forces. Winds blowing on the sea surface displace water particles from their equilibrium position. These waves are called **wind-induced waves** or simply **wind waves** (see

Sect. 3.5). Earthquakes or eruptions of underwater volcanos move the sea bed which in turn displaces the sea surface. Such wave is known as **tsunami**. More details are given in Sect. 3.6. Land masses sliding into reservoirs and bays, or meteorological disturbances can create large standing waves known as **seiches** (see Sect. 3.7).

In general, surface water waves are affected by both gravitational force and surface tension. For very short waves (wavelength less than about 1.7 cm), surface tension is the dominant restoring force. These waves are called **capillary waves**, and are important in the context of processes controlling interactions between the atmosphere and ocean, and in remote sensing of the ocean surface.

In most oceanographic and other applications, the interest lies with longer surface waves where the dominant maintaining force is gravity, and hence, they are known as **gravity waves**. A summary of wave types, their causes and energy contents is given in Fig. 3.1.

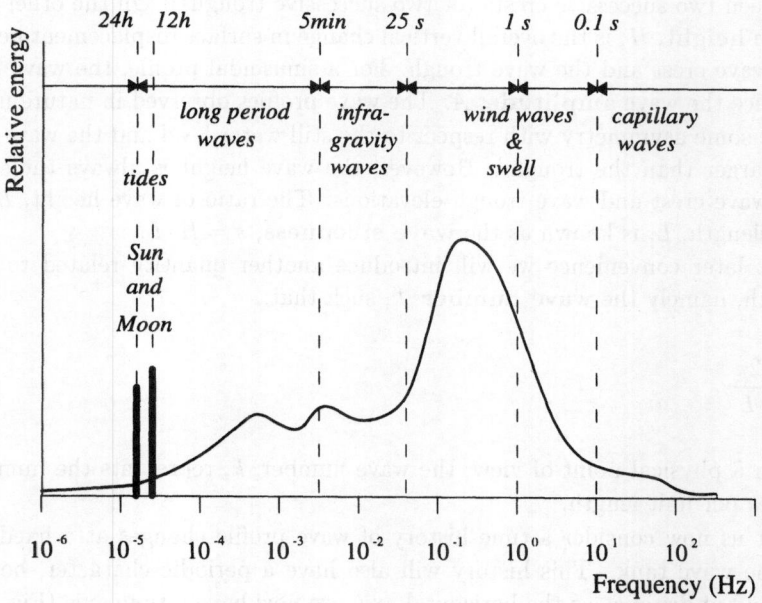

Fig. 3.1: Types of surface waves and relationship between wavelength, wave frequency, nature of displacement forces, and the relative amounts of energy in each type of wave (adapted from Massel, 1996a)

3.4   Basic Wave Characteristics

3.4.1   Definition of Surface Wave Dimensions

In general, there are many forms of wave type oscillations of water mass in the ocean. We will start here with the simple form of periodic waves. The water particle motion resulting from the restoring force acting during one wave cycle provides the displacing force for the next cycle. Such alternate displacements and restorations establish oscillatory motion. The simplest form of such motion is the sinusoidal form.

Consider, for example, a narrow wave tank equipped with a wave-maker at the end of the tank. A 'snapshot' taken through the glass wall of the tank, or recording water oscillations at a given point, provides a picture of wave motion in the tank. In particular, the **surface displacement**, ζ, of water particles from the undisturbed water level varies with distance at a fixed instant of time (Fig. 3.2a), as well as varying with time at a fixed point in the tank (Fig. 3.2b). A simulation given in computer program D.31 (see Appendix D) clarifies the difference between representations of wave motion in both time and space.

The basic horizontal wave dimension is a **wavelength**, L. This is the distance between two successive crests (or two successive troughs). On the other hand, **wave height**, H, is the overall vertical change in surface displacement between the wave crest and the wave trough. For a sinusoidal profile, the wave height is twice the wave **amplitude**, A. The wave profiles observed in nature usually show some asymmetry with respect to the still water level and the wave crests are larger than the troughs. However, the wave height is always the sum of the wave crest and wave trough elevations. The ratio of wave height, H, and wavelength, L, is known as the **wave steepness**, $s = H/L$.

For later convenience we will introduce another quantity related to wavelength, namely the **wave number**, k, such that:

$$k = \frac{2\pi}{L}. \tag{3.1}$$

From a physical point of view, the wave number, k, represents the number of waves per unit length.

Let us now consider a time history of wave profile changes at a fixed point in the wave tank. This history will also have a periodic character, however, instead of distance as the horizontal axis, we now have a time axis (Fig. 3.2b).

The time interval between two successive crests (or two successive troughs) passing a fixed point is known as the **wave period**, T. The number of crests (or number of troughs) which pass a fixed point per second is called the **frequency**, f. Thus,

$$f = \frac{1}{T}. \tag{3.2}$$

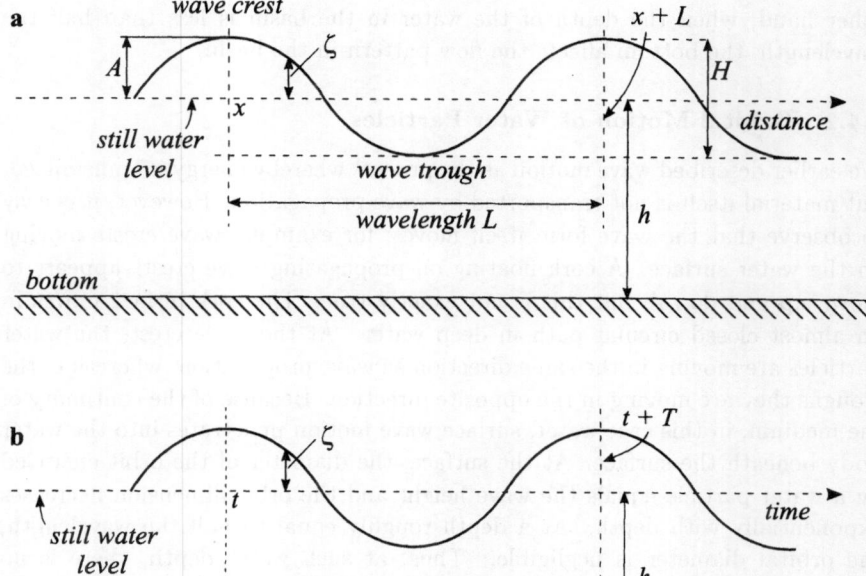

Fig. 3.2: Vertical profile of idealized sinusoidal surface wave: **a** at a given instant of time, **b** at a given location

The frequency, f, is measured in hertz, Hz, and 1 Hz = 1 cycle/s. Apart from frequency, the **angular frequency**, ω, is used in this book. It is expressed as the number of radians per second. We note that 1 radian = $360°/2\pi \sim 57.3°$. Therefore, the wave motion with frequency, f, has an angular frequency, ω, equal to:

$$\omega = 2\pi f = \frac{2\pi}{T}. \tag{3.3}$$

The profile of real wind induced waves is not necessary regular and periodic, but is usually very irregular and random. We will discuss random, irregular waves in Sect. 3.5 and 4.3.

In this book we will frequently distinguish between deep and shallow water depths, in the context of surface waves. Water is considered 'deep' when water depth is larger than half of the wavelength. Otherwise water is treated as 'shallow'. Thus, if the water basin is deeper than one half wavelength, the water at the bottom is virtually stationary whether or not the bottom is there, as the presence or absence of a solid bottom makes little difference. On the other hand, when the depth of the water in the basin is less than half the wavelength, the bottom affects the flow pattern in the basin.

3.4.2   Orbital Motion of Water Particles

We earlier described wave motion as the process whereby energy is transmitted, but material itself is not transported by wave propagation. However, it is easy to observe that the wave form itself moves; for example, wave crests moving on the water surface. A cork floating on propagating wave crests appears to move 'up and down', but this is not exactly true. The cork in fact moves in an almost closed circular path in deep water. At the wave crest, the water particles are moving in the same direction as wave propagation, whereas in the troughs they are moving in the opposite direction. Because of the continuity of the medium, in this case water, surface wave motion penetrates into the water body beneath the surface. At the surface, the diameter of the orbit encircled by a water particle equals the wave height and the orbit dimension decreases exponentially with depth. At a depth roughly equal to half the wavelength, the orbital diameter is negligible. Thus, at such water depth, there is no displacement of the water particles due to surface waves (see Fig. 3.3).

In shallow water, the particle orbits are elliptical with both axes decreasing with depth. At the sea bottom, only horizontal motion exists (Fig. 3.4). The relationships between the intensity of wave attenuation and submergence will be developed in Sect. 4.2.

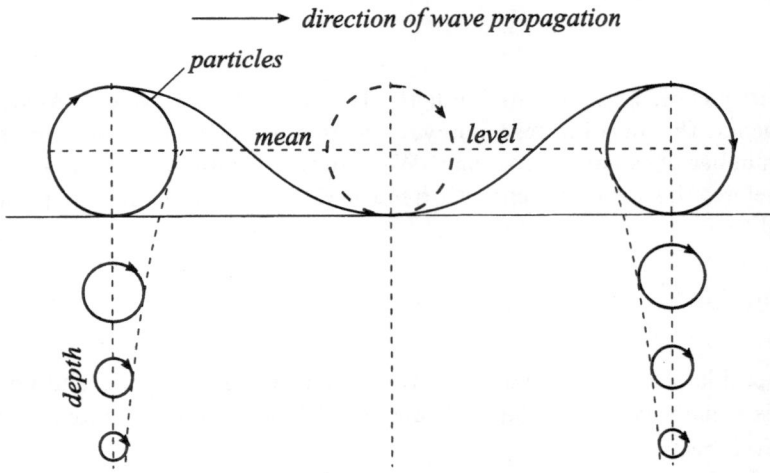

Fig. 3.3: Circular pattern of particle motion in deep water; radius of orbits decreases exponentially with submergence

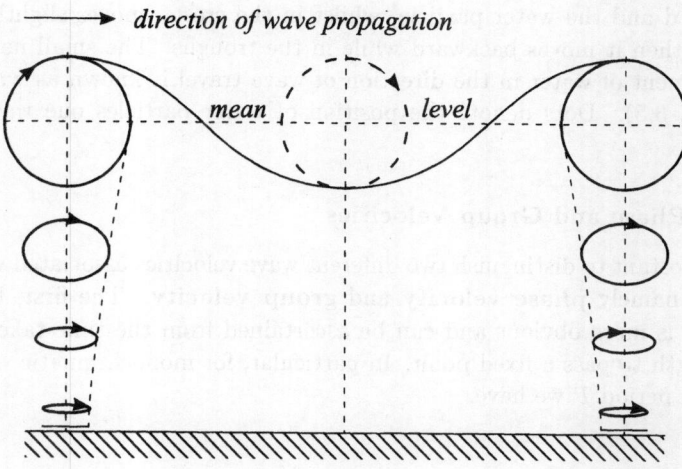

Fig. 3.4: Elliptical pattern of particle motion in shallow water; axes of ellipses decrease with submergence

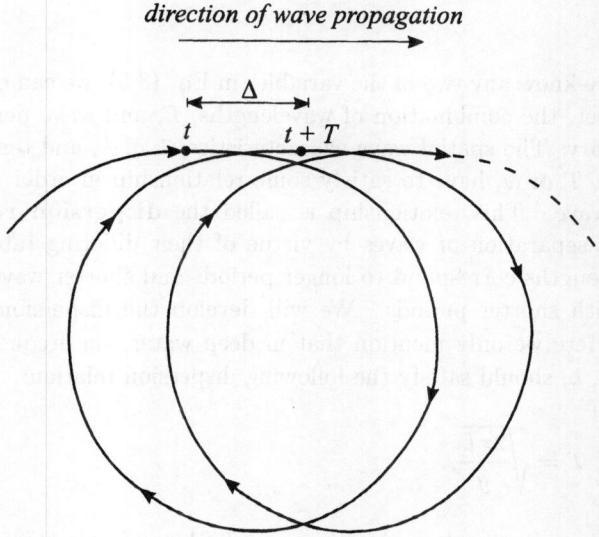

Fig. 3.5: Forward displacement of particle in deep water; Δ denotes a particle drift after one wave period

The circular (or elliptical) orbits of water particles are only an approximate representation of real particle motion. There is always a small net component of forward motion, particularly in waves of large amplitude. The orbits are not closed and the water particle, whilst in the crests, moves slightly further forward then it moves backward while in the troughs. The small net forward displacement of water in the direction of wave travel is known as **wave drift** (see Fig. 3.5). Dots denote the position of water particles one wave period apart.

3.4.3 Phase and Group Velocities

It is important to distinguish two different wave velocities associated with wave motion, namely **phase velocity** and **group velocity**. The first, the phase velocity, is more obvious and can be ascertained from the time taken for one wavelength to pass a fixed point. In particular, for monochromatic wave train with the period T we have:

$$C = \frac{L}{T}. \tag{3.4}$$

So velocity, C, is a measure of the speed at which a particular wave phase (for example the wave crest) is propagating on the sea surface. Combining Eqs. (3.3), (3.4) and (3.1) we obtain:

$$C = \frac{L}{T} = \frac{\omega}{k}. \tag{3.5}$$

Basically, if we know any two of the variables in Eq. (3.5), we can calculate the third. However, the combination of wavelengths, L, and wave period, T, can not be arbitrary. The spatial wave characteristics, L or k, and temporal wave characteristic, T or ω, have to satisfy some relationship in order to represent the surface wave. This relationship is called the **dispersion relation**. It expresses the separation of waves by virtue of their differing rates of travel; greater wavelengths correspond to longer periods and shorter wavelengths are associated with shorter periods. We will develop the dispersion relation in Sect. 4.2.2. Here we only mention that in deep water, the frequency, ω, and wave number, k, should satisfy the following dispersion relation:

$$\omega = \sqrt{gk} \quad \text{or} \quad T = \sqrt{\frac{2\pi L}{g}}, \tag{3.6}$$

in which g is gravitational acceleration, $g = 9.81 \text{ m/s}^2$. Combining Eq. (3.5) and (3.6) we find that in deep water the phase velocity C is:

$$C = \sqrt{\frac{gL}{2\pi}}. \tag{3.7}$$

Group velocity is a quantity not as easily measured as the phase velocity. To understand the relationship between phase velocity and group velocity, let us examine a band of ripples created by tossing a stone into a still pond. The band gets wider with increasing distance from the original disturbance. Ripples of greater wavelength progressively travel faster than shorter ones due to the dispersion effect. More careful observation shows that each individual ripple travels faster than the band of ripples. The velocity of the band is called the **group velocity**, which is about half the phase velocity of the individual ripples which travel through that band. In order to justify this relation, we will follow arguments by Barber (1969), and discuss an idealized example of two waves of the same amplitude but with a relatively small difference in wavelength, say ΔL (see Fig. 3.6a). Superposition of these two waves results in group wave pattern as shown in Fig. 3.6b.

If we look at two wave crests immediately following each other, one lags by a distance L, the other by a distance $L + \Delta L$, so that the crest of the longer train is a distance ΔL behind the crest of the other. The train of longer wavelength travels at a different speed, say $C + \Delta C$, instead of C. The lag, ΔL, is made up in a short time Δt:

$$\Delta t = \frac{\Delta L}{\Delta C}. \tag{3.8}$$

Both wave trains move forward during this time and the place where the crests coincide moves back, relative to them, by one whole wavelength. The highest wave is located in the middle of the wave group, which advances a shorter distance than the waves themselves. This distance is:

$$\Delta x = C \Delta t - L. \tag{3.9}$$

The first term is the distance travelled by the wave crest during the time Δt and the second term is the one wavelength the group has dropped back. Therefore, the velocity of the wave group becomes:

$$C_g = \frac{\Delta x}{\Delta t} = C - \frac{L}{\Delta t} = C - L\left(\frac{\Delta C}{\Delta L}\right). \tag{3.10}$$

For small ΔC and ΔL, Eq. (3.10) becomes:

$$C_g = C - L\frac{dC}{dL}. \tag{3.11}$$

Therefore, differentiating and substituting into Eq. (3.11) we finally obtain:

$$C_g = \frac{1}{2}C, \tag{3.12}$$

showing that the groups travel at only half of the speed of the individual waves.

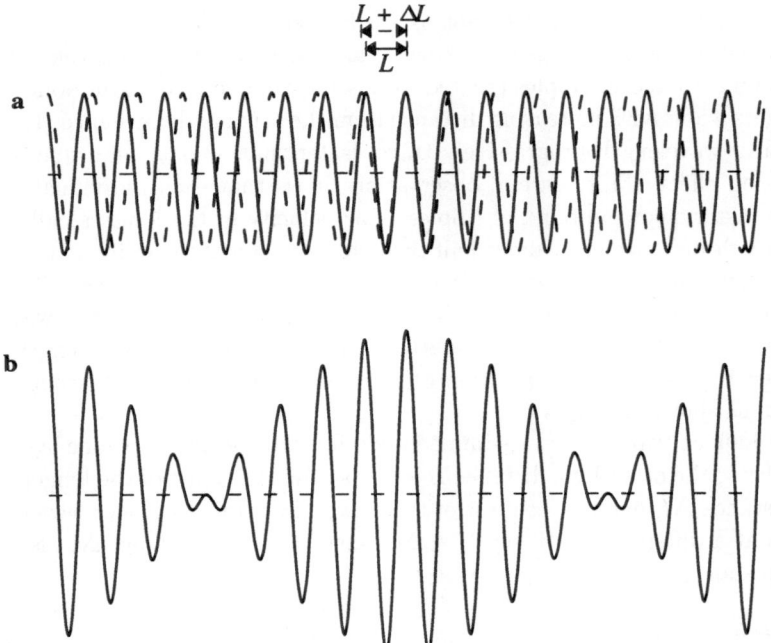

Fig. 3.6: The interference of waves: **a** two elementary waves, **b** resulting wave group

When water becomes shallower, the water depth becomes more important than the wavelength in determining wave velocity. As a result, wave velocity approaches the group velocity. In very shallow water, say when water depth $h < L/20$, all waves travel at the same depth-controlled velocity, $C_g = C = \sqrt{gh}$. Therefore, there is no wave-wave interaction and group velocity can be regarded as equal to phase velocity.

More rigorous derivation of the formulas for group velocity in deep- and shallow-waters is given in Sect. 4.2.2.

It was shown above that specific points of wave form, for example the wave crests, are transported at phase velocity. Let us now ask what quantity is transported at group velocity? In order to answer this question let us look again at Fig. 3.6b. The areas between groups are regions of minimal disturbance. Hence, they are the regions of minimal energy and no energy is transmitted across these regions. Energy is constrained within the wave group, and it is propagated at the group velocity. The rate at which energy is propagated is called the **energy flux**, and it is a product of wave energy, E, and group velocity, C_g:

$$F = C_g E. \tag{3.13}$$

In the SI system of dimensions, the energy is expressed in [N $\cdot$ m]; therefore the dimension of energy flux (F) will be [N$\cdot$ m$^2\cdot$ s^{-1}] (see also Appendix B).

3.5 Wind-Generated Waves

3.5.1 How Does Wind Generate Waves?

The nature of external forces acting on ocean water dictates what types of waves can be induced in the ocean. The most obvious cause of surface waves is the action of wind. The ancient Greeks were well aware of the interaction between the atmosphere and the sea surface. Aristotle (384–322 B.C.) realized that wind acting on the sea surface played a very important role in the development of waves. Pliny (AD 23–79) observed that oil poured upon waves calmed them. However, from the time of Aristotle to the Renaissance of science in the 'Golden Age of Discovery' in the late fifteenth century, very little progress was made towards understanding the generation and growth of waves. It was not until the nineteenth and twentieth centuries that more fundamental knowledge of what causes waves, and how they behave, was accumulated.

Historically, many approaches have been utilized to describe the basic mechanism for the transfer of wind energy to surface waves. Some of these approaches today possess only historical value, while others are still utilized in oceanographic practice. Present understanding of the wave generation mechanism is based on two models proposed by Phillips (1957) and Miles (1957, 1962).

The Phillips model is applicable in the early stage of wave generation while the Miles' predicts further wave growth. To explain these models, consider the sequence of events when, after a period of calm weather, wind starts to blow, increasing to a gale, and continues to blow at constant gale force for some considerable time. At the beginning, no significant wave growth occurs until the wind speed exceeds 1 m/s. Then, small steep waves form as the wind speed increases. At this initial stage, wind energy is transmitted to the water by pressure fluctuations. The natural flow of wind is always turbulent. The water pressure fluctuates randomly and the air particles start to move in random eddies throughout the air above the water surface, although the net direction of movement remains in the direction of the main flow. Random pulses of air pressure induce a resonant response of the water surface. Waves continue to grow in size and their speed increases. The 'resonance model' of Phillips (1957) assumes that there are no feedback reactions of the growing waves on the air layer above water surface.

However, growing waves start to disturb the shear airflow and to induce extra pressure. The shear flow of fluid, gas or water, is a flow where successive layers of fluid move with slightly different speeds. Each successive layer therefore shears over the layer beneath. The shearing air motion, modulated by the motion of the water surface, leads to a resonant coupling between the surface motion and the horizontal variation in the air pressure field. This coupling is a key element in the Miles' generation model, sometimes called a 'shear model'. These two wave generation models are compatible in the process of wave formation. At the initial stage, wave energy grows linearly in time, however, this growth becomes exponential for the later stage of wave generation (Fig. 3.7).

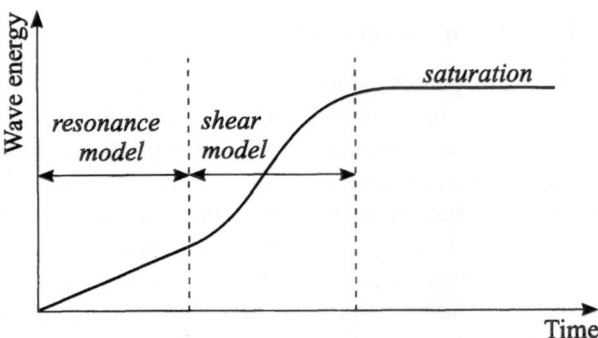

Fig. 3.7: Evolution of wave energy during wave growth

Wave growth does not continue infinitely, but is restricted by processes associated with wind action on the sea surface. Firstly, the phase velocity of waves increases gradually and finally approaches the wind velocity. When phase velocity becomes equal to or greater than the wind velocity, wind does not build the waves further (Massel, 1996a). Waves are at a saturation stage. The second process which contributes to the restriction of wave energy is wave breaking, which takes the form of **white caps** in deep water. A white cap is the tip of the wave crest which is breaking because it is being driven forward by the wind faster than the wave itself is travelling. The energy dissipated during white-capping is converted into forward momentum of the water itself, reinforcing the surface current. Wave breaking in shallow water is more complex than in deep water and we will examine this in more detail in the next chapter.

At the end of this section we point out the difference in the definition of directions of waves and wind. It is common in meteorology and oceanography to identify wind direction as the direction *from* which wind is coming. Thus, a northerly wind is a wind coming from the North and blowing to the South. In contrast, wave direction is identified as the direction *in which* waves are propagating. The same convention is used for ocean currents.

3.5.2   Fetch and Duration Limited Wave Growth

Surface waves resulting from the action of turbulent airflow are also very chaotic and randomly distributed on the sea surface. With some approximation we can say that the observed sea surface is a result of the superposition of many elementary sinusoidal waves, each of different amplitude and phase, which move into and out of phase with and across each other. A simple illustration of such a superposition is given in Fig. 3.8, where the sum of 13 elementary sinusoidal waves produces the final wave profile. At the bottom of this figure, a bar chart displays the energy of each component.

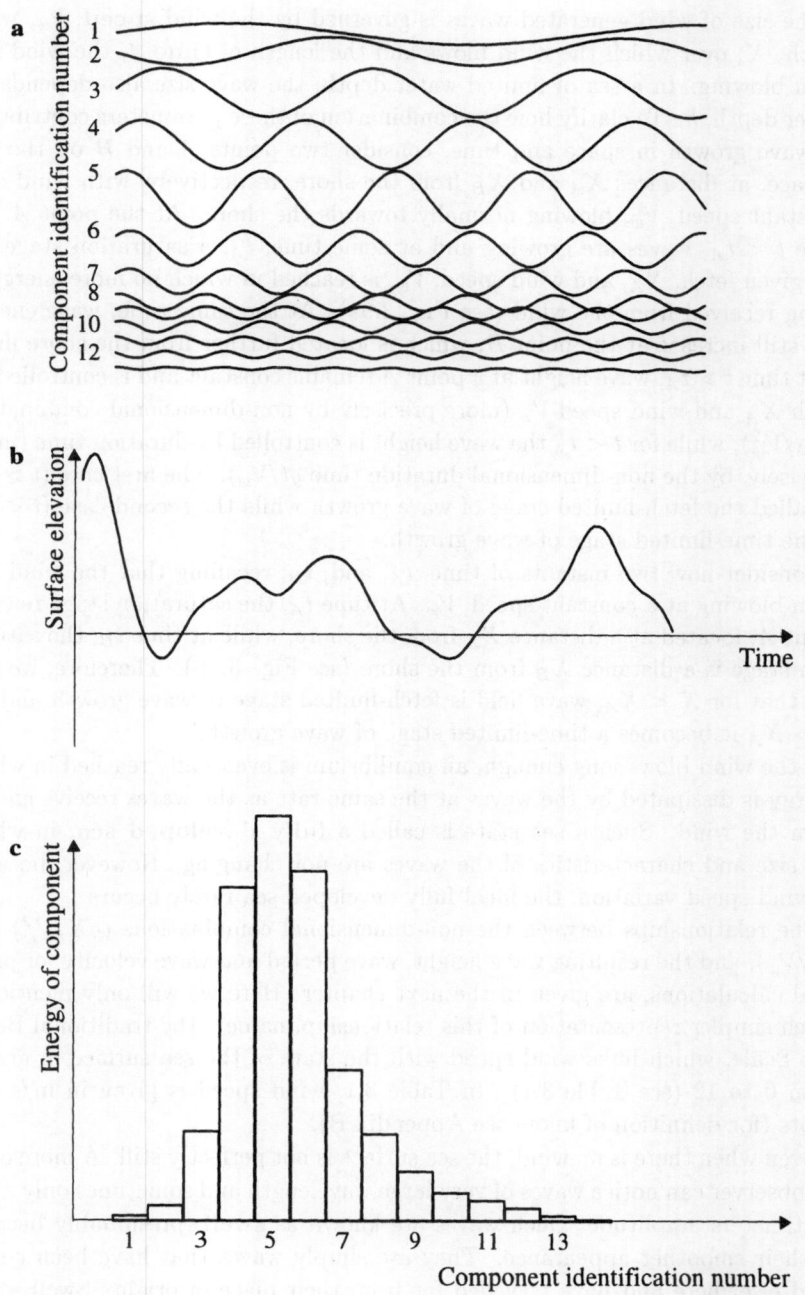

Fig. 3.8: Observed wind-induced surface waves as a result of superposition of many elementary waves: **a** elementary components, **b** final wave profile, **c** relative energy of components

The size of wind generated waves is governed by the wind **speed**, V_w, wind **fetch**, X, over which the wind blows and the length of **time**, t, the wind has been blowing. In a sea of limited water depth, the wave size also depends on water depth, h. To clarify how the combination of these parameters contributes to wave growth in space and time, consider two points A and B on the sea surface, at distances X_A and X_B from the shore, respectively, with wind of a constant speed, V_w, blowing normally towards the shore. At the point A, for time $t < t_A$, waves are growing and at some time, t_A, a saturation stage for the given fetch, X_A, and wind speed, V_w, is reached in which no more energy is being received from the wind (see Fig. 3.9a). At the same time, wave energy can still increase at the point B, which is located further from the shore line.

At time $t > t_A$, wave height at a point A remains constant and is controlled by fetch X_A and wind speed V_w (more precisely by non-dimensional combination $g X_A / V_w^2$), while for $t < t_A$ the wave height is controlled by duration time (more precisely, by the non-dimensional duration time $g t / V_w$). The first case $(t > t_A)$ is called the fetch-limited stage of wave growth while the second case $(t < t_A)$ is the time-limited stage of wave growth.

Consider now two instants of time, t_A, and, t_B, recalling that the wind has been blowing at a constant speed, V_w. At time t_A, the saturation stage reaches point A, located at a distance X_A from the shore, while at time t_B, the saturation stage is a distance X_B from the shore (see Fig. 3.9b). Therefore, we can say that for $X < X_A$, wave field is fetch-limited stage of wave growth and for $X > X_A$ it becomes a time-limited stage of wave growth.

If the wind blows long enough, an equilibrium is eventually reached in which energy is dissipated by the waves at the same rate as the waves receive energy from the wind. Such a sea state is called a **fully developed sea**, in which the size and characteristics of the waves are not changing. However, because of wind speed variation, the ideal fully developed sea rarely occurs.

The relationships between the non-dimensional combinations $(g X / V_w^2)$ and $(g t / V_w)$, and the resulting wave height, wave period and wave velocity for practical calculations, are given in the next chapter. Here we will only mention a much simpler representation of this relationship, namely the traditional Beaufort Scale, which links wind speed with the state of the sea surface in grades from 0 to 12 (see Table 3.1). In Table 3.1, wind speed is given in m/s and knots (for definition of knots see Appendix B).

Even when there is no wind, the sea surface is not perfectly still. A more careful observer can notice waves of very large wavelength and sometimes only a few centimeters amplitude. Such waves are known as **swell**, presumably because of their smoother appearance. They are simply waves that have been generated elsewhere and have travelled far from their place of origin. Swell waves are remarkably conservative; they can travel long distances without substantial attenuation.

There is some experimental evidence that swell can travel nearly halfway around the world, starting east of Australia and moving south, past New

Table 3.1: The Beaufort Wind Scale (adapted from Bearman, 1997)

B°	Name	Wind speed		State of the sea-surface	H
		knots	m/s		(m)
0	Calm	< 1	0.0– 0.2	Sea like a mirror	0
1	Light air	1– 3	0.3– 1.5	Ripples with appearance of waves;	0.1-0.2
2	Light breeze	4– 6	1.6– 3.3	Small wavelets;	0.3-0.5
3	Gentle breeze	7–10	3.4– 5.4	Large wavelets; crests begin to break;	0.6-1.0
4	Moderate breeze	11–16	5.5– 7.9	Small waves becoming longer; frequent white horses;	1.5
5	Fresh breeze	17–21	8.0–10.7	Moderate waves taking longer form;	2.0
6	Strong breeze	22–27	10.8–13.8	White foam crests extensive everywhere;	3.5
7	Moderate gale	28–33	13.9–17.1	Sea heaps up and white foam from breaking waves begins to be blown in streaks;	5.0
8	Fresh gale	34–40	17.2–20.7	Moderately high waves of greater length; edges of crests break into spindrift;	7.5
9	Strong gale	41–47	20.8–24.4	High waves, sea begins to roll;	9.5
10	Whole gale	48–55	24.5–28.4	Sea-surface takes on white appearance as foam in great patches is blown in very dense streaks;	12
11	Storm	56–64	28.5–32.7	Exceptionally high waves; sea covered with white patches of foam, visibility reduced;	15
12	Hurricane	> 64	>32.7	Air filled with foam and spray, visibility greatly reduced	>15

B° denotes the Beaufort wind scale.

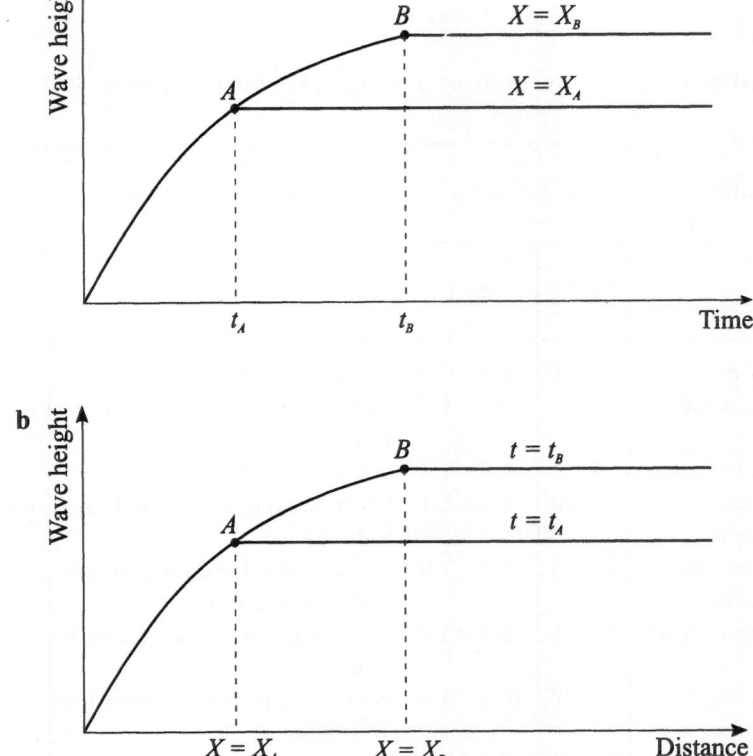

Fig. 3.9: Wave growth in space and time: **a** duration-limited wave growth, **b** fetch-limited growth

Zealand and across the Pacific Ocean to California. It is thought to take the swell about two weeks to travel this distance (Barber, 1969). Swell waves have very little effect on the size and progress of locally generated wind waves. They are able to pass through wind waves without interaction (Massel, 1996a).

3.5.3   Parameters and Functions of Wind-induced Waves

Consider a typical wave record at a given location as shown in Fig. 3.10. This figure illustrates a set of various waves which occur over a short time. However, for many applications, it is necessary to identify particular waves and their associated height and period.

In Fig. 3.2, the wave period, T, was defined as a time interval between two successive crests. However, in case of random waves it is more practical to define particular wave periods as the time interval between two successive points at which the sea surface crosses the mean sea level, when rising upwards. These

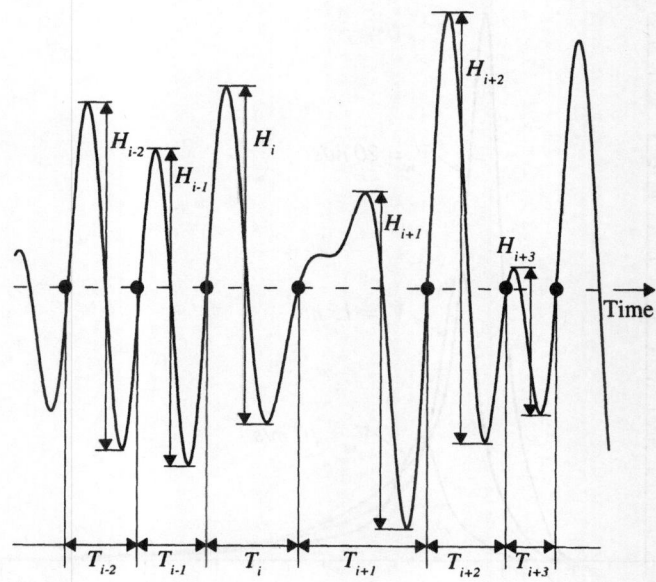

Fig. 3.10: A typical wave record of sea surface variations at a given location

points have been marked in Fig. 3.10 with closed circles, and they separate
wave periods of successive waves. The corresponding wave periods, T_i, and
wave heights, H_i, are noted. It should be noted that the definition of an
individual wave height entirely depends on the choice of the trough occurring
before or after the crest. In Fig. 3.10, a wave height is defined as zero-up-
crossing wave height and the wave crest occurs before the wave trough. Other
possible definitions of the wave height are discussed by Massel (1996a).

Because of the large variation of waves in a wind-induced wave train, there
is a need to define some statistical measures which characterize the wave train.
An obvious candidate for such characteristics is the **mean wave height**, $\bar{H}$.
It is simply an arithmetic mean of all wave heights in a given record, *i.e.*:

$$\bar{H} = \frac{1}{N} \sum_{i=1}^{i=N} H_i. \tag{3.14}$$

Usually, a record of about 20 minutes duration provides good representation
of the sea state; such records generally contain more than 100 waves ($N > 100$).
In addition to the mean wave height, $\bar{H}$, the **significant wave height**, H_s, is
frequently used in oceanographic and engineering practice. The concept of the
significant wave height was introduced by Sverdrup and Munk (1947). They
defined significant wave height, H_s, as the average of the highest one-third of
wave heights. This wave height is close to the mean wave height estimated by

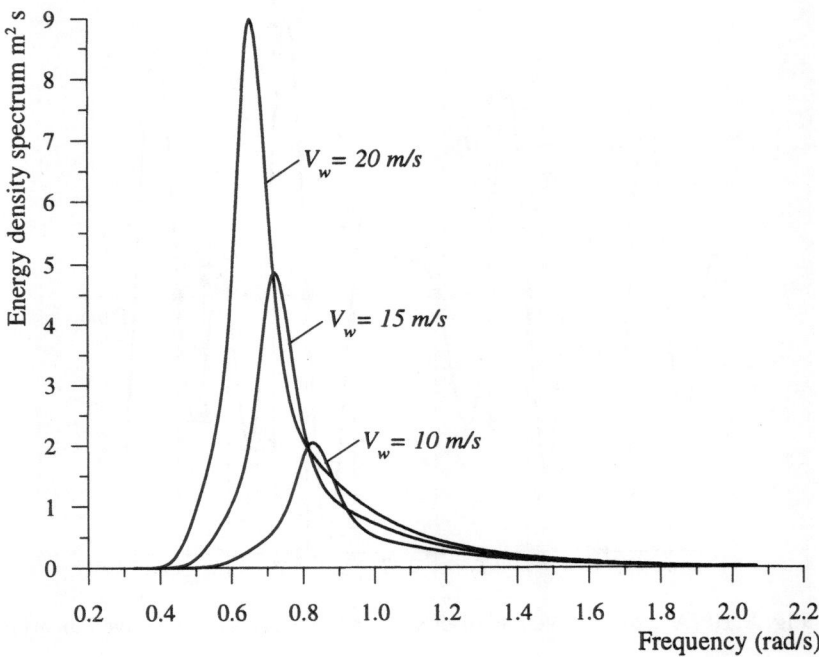

Fig. 3.11: Typical frequency spectrum of surface waves

the human eye. In deep water there is a simple relation between significant wave height, H_s, and mean wave height, $\bar{H}$:

$$H_s = 1.6\,\bar{H}. \tag{3.15}$$

In any wave record, there will also be a maximum wave height, H_{max}. The H_{max} value for a given period of time is very important when designing many coastal and offshore engineering structures, such as breakwaters, jetties, coastal revetments and oil platforms.

In the same way as mean wave height, we can determine the **mean wave period** $\bar{T}$:

$$\bar{T} = \frac{1}{N} \sum_{i=1}^{i=N} T_i. \tag{3.16}$$

When examining Fig. 3.10, it becomes clear that waves of some periods have larger wave heights than other. Therefore, the question arises as to which wave period, or which wave frequency is associated with the highest wave energy? The answer to this question is provided by the **frequency spectrum** $S(\omega)$. The frequency spectrum presents the distribution, with respect to frequencies, of the energy supplied by wind to the ocean surface (Fig. 3.11). The area under

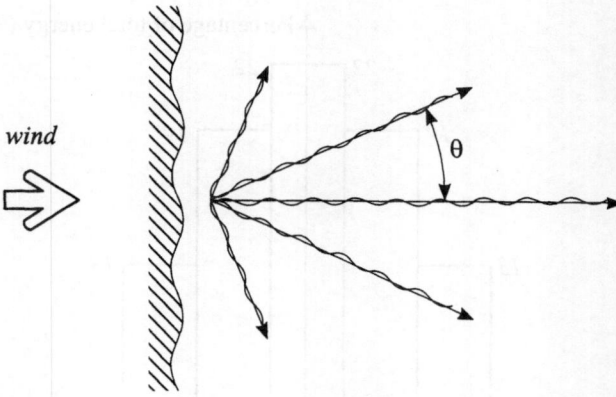

Fig. 3.12: Waves are propagated from the shoreline in various directions; lengths of arrows are roughly proportional to amount of energy transported in a given direction

the spectrum curve is a measure of the total wave energy per unit sea surface, *i.e.*:

$$E = \int_0^\infty S(\omega)d\omega. \tag{3.17}$$

It is clear that for the same fetch, X, but different wind speed, V_w, the wave energy will be different. This is illustrated in Fig. 3.11 for three different wind speeds; 10 m/s ($\approx$ 20 knots), 15 m/s ($\approx$ 30 knots) and 20 m/s ($\approx$ 40 knots), but for the same fetch of 200 km. It is interesting to note the shift of the peak frequency, ω_p, towards lower frequencies when wind speed increases. This means that for the same fetch, higher wind speeds generate longer waves. For the case shown in Fig. 3.11, the wavelength associated with the peak frequency is 91 m, 119 m, and 145 m, respectively. Also the wave energy, proportional to total area under curve $S(\omega)$, is greater for highest wind speed.

As the dimension of energy E is [N· m], the dimension of spectrum, $S(\omega)$, should be [N· m· s]. If we take into account that $E \sim \rho g H^2$ and drop the constant ρg, the dimension of energy becomes [m²] and subsequently the dimension of spectrum $S(\omega)$ is [m²·s], which is frequently used in professional oceanographical literature.

When observing surface ocean waves, we notice their multidirectionality; waves travel in many directions, resulting in an apparently confused sea. Consider the simple case of a long coastline extending in the north-south direction, with a wind of constant speed blowing from the shore normal to shoreline (Fig. 3.12). Due to the continuity of water mass, and due to the interactions between waves, waves do not propagate only in the direction in which the wind is blowing, but spread in the sector of ± 90° from the wind direction. However, the amount of energy carried by waves in a particular direction depends on the angle, θ, between the wind and wave directions. The highest portion of energy

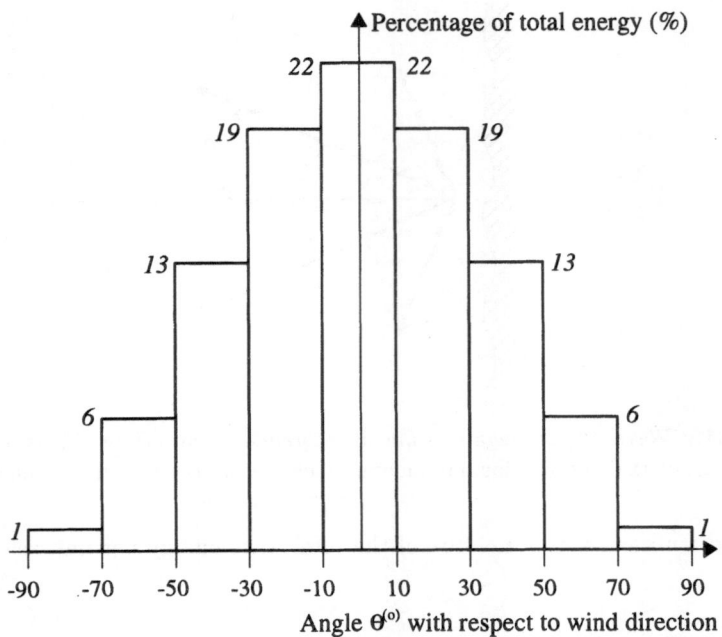

Fig. 3.13: Percentage of energy contained in the angle intervals $\Delta\theta = 20°$; wind is blowing in the direction $\theta = 0°$

propagates along the wind direction, while along the shoreline ($\pm 90°$ from the wind direction) the wave energy is negligible. One of the most popular formula describing the directional partition of wave energy is (Massel, 1996a):

$$D(\theta) = \frac{2}{\pi} \cos^2\theta. \tag{3.18}$$

This formula gives the distribution of unit wave energy among wave directions. Therefore, the obvious relationship should be satisfied:

$$\int_{-\pi/2}^{\pi/2} \frac{2}{\pi} \cos^2\theta \, d\theta = 1. \tag{3.19}$$

Let us calculate the percentage of wave energy which is contained within a particular angular sector. As was shown in Fig. 3.13, in the sector $\pm 10°$ from the wind direction, 22% of wave energy is propagated. About 86% of the energy of the sea-surface propagates within an angle $50°$ either side of the wind direction. More detailed studies indicate that the directional energy distribution depends not only on the angle, θ, but also on the wave frequency. Waves with frequency close to the peak frequency propagate more closely to the wind direction than the waves of the higher or lower frequencies (Massel, 1996a).

3.5.4   Geography of Waves

Wind and wave climate varies considerably between various regions of the World Ocean. Peak wave conditions occur at high latitudes ($\pm 50°$). In the Southern Ocean, the almost constant strong winds acting over large fetches generate consistently high waves, which propagate over very long distances. In contrast, the North Atlantic Ocean has a more variable sea state. In all oceans, the intensity of wave motion decreases from high latitudes towards the Equator. In the tropical zones, the frequency of storm winds is 2-3 times smaller than in higher latitudes. However, the tropical regions are the locations where cyclones (or hurricanes or typhoons) are generated. Due to the complicated pattern of the wind field, the resulting storm waves are usually superimposed upon swell propagating from various directions. Some parameters of extreme wind speed and wave heights, occurring once in 30 or 50 years ($T_R = 30$ or 50 years), in various regions of the World Ocean, are given in Table 3.2, based on data reported by Davidan *et al.* (1985). However, the wind speed during tropical cyclones was omitted from the table. Confidence intervals with confidence level of 95% are added in parentheses.

The Atlantic Ocean extends from north to south through all climatic zones. The number of atmospheric depressions in the Atlantic Ocean is higher during winter (October–March) and decreases during summer (Massel, 1982). During tropical cyclones the wind speed exceeds 40 m/s, reaching 70 m/s and more, and waves can be as high as 20 m. During tropical cyclone *Camille* (1969) in the Gulf of Mexico, a wave height of 23.6 m was recorded. The *S.S. Majestic*'s officers observed waves from 18.3 m to 27.4 m near $48°30'N$ and $21°05'W$ on December 29, 1922. The passenger ship *Michelangelo*, during a North Atlantic crossing, was struck by a wave that collapsed the superstructure and broke heavy windows at 24.6 m above the water line (Massel, 1996a).

The Pacific Ocean is the largest ocean with the greatest average, as well as the greatest observed, depth. Especially in the Southern Pacific, the large, almost unlimited wind fetches create conditions for generation of large waves. For example, on April 2, 1966, the *S.S. Ob*'s officers measured a wave of 24.9 m using the stereophotogrametry method. In 1921 Captain Wilson of the Blue Funnel Line, while en route from Yokohama to Puget Sound, Washington, recorded waves that were higher than 21.3 m. The *U.S.S. Ramapo*, a navy tanker, encountered a wave of 33.5 m on February 7, 1933, between Manila and San Diego.

There are two distinct regions of the Indian Ocean, characterized by different climates. The northern part is dominated by the Asian summer monsoon, while in the southern part, between 40°S and the Antarctic continent, the powerful Westerlies are located. The swell generated in the Southern Ocean propagates towards lower latitudes, resulting in a more energetic wave climate in the Indian Ocean. The waves are consistently larger than those in either the South Pacific or South Atlantic Oceans.

Table 3.2: Extreme wind speed and wave heights in various regions of the World
Ocean (adapted from Massel, 1996a)

Region	Extreme value		
	Wind speed (m/s)		Wave height (m)
	$T_R = 30$ y	$T_R = 50$ y	$T_R = 30$ y
Midlatitude zones of the North Atlantic Ocean	48 (42–54)	53 (44–62)	20 (17–23)
Midlatitude zones of the South Atlantic and Indian Ocean	50 (42–58)	55 (43–67)	23 (19–27)
Midlatitude and subtropical Pacific; Midlatitude South Pacific	48 (40–56)	55 (43–67)	21 (18–24)
Tropical zone of the North Atlantic Ocean	39 (31–47)	44 (34–54)	16 (13–19)
Tropical zone of the South Atlantic and Indian Ocean	32 (28–36)	36 (30–42)	14 (12–16)
Tropical zone of Pacific and tropical monsoon zone in Indian Ocean	33 (29–37)	36 (30–42)	12 (10–14)
Equatorial zone	32 (28–36)	35 (29–41)	12 (10–14)

Significant monthly variations are observed in the Westerlies. The highest
wind speeds are found in a broad band between Africa and Australia, where the
largest waves also occur (Young, 1994; Young and Holland, 1996). On the other
hand, in the western region of the Indian Ocean, close to the African coastline,
between Durban and Port Elizabeth, giant waves (freak waves) sometimes occur, even during relatively calm weather. Their front slope is very steep and
the heights of the waves are typically 15–18 m. During a 22 year period (1952–
1973) in this region, eleven ships were sunk or seriously damaged (Davidan and
Lopatukhin, 1978). Two particularly unfortunate cases were the *World Glory*,
which broke in two and sank in June 1968, and the *Neptune Sapphire*, which
lost 60 m of its bow section in August 1973. It can only be speculated that
giant waves may account for many of the ships which have been lost without

trace off this coast. All except one of the eleven incidents involved vessels riding on the rapid Agulhas Current. This current, generated in the Passat region of the Indian Ocean, flows in a southwest direction along the African coast. The typical current speed is of the order of 2.0–2.5 m/s, and its width is about 90–170 km. Energetic swell, generated at higher latitudes, propagates against this current resulting in very high and steep waves. Interaction of waves propagating against Agulhas Current is discussed in Sect. 7.7.2. For more general elaboration see also Massel (1989).

Very comprehensive atlas on wind and wave climate has recently been published by Young and Holland (1996). The atlas provides quantitative estimates of the global wind and waves which are based on the data measured by the GEOSAT satellite, for 3 years period. The resulting charts contain the contour fields of wind speeds and wave heights for any region of the globe.

3.6   Tsunamis

Wind-induced waves are very common on the ocean surface. However, in some circumstances waves of unusual character may result. They are the long waves generated by seismic disturbances, or by slumping of submarine sediment masses due to gravitational instability. Such waves are called **tsunamis**. Tsunami is a Japanese word being, in fact, a combination of two words: 'ami' which means wave, and 'tsu' which denotes a particular point at the waterline. Thus, a tsunami is a wave which approaches the shoreline.

In the open ocean, tsunamis often remain undetected because of their great wavelength, of the order of hundreds of kilometers, and its small wave height, usually in the order of one metre. However, the velocity of propagation of tsunami in deep water is very high (exceeding 700 km/hour when the mean ocean depth is used), which can be estimated approximately as:

$$C = \sqrt{gh}, \tag{3.20}$$

in which h is the water depth (for more vigorous derivation of formula (3.20) see Sect. 4.2). When tsunami reaches shallower water, the velocity decreases, but the energy flux in the wave remains the same, resulting in an increase in wave height, sometimes to catastrophic values. Tsunamis occur mostly in the Pacific as that ocean experiences frequent seismic activity. Catastrophic tsunamis are recorded in many historical Japanese, Chinese and American documents. In 1495, a high tsunami attacked the southern coast of the Japanese Honshu Island. A huge hall, containing an 11.3 metres tall bronze statue of Buddha was completely washed away, however the statue is still in place. It should be noted that this statue has a total weight of 800 tonnes and is about 1.5 km from the present shoreline and about 50 m above sea level.

Almost four hundred years later, in August 1883, a gigantic explosion of the volcano Krakatoa, situated between Sumatra and Java in the heavy active

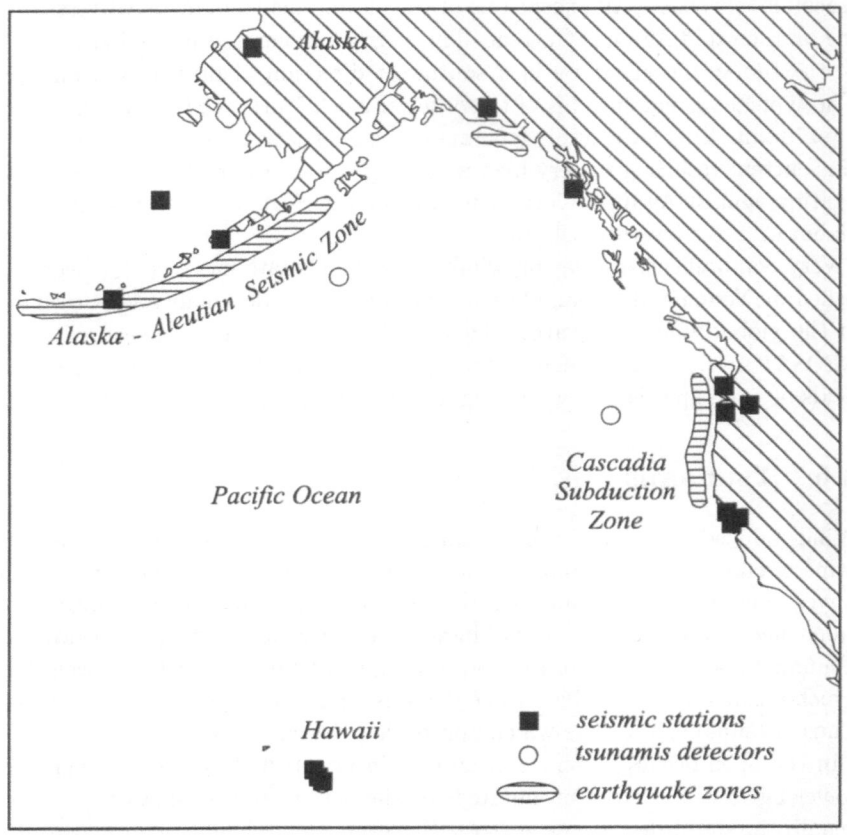

Fig. 3.14: Earthquake zones and location of tsunamis detectors in North-East Pacific (adapted from Bernard, 1998)

seismic zone, generated tsunami with height over 30 m. The gigantic wave swept away thousand of villages on nearby islands, and 36,000 lives were lost. Twenty minutes after the explosion, the harbour of Batavia (Jakarta) was destroyed, and nine hours later 300 river boats were smashed at Calcutta. Halfway around the earth and 32 hours after the eruption, the wave reached the English Channel and was recorded on the tide gauges (Smith, 1973).

Tsunamis continue to strike Pacific coasts. Coastal communities in East Russia, Japan, as well as in Alaska and northern California are particularly threatened by tsunamis generated by local earthquakes. When a large earthquake occurs, the first tsunami may reach nearby coastal communities within 10 minutes of the event.

On October 4, 1994, an earthquake of magnitude Ms 8.0 (Mercalli Intensity Scale) struck the southern region of the Kuril Islands. A tsunami run-up of approximately 1.8 m was reported in Nemuro, Japan, about 90 minutes after the earthquake (Yeh *et al.*, 1995). In the South Kuril Islands, 11 people were killed and 242 were injured. In Hokkaido, Japan, one person was killed and 140 injured. None of the casualties were due directly to the tsunami, in spite of its significant run-up of approximately 10 m on Shikotan Island.

In order to minimize the number of casualties and reduce damage due to tsunamis, three types of tsunami warning systems exist: the Pacific Tsunamis Warning Center, five regional systems (two in the United States and one each in Japan, Russia, and French Polynesia), and local systems in Chile and Japan (Bernard, 1998). The Pacific-wide System can issue a warning about 1 hour prior to arrival of the tsunami (useful for populations located more than about 750 km from the source), regional systems warn in about 10 minutes prior to arrival of the tsunami (useful for 100–750 km from the source) and local systems warn in about 5 minutes prior to arrival of the tsunami (useful for areas situated less than 100 km from the source). Potential earthquake zones and locations of tsunami detectors in the North-East Pacific are shown in Fig. 3.14.

3.7   Seiches

Under specific conditions, very long waves can also be generated in closed water basins such as lakes, bays, estuaries and harbours, and they are known as **seiches**. A seiche is a standing wave, which can be considered as the sum of two progressive waves, travelling in opposite directions. Everybody has experienced a seiche in their domestic bath by setting the water into oscillatory motion by moving a hand to and fro in the water. Figure 3.15 shows an idealized vertical profile of a seiche. At either end of the container, the water level is alternatively high and low, whereas in the middle, the water level remains constant.

For the case illustrated in Fig. 3.15, the length of the container, l, corresponds to half the wavelength, L, of the seiche. At the point where the water level is constant (the **node**), the horizontal flow of water from one end of the container to the other is greatest. Where the fluctuations of water level are greatest (the **antinodes**), there is minimal horizontal movement of the water. The seiche period is the time from peak to peak of the oscillation; it varies with the basin length and depth.

Seiches in natural conditions are much more complicated than that in a simple closed container. The most probable sources of seiche forcing mechanisms are (Massel, 1989):

- passage of barometric fluctuations,
- impact of wind gusts on the water surface,
- release of water forced up on a leeward shore through the lapse of strong onshore winds, and
- seismic oscillations of the earth during earthquakes.

However, the principal source of excitation remains meteorological. After being excited by meteorological disturbances, water surges back and forth until

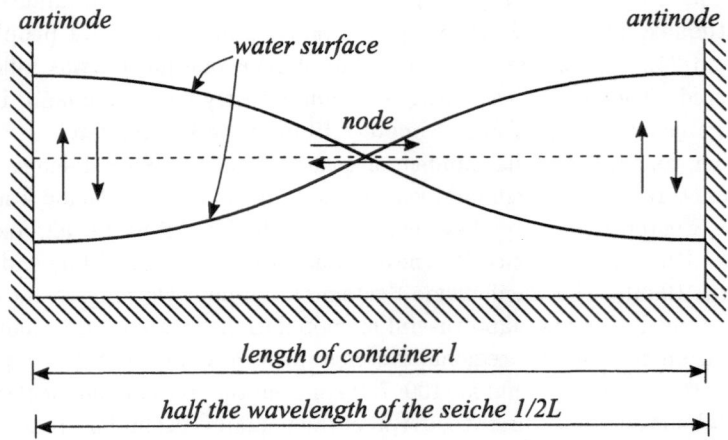

Fig. 3.15: Vertical profile of idealized standing wave in a container

the oscillation is damped out by friction. The main body of water may oscillate longitudinally or laterally at different periods, while water in a separate bay or harbour may oscillate at its own natural period. Coastal seiches are more often forced by direct action of wave energy at the mouth of a gulf or the entrance of a harbour. The resulting sea surface can be unimodal (as in Fig. 3.15) or multimodal. If the water depth in a basin of rectangular shape is constant, the periods of seiches, T, are given by the Merian formula (Massel, 1989):

$$T_n = \frac{2l}{n\sqrt{gh}},\tag{3.21}$$

in which l is the length of the basin, h is the water depth, g is the acceleration due to gravity, and $n = 1, 2, \ldots$ denotes the number of nodes in the basin. The periods, T_n, are known as the resonance periods or natural periods of oscillation.

When the seiche is unimodal ($n = 1$) as is shown in Fig. 3.15, the period of the seiche is given by:

$$T_1 = \frac{2l}{\sqrt{gh}}.\tag{3.22}$$

Therefore, the wavelength of the seiche becomes:

$$L_1 = \sqrt{gh}\, T_1 = 2l.\tag{3.23}$$

When the seiche is bimodal ($n = 2$), we obtain from (3.21):

$$T_2 = \frac{l}{\sqrt{gh}} \quad \text{and} \quad L_2 = l.\tag{3.24}$$

For $n = 3$, Eq. (3.21) gives:

$$T_3 = \frac{2l}{3\sqrt{gh}} \quad \text{and} \quad L_3 = T_3 \sqrt{gh} = \frac{2}{3}l, \tag{3.25}$$

When the number of modes increases, the corresponding wavelength, L_n, decreases and more node points appear on the sea surface.

The Merian formula (3.21) can also be used to provide approximate seiche periods for water basins which are not rectangular in shape. For example, Lake Baikal in Siberia, which is the biggest reservoir of fresh water on the Earth, the Merian formula predicts $T_1 = 292$ min., while the observed period is $T_1 = 278.2$ min.

Seiches in bodies of water (bays, inlets, etc.) which open on to much larger water basins can still be calculated by the Merian formula, assuming that the length, l, is replaced by $2l$ and n is restricted to 1, 3, 5, ... *i.e.*:

$$T_n = \frac{4l}{n\sqrt{gh}}. \tag{3.26}$$

For more complex basins, a modified Merian formula or some analytical models can be used. The reader should consult Wilson (1972) for an in-depth discussion.

3.8 Storm Surges

The meteorological conditions may considerably change the height of predicted seiches and complicate the prediction process. A strong wind blowing on shore piles up the water and causes high waters to be higher than predicted, while winds blowing off the land will have the reverse effect. However, the effect of wind stress on sea level is very variable and depends largely on the topography of the area. This phenomenon is called **storm surge**. A reduction in atmospheric pressure of 10 hectopascals (hPa) from the average level can cause a difference in the sea level of about 0.1 m. This depression of the water surface below areas of high atmospheric pressure, and its elevation below low atmospheric pressure, is often described as the *inverted barometer effect*. The water level does not adjust itself immediately to the change of the pressure over the considerable area.

Storm surges may be expected to occur along coasts with relatively shallow waters affected by storms. In some areas the storm surges have very disastrous effects. In North America they frequently occur in the Gulf of Mexico and along the east coast, due to hurricanes. In Asia, the Bay of Bengal is particularly affected by storm surges causing flooding. In Europe, a comprehensive documentation of storm surges is related particularly to the North Sea and the Baltic Sea (Perry and Walker,1977). For example, in 1953 one of the biggest

storm surges observed in the North Sea caused serious flooding and the loss
of lives in the Netherlands and England. The deep depression, travelling east-
wards the north of the British Isles and then south-eastwards into the North
Sea, raised the water level in the southern part of the North Sea. The amplitude of storm surge on the eastern coast of England reached about 2.75 m,
while in some places in the Dutch coast it was greater than 3.95 m.

3.9 Rossby Waves

Fluid moving on the Earth's surface possesses vorticity about a vertical axis as
a result of the Earth's rotation at any location other than the Equator. This is
the planetary vorticity, and its magnitude is equal to the Coriolis parameter, f.
Planetary vorticity has to be distinguished from a small scale of fluid element
discussed in Sect. 2.6 and Appendix C.1. Following Mann and Lazier (1996),
let us consider the eastward flowing current. Water element, besides moving
east, is rotating around to the left (in the Northern Hemisphere) with the
rotation of the Earth at a rate of $f = 2\omega_E \sin \phi$ (see Sect. 5.3.1). When water
is deflected northward, the vertical component of the rotation increases and
water is moving faster as the Coriolis parameter increases.

In the reference frame attached to the Earth, water coming from south is
rotating to the right or back toward its initial latitude and it can even pass
this latitude arriving in the region where the rate of rotation f is less than the
water in the current. Then the water appears to turn to the left and relative to
the local vertical and again head backward toward its original latitude. These
large, slow, horizontal oscillations in the open ocean are known as **Rossby
waves**. Their amplitudes are of hundreds of kilometres, their wavelengths can
be as long as hundreds to thousands of kilometres, and periods are of the order
of days.

Using the similar arguments as above, it can be shown that the Rossby waves
cannot be formed on a westward flowing current. However, they can be formed
when the vorticity is changed due to change in depth. The explanation of these
topographic Rossby waves is given by Tomczak and Godfrey (1994) and comprehensive mathematical treatment of the Rossby waves is done by Pedlosky
(1979). As will be shown in Chap. 7, the Rossby waves play important role in
the low-frequency variations in the Equatorial Pacific that lead to El Niño.

4 How to Determine Wave Parameters

4.1 Introduction

In Chap. 3 various types of surface waves were described. Each of them possess their own properties, and different techniques should be used for prediction of these properties. However, some characteristics of all waves are common and can be used to distinguish between various wave types. The most important characteristics are relative wavelength (L/h), relative wave height (H/h) and wave steepness (H/L).

Depending on relative wavelength, waves can be classified as short, long or intermediate:

$$\left.\begin{array}{lll} \text{long waves (shallow water waves)} & \text{when} & \dfrac{L}{h} > 10 \text{ (or 20)} \\[3mm] \text{waves on intermediate water depth} & \text{when} & 2 < \dfrac{L}{h} < 10 \text{ (or 20)} \\[3mm] \text{short waves (deep water waves)} & \text{when} & \dfrac{L}{h} < 2 \end{array}\right\} . \qquad (4.1)$$

In shallow water areas, the wavelength is much larger than water depth and therefore long waves are sometimes known as shallow water waves.

The relative wave height (H/h) and wave steepness (H/L) provide some guide as to what kind of method should be used to calculate various wave parameters. The complexity of prediction methods strongly depends on the boundary conditions which should be satisfied in the calculations. Boundary conditions usually describe the continuity (or lack of such continuity) of physical quantities such as pressure or velocity, when crossing the boundary between different media, e.g. when going from air into water or from water on the sea bottom.

When the surface upon which the boundary conditions have to be imposed is simple, the calculations simplify considerably. In wave analysis, the most difficult boundary conditions are those at the sea surface. This results from the fact that the displacement of the sea surface, ζ, is unknown a priori and has to be found in the calculations. However, when relative wave height (H/h) and wave steepness (H/L) are small, we can assume that the sea surface does

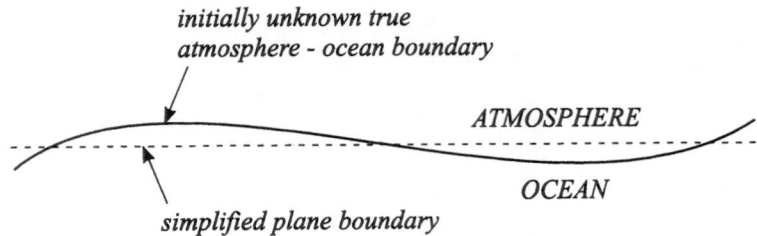

initially unknown true
atmosphere - ocean boundary

ATMOSPHERE

OCEAN

simplified plane boundary

Fig. 4.1: 'True' and 'simplified' atmosphere-ocean boundary

not differ much from the still water plane ($z = 0$), and the boundary conditions can be applied on that plane instead of on the initially unknown wave surface (see Fig. 4.1).

For small surface disturbances, other wave properties such as wave-induced pressure or wave-induced velocities, are also small. If they are of magnitude x, it means that their square (x^2), cube (x^3) and other higher powers (x^n) are even smaller, and can be neglected in calculations. Hence, all resulting quantities will be linearly proportional to the wave amplitude (or wave height), e.g. wave pressure $= \alpha\times$ wave amplitude, where α is some proportionality coefficient. Wave theories which are based on such proportionality relationships are called linear or **small amplitude wave** theories. They have to be distinguished from the higher order, or nonlinear theories, in which all quantities are expressed in terms of wave amplitude (or wave height) to powers higher than one. A brief summary of the theory of small amplitude waves is given in Appendix C.5.

4.2   Wave Parameters Based on Small Amplitude Wave Theory

4.2.1   Introductory Remarks

Consider the wave motion of permanent form on a water of mean depth, h (Fig. 4.2). We assume that wave height, H, and wave period, T, are known. The wave height, H, and wave period, T, are such that the wave steepness $s = H/L \ll 1$ and $H/h \ll 1$. Therefore, the wave parameters correspond to a small amplitude wave. Figure 4.2 illustrates the picture which is obtained when looking through the glass wall of a narrow wave channel with waves propagating in the direction of the positive x-axis. Assuming that waves are uniform in the y direction (crest line is perpendicular to the plane O, x, z), at each point in the water body there are five unknown quantities, induced by wave motion: horizontal water velocity $u(x, z, t)$, vertical water velocity $w(x, z, t)$, pressure $p(x, z, t)$, displacement of the sea surface $\zeta(x, t)$ and wavelength, L. Therefore, to determine these five unknowns, we need at least five equations. When we use some properties of water and wave motion itself, the problem of determining

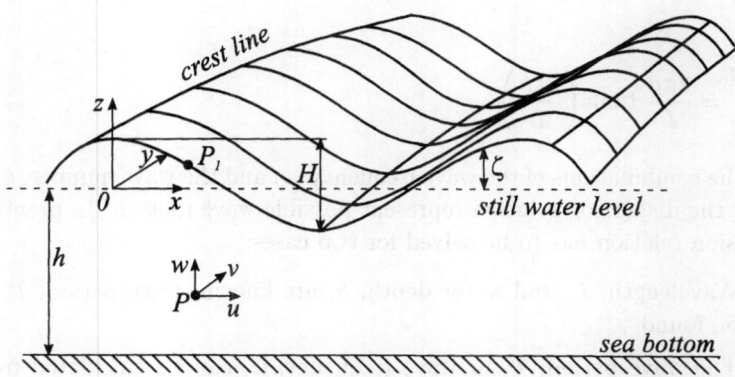

Fig. 4.2: Regular wave propagating on constant water depth

these quantities is simplified. Full derivation of this type of solution is given in the Appendix C.5. However, for convenience of further applications, the final workable formulas are collected and briefly discussed below.

4.2.2 Practical Calculation Formulas for Constant Water Depth

Appendices C.4 and C.5 show that a practical solution for regular waves propagating on a constant water depth can be obtained relatively simply, assuming that water is incompressible and wave motion is irrotational. The final solution is given in the form of velocity potential $\phi(x, z, t)$. The function ϕ is a basic function for calculation of all practical wave parameters, and it takes the form (see Eq. C.115):

$$\phi(x, z, t) = \frac{gH}{2\omega} \frac{\cosh k(z + h)}{\cosh kh} \sin(kx - \omega t). \tag{4.2}$$

The dimension of velocity potential is [m^2/s]. The term $\cosh k(z + h)/\cosh kh$ describes the attenuation of wave motion with submergence, z, below the water surface. It is equal to 1 at the sea surface ($z = 0$), while at the sea bottom ($z = -h$) it becomes $1/\cosh kh < 1$. The term $\sin(kx - \omega t)$ expresses the periodic dependence of the potential on horizontal distance, x, and time, t. The periodicity scale in space is the wavelength, $L = 2\pi/k$, and the periodicity scale in time is the wave period, $T = 2\pi/\omega$. Below, the formulas for calculation of various wave parameters are listed. For interested readers, more discussion on these formulas is given in Appendix C.5.

Wavelength, L, and Wave Period, T. The relationship between wavelength, L, and wave period, T, is given by the dispersion relation. Substituting the velocity potential (C.115) in the boundary condition (C.98) yields:

$$\omega^2 = gk \tanh(kh) \quad \text{or} \quad \frac{\omega^2 h}{g} = kh \tanh(kh), \tag{4.3}$$

or,

$$\left(\frac{2\pi}{T}\right)^2 = \frac{2\pi g}{L} \tanh\left(\frac{2\pi h}{L}\right). \tag{4.4}$$

Only the combinations of the wave frequency, ω, and the wave number, k, which satisfy the dispersion relation represent possible wave motion. In practice, the dispersion relation has to be solved for two cases:

- Wavelength, L, and water depth, h, are known; wave period, T, has to be found.

  The wave period, T, or wave frequency, ω, can be calculated from the above equations in a straightforward manner:

$$T = \frac{2\pi}{\sqrt{\dfrac{2\pi g}{L} \tanh\left(\dfrac{2\pi h}{L}\right)}}, \tag{4.5}$$

  or

$$\omega = \sqrt{gk \tanh(kh)}. \tag{4.6}$$

- Wave period, T, or wave frequency, ω, and water depth, h, are known; wavelength, L, has to be found.

  The determination of wave number, k, or wavelength, L, is not as straightforward as in the first case. In order to clarify the nature of the solution, let us rewrite the second equation in (4.3) in the form:

$$f_1 = \frac{\omega^2 h}{g}\left(\frac{1}{kh}\right) = \tanh(kh) = f_2. \tag{4.7}$$

If the functions $f_1(kh) = (\omega^2 h/g)(1/kh)$ and $f_2(kh) = \tanh(kh)$ are plotted, as illustrated in Fig. 4.3, then the inter-section of these curves at point P denotes the value of (kh) which is a solution of the dispersion relation (4.7).

The solution of Eq. (4.3) can be obtained much more easily for extreme water depths, *i.e.* for very deep water and very shallow water. For very large water depth $\omega^2 h/g \gg 1$, the $\tanh(kh) \to 1.0$. Using this value in Eq. (4.3), we obtain:

$$kh = \frac{\omega^2 h}{g}, \tag{4.8}$$

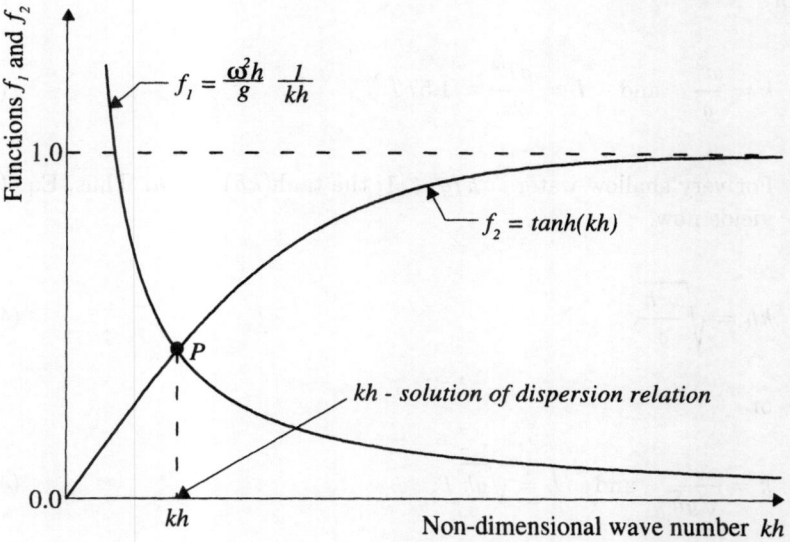

Fig. 4.3: Graphical interpretation of solution of the dispersion relation

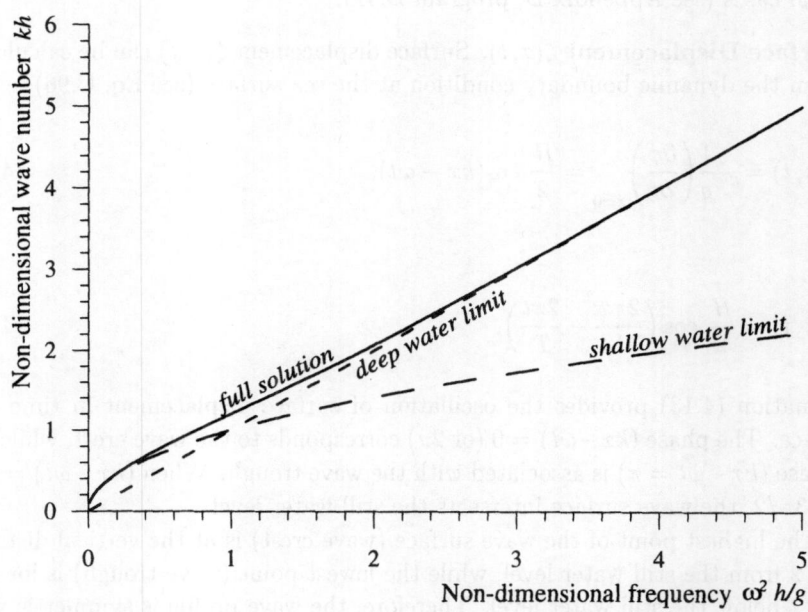

Fig. 4.4: Dispersion relation $kh = f(\omega^2/g)$

or

$$k = \frac{\omega^2}{g} \quad \text{and} \quad L = \frac{gT^2}{2\pi} \approx 1.57\,T^2. \tag{4.9}$$

For very shallow water $\omega^2 h/g \ll 1$, the $\tanh(kh) \to kh$. Thus, Eq. (4.3) yields now:

$$kh = \sqrt{\frac{\omega^2 h}{g}}, \tag{4.10}$$

or

$$k = \frac{\omega}{\sqrt{gh}} \quad \text{and} \quad L = \sqrt{gh}\,T. \tag{4.11}$$

The dispersion relation (4.3) is shown in Fig. 4.4 in the form of the function $kh = f(\omega^2 h/g)$. The asymptotes for both extreme cases, deep and shallow water, are also marked. A computer program is provided to solve Eq. (4.3) for both cases (see Appendix D, program D.41).

Surface Displacement $\zeta(x,t)$. Surface displacement $\zeta(x,t)$ can be calculated from the dynamic boundary condition at the sea surface (see Eq. C.96), e.g.:

$$\zeta(x,t) = -\frac{1}{g}\left(\frac{\partial\phi}{\partial t}\right)_{z=0} = \frac{H}{2}\cos(kx - \omega t), \tag{4.12}$$

or

$$\zeta(x,t) = \frac{H}{2}\cos\left(\frac{2\pi x}{L} - \frac{2\pi t}{T}\right). \tag{4.13}$$

Equation (4.13) provides the oscillation of surface displacement in time and space. The phase $(kx - \omega t) = 0$ (or 2π) corresponds to the wave crest, while the phase $(kx - \omega t = \pi)$ is associated with the wave trough. When $(kx - \omega t) = \pi/2$ or $3\pi/2$, the wave surface intersects the still water level.

The highest point of the wave surface (wave crest) is at the vertical distance $H/2$ from the still water level, while the lowest point (wave trough) is located $H/2$ below the still water level. Therefore, the wave profile is symmetric with respect to the still water level. Equation (4.13) can be rewritten in a slightly different form:

$$\zeta(x,t) = \frac{H}{2}\cos\left[\frac{2\pi}{L}\left(x - t\frac{L}{T}\right)\right] = \frac{H}{2}\cos\left[\frac{2\pi}{L}\left(x - Ct\right)\right]. \tag{4.14}$$

The wave crest appears when the argument of the cos function is equal to zero; when $x - Ct = 0$ or when $x/t = C$. This means that the wave crest and any other point on the wave surface propagate with the phase velocity, C. The surface displacement (4.13) is illustrated in Fig. 4.2. A computer program for calculating wave surface displacement is given in Appendix D (see program D.42).

Phase and Group Velocity. From Eq. (3.5) in Chap. 3 follows that the phase velocity is:

$$C = \frac{L}{T} = \frac{\omega}{k}. \tag{4.15}$$

After substitution for ω from Eq. (4.6), we obtain:

$$C = \sqrt{\frac{g}{k} \tanh(kh)} = \sqrt{\frac{gL}{2\pi} \tanh\left(\frac{2\pi h}{L}\right)}. \tag{4.16}$$

In deep water, when $\tanh(kh) \to 1.0$, Eq. (4.16) simplifies as follows:

$$C = \sqrt{\frac{g}{k}} = \sqrt{\frac{gL}{2\pi}}. \tag{4.17}$$

As $k = \omega^2/g$ in deep water, phase velocity can also be presented as a function of wave frequency (or wave period) independently of the water depth, *i.e.*:

$$C = \frac{g}{\omega} = \frac{gT}{2\pi}. \tag{4.18}$$

For shallow water, when $\tanh(kh) \to kh$, Eq. (4.16) gives:

$$C = \sqrt{gh}. \tag{4.19}$$

In shallow water, phase velocity is totally controlled by water depth, independent of the wavelength.

The group velocity, C_g, is calculated as a derivative of frequency, ω, with respect to wave number, *i.e.*:

$$C_g = \frac{d\omega}{dk}. \tag{4.20}$$

Equation (4.20) results from the theory of kinematics of water particles subjected to the wave motion. For details of derivation the reader should consult, for example, Massel (1989). Using Eq. (4.6) in (4.20) yields:

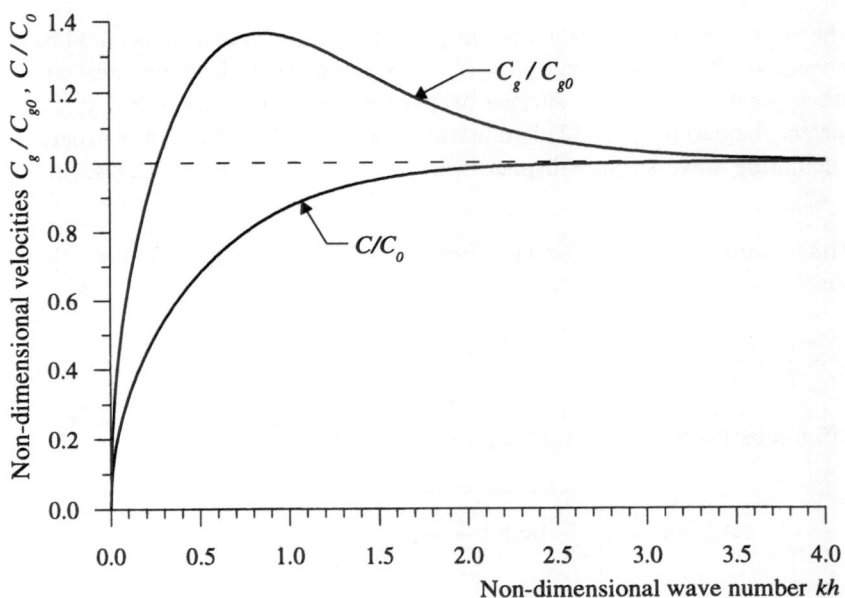

Fig. 4.5: Phase and group velocities, normalized against their values in deep water

$$C_g = m(kh)\, C,\tag{4.21}$$

in which:

$$m(kh) = \frac{1}{2}\left(1 + \frac{2kh}{\sinh(2kh)}\right).\tag{4.22}$$

Hence, the group velocity is proportional to the phase velocity, but the proportionality function, m, depends on the non-dimensional wave number, kh. Again, there are two limiting water depths when group velocity takes simpler forms, *i.e.*:

- deep water when $\dfrac{2kh}{\sinh(2kh)} \to 0$; then

$$m \to \frac{1}{2} \quad \text{and} \quad C_g \to \frac{1}{2}C = \frac{gT}{4\pi},\tag{4.23}$$

- shallow water when $\sinh(2kh) \to 2kh$; then

$$m \to 1 \quad \text{and} \quad C_g \to C = \sqrt{gh}.\tag{4.24}$$

Equation (4.24) confirms a simpler derivation of group velocity which is given in Sect. 3.4.3. The phase and group velocities are presented in Fig. 4.5 as functions of the non-dimensional water depth kh. The velocities have been normalized against their values in deep water, e.g. C/C_0 and C_g/C_{g0}, in which C_0 and C_{g0} are the phase and group velocities at infinite water depth, respectively, i.e.: $C_0 = \sqrt{g/k}$ and $C_{g0} = 1/2\sqrt{g/k}$. Figure 4.5 shows that phase velocity C/C_0 increases monotonically as water depth increases. However, this is not the case for normalized group velocity C_g/C_{g0} which initially increases, when depth increases, and reaches its maximum value at about $kh = 2\pi h/L \approx 1.0$ ($L/h \approx 2\pi$) after which it decreases to 1. The computer program D41 is provided to determine both velocities (see Appendix D).

Orbital Velocity Components. Because of the periodicity of wave movement, water particles move along closed orbits (see Figs. 3.3 and 3.4). Both horizontal and vertical orbital velocity components take the form (see Appendix C.5 for derivation):

• *horizontal velocity component*

$$u(x,z,t) = \frac{gkH}{2\omega} \frac{\cosh k(z+h)}{\cosh(kh)} \cos(kx - \omega t), \qquad (4.25)$$

• *vertical velocity component*

$$w(x,z,t) = \frac{gkH}{2\omega} \frac{\sinh k(z+h)}{\cosh(kh)} \sin(kx - \omega t). \qquad (4.26)$$

Let us now consider a vertical profile located under the wave crest, when $kx - \omega t = 0$. Hence, Eqs. (4.25) and (4.26) give:

$$\left. \begin{array}{rcl} u & = & \dfrac{gkH}{2\omega} \dfrac{\cosh k(z+h)}{\cosh(kh)} \\[2mm] w & = & 0. \end{array} \right\} . \qquad (4.27)$$

Under the wave crest, the vertical velocity component vanishes, as expected, while the horizontal component is non-zero, attenuating from the surface to the sea bottom. At the sea surface ($z = 0$), the velocity, $u = gkH/2\omega$, and at the sea bottom ($z = -h$), $u = (gkH/2\omega)/\cosh kh$. The schematic representation of orbital velocity components for various wave phases is given in Fig. 4.6. Four phases on the wave profile have been selected and the corresponding orbital velocity vectors at some submergence level are indicated. Both deep and shallow water cases are considered. In deep water, particles move along circular paths. At the wave crest phase (point B), only horizontal orbital velocity exists, while at the inter-section of the surface with the still water level (points A and C) water particle moves up and down, respectively. In shallow water, the picture

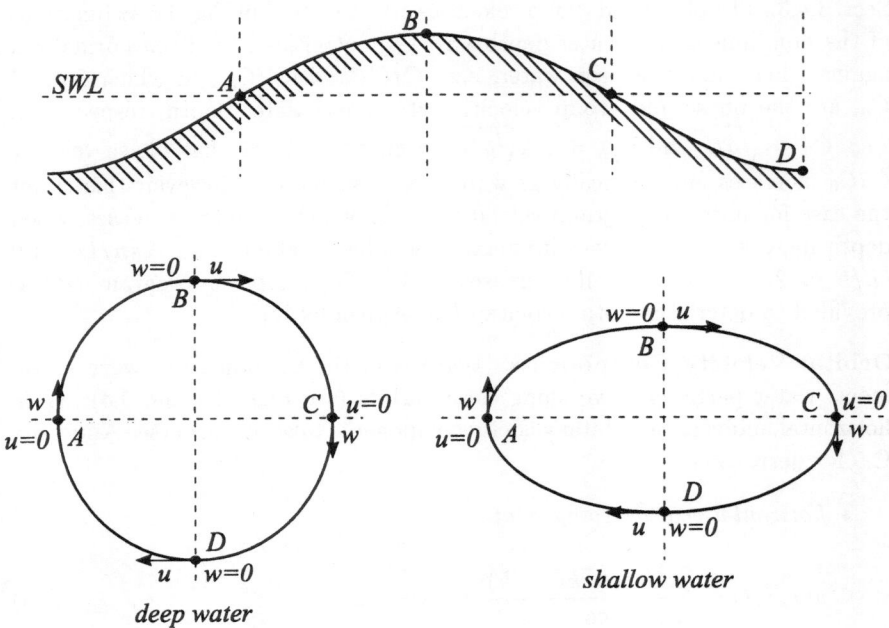

Fig. 4.6: Schematic representation of orbital velocity components for various wave phases at a given submerged point for deep and shallow waters

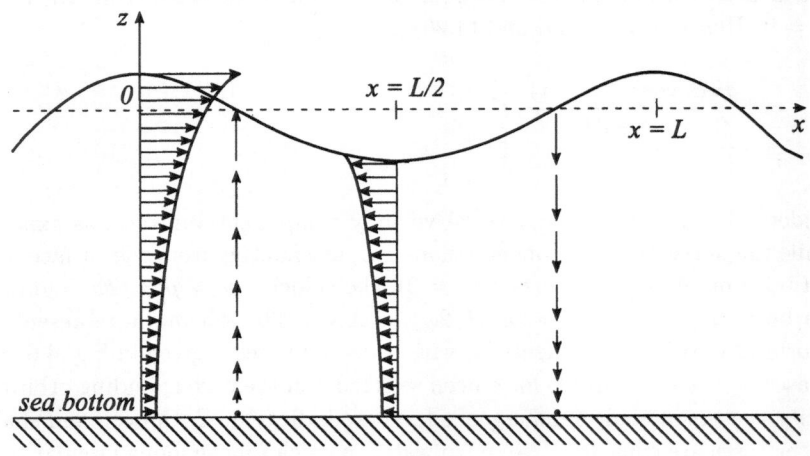

Fig. 4.7: Vertical profiles of the wave-induced velocities

is similar but now the particle path is elliptical. More discussion on orbital velocities is given in Appendix C.5.4.

The vertical profiles of orbital velocities u and w, for four phase positions and for $t = 0$, are shown in Fig. 4.7. The highest horizontal orbital velocities are those under the wave crest; they are directed parallel to the direction of wave propagation. In that part of the wave, when surface displacement is zero, horizontal orbital velocity is also zero. For the profile at the wave trough, the direction of orbital velocity is the reverse to the direction of wave propagation and its magnitude at a given level is the same as for the wave crest profile. At profiles where surface displacement vanishes, $i.e.$ for $x = L/4$ and $x = 3L/4$, only vertical orbital velocity exists, with values attenuating with submergence; at the sea bottom, vertical velocity is obviously zero because of zero bottom permeability. In Appendix D, the computer program D.43 is given to calculate the velocities at wave crest and wave trough and for arbitrary wave parameters.

Orbital Acceleration Components. The particle acceleration at a given point (x, z) and time, t, is obtained by differentiation of the particle velocities with respect to time, $i.e.$:

$$a_x(x, z, t) = \frac{\partial u}{\partial t} = \frac{gkH}{2} \frac{\cosh k(z + h)}{\cosh(kh)} \sin(kx - \omega t), \qquad (4.28)$$

$$a_z(x, z, t) = \frac{\partial w}{\partial t} = -\frac{gkH}{2} \frac{\sinh k(z + h)}{\cosh(kh)} \cos(kx - \omega t), \qquad (4.29)$$

or:

$$a_z(x, z, t) = \frac{gkH}{2} \frac{\sinh k(z + h)}{\cosh(kh)} \cos(kx - \omega t + \pi). \qquad (4.30)$$

Water Particle Displacement. As was shown in Figs. 3.3 and 3.4, water particles under the influence of wave motion move along closed circular or closed elliptical paths. The shape and length of these paths can be found by integrating the particle velocities, u and w, with respect to time. The resulting path equation take the form (Massel, 1989):

$$\left(\frac{\xi}{A}\right)^2 + \left(\frac{\eta}{B}\right)^2 = 1. \qquad (4.31)$$

Equation (4.31) is the equation of an ellipse where ξ and η are horizontal and vertical displacements of a water particle at a point of submergence $(-z)$, and horizontal semi-axis A and vertical semi-axis B (see Fig. 4.8) are given by:

$$\left. \begin{array}{rcl} A & = & \dfrac{H}{2} \dfrac{gk}{\omega^2} \dfrac{\cosh k(z + h)}{\cosh kh} \\[3mm] B & = & \dfrac{H}{2} \dfrac{gk}{\omega^2} \dfrac{\sinh k(z + h)}{\cosh kh} \end{array} \right\}. \qquad (4.32)$$

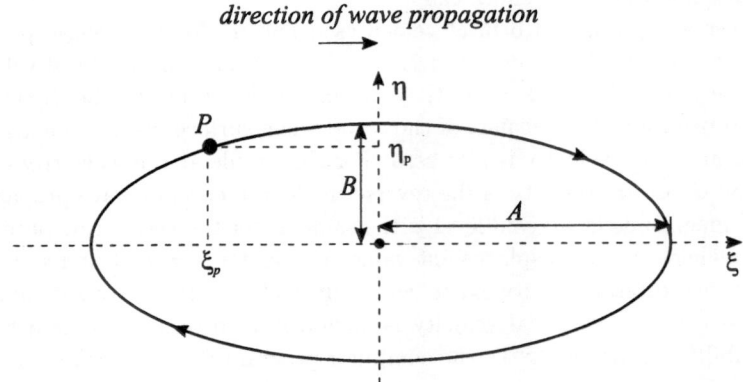

Fig. 4.8: Elliptical path of water particle in waves

At the sea bottom $(z = -h)$, the vertical semi-axis B vanishes and the water particle moves horizontally back and forward along the sea bottom with amplitude $A = (H/2)(gk/\omega^2)/\cosh kh$.

In deep water, when $h \to \infty$, both semi-axes of the water particle path are equal, *i.e.*:

$$A = B = \frac{H}{2} \frac{gk}{\omega^2} e^{kz}, \tag{4.33}$$

and Eq. (4.31) becomes:

$$\xi^2 + \eta^2 = A^2, \tag{4.34}$$

the equation of a circle with radius A. For points deeper than half the wavelength, $z < -L/2$, the radius $A \approx 0$, and surface waves are unable to induce any movement of water particles.

Wave Induced Pressure. Total pressure at a given point in the water body is determined as follows (for derivation see Appendix C.5.4):

$$p(x, z, t) = \rho_w g (-z + \zeta) - \rho_w g [1 - K_p(z)] \zeta, \tag{4.35}$$

in which:

$$K_p(z) = \frac{\cosh k(z + h)}{\cosh(kh)}, \tag{4.36}$$

and ζ is the surface displacement given by Eq. (4.13).

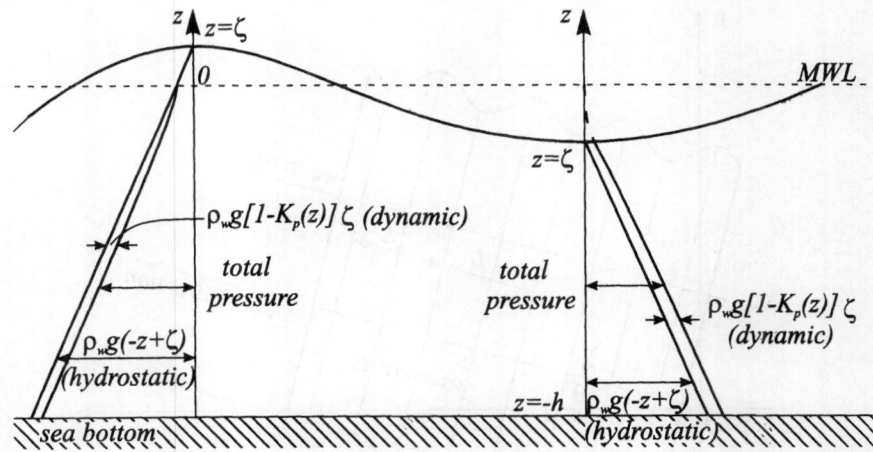

Fig. 4.9: Vertical profiles of wave-induced pressure

The first term on the right-hand side represents a hydrostatic pressure due to submergence and surface displacement. The second term is the dynamic part of the total pressure, and its contribution to the resulting pressure depends on wave phase. A more detailed discussion of wave-induced pressure is given in Appendix C.5.4 and a final vertical profile of wave induced pressure below both a wave crest and a trough is illustrated in Fig. 4.9. The computer program D.44 (Appendix D) calculates pressure for an arbitrary submergence, under wave crest and wave trough.

Wave Energy. From classical physics it is known that a displacement of some mass, m, from a position of equilibrium against a gravitational field results in the generation of potential energy. This principle is used in various facilities to generate energy. Water falling from a reservoir through a turbine eventually generates electricity. As in surface waves, the water mass has been displaced from an equilibrium position, and it possesses a potential energy. Even when there are no waves, potential energy exists because the centre of gravity of the water mass is above the sea bottom which can be considered as a reference surface (see Fig. C.7).

Water particles in wave motion are not only displaced from their equilibrium positions, but they move along circular or elliptical paths. Thus, their velocity is not zero and therefore they possess some kinetic energy. The total wave energy, per unit area, is the sum of the potential and kinetic energy (see Appendix C.5 for derivation):

$$E = E_p + E_k = \rho g \frac{H^2}{8}. \tag{4.37}$$

Waves Propagating in an Arbitrary Direction. Consider a top view of surface waves as in Fig. 4.10. Wave crests are perpendicular to the wave

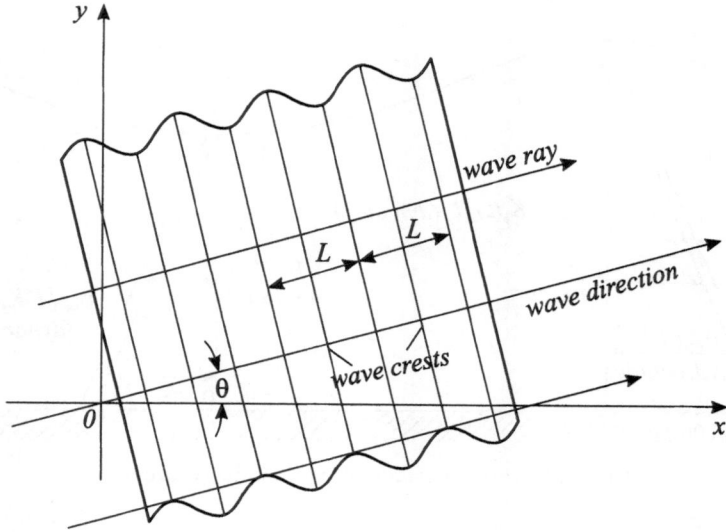

Fig. 4.10: Wave propagating at angle θ against x-axis

direction. The lines parallel to wave direction are called wave rays. Thus, wave rays are inclined at angle θ to the x-axis. Surface displacement, $\zeta(x, y, t)$, can now be expressed as follows:

$$\zeta(x, y, t) = \frac{H}{2} \cos \left[k \left(x \cos \theta + y \sin \theta \right) - \omega t \right]. \tag{4.38}$$

When $\theta = 0$, Eq. (4.38) is identical to Eq. (4.13). All other expressions can be similarly modified, simply replacing kx by $k(x \cos \theta + y \sin \theta)$.

4.2.3   Higher Order Wave Theories

Small amplitude wave theory provides reasonable results in many oceanographic and engineering applications. Also, for the purpose of this book this theory will mostly be used. However, for completeness, it is worthy to briefly outline some higher order nonlinear wave theories.

In particular, in the coastal zones the wavelength, L, is much larger than the water depth, h, and wave height, H, is an appreciable fraction of water depth, h. In such a situation, small amplitude wave theory provides a simple, but only approximate description of the wave motion. To improve the accuracy of predictions, several alternative theories have demonstrated their usefulness in the study of shallow water waves. Generally speaking, these theories belong to two different groups, namely Stokes' theory of short waves of finite height, and theories of long waves. A distinction between these groups depends on a combination of (L/h) and (H/h) values in the form of the so called Ursell parameter (Massel, 1989):

$$U = \left(\frac{H}{h}\right)\left(\frac{L}{h}\right)^2. \tag{4.39}$$

Stokes' theory is applicable when $U < 75$ (Hedges, 1995, suggested $U < 40$), while long-wave theories should be used when $U \geq 75$. Fenton (1990) proposed to demarcate the regions of validity of the two theories using the expression:

$$\frac{L}{h} = 21.5 \exp\left(-1.87\frac{H}{h}\right). \tag{4.40}$$

This demarcation line is based on the careful examination of the accuracy of both theories (demarcation line (4.40) is shown in Fig. 4.19). A derivation of the calculation formulas resulting from the short- and long-wave theories is beyond the scope of this book and interested readers are directed to professional books or articles (Fenton, 1979, 1985; Mei, 1983; Dean and Dalrymple, 1992; Massel, 1989; Hedges, 1995). However, we present some of the most useful formulae here.

Stokes' Theory for Waves of Finite Height. The basic difference between linear and nonlinear solutions of the surface wave problem is in the treatment of the boundary conditions. Instead of simplified conditions (C.93) and (C.97), the exact conditions (C.92) and (C.95) have to be used. In order to satisfy the Laplace equation and these exact boundary conditions, Stokes in 1847 proposed the solution in which a velocity potential ϕ is presented as a summation of many components, *i.e.* (Stokes, 1847; Massel, 1989):

$$
\begin{aligned}
\phi(x, z, t) \approx \ & B_1 \cosh\left[k(z + h)\right] \sin(kx - \omega t) + \\
& + B_2 \cosh\left[2k(z + h)\right] \sin\left[2(kx - \omega t)\right] + \\
& + B_3 \cosh\left[3k(z + h)\right] \sin\left[3(kx - \omega t)\right] + \ldots
\end{aligned}
\tag{4.41}
$$

in which coefficients B_1, B_2, B_3, ... are some functions of wave height, H, wave frequency, ω, wave number, k, and water depth, h (see, for example, Fenton, 1985).

The rate of accuracy of the solution depends on how many terms in Eq. (4.41) are taken into account. If only two terms are taken into account, water surface displacement becomes:

$$\zeta(x, t) = \frac{H}{2}\cos(kx - \omega t) + \frac{kH^2}{16}\left[2 + \cosh(2kh)\right]\frac{\cosh(kh)}{\sinh kh^3}\cos 2(kx - \omega t). \tag{4.42}$$

Note that the surface displacement now depends on wave height in powers one and two. Thus, it is a nonlinear function of wave height. The first term on the

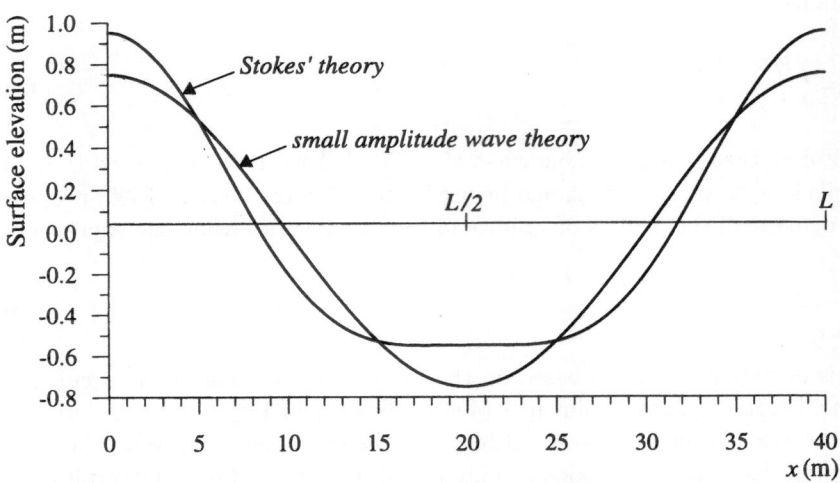

Fig. 4.11: Comparison of wave profile resulting from small amplitude and Stokes' wave theories ($H = 1.5$ m, $L = 40$ m, $h = 5$ m)

right-hand side of equation (4.42) is a displacement resulting from the theory of small amplitude while the second term represents nonlinear modification. Only when wave height becomes very small, can the second term be neglected. The greater accuracy of a nonlinear solution for short waves can be achieved when more terms are used in an expansion (4.41) (see, for example, Fenton, 1985). However, it should be noted that Stokes' theory should not be used for long waves when the corresponding Ursell parameter, U, becomes greater than 75. The computer program for calculating nonlinear wave profiles according to Eq. (4.42) is given in Appendix D (program D.42). In Fig. 4.11, a comparison of wave profiles resulting from Stokes' and small amplitude wave theory is shown for the $L/h = 8$ and $H/h = 0.3$; thus $U = 19$. The resulting Stokes' wave profile is much more peaked at the wave crest and flatter at the troughs than the small amplitude sinusoidal wave form as commonly observed in the sea.

Long-wave Theories. In coastal regions, wavelength, L, is generally much larger than the water depth, h. Moreover, the wave height, H, is an appreciable fraction of h so the Ursell parameter, U, becomes too large to allow Stokes theory to be used. There are a number of nonlinear solutions which can be used successfully for long waves, such as the Boussinesq equation, the Korteweg-de Vries equation, and the solitary wave equation (for details see Fenton, 1979, 1990; Mei, 1983; Dean and Dalrymple, 1992). In particular, the Korteweg-de Vries equation provides a profile in terms of the 'elliptic function' cn, and such a profile is named the **cnoidal wave**. A cnoidal wave is characterized by a very sharp, narrow crest and a long, shallow trough. Cnoidal waves will not be discussed here and interested readers should consult Fenton (1979). When

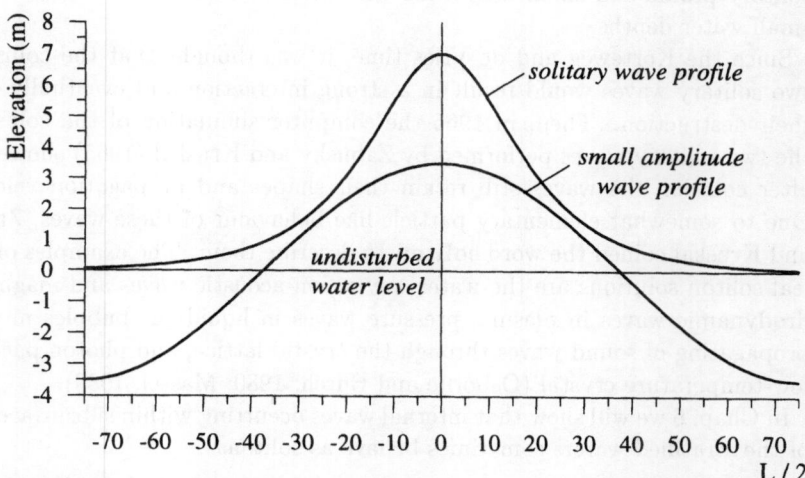

Fig. 4.12: Comparison of solitary wave profile with a profile resulting from small amplitude wave theory ($H = 7$ m, $L = 150$ m, $h = 10$ m)

the ratio $L/h \rightarrow \infty$, the cnoidal wave profile approaches the so called **solitary wave profile**, located above the undisturbed level (see Fig. 4.12).

The first documented observation of a solitary wave was made by Scott Russell in the 19th century. However, a model differential equation describing solitary wave behaviour was not developed until 1895, when Korteweg and de Vries approximated the Navier Stokes equations for the small finite-amplitude waves in a shallow channel. This equation is known as K-dV equation (Massel, 1989). The profile of a solitary wave can be determined from the formula:

$$\zeta(x,t) = \frac{H}{\cosh^2 \left[\sqrt{\frac{3H}{4h^3}} (x - Ct) \right]}, \tag{4.43}$$

in which the phase speed C is given by:

$$C = \sqrt{\left(1 + \frac{H}{h}\right) gh} \approx \sqrt{gh} \left[1 + \frac{1}{2} \left(\frac{H}{h}\right) \right]. \tag{4.44}$$

The solitary wave profile is frequently used to represent motion in extremely shallow water. For example, the shape of tsunami approaching a coastline is very close to the solitary wave shape. A computer program for determining a solitary wave profile is given in Appendix D (program D.42). In Fig. 4.12, a solitary wave profile is compared with a profile resulting from the small amplitude wave theory for $L/h = 15$ and $H/h = 0.7$; thus $U = 157$. In this very shallow water region, a sinusoidal wave form is completely different to the solitary profile and small amplitude wave theory should not be used for very small water depths.

Since the Korteweg and de Vries time, it was thought that the collision of two solitary waves would result in a strong interaction and eventually end in their destruction. Then, in 1965 the computer simulation of the collision of the two solitary waves performed by Zabusky and Kruskal (1965) showed that after collision the waves still retain their shapes and propagation velocities. Due to somewhat elementary particle-like behaviour of these waves, Zabusky and Kruskal coined the word **soliton** to describe them. The examples of physical soliton solutions are the water waves, ion-acoustic waves and magnetohydrodynamic waves in plasma, pressure waves in liquid-gas bubbles mixtures, propagating of sound waves through the crystal lattice, and photon packets in low-temperature crystal (Osborne and Burch, 1980; Massel, 1989).

In Chap. 6 we will show that internal waves occurring within subsurface layers of the stratified waters sometimes behave as solitons.

4.2.4 Wave Shoaling and Refraction

In previous sections we have dealt only with waves propagating on water of constant depth. However, the sea bottom very rarely is horizontal. In the deep ocean, when water depth is larger than half of the wavelength, depth changes (even by a few hundreds of meters) do not influence the surface waves. On the other hand, in the coastal zone the shoaling water depth effects the phase speed of waves (Eq. 4.16), and as was demonstrated in Fig. 4.5, the phase velocity increases with the local water depth. Therefore, parts of the wave crest lying over deeper water travel faster than the parts of the same crest lying over shallow water. In the course of propagation, such a wave front gradually turns towards the shallow water (see Fig. 4.13).

Let us consider a wave crest at some time instant, t, and two points, A and B, located on that crest, at water depths h_3 and $h < h_3$, respectively. After some time, Δt, point A moves for distance l_A along ray 1, while point B moves some distance, l_B, along ray 2. The term **wave ray** has been taken from optics. Rays are the lines which can be drawn perpendicularly to the wave crests and which indicate the direction of wave movement. Because water depth at point A is larger than depth at point B, the velocity of point A is greater than velocity of point B. Therefore, the distance l_A will be greater than the distance l_B, and the wave crest which joins points A and B gradually becomes parallel to the shoreline. This is in agreement with the common observation on beaches that the crests end up almost parallel to the shoreline, even when they are approaching the coast at an oblique angle from the sea.

This phenomenon is known as **refraction**. Refraction of waves over decreasing water depth can be described by a relationship similar to Snel's law, which describes refraction of light rays through materials of different refractive indices (Willebrord Snel van Royen was a Dutch scientist who lived 1580–1626).

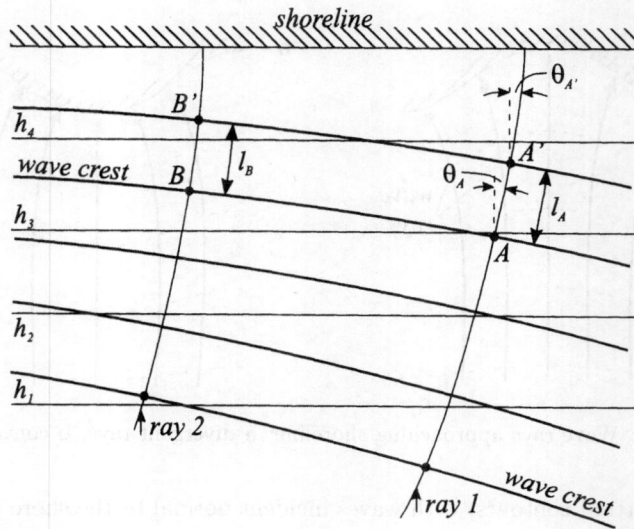

Fig. 4.13: Wave crests refracting in the shallow water zone

Prior to formulating Snel's law for surface waves, we define wave direction by an angle θ between the wave ray and a line normal to local depth contour (see Fig. 4.13). Hence, Snel's law for two points A and A' located on ray 1 can be represented as follows:

$$\frac{\sin \theta_A}{C_A} = \frac{\sin \theta_{A'}}{C_{A'}} = \text{Const}, \tag{4.45}$$

in which C_A and $C_{A'}$ are the respective phase velocities at points A and A'. Exactly the same relationship can be obtained using the angles between wave crests and bottom contours.

When depth contours vary arbitrarily in the coastal zone (see, for example, Fig. 4.14), the relationship between angles θ and velocities C takes the form (Massel, 1989):

$$\frac{\partial \left(\dfrac{\sin \theta}{C} \right)}{\partial x} - \frac{\partial \left(\dfrac{\cos \theta}{C} \right)}{\partial y} = 0, \quad \text{or} \quad \frac{\partial \left(k \sin \theta \right)}{\partial x} - \frac{\partial \left(k \cos \theta \right)}{\partial y} = 0. \tag{4.46}$$

Equation (4.46) is usually solved for the unknown angle, θ, and phase velocity, C (or wave number k), using computer programs.

Depending on bottom contour configurations, wave rays can converge or diverge (Fig. 4.14). The distance between two adjacent wave rays is an indication of wave height. To explain this relationship, let us consider a simpler case of a

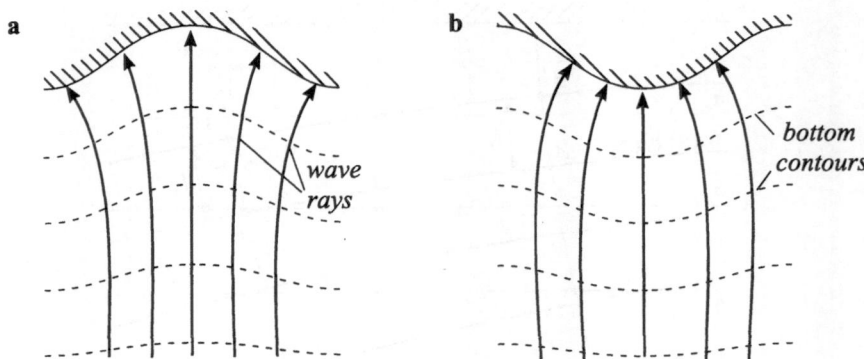

Fig. 4.14: Wave rays approaching shoreline: **a** divergent rays, **b** convergent rays

parallel bottom contours, with waves incident normal to the shore and assume they are not breaking (Fig. 4.15). The bottom slope is considered to be very gentle, say < 0.05. As was shown in Chap. 3, wave energy is transported in the direction of wave propagation with the group velocity, C_g, and the product of wave energy, E, and velocity, C_g, is the energy flux $F = E\,C_g$. If there are no energy losses due to breaking or bottom friction, it is reasonable to assume that energy flux remains constant, *i.e.*:

$$E\,C_g = \text{Const.} \tag{4.47}$$

Applying this principle to waves approaching coastline, for the two water depths, h_1 and h_2, we obtain:

$$(E\,C_g)_{h=h_1} = (E\,C_g)_{h=h_2}. \tag{4.48}$$

As wave energy is proportional to the square of wave height (see Eq. 4.37), Eq. (4.48) becomes:

$$\left(H_1^2\,C_g\right)_{h=h_1} = \left(H_2^2\,C_g\right)_{h=h_2}, \tag{4.49}$$

and

$$H_2 = H_1\sqrt{\frac{C_{g1}}{C_{g2}}}. \tag{4.50}$$

Therefore, when wave height, H_1, at water depth, h_1, is known, wave height at another water depth, h_2, can be found, as the group velocities, C_{g1} and C_{g2}, follow from the formula (4.21) for given water depths, h_1 and h_2. For very

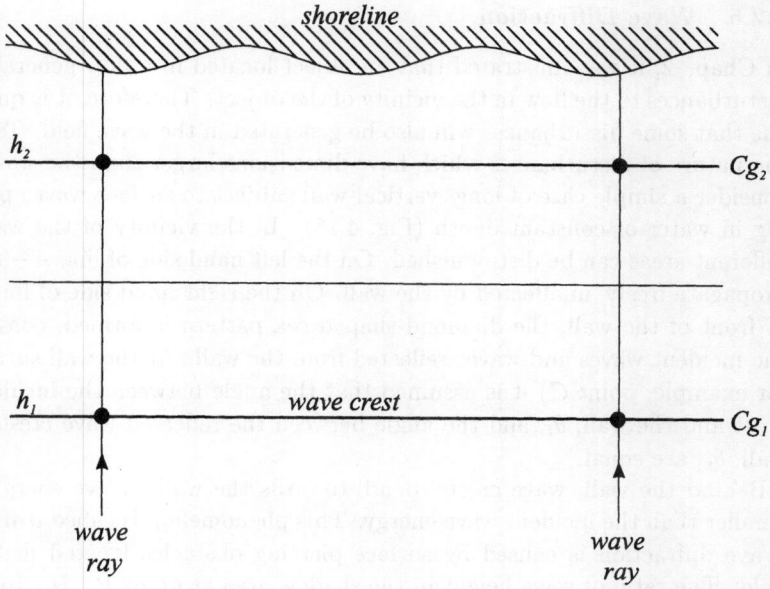

Fig. 4.15: Wave approaching normally to shoreline with energy flux being constant

shallow waters, group velocity $C_g = \sqrt{gh}$ (see Eq. 4.24). Hence, Eq. (4.50) becomes Green's law (Massel, 1989):

$$H_2 = H_1 \left(\frac{h_1}{h_2}\right)^{1/4}. \qquad (4.51)$$

When waves approach at some angle to a coastline with parallel bottom contours, as is shown in Fig. 4.13, Eq. (4.50) has to be slightly modified as follows:

$$H_2 = H_1 \sqrt{\frac{C_{g1}}{C_{g2}}} \sqrt{\frac{\cos \theta_1}{\cos \theta_2}}. \qquad (4.52)$$

Using the arguments developed above, we can say that conservation of energy flux results in higher waves on headlands than in bays (Fig. 4.14).

The computer program for wave shoaling on a slope with parallel isobaths is given in Appendix D (Program D.45). The relationships between wave height, group velocity and angle, developed above, are valid for gentle slopes only, say smaller than 1/10. When the bottom slope becomes steeper, transition of wave parameters is very rapid and an additional mechanism, known as wave diffraction starts to play an important role.

4.2.5   Wave Diffraction

In Chap. 2, it was illustrated that an object located in a flow generates some
disturbances to the flow in the vicinity of the object. Therefore, it is quite obvious that some disturbances will also be generated in the wave field. To explore
the nature of disturbances which have dimensions larger than the wavelength,
consider a simple case of long, vertical wall subject to surface waves propagating in water of constant depth (Fig. 4.16). In the vicinity of the wall, three
different areas can be distinguished. On the left hand side of line $A - B$, waves
propagate freely, unaffected by the wall. On the right hand side of line $A - B$,
in front of the wall, the diamond-shaped sea pattern is formed, consisting of
the incident waves and waves reflected from the wall. At the wall surface (see,
for example, point C) it is assumed that the angle between the incident wave
crest and the wall, θ_i, and the angle between the reflected wave crest and the
wall, θ_r, are equal.

Behind the wall, wave crests 'bend' towards the wall. Wave energy here is
smaller than the incident wave energy. This phenomenon is called **diffraction**.
Wave diffraction is caused by surface piercing obstacles located in the wave
field. The ratio of wave height in the shadow area at point P , H_p, to incident
wave height, H_i, is known as the diffraction coefficient (see Fig. 4.16):

$$K_d = \frac{H_p}{H_i}. \tag{4.53}$$

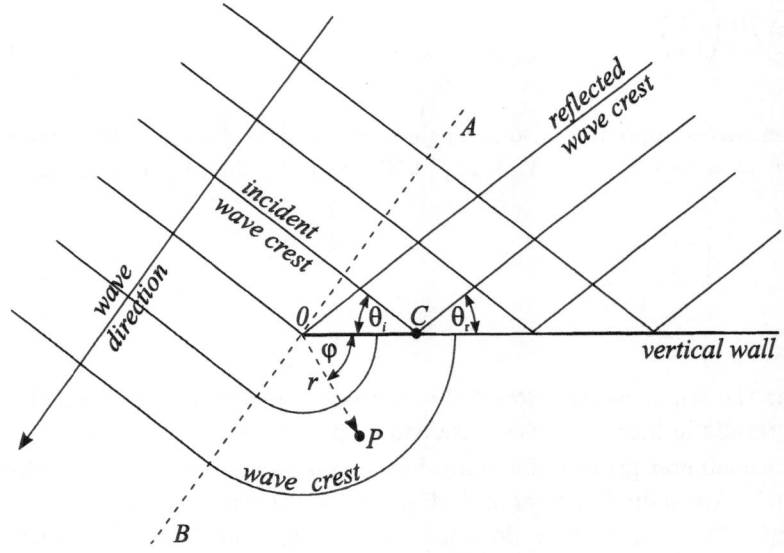

Fig. 4.16: Wave field in vicinity of vertical wall

The diffraction coefficient depends on the distance, r, between a given point and the tip of the wall (point O) and angle, φ. For an obstacle of simple shape such as a long vertical wall, the values of the diffraction coefficient, K_d, are available in graphical form (SPM, 1984). More complicated obstacles require more advance techniques (Mei, 1983; Massel, 1989). We will again discuss the diffraction phenomenon while discussing wave behaviour on steep offshore islands and coral reefs.

4.2.6   Wave Breaking

Waves approaching a shoreline over decreasing water depth gradually change their height and length. Wavelength decreases because period is constant but phase speed is smaller, while amplitude increases to ensure constant energy flux (see Green's law 4.51). Waves approach their limiting slope and eventually lose their stability and break. Wave breaking on a beach is one of the most significant physical phenomena in the coastal zone. Its significance arises from the fact that during breaking, most of the energy transmitted with the waves is dissipated and partly transmitted to the nearshore currents which may cause sediment transport both in the on-offshore and along shore directions. Some wave energy is reflected back out to sea, the amount depending upon the slope of the beach – the shallower the angle of the beach slope, the less energy is reflected. Finally, a small amount of wave energy is dissipated as heat in the final small-scale mixing of foaming water and sand.

Breaking waves are not limited to the nearshore zone. During storms, waves break at sea in the form of **white-capping**, *i.e.* breaking of the tip of the wave crest as it is driven forward by the wind faster than the wave itself is travelling. The initiation of wave breaking in deep water is controlled by wave slope, $s = H/L$. Stokes (1847) predicted that limiting wave steepness is:

$$s = \frac{H}{gT^2} = 0.027. \tag{4.54}$$

However, available field and laboratory data, as well as theoretical analysis (Massel, 1998), indicate that a limiting steepness in deep water is lower than Stokes' limiting steepness. For the commonly observed probability of breaking (from 3 to 15%), the limiting wave steepness during a storm is of the order of 0.005.

Breaking on a beach is more complicated than breaking in deep water. Shallow water breakers can be classified into four major types (Massel, 1989):

1. *Spilling.* White foam and turbulence appears at the wave crest and spills down the front face of the wave (Fig. 4.17a). Spilling breakers usually start some distance from shore and dissipate their energy gradually. Such waves are characteristic of a gently sloping shoreline.

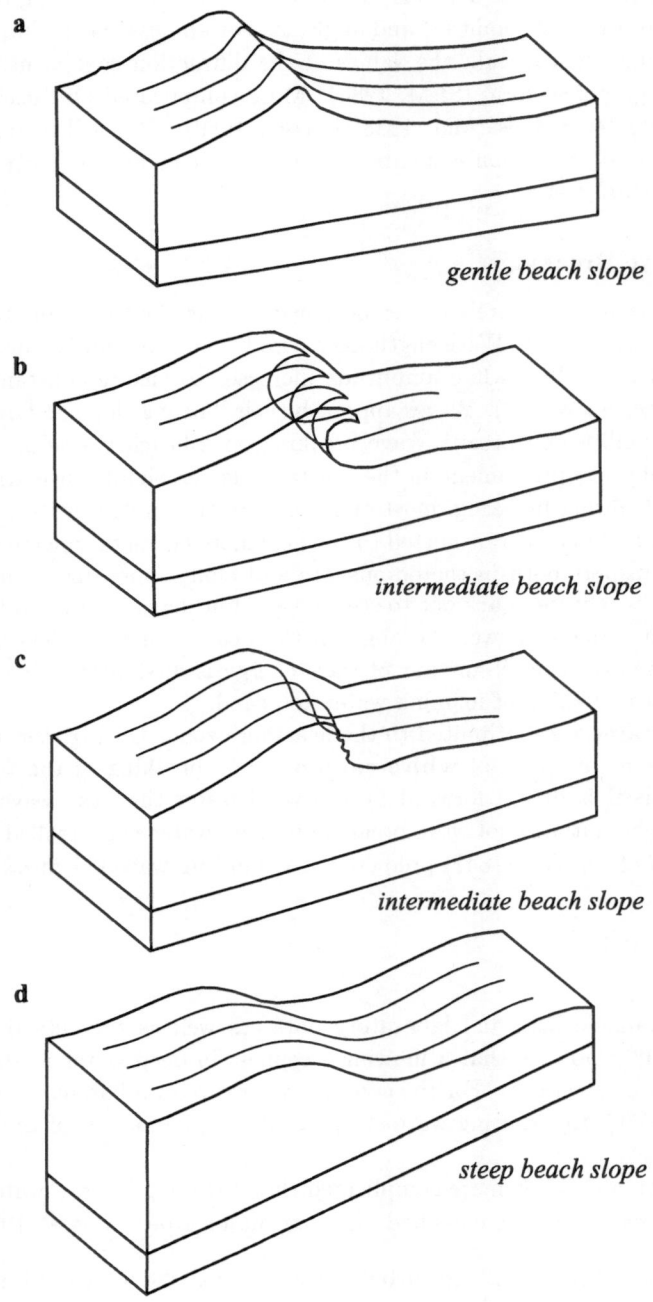

Fig. 4.17: Types of shallow water breakers: **a** spilling breaker, **b** plunging breaker, **c** collapsing breaker, **d** surging breaker

2. *Plunging.* The whole front face of the wave steepens until vertical; the crest curls over the front face and falls into the base of the wave (Fig. 4.17b). The classic form of plunging breakers, much beloved by surf-riders, is arched, with a convex back and a concave front. This type of breaker appears on beaches of relatively gentle slope where long swells, generated by distant storms, are present.

3. *Collapsing.* The lower part of the wave's front face steepens until vertical, and this front face curls over as an abbreviated plunging wave (Fig. 4.17c). Minimal air pocket and usually no splash-up are observed. Such breakers occur on beaches with moderately steep slopes.

4. *Surging.* The wave slides up the beach with little or no bubble production. The water surface remains almost plane, except where ripples may be produced on the beach face (Fig. 4.17d). Surging breakers are found on very steep beaches.

Laboratory experiments have shown that breaker types can be distinguished depending on the so called 'surf similarity parameter' ξ (Massel, 1989):

$$\xi = \frac{\beta}{\sqrt{\frac{H_b}{L_0}}}, \tag{4.55}$$

where β is the beach slope, H_b is the wave height at breaking point, and L_0 is the deep water wavelength. The following values of parameter ξ are associated with particular breakers types:

$$\left.\begin{array}{rl} \text{surging or collapsing}: & \xi > 2.0 \\ \text{plunging}: & 0.4 < \xi < 2.0 \\ \text{spilling}: & \xi < 0.4 \end{array}\right\}. \tag{4.56}$$

The breaking wave height, H_b, in Eq. (4.55) can be found from the formula developed by Goda (1985):

$$H_b = 0.18 L_0 \left\{ 1 - \exp\left[-1.5\frac{\pi h}{L_0}\left(1 + 15\beta^{4/3}\right)\right]\right\}. \tag{4.57}$$

Formula (4.57) is presented in graphical form in Fig. 4.18.

In Fig. 4.18, a limiting wave height for waves propagating over a horizontal bottom is added. Horizontal sea bottoms usually occur at sea banks or at the top of coral reefs. In current oceanographic and engineering practice, it is assumed that for a horizontal bottom the H/h ratio is equal to 0.78, what is consistent with the value predicted from solitary waves theory. However, Tucker *et al.* (1983) measured waves using two waverider buoys, seaward and landward of a relatively flat offshore bank of the east coast of England. Minimum water depth over the bank was approximately 4.5 m at mid-tide level. The measured

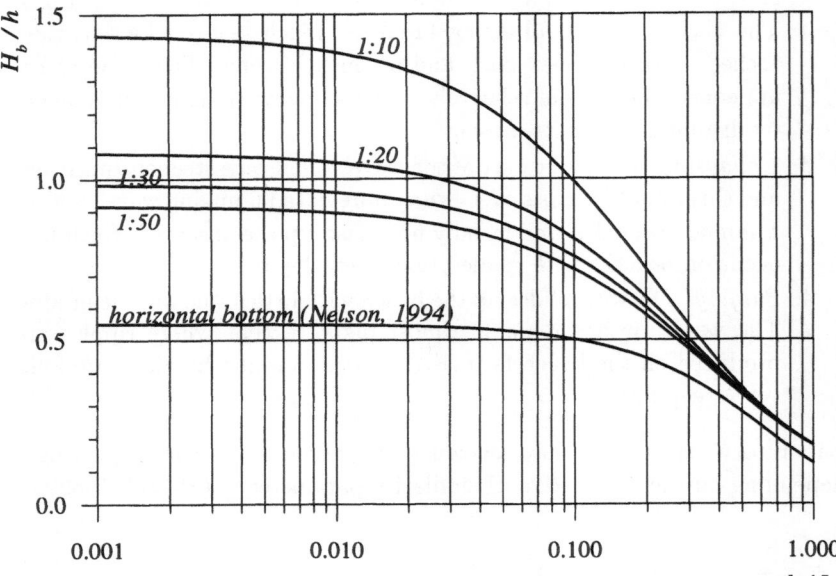

Fig. 4.18: Breaking wave height and bottom slope relationship

saturation level corresponding to a breaking wave showed the ratio of maximum wave height to water depth of about 0.5.

Laboratory experiments on random waves propagating over a horizontal bed, reported by Riedel and Byrne (1986), showed that the limiting ratio $(H/h)_{max}$ of 0.55 applies equally well to wind-induced and regular waves. Nelson (1994) reviewed all existing laboratory and field data for shallow water waves and concluded that the upper limit value for the ratio H/h is 0.55, which is a value considerably less than 0.8, used in current oceanographic and engineering practice. Using experimental data sets, he proposed an envelope curve for both transitional and shallow water waves, in the form:

$$\left(\frac{H}{h}\right)_{max} = \frac{F_c}{22 + 1.82F_c},$$ (4.58)

in which:

$$F_c = \left(\frac{H}{h}\right)^{1/2} \left(T\sqrt{\frac{g}{h}}\right)^{5/2}.$$ (4.59)

For practical calculations, it will be more convenient to express the ratio H/h as a function of some independent variable, say h/L_0 ($L_0 = g/2\pi T^2$). Thus we have (Massel, 1996b; see also Fig. 4.18):

$$\frac{H_{max}}{h} = \left[\frac{\sqrt{1 + 0.01504h_*^{-2.5}} - 1}{0.1654h_*^{-1.25}}\right]^2,$$ (4.60)

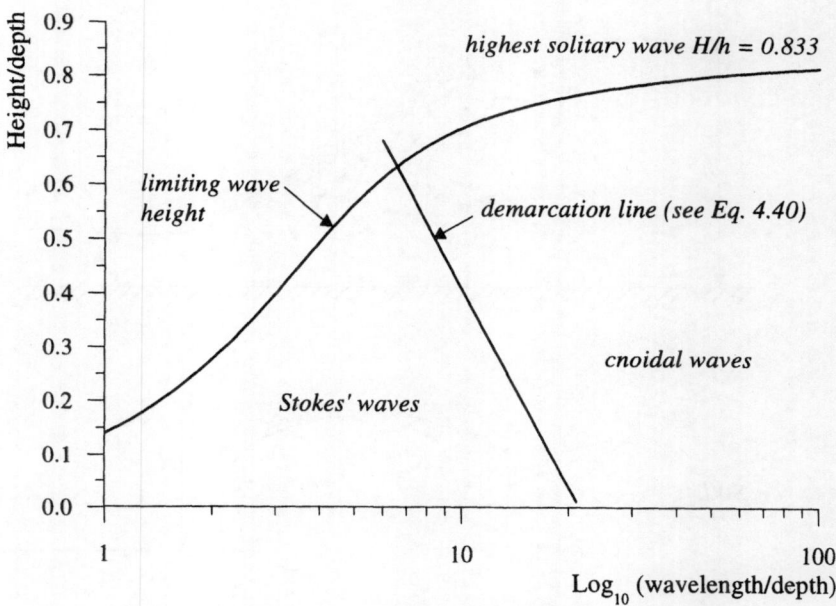

Fig. 4.19: Regions of validity of the Stokes' and cnoidal wave theories

in which:

$$h_* = \frac{h}{gT^2}. \tag{4.61}$$

Most of the experimental data reported by Nelson are the results of experiments in wave flumes. However, laboratory studies of surface waves are complicated due to the contamination contributed to the wave motion by wavemakers, as the simple harmonic motion produces a wave train not only with the wave-maker frequency, but also with it's higher harmonic. The sinusoidal motion of the generator does not match the water particle motion required by the wave. The theoretical explanation of the observed limiting wave height and influence of mechanical generation was proposed recently by Massel (1996b).

The highest waves possible, $H_{\max}$, have been determined in many theoretical and computational studies. In Fig. 4.19, the limiting wave height $H_{\max}/h$ is shown according to the approximation given by Fenton (1990):

$$\frac{H_{\max}}{h} = \frac{0.141063\left(\frac{L}{h}\right) + 0.0095721\left(\frac{L}{h}\right)^2 + 0.0077829\left(\frac{L}{h}\right)^3}{1 + 0.0788340\left(\frac{L}{h}\right) + 0.0317567\left(\frac{L}{h}\right)^2 + 0.0093407\left(\frac{L}{h}\right)^3}. \tag{4.62}$$

As was pointed out by Fenton (1990), there are enough instabilities (for example, mechanical generation discussed above) at work that real waves propagating over a flat bed cannot approach the theoretical limit given by Eq. (4.62).

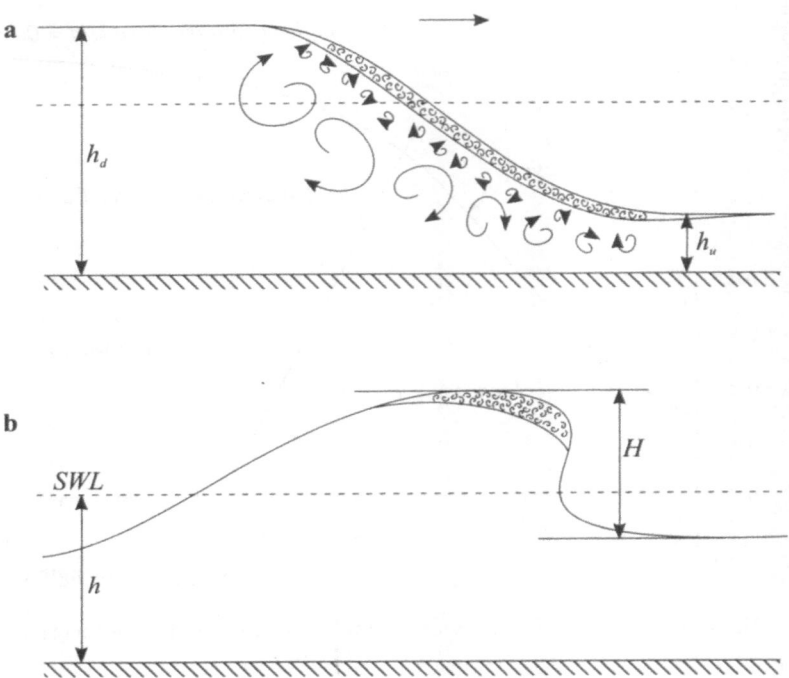

Fig. 4.20: Periodic bore model: **a** classic bore propagation, **b** approximation of wave breaking by bore shape breaker

The area between the water depth contour at which waves start to break and the shoreline is known as the **surf zone**. Prediction of wave parameters in the surf zone is highly complex and the present understanding of wave-breaking processes does not allow for adequate substitution for empirical observations. The development of rational prediction models is just beginning (Battjes and Jansen, 1978; Thornton and Guza, 1983; Dingemans, 1997). Therefore, it will not be discussed here. Interested readers can find detailed information in the suggested books and referred articles.

Here we will only briefly describe the basic physical processes involved in breakers propagating in the surf zone. When waves start to break, energy gradually decreases and the rate of this attenuation depends on wave parameters as well as on the geometry of the sea bottom. To allow some quantitative estimation of the amount of energy which is lost during breaking, the similarity of a breaker to the so called ' bore' is utilized (Battjes and Jansen, 1978). A bore is a pattern of fluid motion which occurs when two volumes of fluid with different surface levels are joined with a very short transition interface (Fig. 4.20). Bores, also called tidal bores, are observed at some tidal rivers during high tides. They move upstream as a rolling wall of water with a speed faster than that of a normal tidal current.

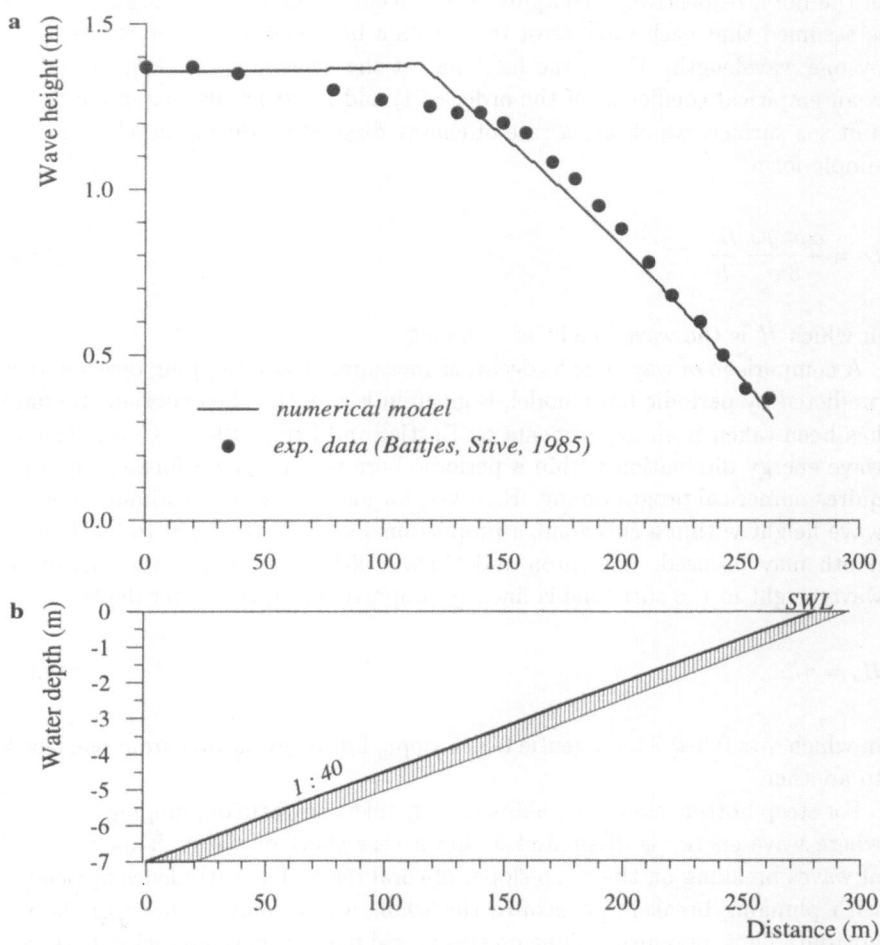

Fig. 4.21: Wave height decay over plane beach. Comparison of periodic bore model with experimental data: **a** wave height distribution, **b** water depth (from Massel, B., 1998, with permission)

Most tidal bores are relatively small, of the order of 0.5 m, but some can be exceptionally high. The Severn River undular bore in England is some 1–2 m high, whereas the Amazon bore reaches about 5 m.

The conditions of conservation of mass and momentum across the bore provide a rate of energy loss in the bore (Massel, 1989):

$$\Delta E = \frac{1}{4}\rho_w g \frac{(h_d - h_u)^3}{h_d h_u}, \tag{4.63}$$

in which h_d and h_u are water levels in the downstream and upstream regions

of the bore, respectively. To apply this concept to breakers in the surf zone, it
is assumed that each wave crest represents a bore and each bore is separated
by one wavelength. Using the fact that at the wave crest $h_d - h_u \approx \alpha H$, α
is an empirical coefficient of the order of 1, and referring dissipated energy to
unit sea surface, we obtain a rate of energy dissipation during breaking in the
simple form:

$$D = \frac{\alpha \rho_w g \omega}{8\pi} \frac{H^3}{h},$$ (4.64)

in which H is the wave height at breaking.

A comparison of wave height decay, as measured over a slopping beach and as
predicted by periodic bore model, is given in Fig. 4.21. The experimental data
has been taken from experiments by Battjes and Stive (1985). Calculation of
wave energy dissipation within a periodic bore is not straightforward, and requires numerical programming. However, for an approximate estimation of the
wave height within a surf zone, a proportionality of wave height to local water
depth may be used. Thornton and Guza (1983) showed that the significant
wave height in the surf zone is linearly proportional to the water depth:

$$H_s = \gamma h,$$ (4.65)

in which $\gamma \approx 0.6–0.7$ for a gentle beach slope, but likely to vary from one beach
to another.

For steep bottom slopes, breaking usually takes the form of plunging breakers,
where wave energy is dissipated within a very short distance. This is typical
of waves breaking on the steep slopes of coral reefs. The turbulence associated
with plunging breakers penetrates the whole water column, down to the sea
bottom. Thus, organisms living on steep coral reefs or on steep rocky shores are
very likely to experience turbulence with every breaking wave (Massel, 1994).
The physics of wave propagation on coral reefs and biological consequences of
this intensive turbulent motion will be discussed in Chap. 14.

4.2.7 Boundary Layer Induced by Waves

In Sect. 2.5 we considered bottom boundary layers induced by unidirectional,
steady flow. In the coastal zone, where water depth is usually smaller than
half of the wavelength, wave motion attenuates slowly and non-zero horizontal
velocity exists at the sea bottom. This velocity oscillates back and forth along
the direction of wave propagation. Thus, Eq. (4.25) gives, for $z = -h$ (sea
bottom) and at a point $x = 0$:

$$u(x, -h, t) = \frac{gkH}{2\omega} \frac{1}{\cosh kh} \cos \omega t.$$ (4.66)

Assuming that the bottom is flat and smooth, and flow is laminar, the velocity distribution in the boundary layer has been found by Stokes solving the viscous equation of motion in the form (Fredsøe and Deigaard, 1992):

$$u(z_1, t) = u_0 \cos(\omega t) - u_0 \exp\left(-\frac{z_1}{\delta_s}\right) \cos\left(\omega t - \frac{z_1}{\delta_s}\right), \tag{4.67}$$

in which z_1 is the distance from the bed, $z_1 = h + z$, and

$$u_0 = \frac{gkH}{2\omega} \frac{1}{\cosh(kh)}, \tag{4.68}$$

while δ_s is the boundary layer thickness defined by Stokes as:

$$\delta_s = \sqrt{\frac{2\nu}{\omega}}. \tag{4.69}$$

The quantity δ_s is sometimes called the 'Stokes length'. There are a few other proposals for δ_s definition. Sleath (1987) defined the top of the boundary layer as the level where velocity decreases to a certain small fraction of u_0, say 0.05. Then:

$$\delta_1' = 3\delta_s. \tag{4.70}$$

Jonsson (1980) suggested a different type of definition for δ_s, as the distance from the bed to a level where velocity, u, equals u_0, i.e.:

$$\delta_1'' = \frac{\pi}{2}\delta_s. \tag{4.71}$$

For example, for a wave with 6 s period, boundary layer thickness δ_s, δ_1' and δ_1'' are 1.4 mm, 4.3 mm and 2.2 mm, respectively. An intercomparison and discussion of these and other definitions of boundary layers are given by Nielsen (1992).

When $z_1 = 0$ ($z = -h$), Eq. (4.67) predicts velocity $u(0, t) = 0$, as expected. At levels above the sea bottom, when z_1 is large, velocity $u(z_1, t)$ approaches ambient velocity, u_0, because the term $\exp(-z_1/\delta_s) \to 0$.

Very close to the bottom, the term z_1/δ_s in Eq. (4.67) introduces a phase shift in the velocity, relative to ambient velocity $u_0 \cos(\omega t)$. For example, when $z_1/\delta_s = \pi$, the velocity in the boundary layer is exactly out of phase with velocity due to wave motion alone. The comparison of the velocity distribution (2.48) and (4.67) indicates that the velocity decay with a distance in an oscillatory flow is much greater than for unidirectional flow due to creating a vorticity with different sign, which cancel each other during wave period.

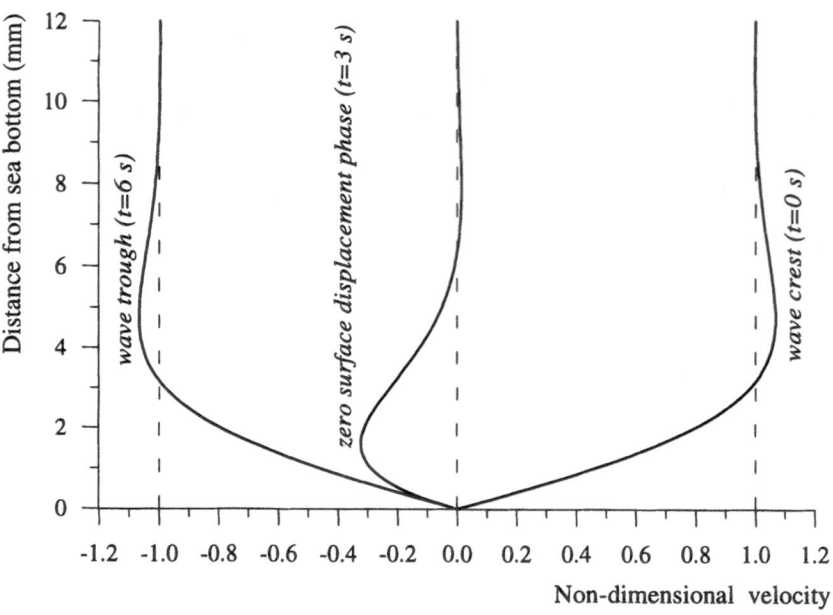

Fig. 4.22: Non-dimensional velocity $u(z,t)/[u_0 \cos(\omega t)]$ as a function of distance from sea bottom ($T = 12$ s)

In Fig. 4.22, non-dimensional velocity $u(z,t)/[u_0 \cos(\omega t)]$ is presented as a function of distance from the sea bottom for a wave period of $T = 12$ s and three wave phases (wave trough, wave crest and zero surface displacement phase). Thickness, δ_s, of the boundary layer now equals 2 mm. Under the wave crest (at $t = 0$ s) velocity within the boundary layer increases from zero at the bottom to its maximum value $1.067\, u_0$ at $z \approx 4.7$ mm, and then asymptotically approaches u_0 (or non-dimensional value $= 1.0$). When the wave surface intersects the still water level (at $t = 3$ s for $T = 12$s period wave), horizontal velocity above the boundary layer is exactly zero (see Eq. 4.25). However, in the boundary layer, velocity is not equal to zero, but is negative along the whole vertical profile with the smallest value, $-0.322 u_0$, at $z \approx 1.5$ mm above bottom. Finally, under the wave trough ($t = 6$ s), velocity is negative, and at $z = 4.7$ mm velocity reaches its largest negative value $1.067\ u_0$.

If we proceed further and expand the Stokes' solution to a second approximation, we find that the water movement in the bottom boundary layer is not strictly oscillatory as was given by Eq. (4.67). Longuet-Higgins (1953) discovered that there is a non-zero net transport in the direction of wave motion varying with distance from the bed:

$$u(z_1) = \frac{\pi^2 H^2}{4LT \sinh^2(kh)}\, [5 - 8\exp(-\xi)\cos\xi + 3\exp(-2\xi)], \qquad (4.72)$$

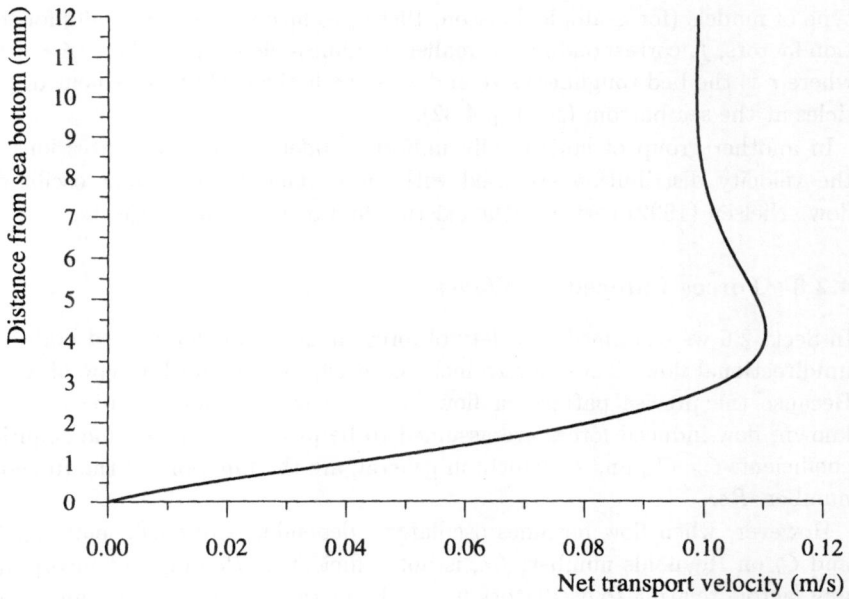

Fig. 4.23: Vertical distribution of net transport velocity in the boundary layer ($H = 2$ m, $T = 12$ s, $h = 10$ m)

where $\xi = z_1/\delta_s$. This transport reaches a maximum velocity:

$$u_{max} = 1.057 \frac{\pi^2 H^2}{LT \sinh^2(kh)}, \tag{4.73}$$

at level $z_1 = 2.3\delta_s$. When z_1 increases, transport velocity tends to value $5\pi^2 H^2/[4LT \sinh^2(kh)]$. In Fig. 4.23, a vertical distribution of net transport velocity is given for waves with $H = 2$ m, $T = 12$ s and water depth $h = 10$ m. Maximum velocity of ~ 0.12 m/s is reached at about 4.5 mm above the sea bottom.

Natural sea beds never are perfectly flat and smooth, and flows tend to be turbulent. Turbulent boundary layers in unidirectional flow have been examined in Sect. 2.5. In general, results of this analysis are qualitatively correct also for oscillatory turbulent boundary layers, but their quantitative prediction is much more complex. These prediction models fall into two broad categories; namely horizontally uniform (in sense of bed roughness) models and models which take into account the horizontal variability between crests and troughs of the bed roughness elements (bed ripples).

For a horizontal sea bottom, some models assume that the velocity distribution is at all times logarithmic (as in the case of unidirectional flow) throughout a boundary layer thickness which may be constant or time-dependent. Such type of models (for example Jonsson, 1966) produce very good results for friction factors, f, corresponding to smaller roughness elements, when $r/A < 0.03$, where r is the bed roughness size and A is the horizontal displacement of particles at the sea bottom (see Eq. 4.32).

In another group of horizontally uniform models, empirical expressions for the velocity distribution are used with close analogy to laminar oscillatory flow. Nielsen (1992) reviewed the existing literature on this subject.

4.2.8 Forces Induced by Waves

In Sect. 2.6 we examined a variety of forces imposed on submerged bodies by unidirectional flow. These forces included drag, inertia and lift type of forces. Because the precise pattern of flow around most submerged objects is not known, flow-induced forces are assumed to be proportional to some empirical coefficients C_d, C_m and C_l which, in general, are the functions of the Reynolds number, Re.

However, when flow becomes oscillatory, dependence of coefficients C_d, C_m and C_l on Reynolds number, Re, is not sufficient as Re does not incorporate any factors relating to oscillatory flow. Therefore, for complete specification of the flow pattern, especially its periodicity, another parameter is required. This dimensionless number is known as the Keulegan-Carpenter number K, or the period parameter (Sarpkaya and Isaacson, 1981):

$$K = \frac{u_m T}{L_b}, \tag{4.74}$$

in which u_m is the maximum velocity of the oscillation flow, T is the wave period and L_b is the characteristic length of an object along the direction of flow. The physical meaning of the Keulegan-Carpenter number, K, is quite simple. In wave motion, the direction of flow changes every half wave period. Thus, flow past the object travels in one direction, stops, and then flows back in the other direction. When the distance that the water moves when flowing in one direction is greater than the length of the object, the flow pattern is similar to that for unidirectional flow. In this case, the Reynolds number, Re, controls fluid separation from the surface, and there is sufficient time for the formation of a wake of vortices behind the body, which results in a reduction of pressure (see Sect. 2.6). If, however, the distance travelled by the fluid in one direction during half of wave period is smaller than the length of the object, there is not enough time for separation to occur, before the fluid again changes direction. The distance traveled in one direction is roughly equal to $u_m T$, and it has to be compared with the characteristic length of an object, L_b. This is the significance of the Keulegan-Carpenter number, K. When K is large, the fluid moves many times the length L_b, and when $K > 30$, the flow may be expected to mimic steady flow. For smaller values of K, the flow pattern is affected by changes in flow direction.

Let us consider a typical wave period, $T = 10$ s, and maximum horizontal velocity $u_m = 1$ m/s. Thus, for marine organisms with characteristic lengths between 1 mm and 10 cm, the Keulegan-Carpenter number, K, varies between 100 and 10^4. For such large values of K, the oscillatory nature of the flow does not effect the flow around the organisms, and the drag, lift and inertia coefficients are nearly independent of K (Sarpkaya and Isaacson, 1981).

Resulting total wave-induced force can be found from Morison's equation (2.90) for C_d and C_m coefficients being, in general, the functions of the Reynolds number, Re (see, for example, Fig. 2.24). Because in oscillatory flow, the direction of flow changes every wave period, the expression for drag force (2.62) should be:

$$F_d = \frac{1}{2}\rho_w C_d S u_m^2 \cos(\omega t)\left|\cos(\omega t)\right|, \tag{4.75}$$

in which u_m is the maximum velocity, ω is the frequency of wave motion, $i.e.$ $\omega = 2\pi/T$, and S is the projected area of the body on the flow direction. Symbol $|\ |$ denotes an absolute value. For example, if the value of x is positive, then $|x| = x$, but if this value is negative, absolute value $|x| = -x$. Therefore, in each case, the $|x|$ always has a positive value.

Let us represent horizontal acceleration as (see Eq. 4.28):

$$a(t) = a_m \sin(\omega t), \tag{4.76}$$

where a_m is the maximum acceleration. From Eq. (4.25) and (4.28) it follows that $a_m = \omega u_m$. Thus, the inertia force in Eq. (2.86) takes a form:

$$F_i = \rho_w C_m V a_m \sin(\omega t) = \rho_w C_m V u_m \omega \sin(\omega t), \tag{4.77}$$

where V is the volume of the body.

Let us now form a ratio of the maximum inertia and drag forces, $i.e.$:

$$\frac{F_i}{F_d} = \frac{\rho_w C_m V u_m \omega}{\frac{1}{2}\rho_w C_d S u_m^2} = 2\frac{C_m}{C_d}\frac{V\omega}{S u_m}. \tag{4.78}$$

The ratio V/S is proportional to the characteristic length of the body in the flow direction, L_b, $i.e.$ $V/S = \alpha L_b$, in which α is a proportionality coefficient. For example, for a sphere of diameter D, we have $V/S = D/6$ ($\alpha = 1/6$), and for a vertical cylinder of diameter D and unit height, $V/S = \pi D/4$ ($\alpha = \pi/4$). Using this in Eq. (4.78), we obtain:

$$\frac{F_i}{F_d} = 2\frac{C_m}{C_d}\frac{\alpha L_b \omega}{u_m} = 2\alpha\frac{C_m}{C_d}\omega T\left(\frac{\omega T}{L_b}\right)^{-1} = 4\pi\alpha\frac{C_m}{C_d}\frac{1}{K}. \tag{4.79}$$

Equation (4.79) indicates that inertia force is dominant only for very small values of the Keulegan-Carpenter number K. As C_m, C_d and α are all of the order of one, for $K > 100$, wave-induced force on a body is totally dominated by drag force, and the oscillatory character of the flow can be neglected in calculation of forces.

When exploring wave-induced forces on submerged objects, another spatial scale has to be taken into account. This is the wavelength and its relation to the characteristic length of the object. When an object spans a significant fraction of a wavelength, the incident waves undergo significant scattering or diffraction. The Morison equation is not applicable in such a situation because this equation is based on the assumption that the kinematics of the undisturbed flow close to the object do not change in the incident wave direction. Since flow velocity and acceleration do in fact vary with a wavelength, L, the assumption implicit in the use of the Morison equation is that the ratio L_b/L is small.

For marine plants and animals, even as large as whales, the condition $L_b/L \ll 1$ is usually satisfied. Therefore, we will not proceed further with the problem of calculating wave forces on large bodies when this condition is violated. Let us only note here that for offshore concrete gravity petroleum structures, with typical diameter of the order of 200 m and columns with diameters of the order of 20 m, wave scattering becomes important when determining wave forces. For example, for a vertical column when $D/L > 0.2$ (D is column diameter, and L is wavelength), K does not exceed 2.2 and usually is less than 1. This means that for large bodies, inertia force dominates over drag force (see Eq. 4.79). Special numerical techniques, known as diffraction techniques, have been developed to deal with objects of large dimensions.

For further details, the reader should consult Sarpkaya and Isaacson (1981) and Massel (1981).

4.3 Statistical and Spectral Properties of Waves

4.3.1 A Brief Orientation

Looking at ocean surface waves, one can notice both their randomness and their quasi-regularity. Wave profiles are constantly changing with time and in a random fashion. Consequently, the properties of waves are not readily defined on a wave-by-wave basis. The fundamental property of surface waves induced by wind is their irregularity. The prediction of wave parameters can be achieved through stochastic analysis only, which can be developed in two basic domains, *i.e.* **probability and frequency domains**.

In the probability domain, particular wave parameters, such as surface displacement at a given time, wave height, wave period, etc. are considered as elementary random events. This approach is easy to understand when dealing with digitized data. The digitized data of a particular parameter forms a set of random realizations of a random variable, when the time sequence of the parameter is not taken into account. The final results in this approach are expressed in terms of probability density functions, distributions functions and some statistical values.

Frequency analysis mainly deals with an evaluation of the distribution of wave energy among various frequencies and directions. There are two main methods for the development of the frequency spectrum. The traditional method is based on the Fourier Transform of the correlation function while in the second method, the time series is simply transferred into its Fourier components. This method is known as Fast Fourier Transform (FFT). We examine these techniques in more detail in Chap. 9.

4.3.2 Outline of Waves Statistics

Let us assume that a wave record of sea surface oscillation at a given location is available (see, for example, Fig. 3.10). If the sea surface is so irregular, how can we determine the probability of the occurrence of waves with a particular height, say 1 m or 3 m above mean sea level? What is the most frequently observed surface displacement? To answer these and other questions we have to know the probability density distribution of the surface displacement. In most cases, the probability density distribution of the ordinates of ocean surface waves takes the form of the so called normal distribution or Gaussian function (Massel, 1996a):

$$f(\zeta) = \frac{1}{\sqrt{2\pi}\sigma_\zeta} \exp\left[-\frac{\left(\zeta - \bar{\zeta}\right)^2}{2\sigma_\zeta^2}\right], \tag{4.80}$$

in which $\bar{\zeta}$ is the mean value of surface displacement; usually $\bar{\zeta} = 0$ and σ_ζ is a standard deviation of the distribution defined as:

$$\sigma_\zeta = \left\{\frac{1}{N-1} \sum_{i=1}^{N} \left(\zeta_i - \bar{\zeta}\right)^2\right\}^{1/2}, \tag{4.81}$$

in which N is a total number of sea surface ordinates taken into account in the summation. In oceanographic practice it is common to take records of about 20 minutes duration, sampled every 1/4 or 1/2 s, depending on the instrument used. Therefore, the value of N in (4.81) varies between 2 400 and 4 800 readings for 1/2 s and 1/4 s sampling, respectively. More details on measurement techniques and data collection procedures can be found in Chap. 9.

The function (4.80) is illustrated in Fig. 4.24, for a wave record with $\sigma_\zeta = 0.5$ m and $\bar{\zeta} = 0$. The probability density distribution is a symmetrical curve with maximum at $\zeta = \bar{\zeta} = 0$ which satisfies the following condition:

$$\int_{-\infty}^{\infty} f(\zeta)d\zeta = 1. \tag{4.82}$$

This condition expresses the fact that the probability of occurrence of all cases should be equal to 1.

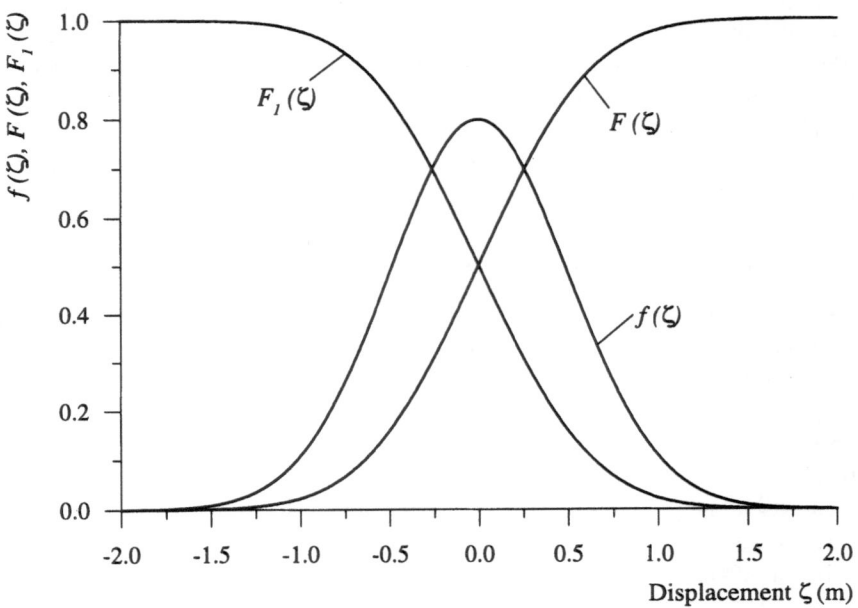

Fig. 4.24: Theoretical probability density distribution f, cumulative probability distribution F and probability wave exceedence distribution F_1, for wave record with $\sigma = 0.5$ m and $\bar{\zeta} = 0$

On the other hand, if we would like to determine the probability of the surface displacement ζ, lying in the band between some values ζ_1, and ζ_2, *i.e.* $\zeta_1 < \zeta < \zeta_2$, we have to integrate function $f(\zeta)$ for respective values of ζ, *i.e.*:

$$\text{Prob}(\zeta_1 < \zeta < \zeta_2) = \int_{\zeta_1}^{\zeta_2} f(\zeta)d\zeta. \tag{4.83}$$

In many practical applications, in place of the probability density distribution, the cumulative probability distribution $F(\zeta)$ is often used. Function $F(\zeta)$ defines the probability (not probability density!) that some values of ζ are smaller than ζ_0, $\zeta < \zeta_0$; thus:

$$F(\zeta_0) = Prob(\zeta < \zeta_0) = \int_{-\infty}^{\zeta_0} f(\zeta)d\zeta. \tag{4.84}$$

If $\zeta_0 \to \infty$, $F(\zeta_0) \to 1$, which is expected, as all events are involved in the integration. Occasionally, a complementary probability distribution $F_1(\zeta)$ is used, *i.e.*:

$$F_1(\zeta_0) = 1 - F(\zeta_0) = Prob(\zeta > \zeta_0) = \int_{\zeta_0}^{\infty} f(\zeta)d\zeta. \tag{4.85}$$

It should be noted that probability distributions $F(\zeta)$ and $F_1(\zeta)$ are dimensionless, while the probability density distribution, $f(\zeta)$, possesses a dimension depending on the dimension of the variable ζ. For example, for ζ which denotes surface displacement, a dimension of $f(\zeta)$ is m^{-1}. For completeness, the functions F and F_1 also are illustrated in Fig. 4.24.

In a similar way, we can determine the probability distribution functions for other wave parameters. However, here we restrict ourselves to the wave height only. A more thorough explanation of wave statistics can be found in professional literature (for example, Phillips, 1977; Massel, 1996a).

The probability density distribution for wave height H takes the form:

$$f(H) = \frac{H}{4\sigma_\zeta^2} \exp\left(-\frac{H^2}{8\sigma_\zeta^2}\right), \tag{4.86}$$

in which σ_ζ is the standard deviation of the surface displacements. The function (4.86) is known as a Rayleigh distribution. Using the relationships between the standard deviation, σ_ζ, and the mean wave height, $(\bar{H})$, or the so called 'root-mean-square' wave height (H_{rms}), function $f(H)$ can be represented in different ways, $i.e.$:

$$f(H) = \frac{\pi}{2} \frac{H}{\bar{H}^2} \exp\left[-\frac{\pi}{4}\left(\frac{H}{\bar{H}}\right)^2\right], \tag{4.87}$$

or:

$$f(H) = \frac{2H}{H_{rms}^2} \exp\left(-\frac{H^2}{H_{rms}^2}\right), \tag{4.88}$$

where:

$$\bar{H} = \sqrt{2\pi}\sigma_\zeta \approx 2.5\sigma_\zeta, \tag{4.89}$$

and

$$H_{rms} = 2\sqrt{2}\sigma_\zeta = \frac{2}{\sqrt{\pi}}\bar{H}, \tag{4.90}$$

where the root-mean-square wave height is defined as:

$$H_{rms} = \left[\frac{1}{N}\sum_{i=1}^{N}H_i^2\right]^{1/2}. \tag{4.91}$$

The function (4.87) is illustrated in Fig. 4.25 for $\bar{H} = 1.0$.

Table 4.1: Some wave height relations based on the Rayleigh distribution. Adapted from Massel (1989)

Characteristic height		$\frac{H}{H_{rms}}$	$\frac{H}{\sigma_\zeta}$	$\frac{H}{H_s}$
Root-mean-square height	H_{rms}	1.0	$2\sqrt{2}$	0.706
Median height	$H(P = 1/2)$	$(\ln 2)^{1/2}$	$(8\ln 2)^{1/2}$	0.588
Mean height	$\bar{H} = H_1$	$\sqrt{\pi/2}$	$\sqrt{2\pi}$	0.626
Significant height	$H_s = H_{1/3}$	1.416	4.005	1.000
Average of tenth highest	$H_{1/10}$	1.80	5.091	1.271
Average of hundredth highest	$H_{1/100}$	2.359	6.672	1.666

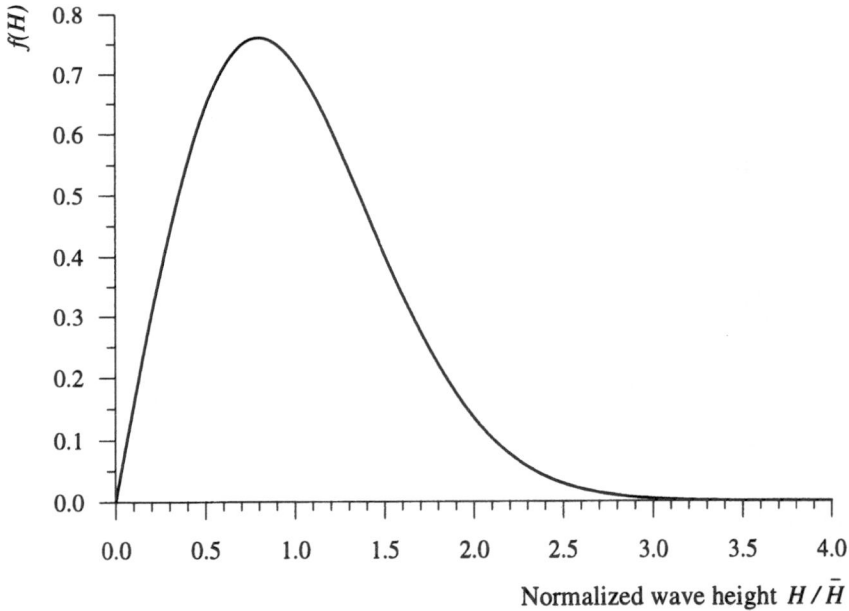

Fig. 4.25: Probability density distribution of wave height ($\bar{H} = 1.0$ m)

The highest probability of occurrence is associated with a wave height slightly smaller than the mean wave height. If the wave height obeys the Rayleigh distribution (4.86), then the significant wave height (H_s), defined in Chap. 3, becomes (Massel, 1996a):

$$H_s = 4.004\sigma_\zeta = \sqrt{2}H_{rms}. \tag{4.92}$$

Substituting Eq. (4.90) into (4.92) yields:

$$H_s \approx 1.6\bar{H}. \tag{4.93}$$

In Table 4.1 the relationships between various characteristic wave heights are given. For example, the mean height of the highest 100 waves is about 1.67 times higher than the significant height.

4.3.3 Spectral Properties of Wind-Induced Waves

In order to clarify the fundamentals of spectral analysis of time series of random, irregular ocean surface waves, let us begin with analysis of time series for $0 < t < T$ (see Fig. 4.26a). As is shown in many basic mathematical text books (for example, Hildebrand, 1965), any irregular function can be represented as a summation of many sinusoidal curves. The basis for such representation is the Fourier series, named after Joseph Fourier (1768-1830):

$$f(t) = \frac{1}{2}a_0 + \sum_{n=1}^{n=N} \left[a_n \cos\left(n\frac{2\pi}{T}t\right) + b_n \sin\left(\frac{2\pi}{T}t\right) \right], \tag{4.94}$$

or

$$f(t) = \frac{1}{2}a_0 + \sum_{n=1}^{n=N} \left[a_n \cos\left(n\omega t\right) + b_n \sin\left(n\omega t\right) \right]. \tag{4.95}$$

The frequencies of these curves are multiples of the basic frequency $\omega = 2\pi/T$. Therefore, we have curves with frequencies 2ω, 3ω, ... , $n\omega$. By analogy to acoustics and music we called them harmonics. Accuracy of the representation (4.95) depends on the complexity of the function $f(t)$ and the number of harmonics, N. For convergence of the series to the function $f(t)$ it is sufficient that functions $f(t)$ and $f'(t)$ are continuous except at a finite number of points in the interval $(0, T)$. Fourier series not only represent the function on the interval $(0, T)$ but also give the periodic extension of the function f outside this interval.

It is convenient for further analysis to rewrite Eq. (4.95) in the more compact form:

$$f(t) = \frac{1}{2}a_0 + \sum_{n=1}^{n=N} c_n \cos\left(n\omega t + \epsilon_n\right), \tag{4.96}$$

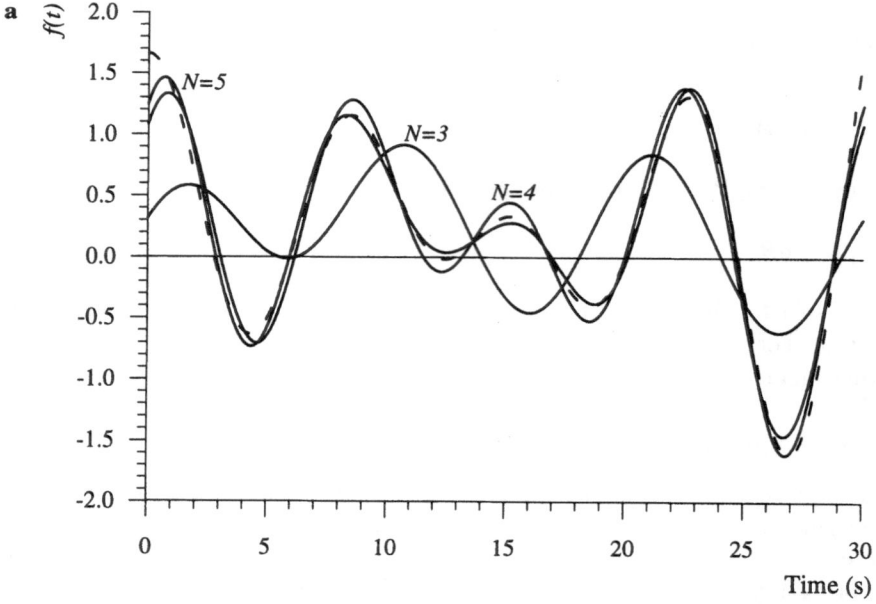

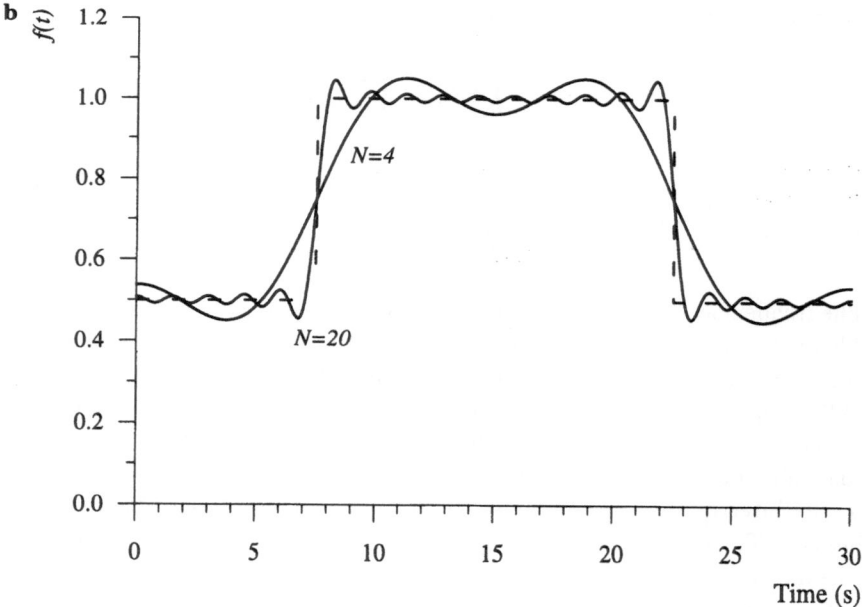

Fig. 4.26: Harmonic analysis of a given irregular sea surface variation in time interval (0, 30 s): a wave type profile, b box-type profile; dashed lines denote given shape, solid lines denote Fourier series fit to given shape; the parameter N denotes the number of terms in the Fourier series

in which:

$$c_n = \sqrt{a_n^2 + b_n^2},$$
(4.97)

$$\epsilon_n = \arctan\left(\frac{b_n}{a_n}\right).$$
(4.98)

The term $1/2(a_0)$ in Eq. (4.96) denotes the mean value of function $f(t)$. According to Eq. (4.96), the function $f(t)$ is a summation of N regular sinusoidal functions with amplitudes c_n and phases ϵ_n. Equation (4.98) suggests that phases ϵ_n are angles within the range $(0, 2\pi)$. The frequency of the first component is equal to $2\pi/T$, the second one has a frequency $2 \cdot 2\pi/T$, the third $3 \cdot 2\pi/T$, etc.

The method to find the amplitudes of harmonics a_0, a_n, b_n, and c_n, for all n as well as phases ϵ_n is known as **harmonic analysis**. For a given number of harmonics, N, harmonic analysis provides the best approximation of function $f(t)$ by Fourier series (4.94), minimizing the mean squared error, ϵ, such that:

$$\epsilon = \frac{1}{T}\int_0^T \left\{ f(t) - \left[\frac{1}{2}a_0 + \sum_{n=1}^{n=N}(a_n\cos(n\omega t) + b_n\sin(n\omega t))\right]\right\}^2 dt = min.$$
(4.99)

A computer program for harmonic analysis for an irregular function defined in the time interval $(0, T)$ is given in Appendix D (Program D.46). Applying this program, the harmonic analysis of the function given in Fig. 4.26a was carried out, with N equal to 3, 4 and 5 being used to approximate the given wave-type function over the time interval of 30 s. For N equals 4 and 5, an accuracy of approximation is very good.

Fourier series analysis is a very powerful tool for representing various irregular shapes, not necessary wave patterns. In Fig. 4.26b, the Fourier series representation (4.94) was used to fit a box-type profile. The N=4 and N=20 harmonics have been applied in calculations. The (4.94) representation with $N = 20$ harmonics provides a very good approximation of the given box-type shape.

We now extend the concept of harmonic analysis to the case of ocean surface waves. As was shown in Fig. 3.8, the observed ocean surface can be regarded as a summation of many elementary waves of various amplitudes and frequencies, propagating in various directions. Thus, for displacement at a given point $P(x, y)$, we can write:

$$\zeta(x, y, t) = \sum_{n=1}^{n=N} c_n \cos\left[k_n(x\cos\theta_n + y\sin\theta_n) - \omega_n t - \epsilon_n\right],$$
(4.100)

in which wave numbers, k_n, are related to wave frequencies, ω_n, by Eq. (4.3) and ϵ_n are the phase angles from the range $(0, 2\pi)$. The resemblance of representation (4.100) to representation (4.96) becomes clear if we assume $x = y = 0$ and neglect the offset a_0. However, there is a fundamental difference between both representations. In contrast to Eq. (4.96), the amplitudes c_n and phases ϵ_n in Eq. (4.100) are **random functions**. The wave amplitudes, c_n, obey some probability density distribution, like that given by Eq. (4.87). For the phase, ϵ_n, it is usually assumed that it has a uniform distribution in $(0, 2\pi)$ range:

$$f(\epsilon) = \frac{1}{2\pi} \quad \text{for} \quad 0 < \epsilon < 2\pi. \tag{4.101}$$

Equation (4.101) denotes that any angle in the band $(0, 2\pi)$ has an equal opportunity of occurrence. In other words, elementary wave components can be superimposed on each other with arbitrary phase angles, ϵ_n, from a range $(0, 2\pi)$, to form the final wave displacement, ζ.

A given record of wave oscillations, at a given point, is only one possible realization of random displacement $\zeta(x, y, t)$ among an infinite number of other possibilities. Therefore, each other record will provide a different realization function $\zeta(x, y, t)$ and the harmonic analysis, given above for a real function, does not provide an explicit representation of the function $\zeta(x, y, t)$, and can not be applied to this function in a straightforward manner.

In contrast to the real function, the number of components (N) is very large, in fact $N \to \infty$. Therefore, in the analysis of random surface displacement $\zeta(x, y, t)$, we are usually not interested in reproduction of the function $\zeta(x, y, t)$ but in determination of the frequency and directional spectra. As we mentioned in Sect. 3.5.3, the frequency spectrum $S(\omega)$ represents the distribution of wave energy among frequencies, while the directional spectrum $D(\theta)$ provides the energy distribution among wave directions. The techniques of determination of spectra $S(\omega)$ will be briefly outlined in Chap. 9. This subject is also described in more detail in other books (see, for example, Kinsman, 1965; Massel, 1996a). The computer program for spectral analysis is given in Appendix D (Program D.47).

Although wave generation conditions throughout the ocean can be quite different, field observations and theoretical analysis have shown that the frequency spectra $S(\omega)$ can be represented in some typical forms. The most popular representations of the ocean wave spectra are (Massel, 1996a):

- deep water ocean: Pierson-Moskowitz spectrum (Pierson and Moskowitz, 1964):

$$S(\omega) = \alpha g^2 \omega^{-5} \exp\left[-\frac{5}{4}\left(\frac{\omega}{\omega_p}\right)^{-4}\right], \tag{4.102}$$

- sea of finite water depth: JONSWAP spectrum (Hasselmann *et al.*, 1973):

$$S(\omega) = \alpha g^2 \omega^{-5} \exp\left[-\frac{5}{4}\left(\frac{\omega}{\omega_p}\right)^{-4}\right] \gamma^\delta, \qquad (4.103)$$

in which:

$$\gamma = 3.3, \quad \delta = \exp\left[-\frac{(\omega - \omega_p)^2}{2\sigma_0^2 \omega_p^2}\right], \qquad (4.104)$$

where ω_p is a peak frequency and σ_0 is a shape parameter such that:

$$\sigma_0 = \begin{cases} 0.07 & \text{for} \quad \omega < \omega_p \\ 0.09 & \text{for} \quad \omega > \omega_p. \end{cases} \qquad (4.105)$$

The spectrum (4.102) has been obtained by Pierson and Moskowitz (1964) using field data collected in the Atlantic Ocean and applying some theoretical results of Phillips (1958) and Kitaigorodskii (1962). The JONSWAP spectrum is based on an extensive wave measurement program (Joint North Sea Wave Project) carried out in 1968 and 1969 in the North Sea (Hasselmann *et al.*, 1973). In particular, the JONSWAP experiment suggests that the so called Phillips' constant, α, in Eq. (4.102) and (4.103), and peak frequency, ω_p, are the following functions of wind speed, V_w, and wind fetch, X:

$$\alpha = 0.076\left(\frac{gX}{V_w^2}\right)^{-0.22}, \qquad (4.106)$$

$$\omega_p = 7\pi \frac{g}{V_w}\left(\frac{gX}{V_w^2}\right)^{-0.33}. \qquad (4.107)$$

The spectra (4.102) and (4.103) are shown in Fig. 4.27. A wind speed $V_w = 20$ m/s and fetch $X = 200$ km have been used in the calculations. This very high wind speed (20 m/s) and long fetch (200 km) results in very high waves, with significant wave height $H_s \approx 4.5$ m. The sea is fully developed and the phase speed of the waves is equal to about 75% of wind speed. For such an almost fully developed sea state, the Pierson-Moskowitz spectrum (4.102) is applicable. The JONSWAP spectrum (4.103) is more useful for fetch limited conditions when phase velocity is still much lower than wind speed. This different range of applicability of both spectra is clearly seen in Fig. 4.27 in which the JONSWAP spectrum has an enhanced peak, with a contrast to a much broader Pierson-Moskowitz spectrum.

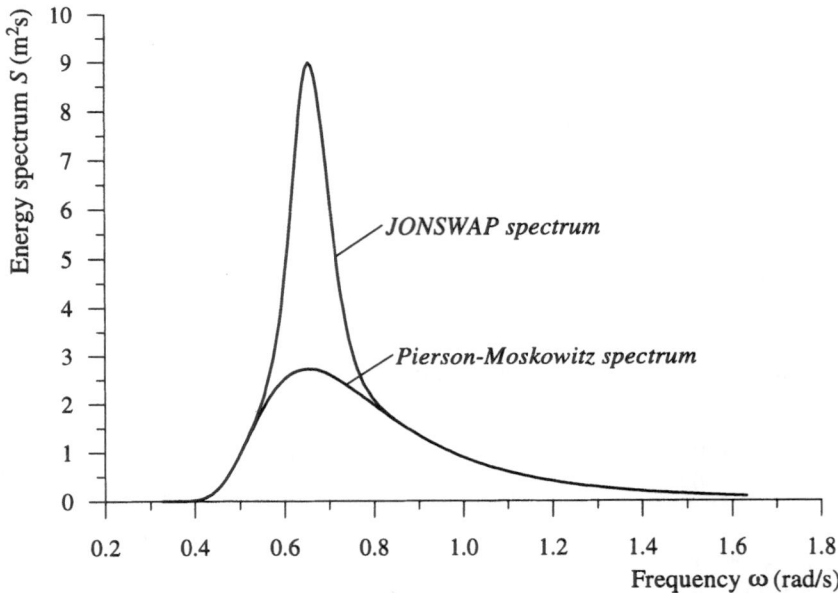

Fig. 4.27: Pierson-Moskowitz and JONSWAP spectra for wind fetch $X = 200$ km and wind speed $V_w = 20$ m/s

For high frequencies, both spectra approach the same universal shape:

$$S(\omega) \rightarrow \alpha g^2 \omega^{-5}. \tag{4.108}$$

Spectrum (4.108) is sometimes called the saturation range spectrum or the Phillips' spectrum, after Phillips, who developed this spectrum in 1958 (Phillips, 1958). This spectrum reflects the existence of some balance between energy supplied by wind and that lost by dissipation due to 'white caps'.

Observed ocean wave spectra sometimes possess a very complicated pattern where more than one peak in the frequency spectrum is observed. Various mechanisms can be responsible for such behaviour. The obvious one is the superposition of various wave systems approaching the observation point. For example, as well as locally generated waves, the swell from other remote generating areas can be recorded at a given point. In a finite water depth or in a shallow water zone, the complicated processes within the wave field also manifest themselves in the form of additional peaks at higher harmonics of main frequency. The most common are spectra with more gentle decreasing of wave energy at higher frequencies or even with double peaks; one in the lower frequency part, the other in the higher frequency part.

Multipeak spectra cannot be represented by the previous models. The simplest way to represent the entire spectral shape is to decompose the spectrum

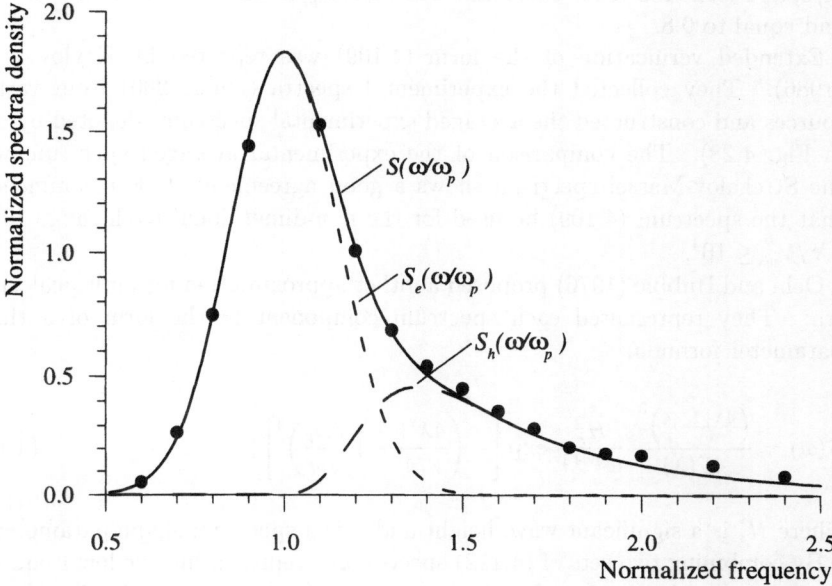

Fig. 4.28: Comparison of the Strekalov and Massel spectrum (solid line), with average experimental spectrum (dots); dashed lines denote two spectral components (adapted from Massel, 1996a)

into two parts: the main energetic component and the high-frequency component. Strekalov and Massel (1971) proposed such a decomposition in the non-dimensional form (see Fig. 4.28):

$$S\left(\frac{\omega}{\omega_p}\right) = S_e\left(\frac{\omega}{\omega_p}\right) + S_h\left(\frac{\omega}{\omega_p}\right),$$

(4.109)

in which:

$$\tilde{S}_e\left(\frac{\omega}{\omega_p}\right) = A\,\exp\left[-B\left(\frac{\omega}{\omega_p} - 1\right)^2\right],$$

(4.110)

$$\tilde{S}_h\left(\frac{\omega}{\omega_p}\right) = C\left(\frac{\omega}{\omega_p}\right)^{-n}\exp\left[-7.987\left(\frac{\omega}{\omega_p}\right)^{-m}\right].$$

(4.111)

The spectrum components $\tilde{S}_e\left(\omega/\omega_p\right)$ and $\tilde{S}_h\left(\omega/\omega_p\right)$ are normalized using peak frequency ω_p and variance σ_ζ^2. The free parameters are: $A = 1.835$, $B = 22.222$, $C = 4.211$, $n = 5$, and $m = 8$. These values of the free parameters were based on the experimental data from the Kaspiyan Sea, and wind speed $7 \leq U \leq 15$

m/s, and fetch $100 < X < 300$ km. The ratio ω_p/ω was assumed to be constant and equal to 0.8.

Extended verification of the form (4.109) was reported by Krylov *et al.* (1986). They collected the experimental spectra (about 200) from various sources and constructed the averaged experimental spectrum (denoted by dots in Fig. 4.28). The comparison of the experimental averaged spectrum with the Strekalov-Massel spectrum shows a good agreement. It is recommended that the spectrum (4.109) be used for the non-dimensional fetch range $10^2 \leq gX/V_w^2 \leq 10^4$.

Ochi and Hubble (1976) proposed another approximation for multipeak spectra. They represented each spectrum component in the form of a three-parameter formula:

$$S(\omega) = \frac{\left(\frac{4\lambda+1}{4}\omega_p^4\right)^\lambda}{4\Gamma(\lambda)} \frac{H_s^2}{\omega^{4\lambda+1}} \exp\left[-\left(\frac{4\lambda+1}{4}\right)\left(\frac{\omega_p}{\omega}\right)^4\right], \qquad (4.112)$$

where H_s is a significant wave height and λ is a spectrum shape parameter.

By combining two sets of (4.112) spectra, one representing the low-frequency component and the other the high-frequency component, they finally obtained the following six-parameter spectral representation:

$$S(\omega) = \sum_{j=1}^{2} \frac{\left[(4\lambda_j+1)\,\omega_{pj}^4/4\right]^{\lambda_j}}{4\Gamma(\lambda_j)} \frac{H_{sj}^2}{\omega^{4\lambda_j+1}} \exp\left[-\left(\frac{4\lambda_j+1}{4}\right)\left(\frac{\omega_{pj}}{\omega}\right)^4\right], \qquad (4.113)$$

in which $j = 1$ and 2 represents the lower and higher frequency components, respectively, and Γ is the gamma function. The parameters of spectrum (4.113) should be determined numerically to best fit the observed spectra.

5 Tides

5.1 Introduction

Rhythmic variations of sea level, known as tides, have been observed by sailors and coastal communities for thousands of years. From the earliest time it has been realized that there is some connection between tides and the motion of the Moon and Sun. However, it was only in the seventeenth century that Isaac Newton (1642-1723) and Pierre-Simon Laplace (1749-1827) provided the theoretical explanation of the nature of tides and opened the door for tidal prediction.

The keystone for understanding tides is the law of gravitational attraction between the Earth, the Moon and the Sun. The simplest explanation possible is the hypothetical example when the continents on the Earth are neglected and the Earth is assumed to be a perfectly smooth sphere, completely covered by water. The water is acted upon by the same forces that act on the solid Earth. The prediction model associated with this case is known as the equilibrium model of tides, as tides in this model result from the equilibrium of gravitational forces. However, the equilibrium model cannot explain many aspects of tides, especially the varying tidal amplitudes in many locations on the Earth. A substantial improvement in tidal prediction has been achieved by considering tides, in a dynamic way, as waves. In fact, tides are the longest oceanic waves with periods of the order of 12 hours. This wave type approach, expanded and revised many times since Laplace's first formulation, is now known as the dynamic model of tides.

Complexity of land contours on the Earth and complicated bathymetry cause great difficulties in the prediction of tides at some points on the Earth. However, the tides can still be predicted using the harmonic analysis of a sufficiently long record of water level fluctuation. Harmonic analysis was already mentioned in Chap. 4, as the technique for representing an arbitrary, irregular time series in the form of a summation of many sinusoidal and cosinusoidal curves with various amplitudes and phase lags. In this chapter we describe the nature of tidal generating forces and the available prediction techniques. The significance of tide variations and tidal induced currents for life in the ocean will be discussed in Part III of the book.

5.2   Tide Generating Forces and Equilibrium Theory

5.2.1   The Earth-Moon System

To determine the forces needed for generation of tides we first have to consider the basic mechanisms involved in the process. Ocean water partly covers the Earth which moves in space, and in the same time rotates around its own axis. The rotating Earth is attracted by the Moon and the Sun. The forces which generate tides result from the balance of these two mechanisms. Initially, for simplicity, we will neglect the attraction induced by the Sun and consider the Earth-Moon System first.

Nature of Gravitational Force. Gravitational force, or gravity, is the most obvious of nature's forces. It keeps us on the ground, and it controls the behaviour of the Universe. Isaac Newton was the first to realize that all bodies with mass attract each other. In 1687, in his *Principia Mathematica*, he laid the conceptual foundation of universal gravitation, discovering that the gravitational force, F_g, between two bodies is proportional to the product of their mass and inversely proportional to the square of the distance between them. Applying Newton's law to the Earth and Moon, we obtain (Godin, 1972):

$$F_g = G \, \frac{m_E \, m_M}{R_{EM}^2}, \tag{5.1}$$

in which m_E and m_M are the masses of the Earth and Moon, respectively, R_{EM} is the distance from the centre of the Earth to the centre of the Moon, and G is the universal gravitational constant ($G = 6.67 \times 10^{-11}$ Nm^2/kg^{-2}). For convenience, in Table 5.1 some relevant quantities for the Earth, Moon and Sun have been listed.

Let us now demonstrate a practical application of Newton's law of universal gravitation by calculating how the apparent weight of an astronaut's body will change when he travels from the Earth to the Moon. For simplicity we assume that the astronaut's weight on Earth is 75 kg (weight) in the metric system. Using the relationship given in Appendix B, we find that weight in SI unit will be $W_E = 75$ kg (weight) = 75× 9.806 N ≈ 735.45 N. Therefore, the mass of the body is $m_b = W_E/g = 75$ kg. This simple derivation is given here to stress the difference between weight and mass, and between former and new unit systems. These differences are discussed in detail in Appendix B.

The gravitational force acting on the astronaut's body on Earth can be calculated from Eq. (5.1):

$$F_E = G \, \frac{m_b \, m_E}{R_E^2}, \tag{5.2}$$

in which m_b is the mass of the astronaut, m_E is the Earth's mass, and R_E is the distance between centres of mass of the astronaut and the Earth. We simply take R_E as the mean Earth radius. The astronaut's mass will not change as

Table 5.1: Some quantities related to the Earth, Moon and Sun

Quantity	Value
Equatorial diameter of the Earth	12 755 km
Polar diameter of the Earth	12 714 km
Diameter of the Moon	3 476 km
Diameter of the Sun	7 392 530 km
Mass of the Earth	5.98×10^{24} kg
Mass of the Moon	7.34×10^{22} kg
Mass of the Sun	1.98×10^{30} kg
Average distance between the centres of the Earth and the Moon	384 329 km
Average distance between the centres of the Earth and the Sun	149 360 000 km

he walks on the Moon, however, his apparent weight will be different from what is on the Earth. Equation (5.2) provides the following relationship for gravitational force acting on the astronaut on the Moon:

$$F_M = G\,\frac{m_b\,m_M}{R_M^2}, \tag{5.3}$$

where m_M and R_M are the Moon mass and radius, respectively. As the ratio of the gravitational forces is equal to the ratio of the respective weights, from Eq. (5.2) and (5.3) we obtain:

$$\frac{W_M}{W_E} = \frac{F_M}{F_E} = \left(\frac{m_M}{m_E}\right)\left(\frac{R_E}{R_M}\right)^2. \tag{5.4}$$

in which W_M is the apparent weight of the astronaut on the Moon. After substituting for masses and radii in Eq. (5.4), we find that the astronaut's weight on the Moon will be about one sixth of that on the Earth.

Newton's gravitational laws also provide an opportunity to determine the gravitational acceleration associated with the gravitational force on the Earth. As the gravitational force, F_E, acting on any body located on the Earth's surface is equal to its weight, it follows that the following equation should be true:

$$G\,\frac{m_b\,m_E}{R_E^2} = m_b\,g, \tag{5.5}$$

in which m_b is the mass of the body. Thus:

$$g = G \frac{m_E}{R_E^2}. \tag{5.6}$$

Substituting all necessary quantities into Eq. (5.6) we obtain a gravitational acceleration $g = 9.805$ m/s^2 when using the Earth's equatorial radius, and $g = 9.870$ m/s^2 when using the Earth's polar radius (see Appendix B).

It is apparent that the gravitational force will not be the same at all points on the Earth's surface, because not all these points are at the same distance from the Moon. Points on the Earth nearest to the Moon will experience greater gravitational pull from the Moon than will the points on the opposite side of the Earth.

Centrifugal Force. The gravitational force, F_g, between the Earth and the Moon has to be balanced by another force, otherwise the Earth-Moon system would become unstable and the Moon would collide with the Earth or vice versa. The balancing force results from the fact that the Earth and the Moon are not motionless bodies. If we forget for a moment that the Earth rotates upon its own axis, then both the Moon and Earth are mutually revolving around the common centre of mass with a period of 27.3 days. The orbits of motion are slightly elliptical but for simplicity we will treat them as circular. As the Earth's mass is about 81 times larger than the mass of the Moon (see Table 5.1), the common centre of mass for the Earth-Moon system lies about 4660 km from the centre (Fig. 5.1) within the Earth.

As the Earth-Moon system revolves around a common centre of mass, all the points within and upon the Earth follow circular paths having exactly the same radius (see radii of points A, B and C in Fig. 5.2). As the angular velocities and radii of the circular paths travelled are the same for the all points, each of these points experiences equal acceleration and equal centrifugal force, F_c.

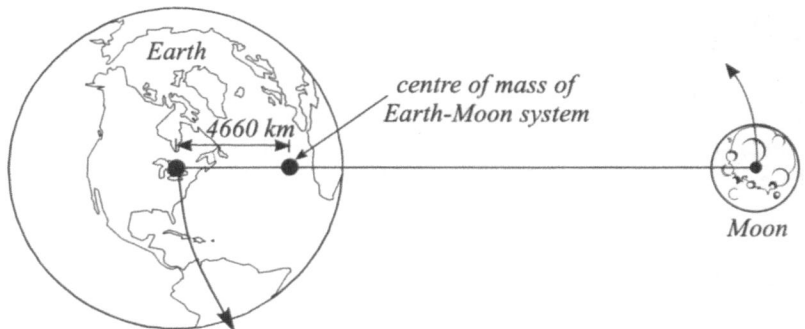

Fig. 5.1: Rotation of the Earth and Moon about a common centre of mass

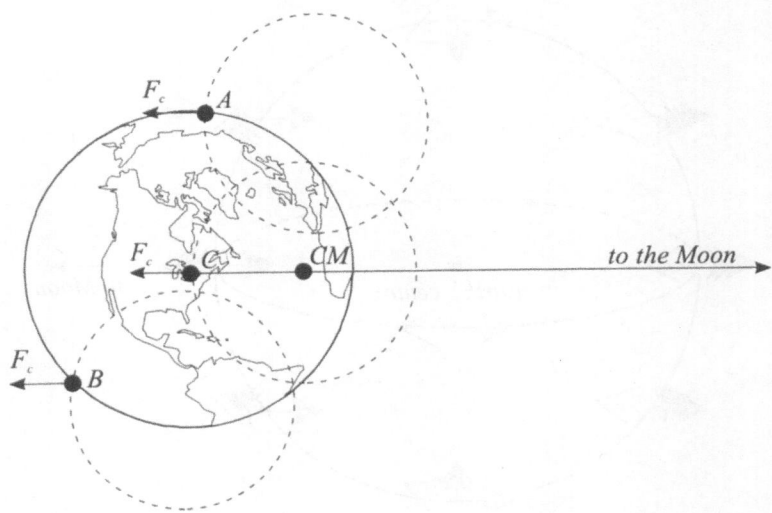

Fig. 5.2: Points A, B and C move in circular paths of the same radii; CM is the centre of mass of the Earth-Moon system, and F_c denotes the centrifugal force

The vectors of the centrifugal force are directed parallel to the line of the centres of the Earth and the Moon.

We note that acceleration of a particle travelling along a circular path with radius r and with angular frequency ω is:

$$a_c = r\omega^2,$$ (5.7)

where $\omega = 2\pi/T$, and T is the period of rotation. As the frequency ω is related to the tangential velocity u_t by:

$$u_t = \omega r,$$ (5.8)

the acceleration a_c becomes:

$$a_c = \frac{u_t^2}{r}.$$ (5.9)

Thus, the centrifugal force F_c is:

$$F_c = m\,a_c = m\,r\,\omega^2 = m\,\frac{u_t^2}{r},$$ (5.10)

where m is the mass involved.

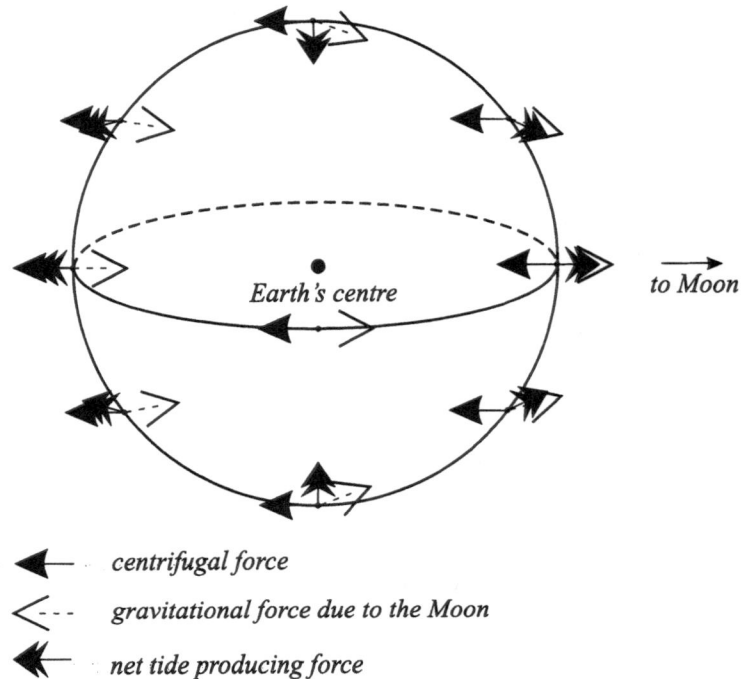

Fig. 5.3: Relative strengths and directions of gravitational, centrifugal and tide-producing forces (adapted from Bearman, 1997)

The total centrifugal force within the Earth-Moon system exactly balances the gravitational force induced by mutual attraction of the Earth and the Moon. The combined effect of the gravitational and centrifugal forces is known as the **tide-producing force**. This force changes its magnitude and direction depending upon the position on the Earth's surface with the respect to the Moon. Tide-producing force may be directed into, parallel to, or away from the Earth surface. In Fig. 5.3, the relative strengths and directions of the gravitational and centrifugal forces are shown for a given position of the Earth and the Moon.

Tide-Producing Force. Until now the gravitational attraction of the water on the surface of the Earth was neglected. In order to find the influence of the local gravitational force, we will consider a point P on the Earth, initially covered with a uniform layer of water (Fig. 5.4). To facilitate a comparison of all acceleration components, we will consider accelerations rather than forces. In particular, acceleration acting normal to the Earth's surface (acceleration is positive when directed away from the Earth's centre) is given by:

$$a_n(P) = -g - a_c \cos\theta + a_a \cos(\varphi + \theta) = -g - \frac{F_c}{m_E}\cos\theta + G\frac{m_M}{R_{EM}^2}\cos(\varphi + \theta), \quad (5.11)$$

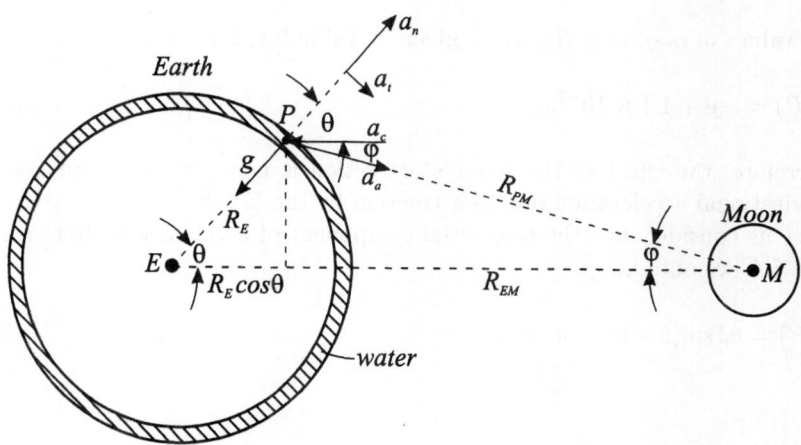

Fig. 5.4: Vectors of the three accelerations acting on a given particle P: gravity, g, centrifugal acceleration, a_c, and attraction acceleration, a_a

where g is the acceleration due to attraction by the Earth, a_c is the acceleration due to centrifugal force, a_a is the acceleration due to attraction by the Moon and the angles φ and θ are defined in Fig. 5.4.

The gravitational force exerted by the Moon on a point at the Earth's centre (point E in Fig. 5.4) is exactly equal and opposite to the centrifugal force. Thus:

$$F_c = G \frac{m_E\, m_M}{R_{EM}^2}. \tag{5.12}$$

Substituting Eq. (5.12) into Eq. (5.11) we obtain:

$$a_n(P) = -g - G \frac{m_M}{R_{EM}^2} \cos\theta + G \frac{m_M}{R_{PM}^2} \cos(\varphi + \theta). \tag{5.13}$$

Using Eq. (5.6) in Eq. (5.13) gives:

$$a_n(P) = -g + g \left(\frac{m_M}{m_E} \right) \left(\frac{R_E}{R_{PM}} \right)^2 \left[\left(\frac{R_{EM}}{R_{PM}} \right)^2 \cos(\varphi + \theta) - \cos\theta \right]. \tag{5.14}$$

A maximum contribution of the Moon attraction and centrifugal force is when $\theta = 0°$ or $\theta = 180°$, *i.e.* when the Moon is straight overhead, or beneath on the opposite side of the Earth. For example, for $\theta = 0°$, Eq. (5.14) simplifies as follows:

$$a_n(P) = -g + g \left(\frac{m_M}{m_E} \right) \left(\frac{R_E}{R_{PM}} \right)^2 \left[\left(\frac{R_{EM}}{R_{PM}} \right)^2 - 1 \right]. \tag{5.15}$$

For values of m_M, m_E, R_E, R_{EM} given in Table 5.1, Eq. (5.15) yields:

$$a_n(P) = -g + 1.1 \times 10^{-7}g. \tag{5.16}$$

Therefore, the effect of the Moon's attraction is negligible in comparison to gravitational acceleration due to attraction by the Earth.

Let us consider now the tangential component of acceleration. Referring to Fig. 5.4, we obtain:

$$a_t(P) = a_a \sin(\theta + \varphi) - a_c \sin\theta, \tag{5.17}$$

or

$$a_t(P) = g\left(\frac{m_M}{m_E}\right) R_E^2 \left[\frac{\sin(\theta + \varphi)}{R_{PM}^2} - \frac{\sin\theta}{R_{EM}^2}\right]. \tag{5.18}$$

In order to proceed further with the calculations, we expand the expression in parentheses noting that:

$$R_{PM}^2 = R_{EM}^2 - 2R_E R_{EM} \cos\theta \tag{5.19}$$

is the law of cosines, and:

$$\tan\varphi = \frac{R_E \cos\theta}{R_{EM} - R_E \cos\theta}. \tag{5.20}$$

After substitution of Eq. (5.19) and (5.20) into Eq. (5.18) and neglecting higher order terms of R_E/R_{EM}, we obtain:

$$a_t(P) = \frac{3}{2}g\left(\frac{m_M}{m_E}\right)\frac{R_E^3}{R_{EM}^3} \sin 2\theta. \tag{5.21}$$

Note that the tangent attractive acceleration, $a_t(P)$, (and force) varies as the inverse cube of the distance between the Earth and the Moon. This acceleration (as well as the force) has maxima and minima values for $\sin 2\theta = 1$, or for $\theta = 45°$ and $\theta = 135°$. The magnitude of $a_t(P)$ for those angles is $\pm 0.84 \times 10^{-7}g$. Although tangential force, $F_t = ma_t$, is very small compared with the Earth's gravitational force, it is not opposed by any other lateral force. Thus, it is capable of moving the water on the Earth's surface. This force is known as the **tractive force**, and is responsible for the tides. This force vanishes only for $\theta = 0°$, 90° and 180°.

Figure 5.5a,b shows the relative magnitude of the tractive forces and the tidal bulges of water drawn out by tangential acceleration components. For simplicity, it was assumed that the Moon is over the Equator. In such a case,

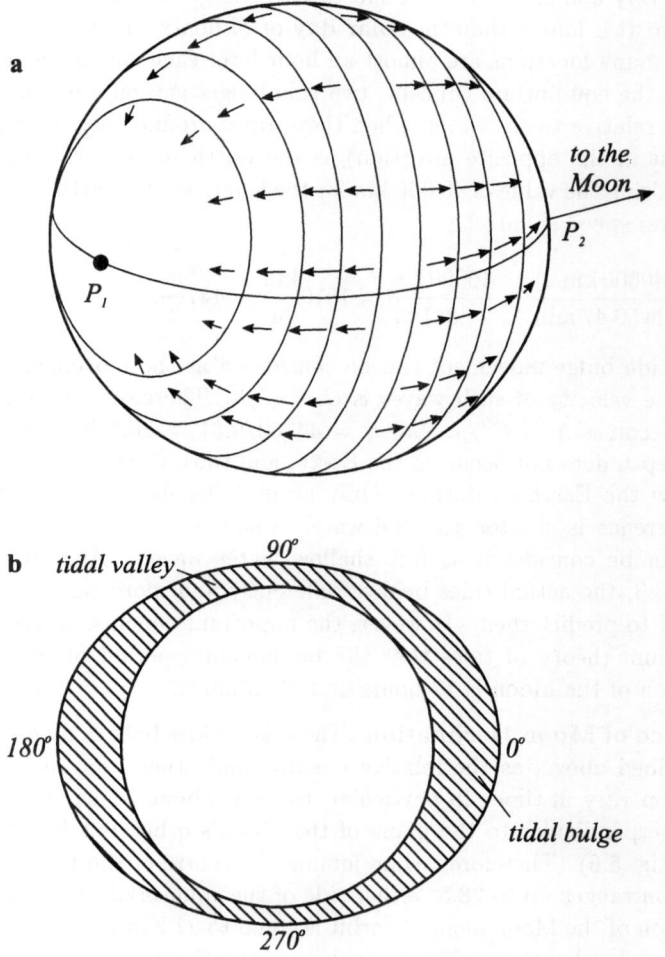

Fig. 5.5: The tractive forces and resulting water bulges: **a** relative magnitude of tractive force vectors, **b** tidal bulges and valleys (adapted from Bearman, 1997)

the tide-producing force causes movement of water towards points P_1 and P_2. Movement of water will continue until the pressure gradient associated with the sloping water surface offsets and balances the tide-producing force. For the Earth completely covered by water, an **equilibrium state** would be reached in the form of an ellipsoid with its two bulges directed towards and away from the Moon. Note that the tide at points P_1 and P_2 is the highest, although the tide-producing forces at these points are minimal.

However in practice, an equilibrium tide cannot occur at low latitudes on Earth. Prior to justifying this, we note that the period of the Earth's rotation with respect to the Moon is 24 hours and 50.47 minutes. This period is called a **lunar day** and is the time for successive passages of the Moon across a given meridian. It is longer than the **solar day** of 24 hours. In other words, the high tides at many locations are almost an hour later each successive day.

Under the equilibrium concept, two tide bulges can only maintain the same position relative to the Moon when they travel around the Earth at the same rate (but in the opposite direction) as the Earth rotates with respect to the Moon. Using the value of 40 000 km for the length of the Earth's circumference, the bulges speed should be:

$$C = \frac{40000\,\text{km}}{24\,\text{hr}\,50.47\,\text{min}} = \frac{40000\,\text{km}}{24.841\,\text{hr}} = 1610\frac{\text{km}}{\text{hr}} = 447\frac{\text{m}}{\text{s}}. \tag{5.22}$$

As the tide bulge movement can be considered as the movement of very long wave, the velocity of such waves is $C = \sqrt{gh}$. Therefore, the required water depth becomes $h = C^2/g$. So, $h = 44762/9.81 = 20.4$ km. However, such water depth does not occur on the Earth, and thus, the tidal bulges are unable to follow the Earth's rotation. Only around the Antarctica, at 60°S, where circumference is shorter and the water is sufficiently deep, the semi-diurnal tides can be considered as free shallow water waves. As will be shown in Sect. 5.3.3, the actual tides behave differently and more complex approach is required to predict them. However, the important conclusion follows from the equilibrium theory of tides that the fundamental period of tides due to the attraction of the Moon is 12 hours and 25 minutes.

Influence of Moon Declination. The actual tides behave differently to what is described above, as the relative position and orientation of the Earth and the Moon vary in time. In particular, the axis about which the Earth rotates is not perpendicular to the plane of the Moon's orbit, but is tilted by about 28.5° (Fig. 5.6). Therefore, a line joining the centre of the Earth with that of the Moon ranges up to 28.5° either side of the equatorial plane. The period of revolution of the Moon along its orbit is equal to 27.2 days. This period should not be confused with the 27.3 day period of the Earth-Moon system's rotation.

Let us now consider a few points on the Earth's surface at various latitudes. In particular, let the latitude of points A and A' be 60°N. An observer at point A would experience a higher tide than that at point A'; 12 hours and 25 minutes later, their position would be reversed. Thus, the water level at points A and A' would be high twice a day and also low twice a day. However, these maxima (minima) will not be equal, *i.e.* there will be a higher high tide and a lower high tide and similarly, a higher low tide and a lower low tide. The daily inequality becomes even larger at the maximum Moon declination of 28.5°, *i.e.* at points B and B'. On the other hand, at the minimum declination, when the Moon is vertically above the Equator, there is no daily variation in tide heights at points C and C'.

When observations at points A, B and C (also A', B' and C') last for an extended period of time, other periodic variations in tide heights will be recorded. These variations are due to eccentricity of the Moon's orbit around the Earth

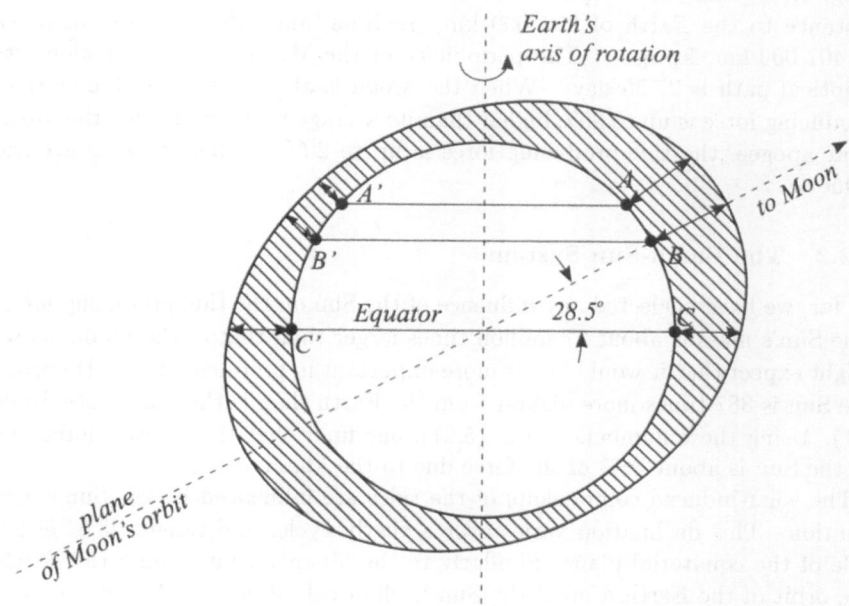

Fig. 5.6: Influence of the Moon's declination on tides

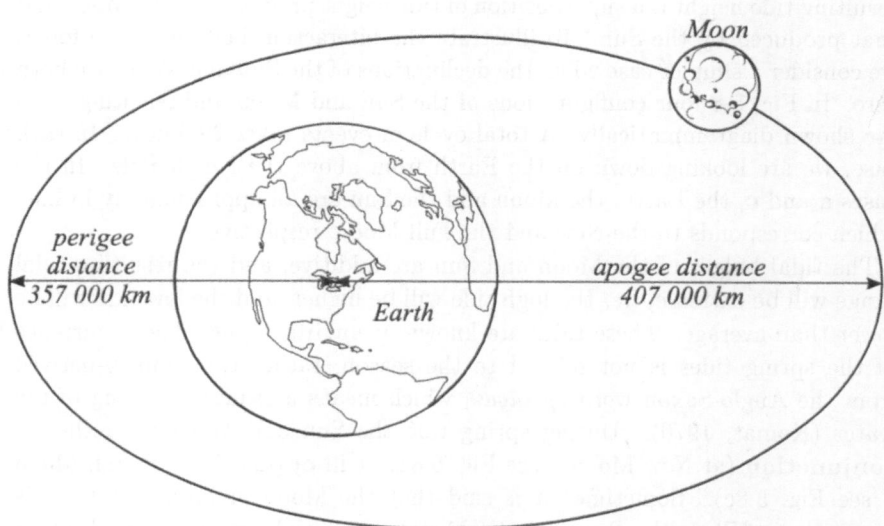

Fig. 5.7: The elliptical path of the Moon around the Earth

(Fig. 5.7). The Moon's path around the Earth is elliptical with the closest distance to the Earth of 357 000 km, **perigee**, and the farthest distance of 407 000 km, **apogee**. The periodicity of the Moon's movement along its elliptical path is 27.55 days. When the Moon is at perigee, the Moon's tide-producing force is up to 20% higher than its average value, and when the Moon is at apogee, the tide-producing force is up to 20% smaller than its average value.

5.2.2   The Earth-Sun System

So far, we have neglected any influence of the Sun on the tide-producing force. The Sun's mass is about 27 million times larger than that of the Moon, so we might expect that it would be far more important in producing tides. However, the Sun is 387 times more distant from the Earth than is the Moon (see Table 5.1). Using these numbers in Eq. (5.21), one finds that the tractive force due to the Sun is about 46% of the force due to the Moon.

The solar-induced components of the tides are influenced by the Sun's declination. This declination varies over a yearly cycle, and ranges 23.5° either side of the equatorial plane. Similarly to the Moon's orbit around the Earth, the orbit of the Earth around the Sun is elliptical. When the Earth is closest to the Sun, it is said to be in **perihelion**, and when the Earth is at maximum distance, it is said to be in **aphelion**. The difference between perihelion and aphelion is only about 4% of the difference between perigee and apogee (Bearman, 1997).

Tide-producing forces due to the Sun and the Moon are additive. Thus, the resulting tide height is a superposition of tide height produced by the Moon and that produced by the Sun. To illustrate the interaction between these forces we consider a simpler case when the declinations of the Sun and Moon are both zero. In Fig. 5.8 four configurations of the Sun and Moon and resulting tides are shown diagrammatically. A total cycle of events takes 29.5 days. In each case, we are looking down on the Earth from above the North Pole. In the cases **a** and **c**, the Earth, the Moon and the Sun are all approximately in line, which corresponds to the New and the Full Moon, respectively.

The tidal bulges of the Moon and Sun are additive, and the resulting tidal range will be extreme, *i.e.* the high tide will be higher, and the low tide will be lower than average. These tides are known as **spring tides**. The occurrence of the spring tides is not related to the season, rather the term is derived from the Anglo-Saxon word *springan*, which means a rising or welling of the water (Komar, 1976). During spring tide the Sun and Moon are either in **conjunction** (at New Moon – see Fig. 5.8a) or in **opposition** (at Full Moon – see Fig. 5.8c). Sometimes it is said that the Moon in both situations is in **syzygy**. When the Sun and the Moon act at right angles to each other (see Fig. 5.8b and 5.8d), it is said that they are in **quadrature**. Tidal range, produced by the Sun and the Moon, is correspondingly smaller than average; these tides are known as **neap tides**.

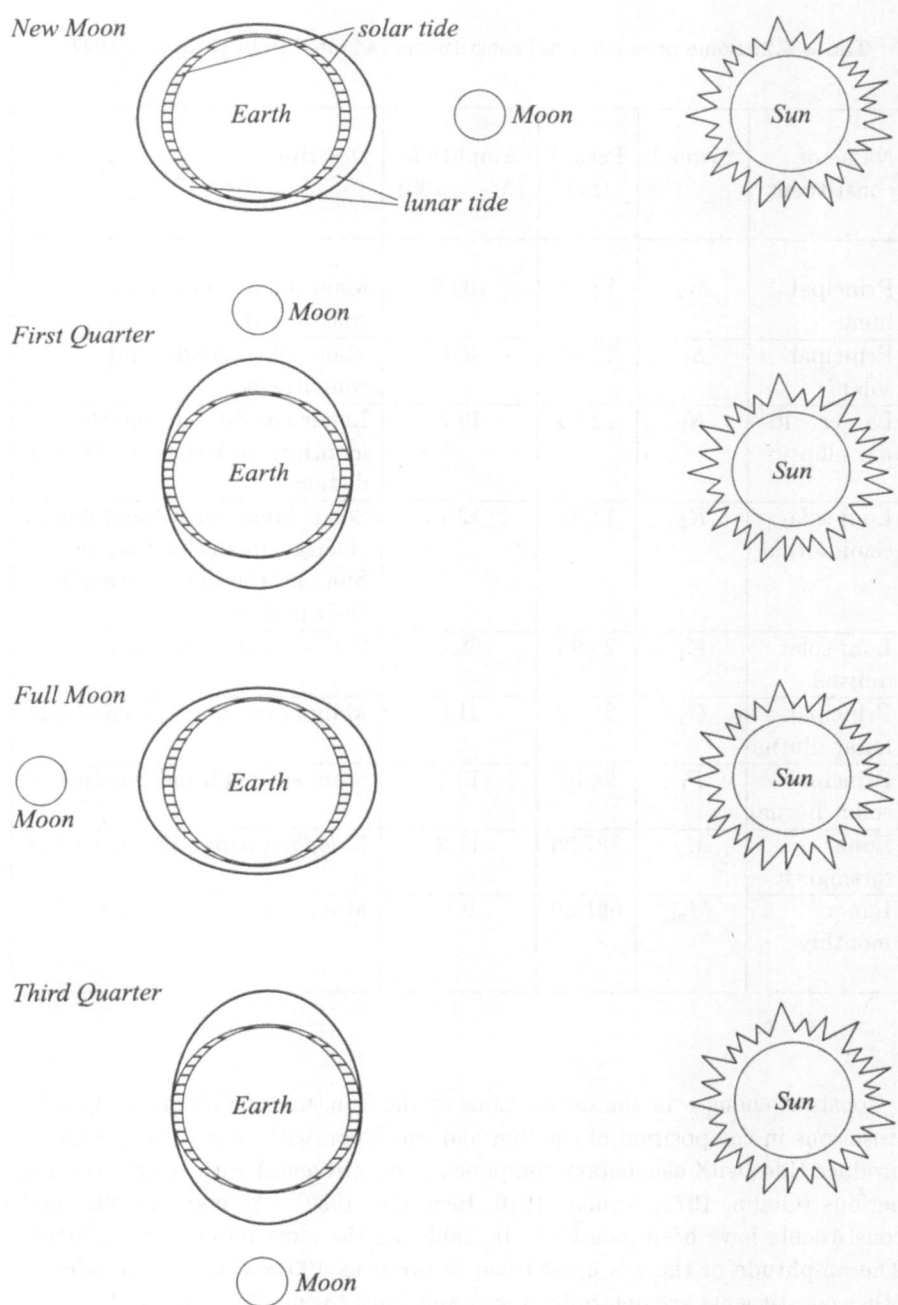

Fig. 5.8: Combined influence of the Moon and the Sun on the tides (adapted from Bearman, 1997)

Table 5.2: Some principal tidal constituents (adapted from Bearman, 1997)

Name of constituent	Symbol	Period (hr)	Amplitude ($M_2 = 100$)	Description
Principal lunar	M_2	12.42	100.0	Main lunar semidiurnal constituent
Principal solar	S_2	12.00	46.6	Main solar semidiurnal constituent
Larger lunar elliptic	N_2	12.66	19.2	Lunar constituent due to monthly variation in Moon's distance
Luni-solar semidiurnal	K_2	11.97	12.7	Solar- lunar constituent due to changes in declination of the Sun and the Moon throughout their orbital cycle
Luni-solar diurnal	K_1	23.93	58.4	Solar- lunar constituent
Principal lunar diurnal	O_1	25.82	41.5	Main lunar diurnal constituent
Principal solar diurnal	P_1	24.07	19.3	Main solar diurnal constituent
Lunar fortnightly	M_f	327.86	17.2	Moon's fortnightly constituent
Lunar monthly	M_m	661.30	9.1	Moon's monthly constituent

Constant changes in the declinations of the Sun and the Moon, and cyclic variations in the position of the Sun and the Moon with respect to the Earth produce tides with elementary components (constituents) with largely varying periods (Godin, 1972; Komar, 1976; Bearman, 1989). As many as 390 tidal constituents have been identified. In Table 5.2 the most important are listed. The amplitude of the M_2 constituent is taken as 100 and the amplitudes of other constituents are determined with reference to the M_2 amplitude.

Constituents responsible for long-term tidal variations have been omitted in Table 5.2. For example, the 5° tilt of the Moon's orbit with respect to the Earth's orbit around the Sun ($28.5° - 23.5° = 5°$) causes the intersection of these orbits to rotate slowly, completing a revolution in 18.6 years.

A specific configuration of the Earth, Moon and Sun produces exceptionally high tides when the Earth is at perihelion, the Moon is at perigee, the Sun and the Moon are in declination. This combination produces the greatest possible tidal range. However, such an occurrence is very rare, and happens about every 1 600 years. The last occurrence took place in about the year A.D. 1400.

5.3   Dynamic Model of Tides

5.3.1   Brief Overall

A comparison of the equilibrium tide theory, developed by Newton, with observed tides immediately shows various discrepancies. High tide often occurs at the wrong time, with the range of the tide not properly predicted by equilibrium theory. For example, the equilibrium theory predicts that high tide at Monterey and at the Golden Gate, in California, should occur at the same time, but in fact high tide at the Golden Gate occurs 45 minutes after Monterey (Denny, 1988). There are many other examples of discrepancies between the equilibrium theory of prediction and observations. According to equilibrium theory, semidiurnal tidal ranges would reach their maximum value of about 10.5 m at equatorial latitudes. In fact, observed tides in the ocean have mean ranges of 0–1 m.

When tides propagate into the relatively shallow waters of the continental shelf, their heights increase. At locations within bays and embayments, the tidal range becomes much higher. For example, in shelf areas of the Bay of Fundy, the Bristol Channel, and the Kimberley Coast (North Australia), the spring tidal ranges exceed 10 m. In Fig. 5.9, an example of such very high tides is shown for the Bay of Le Mont-Saint-Michel (Normandy, France). The Mont-Saint-Michel Benedictine Abbey stands high on the summit of granite rock out in the Bay. The Bay surrounding the rocky island is known for its quicksands and extremely fast rising tides, with the speed of tidal currents exceeding 10 m/s.

There are several reasons for discrepancies between observations and predictions by the equilibrium theory. The most important may be summarized as follows (Pugh, 1987):

1. The Earth is not uniformly covered with the water. The average depth of the oceans is much smaller than the depth of 20 km which is required to allow the tidal bulges to travel as shallow water free waves at the equatorial zone. Only around the Antarctica, at 60°S, can the semidiurnal tides be considered as free shallow water waves.

2. The presence of continents prevents the tidal bulges from propagating around the Earth and complex ocean bathymetry constrains the direction of tidal flows.

3. As we showed in Sect. 3.7, the ocean basins have their own natural modes of oscillation, with many resonant frequencies. These oscillations interact

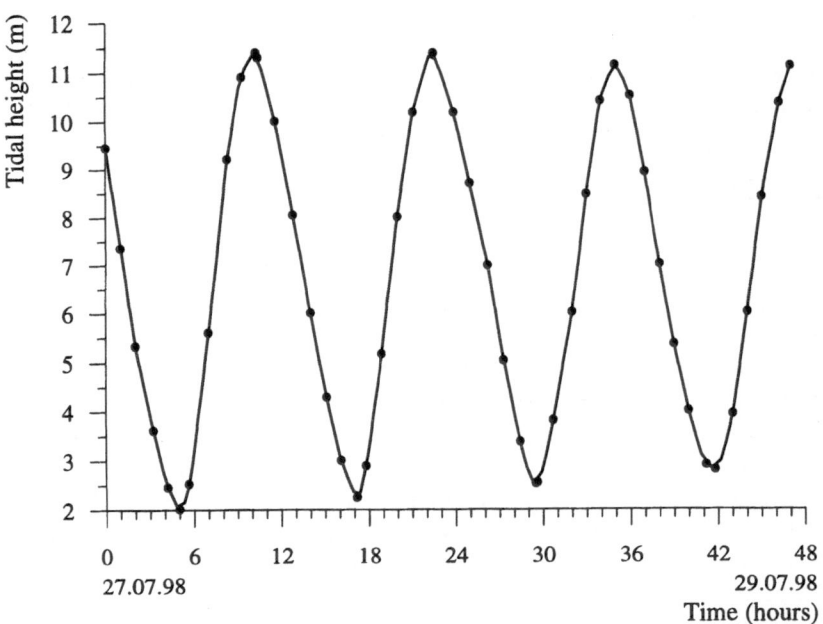

Fig. 5.9: Tides for port Saint-Malo, in proximity of Mont-Saint-Michel, predicted on 27–28 July, 1998

with tidal oscillations, and result in an even more complex set of reso-
nances and local amplifications, not predicted by the equilibrium tidal
model. The most notable example of such amplification is provided by
the Bay of Fundy, in eastern Canada. The bay is about 270 km long and
has an average depth of about 70 m. Using these values in Eq. (3.26), we
can find that the resonant period of the Bay is about 12 hours. Therefore, it is not surprising that very large semi-diurnal tidal oscillations are
observed at the head of the Bay, with the range exceeding 15 m during
spring tides.

4. Equilibrium theory suggests that water responds immediately to the gravitational and attractive forces. This means that water has no inertia,
which is an obvious shortcoming of the theory. Water movement on the
Earth is affected by the rotation of the Earth. Water under the tide-
producing force tries to maintain a uniform direction in absolute space,
but it follows a curved path in the rotating system of coordinates within
which we make observations. This effect is known as the **Coriolis force**.

These examples indicate that equilibrium theory cannot explain all the features of the observed tides. However, before we pursue the matter of improvement of tidal models any further, we must first explain the source and nature
of Coriolis acceleration (Coriolis force).

5.3.2 Coriolis Acceleration

Coriolis acceleration is introduced to allow the application of Newton's second law to the rotating Earth, as originally Newton's law applied only for motion in an unaccelerating coordinate system. A rigorous mathematical development of the Coriolis terms is beyond the scope of this book (for details see, for example, Le Méhauté, 1976). Rather, in this chapter we will describe the physical consequences of Coriolis acceleration. For simplicity, let us consider a fluid particle located on the Earth's surface, in the Northern Hemisphere, and rotating with the Earth. A rotation about the Earth's axis creates a centrifugal force, which for a particle resting on the Earth is balanced by the component of gravity perpendicular to the Earth's axis. Suppose now that this particle moves south, toward the Equator. Therefore, the particle's distance from the Earth's axis will increase and according to Eq. (5.10), centrifugal force experienced by the particle will increase. This larger force cannot be totally balanced by the component of gravity; thus the particle starts to fall behind the rotating Earth. Since the Earth rotates from west to east, the fluid particle describes a path westward (Fig. 5.10a). For example, if corrections for this deflection are not made, an aircraft departing from Stockholm, Sweden, on a flight to Lagos, Nigeria, would drift 4,800 km westward (to the right) and be forced to land in South America rather than in Africa (Pinet, 1992).

Using similar arguments, we find that when the same particle moves northward from the Equator, it's path turns eastward. This means that again the deflection is to the right of the direction of the motion (Fig. 5.10a). In the Southern Hemisphere the deflection of the particle will be to the left (Fig. 5.10a). This is expressed in saying that the Coriolis force deflects tidal flows *cum sole* (with the Sun); in the Northern Hemisphere to the right, clockwise, and in the Southern Hemisphere to the left, anticlockwise (Bearman, 1997).

In Fig. 5.10b, the deflections associated with the eastward or westward movement of the particle in the Northern and Southern Hemispheres are shown. If for example, the particle moves eastward with respect to the Earth along a parallel in the Northern Hemisphere, the particle experiences an additional centrifugal force because of additional tangential velocity. Therefore, in order to be balanced by the component of gravitational force, a particle tends to move further away from the Earth's axis, towards the equator (Fig. 5.10b). In this way, the centrifugal acceleration, u_t^2/r (as well as centrifugal force), remains the same and the balance between centrifugal force and the component of the gravitational force is restored. In the case of particle movement westward, a fluid particle is deflected towards the Pole (Fig. 5.10b). All above deviations are known as Coriolis effects.

To obtain an expression for the Coriolis acceleration, let us consider the simple case of a water particle, P, at rest with respect to the Earth, which is rotating with the frequency ω_E (Fig. 5.11). From Eqs. (5.7) and (5.8) we find that the acceleration of the particle with respect to the axis of rotation is:

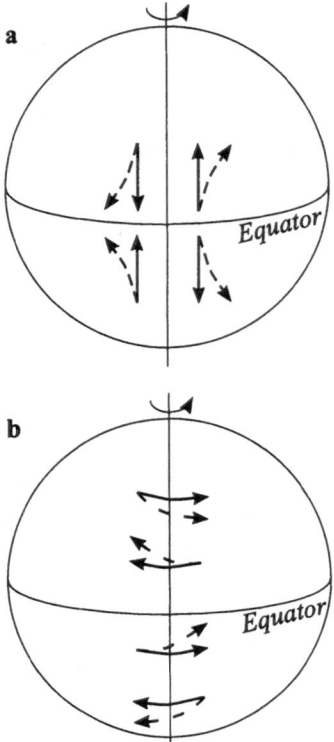

Fig. 5.10: Influence of Coriolis force on particle motion on the rotating Earth:
a particle movement parallel to the Earth's axis, b particle movement normal to the
Earth's axis; solid line denotes initial path of particle, dashed line denotes deflection
of particle path due to Coriolis effect

$$\omega_E^2 r = \frac{u_E^2}{r}, \tag{5.23}$$

in which u_E is the tangential velocity of the Earth at point P.

We now assume that the water particle starts to move with the current eastward, with horizontal velocity u_P over the Earth's surface. Thus, the new acceleration of the particle becomes:

$$a_c = \frac{(u_E + u_P)^2}{r} = \frac{u_E^2 + 2u_E u_P + u_P^2}{r}, \tag{5.24}$$

or:

$$a_c = \omega_E^2 r + 2\omega_E u_P + \frac{u_P^2}{r}. \tag{5.25}$$

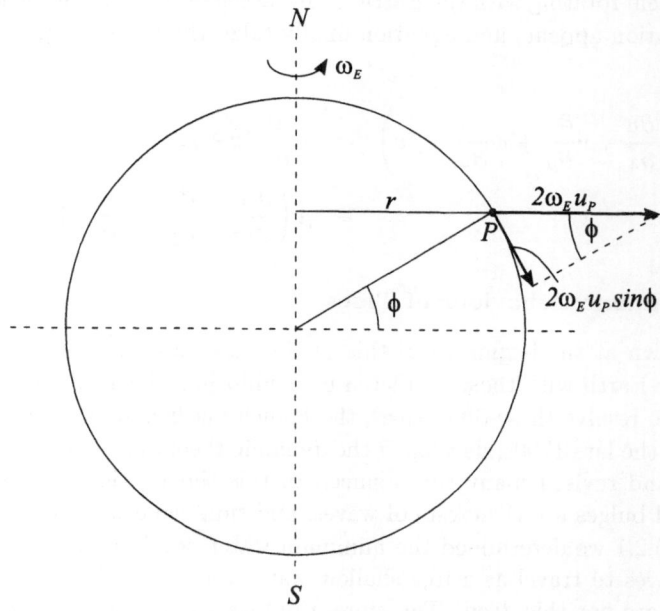

Fig. 5.11: Coriolis acceleration vector at point P

Thus, the eastward moving particle experiences an apparent outward acceleration $2\omega_E u_P + u_P^2/r$. The term u_P^2/r is negligible as the speeds of typical ocean currents are much less than the tangential speed of the Earth. The horizontal component of the acceleration, a_c, on the Earth's surface (see Fig. 5.11) is the required Coriolis acceleration, a_c:

$$a_c = 2\omega_E u_P \sin \phi = f u_P, \tag{5.26}$$

in which:

$$f = 2\omega_E \sin \phi, \tag{5.27}$$

where ϕ is the latitude of the point P. The quantity f is known as the Coriolis parameter, which is dependent on the latitude through the function $\sin \phi$. From Eq. (5.26) follows that at the Equator, Coriolis acceleration vanishes, and reaches its maximum at the Poles. For local particle motion, the effect of Coriolis acceleration is negligible, as the time of the particle motion is much less than the period of rotation of the Earth. However, for tides and ocean currents, which persist for periods of time that are a significant fraction of a day, the Coriolis acceleration and resulting particle deflection cannot be neglected.

In Chap. 2 we have shown that the momentum equation for motion of a fluid along the Ox axis in an unaccelerating coordinate system is given by Eq. (2.22).

For the system rotating with the Earth, a new acceleration, fv, on the left-hand side of equation appears and equation finally takes the form (Pugh, 1987):

$$\rho_w \left(\frac{\partial u}{\partial t} + u\frac{\partial u}{\partial x} + v\frac{\partial u}{\partial y} + w\frac{\partial u}{\partial z} - fv \right) = -\frac{\partial}{\partial x}\left(p + \rho_w g z \right) +$$

$$+ \mu \left(\frac{\partial^2 u}{\partial x^2} + \frac{\partial^2 u}{\partial y^2} + \frac{\partial^2 u}{\partial z^2} \right). \qquad (5.28)$$

5.3.3   Dynamic Behaviour of Tides

As was shown at the beginning of this section, a comparison of the observed tides on the Earth with those predicted by equilibrium theory discloses several conflicts. To resolve these differences, the French mathematician Pierre-Simon Laplace, in the late 1700s, developed the dynamic theory of tides that has been expanded and revised many times since. In this theory, tides are treated as waves; tidal bulges are the crests of waves, and tidal valleys are wave troughs.

In Sect. 5.2.1 we determined the minimum water depth of 20.4 km required for long waves to travel as a free shallow water waves $\left(C = \sqrt{gh} \right)$. However, the oceans are not this deep. Therefore, tidal waves are forced to travel with lower speeds, which do not correspond to water depth. In order to emphasize this fact, the tidal waves are called **forced waves**, as opposed to **free waves** for which the phase velocity is at all times given by $C = \sqrt{gh}$. Thus, an actual tide cannot keep up with the Moon's passage around the Earth, and will be retarded with respect to the situation predicted by equilibrium theory, *i.e.* the tides will be out of phase with the Moon. However, near the South Pole, along the Antarctic Circle (66.5°S), the ocean is deep enough and the distance around the Pole is only 17300 km, which allows the tidal waves to travel as free waves. Therefore, there is no lag and the actual tide can keep up with the theoretical lunar equilibrium tide.

So far we have neglected the presence of the continents. As the tidal wave meets the ocean or basin boundary, it is reflected and moves in the opposite direction. The combination of incident and reflected tidal waves forms a system of standing waves or seiches (see Chap. 3). However, because of the very complex and irregular shape of continents, and the rotation of the Earth, the nodal points and lines have a very complicated shape, and form what is known as the **amphidromic system**. In this system, the crest of the tidal wave, or high water, circulates around an **amphidromic point** which is a point at which there is no tidal motion (from the Greek *amphi*, 'around', and *dromas*, 'running'), once during each tidal period. In each amphidromic system, **co-tidal lines** radiate outwards from the amphidromic points, linking all points where the tide is at the same phase of the cycle. Lines which join locations of the equal tidal range are known as **co-range lines**. Co-range lines form almost concentric paths about the amphidromic points. At these points tidal range is zero and increases with distance from the amphidromic point. Tidal waves

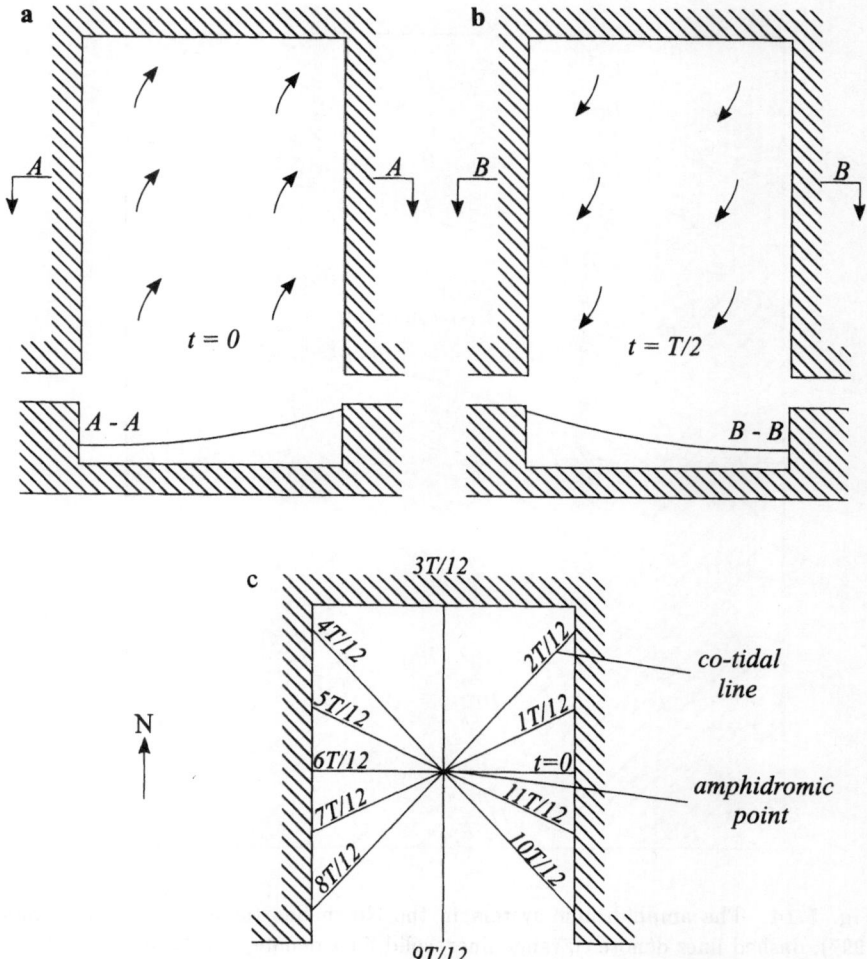

Fig. 5.12: Kelvin waves in rectangular bay in the Northern Hemisphere: **a** flood tide phase, **b** ebb tide phase, **c** total amphidromic system; arrows indicate direction of deflection due to the Coriolis effect; arrows denote the volume flux

within amphidromic systems in the Northern Hemisphere rotate anticlockwise, and clockwise in the Southern Hemisphere.

To illustrate the mechanism of formation of the amphidromic system, we consider a rectangular bay elongated in the north-south direction in the Northern Hemisphere (Fig. 5.12a, b, c). At time $t = 0$, the flood tide enters the bay and is deflected to the right by the Coriolis force. Thus, the water is piled up on the eastern side of the bay. The point with the maximum tidal height moves around the bay anticlockwise. For the ebbing phase of the tide, when the flow is out of the bay, the Coriolis force pushes water to the western boundary (Fig. 5.12b). Because the tidal wave is constrained by a land mass, an amphidromic system

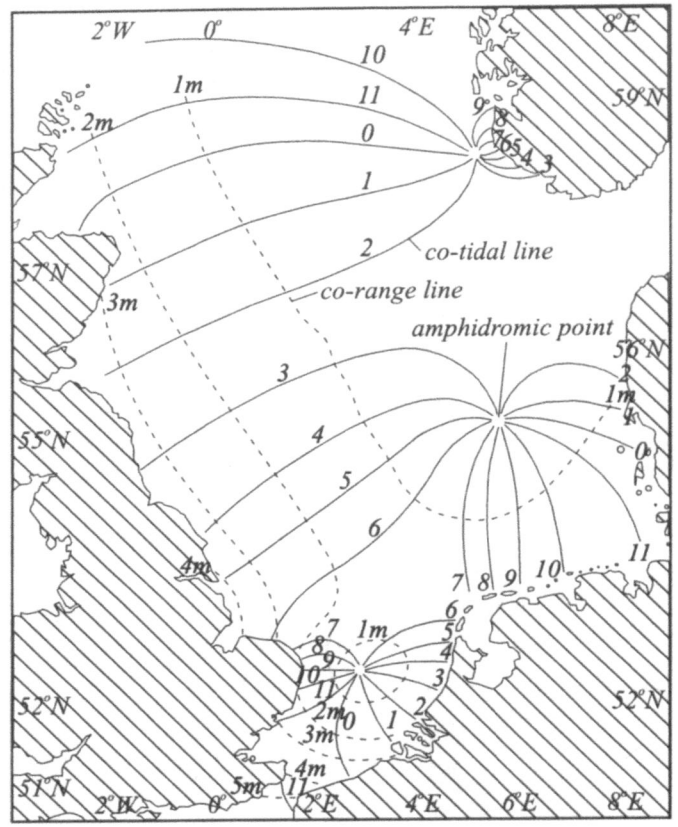

Fig. 5.14: The amphidromic system in the North Sea (adapted from Bearman, 1997); dashed lines denote co-range lines, solid lines denote co-tidal lines

is created with water moving anticlockwise. In Fig. 5.12c, a set of co-tidal lines is shown schematically for one full tide cycle.

A confined amphidromic system, as shown in Fig. 5.12, produces what is known as **Kelvin waves**. Water level along particular co-tidal lines decreases exponentially with increasing distance from the bank, *i.e.*:

$$\zeta \sim e^{-ms}, \tag{5.29}$$

where s is the distance from the bank and attenuation coefficient $m = f/\sqrt{gh}$.

In a real situation on the rotating Earth, the amphidromic systems are much more complicated. Recently, sophisticated numerical models have been developed to predict tides around the Earth. They are based on the Laplace tidal equations, augmented by the energy dissipation due to bottom friction, especially over the shelf areas and in the marginal seas. The accuracy of

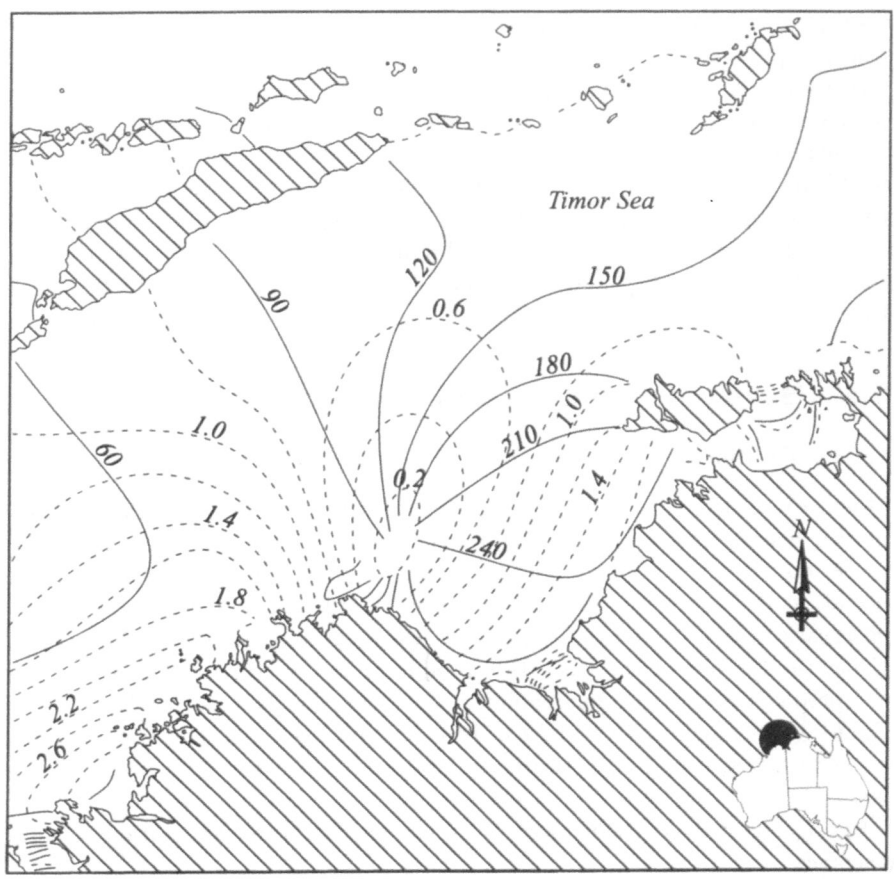

Fig. 5.15: Amphidromic system in Timor Sea, north-west shelf of Australia (data courtesy of L. Mason)

contemporary tidal models has been substantially improved by a combination of the finite element hydrodynamic model and available data, including altimeter data from the TOPEX/POSEIDON satellite (Le Provost *et al.*, 1995, 1998b). This approach is known as data assimilation (Komen *et al.*, 1994).

The present ocean tide prediction models provide very accurate tide heights and a root-mean-square error of prediction is about 3 cm in the deep ocean. On the shelves and in the coastal waters, the error of prediction is slightly higher, at about 10.5 cm. Figure 5.13 (see colour plate p.563) illustrates the amplitudes and phases of the major lunar tide constituent M_2, resulting from the model FES 95.2 developed by Le Provost *et al.* (1998a).

Let us now focus our attention on tides in shelf waters. Probably the best known tides are those in the North Sea (Fig. 5.14). The North Sea is a rather

shallow sea where frictional effects on the tidal motion are significant. As a result, the amphidromic points move away from the source of the tidal energy, which approaches from the Atlantic Ocean from the north and moves counterclockwise around the basin. Due to the effect of friction, the amphidromic points in the North Sea are shifted eastward. Hence, the east coast of Britain has high tides, because of the larger distance from the amphidromic points, while the eastern regions of the North Sea experience smaller tides. In the English Channel, the amphidromic point is situated in the south of England. The tidal waves in the English Channel move in the same direction along both coasts of the Channel. An example of an amphidromic system in the Southern Hemisphere, on the north-west coast of Australia is shown in Fig. 5.15.

5.4   Harmonic Analysis and Prediction of Tides

5.4.1   Prediction of Tides

Modern numerical hydrodynamic models provide accurate prediction of tides in deep water. However, for prediction of tides on the shelf and close to coasts, these models are not yet developed to the stage where they are of practical use at arbitrary points. In such situations, tidal predictions are based on the observations of tides at the desired locations for very long periods of time. Using the periods of the basic tide producing forces (tidal constituents), we can predict the tide at a given point by harmonic analysis, as was described in Chap. 4. As many as 390 tidal constituents have been identified, however not all of these are used in practical calculations. The most important have been listed in Table 5.2.

Hence, for the given time series of duration T, the representation of tides becomes:

$$\zeta(t) = \sum_{n=1}^{n=N} c_n \cos(\omega_n t + \varphi_n), \tag{5.30}$$

in which $\omega_n = 2\pi n/T$; T is the period of observation; N is the number of constituents involved; c_n and φ_n are the amplitudes and phases of particular constituents, respectively. For many points around the Earth, the amplitudes and phases of tidal constituents have been calculated and predicted tides are listed in tide tables. An example of such predictions is shown in Fig. 5.16 for the Mackay area on the east coast of Australia. The prediction is based on long-term observation and 22 constituents were used in the harmonic analysis.

It was shown by Defant (1961) that the nature of the tide at the particular point can be characterized by the form number F, which is the ratio of the sum of the amplitudes of the major constituents:

$$F = \frac{K_1 + O_1}{M_2 + S_2}, \tag{5.31}$$

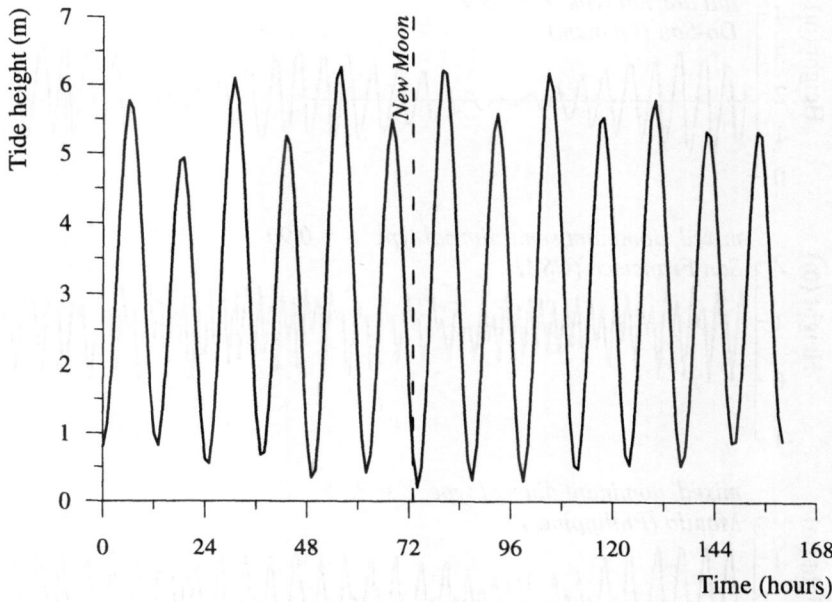

Fig. 5.16: Prediction of tides for Mackay, Australia, for the period of time 24.02.1998, 2:55 am – 2.03.1998, 7:56 pm (Tide Tables, 1998)

where K_1 is the lunar-solar constituent, O_1 is the principal lunar diurnal constituent, M_2 is the principal lunar semidiurnal constituent, and S_2 is the principal solar semidiurnal constituent. Using the form number F, the tides can be divided roughly as follows:

$$F = \begin{cases} 0 & \text{to} \quad 0.25 \quad \text{semidiurnal form} \\ 0.25 & \text{to} \quad 1.50 \quad \text{mixed, predominantly semidiurnal} \\ 1.50 & \text{to} \quad 3.00 \quad \text{mixed, predominantly diurnal} \\ & > \quad 3.00 \quad \text{diurnal form} \end{cases} \qquad (5.32)$$

For tides in the Mackay area (Fig. 5.16), the amplitudes of K_1, O_1, M_2 and S_2 constituents are: 0.390 m, 0.196 m, 1.668 m, and 0.612 m, respectively. Thus, the form number, $F = 0.257$, and tides should be treated as semidiurnal. Other types of tides are illustrated in Fig. 5.17. Tides at Do-San (Vietnam) are example of full diurnal ($F = 18.9$) tides. Between these two extreme cases are the mixed tidal types with domination of semidiurnal (San Francisco, $F = 0.90$) or diurnal (Manila, $F = 2.15$) constituents.

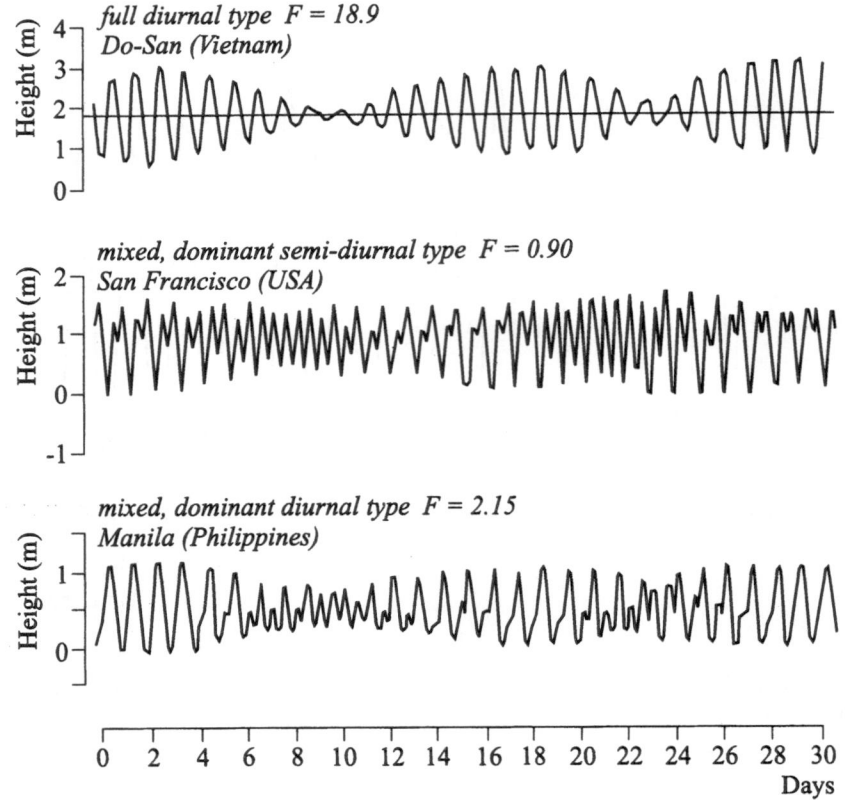

Fig. 5.17: Example of various types of tides with different value of form number F (adapted from Bearman, 1997)

5.4.2   Tidal Tables

At present, a variety of tide tables, in books or in electronic form, are available for many ocean and coastal areas. They are based on long-term observations and numerical predictions. In the tables, the highest and the lowest water levels are usually shown for each day of the year. These levels are referred to some datum, established by the local authorities. Also, for selected ports and locations, the most important characteristic water levels are given, such as (Fig. 5.18):

- H.A.T. (**Highest Astronomical Tide**) and L.A.T. (**Lowest Astronomical Tide**). These are the highest and the lowest levels predicted to occur under normal meteorological conditions and some configuration of the Moon and the Sun. These water levels are not reached every year.
- M.H.H.W. (**Mean Higher High Water**) and M.L.L.W. (**Mean Lower Low Water**). The height of mean higher high water is the mean of the

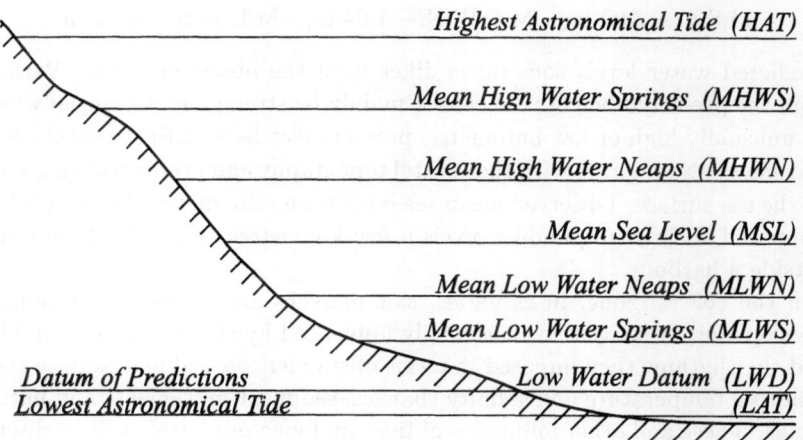

Fig. 5.18: Definition of characteristic tide levels

higher of the two daily high waters over a long period of time. A similar definition applies for the mean lower low water.

- M.L.H.W. (**Mean Lower High Water**) and M.H.L.W. (**Mean Higher Low Water**). The height of mean lower high water is the mean of the lower of the two daily high waters over a long period of time. The definition for mean higher low water is similar.

- M.H.W.S. (**Mean High Water Spring**). This is a long-term average of the heights of two successive high waters during 24 hours of Spring tides, when the range of tide is greatest, at Full and New Moon (approximately once a fortnight).

- M.L.W.S. (**Mean Low Water Spring**). This is a long-term average value of two successive low waters over the same periods as defined for M.H.W.S.

- M.H.W.N.. (**Mean High Water Neaps**). This is the average throughout a year of the heights of two successive high waters when the range of tide is at its minimum, at the time of the first and the last Quarter of the Moon.

- M.L.W.N (**Mean Low Water Neaps**). This is the long-term average value of two successive low waters over the same period of time as defined for M.H.W.N.

- M.S.L. (**Mean Sea Level**). This is the average level of the sea over a long period of time, preferably 18.6 years, or average water level which would exist in the absence of tides.

- L.W.D. (Low Water Datum). This is a local plane which approximates to the mean of the lower low waters, and usually satisfies the criterion that the tide seldom falls below it. For example, for the Mackay area, Australia, these levels are as follows:

H.A.T. – 6.41 m,   M.H.W.S. – 5.28 m,   M.H.W.N. – 4.06 m
M.S.L. – 3.00 m,   M.L.W.N. – 1.94 m,   M.L.W.S. – 0.72 m

Predicted water levels sometimes differ from the observed levels. Variations from the predicted heights are caused mainly by strong and prolonged winds or by unusually high or low barometric pressure (see Sect. 3.8). Coastal currents and waves also interact with the coastal topography and produce local gradients on the sea surface. Observed mean sea level at one site may be higher (or lower) compared with corresponding levels a few kilometres away along the coast or outside a harbour.

In the coastal zone, in estuaries, salt marshes and mangrove swamps, the development of ecosystems is strongly influenced by the changes of tidal levels and the rhythms they imposed in terms of submersion and exposure to the air, resulting temperature and salinity changes, sediment movements and nutrients fluxes. These and other influences of tides on living organisms will be discussed in Chap. 14.

6 Internal Waves

6.1 Introduction and Useful Definitions

The propagation of waves at the sea surface is strongly dependent on the density difference between water and air, and on gravitational acceleration. However, air density is small enough to be ignored in the theoretical analysis and practical calculations of surface waves. In Chap. 1 we have shown that the density of ocean water changes with depth. Therefore, it is quite likely that waves appear along the density gradients within the ocean. Such waves are known as **internal waves**.

While internal waves were known to exist in the ocean in the early 1900's, it is only within the last three decades or so that technology has advanced to a point where large numbers of observations are available. Most internal wave measurement methods are based on recording the profiles of temperature and salinity in the water column. The movement of points of equal temperature, **isotherms**, or the movement of points of equal salinity, **isopycnals**, are the manifestations of the passage of internal waves. Figure 6.1 (see colour plate p.563) shows a time series of density on the Australian North-West Shelf (Burrage *et al.*, 1996). In this figure, water density has been denoted in σ_t units as a form of shorthand. The quantity σ_t is related to density ρ_w by the expression:

$$\sigma_t = \rho_w - 1000. \tag{6.1}$$

Thus, the density $\sigma_t = 25$ kg/m^3 corresponds to the density ρ_w equal to 1025 kg/m^3.

The zone where the vertical density gradient is the greatest is known as a **pycnocline**. For completeness we define here the **thermocline** as a zone where density variations are determined mostly by the temperature variations, and **halocline** where the density variations are controlled mostly by the salinity. Figure 6.2 illustrates typical locations of the pycnocline and thermocline in the tropical zone of the Atlantic Ocean (Miropolskiy and Monin, 1978). The thermocline is about 65 m below the surface, with the pycnocline slightly deeper, at about 70 m. In the same figure, a vertical profile of the Brunt-Väisälä frequency, $N(z)$ is also given. We will discuss this frequency later in this chapter.

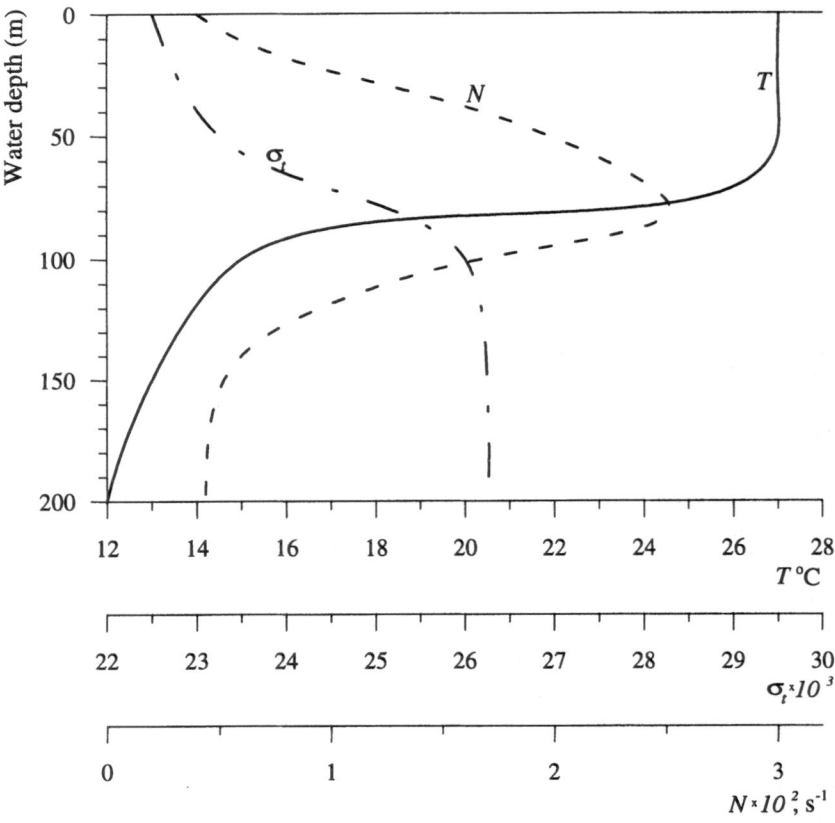

Fig. 6.2: Typical vertical profile of temperature, T, density, σ_t, and Brunt-Väisälä frequency, N, in Atlantic Ocean (adapted from Miropolskiy and Monin, 1978)

We define the isotherm surface as one on which the temperature of the fluid is constant, and the isopycnal surface as one on which the density of the fluid is constant. When the density of fluid is a function of pressure only (*i.e.* $\rho_w = \rho_w(p)$), the isobaric and isopycnal surfaces are parallel to each other. This situation is known as a **barotropic** water mass. If the density is also a function of other parameters (*i.e.* $\rho_w = \rho_w(S, T, p)$), the isobaric and isopycnal surfaces may be inclined to each other. This is known as a **baroclinic** water mass. The barotropic case is most common in deep water, while the baroclinic case is typical for the upper layer of the ocean. As the vertical density gradients in the oceans are mostly very small, little energy is required to move water particles vertically. Therefore, the amplitudes of internal waves can be very large. For example, Bockel (1962) observed vertical displacement of 180 m in the Strait of Gibraltar. In observations of internal waves, both short and long waves can usually be distinguished. Short internal waves are those with periods significantly less than 12 hours; the periods can be as short as

2.5 minutes, but most have periods ranging from 5 to 20 minutes (Roberts, 1975). For example, temperature measurements made in Massachusetts Bay (West Atlantic) indicate that the thermocline can heave up and down with periods of 6 to 8 minutes with vertical displacements of 10 m, occurring at 17.2 m below the surface (Halpern, 1971). In the Andaman Sea, internal waves with periods of about 20 minutes and amplitudes over 40 m were observed. The existence of internal waves with a semi-diurnal period was observed during temperature/salinity measurements on the Australian North-West Shelf. Vertical displacements of isopycnal surfaces were about 30 m, equal to nearly half the water depth (Holloway, 1983).

The phase speed of short-period internal waves varies between 0.1 m/s and 1 m/s, and their steepness is very small, equal to about 0.04. They are not necessary of sinusoidal form; sometimes they are flattened on the crests when the thermocline is shallow and peaked when the thermocline is deep.

Long internal waves with periods around 12 hours are known as semi-diurnal internal tides or baroclinic tides. The amplitudes of internal tides are usually 2–10 m, while their length is about 3×10^4 m. A comprehensive review of internal tide observations was published by Huthnance (1989).

There are a number of hypotheses on how internal waves are generated. There are instances where some generating mechanisms are more likely to occur. For example, it was shown that changes in the topography, e.g. slopes, sills, or seamounts, may produce internal waves (Roberts, 1975; Holloway, 1983; Huthnance, 1989; Holloway et al., 1997). Another possible mechanism for generation of short-period internal waves is forcing provided by long-period waves. The relationship between short-period internal waves and the internal tides has been observed east of the Straits of Gibraltar, the Bay of Biscay and on the California Coast (Roberts, 1975). It is also likely that winds, air-pressure fluctuations, surface swell, as well as ships slowly moving through highly stratified water are able to generate internal waves.

In the next section we will discuss the basic methods of quantitative description of internal waves in a simpler two-layer ocean and in an ocean with density varying continuously with depth. However, we will start with a description of the stability of water masses.

6.2   Stability of Water Masses

Understanding the stability of ocean water is the basic requirement needed to understand the physics of internal waves. In general, a fluid element can be in one of the three states: stable, unstable, or neutral stability. These three states are demonstrated in Fig. 6.3 for a small ball displaced on a concave, convex, and horizontal surface. In the stable state, when the ball is given a small vertical displacement, there is a restoring force which acts on the ball to return it to its original level. On the other hand, in the unstable state, when

Fig. 6.3: States of the particle: **a** stable, **b** unstable, **c** neutral; black dots denote initial position of the particle, white dots denote the particle out of initial position

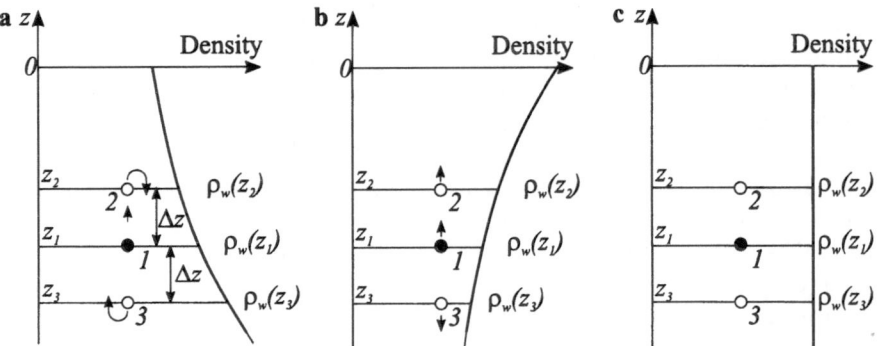

Fig. 6.4: Different profiles of water density: **a** stable, **b** unstable, **c** neutral vertical

the ball is displaced, there is a force moving it further away from its initial position. Finally, for the ball in a neutrally stable state, there is no net force acting to either return the particle its initial position or move it further.

Let us now apply these three stability states to a small element of water in the ocean with some specific vertical density profile, $\rho_w(z)$, as presented in Fig. 6.4. For the moment, we will neglect any dynamic phenomena associated with the particle displacement. In case **a**, it is assumed that water density increases with water depth. Initially at position 1, the forces acting on a small element of water, i.e. weight of a given volume of water and buoyancy force (see Sect. 2.2) are in balance; thus:

$$\rho_w(z_1)Vg - W = 0, \tag{6.2}$$

in which $\rho_w(z)$ is the water density at the level $z = z_1$, V is the volume of the small element of the water under consideration, and W is the weight of this element of water.

Suppose now that due to some external force, the water element is moved to position 2, at the level z_2. The weight of the water element remains the same, however, the buoyancy force changes because change of the density of surrounding water. Thus we have:

$$\rho_w(z_2)Vg - W = \rho_w(z_2)Vg - \rho_w(z_1)Vg = [\rho_w(z_2) - \rho_w(z_1)]\,Vg < 0. \tag{6.3}$$

Because $\rho_w(z_2) < \rho_w(z_1)$, the resulting force is directed downwards and the water element tends to sink back to its original level. A similar consideration for the water element moved to position 3 (level z_3) yields the conclusion that an excess buoyancy force moves the element back to its original position. Thus, case **a** is related to the situation defined as stable. A water element approaching its initial position oscillates around that position, until kinetic energy of the element dissipates.

In case **b**, the water density decreases with depth. Hence, the balance of forces for the water element moved to level z_2 is as follows:

$$\rho_w(z_2)Vg - W = [\rho_w(z_2) - \rho_w(z_1)]\,Vg > 0. \tag{6.4}$$

Thus, the water element will be pushed up, away from its initial position. On the other hand, when the water element is moved from its initial position to position 3 at the level z_3, the buoyancy is not sufficient to balance the element's weight. As a result, the element will sink, moving away from its initial position.

In case **c**, the water density remains constant; thus the balance of forces is satisfied at any position. Such situation is defined as neutral stability.

Consider now a stable profile shown in Fig. 6.4a, and assume that the Earth's rotation is ignored. From Eq. (6.3), it follows that when a small amount of water is displaced from its initial position, the resulting force tends to move it back to the initial position. The movement of the water element is restrained by an inertia force, and the resulting balance of force can be expressed as follows:

$$\rho \frac{d^2\zeta}{dt^2} = g\Delta\rho, \tag{6.5}$$

in which ζ is the vertical water displacement, and $\Delta\rho$ is the resulting change of water density. If we ignore the water compressibility, from a power series for ρ we can write $\Delta\rho$ in the form:

$$\Delta\rho = \zeta\frac{d\rho}{dz}. \tag{6.6}$$

Substituting Eq. (6.6) into Eq. (6.5) yields:

$$\rho\frac{d^2\zeta}{dt^2} + \left(-\frac{g}{\rho}\frac{d\rho}{dz}\right)\zeta = 0, \tag{6.7}$$

or:

$$\frac{d^2\zeta}{dt^2} + N^2(z)\zeta = 0. \tag{6.8}$$

The elementary solution of Eq. (6.8) takes the form:

$$\zeta(t) = \cos(N\,t). \tag{6.9}$$

Equation (6.9) represents the simple harmonic motion of element of the fluid moving up and down. The quantity $N(z)$ is known as the **Brunt-Väisälä frequency** or stability frequency:

$$N(z) = \left\{ -\frac{g}{\rho}\frac{d\rho}{dz} \right\}^{1/2} \tag{6.10}$$

Usually frequency $N(z)$ has a maximum in the thermocline where the density gradient is the greatest, and it decreases both above and below this level, as the water becomes more homogeneous (see Fig. 6.2). Sometimes multiple maxima in the vertical profile of $N(z)$ are observed; they correspond to the seasonal thermoclines.

For the stable static equilibrium in an incompressible fluid (case **a** in Fig. 6.4), $\partial\rho/\partial z < 0$, and $N^2 > 0$. The period $2\pi/N$ varies in the ocean from a few minutes in the thermocline, to several hours in the deep ocean. In some circumstances, such as in lakes and in the deep ocean where $\partial p/\partial z$ is very small, the extra term $-g^2/C_s^2$ (where C_s is the sound speed in the ocean) should be included for the Brunt-Väisälä frequency N^2.

In case **b** (Fig. 6.4), when $\partial\rho/\partial z > 0$, and $N^2 < 0$, Eq. (6.8) has no solution of the wave type. In fact, the only solution is:

$$\zeta(t) = \cosh\left[\left(\frac{g}{\rho}\frac{d\rho}{dz}\right)^{1/2} t \right], \tag{6.11}$$

and as t increases, the displacement goes exponentially to infinity. When $N^2 = 0$, the water mass is uniform in terms of density.

As will be shown in the following sections, the frequency N^2 plays a fundamental role in determining internal wave patterns.

6.3   Internal Waves in Two-Layer Ocean

In order to shed some light on the nature of internal waves, we first consider a simple case of a two-layer ocean, which is a good approximation when the thermocline and pycnocline change very sharply with water depth. We consider the situation with an upper layer of depth h_1 and density ρ_1, and a lower layer of depth h_2 and density ρ_2, with an internal wave of length L_i at the interface between the layers (Fig. 6.5). It should be noted that at the level $z = -h_1$, the gradient of water density becomes infinitely large; thus the Brunt-Väisälä frequency, $N(z)$, cannot be defined at that level. The period T_i and phase

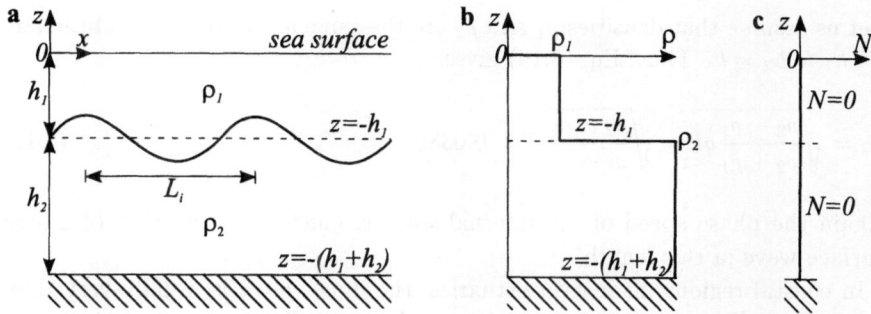

Fig. 6.5: Two-layer ocean model: **a** two layers of fluid of different densities, **b** vertical density profile, **c** vertical profile of frequency

velocity of the internal wave C_i, can be found from the following dispersion relation (Pond and Pickard, 1983):

$$\omega^2 - \frac{(\rho_2 - \rho_1)gk}{\rho_2 \coth(kh_2) + \rho_1 \coth(kh_1)} = 0, \tag{6.12}$$

and

$$C_i = \frac{\omega}{k} = \sqrt{\frac{g}{k} \frac{\rho_2 - \rho_1}{\rho_2 \coth(kh_2) + \rho_1 \coth(kh_1)}}, \tag{6.13}$$

in which $\omega = 2\pi/T_i$ and $\coth(x) = 1/\tanh(x)$.

For the deep water waves (h_1/L_i and h_2/L_i are greater than $1/2$), the phase velocity (6.13) simplifies as follows:

$$C_i = \sqrt{\frac{g}{k} \frac{\rho_2 - \rho_1}{\rho_1 + \rho_2}}. \tag{6.14}$$

Assuming, for example, that $\rho_1 = 1020$ kg/m^3 and $\rho_2 = 1023$ kg/m^3, Eq. (6.14) gives:

$$C_i = 0.038\sqrt{\frac{g}{k}} = 0.038C, \tag{6.15}$$

where C is the phase velocity of the surface wave of this length (see Eq. 4.17). Thus, the phase speeds of internal waves are much smaller than those of surface gravity waves.

On the other hand, for long, shallow water waves, when L_i/h_1 and $L_i/h_2 > 20$, the phase velocity becomes:

$$C_i = \sqrt{\frac{g(\rho_2 - \rho_1)h_1 h_2}{\rho_2 h_1 + \rho_1 h_2}}. \tag{6.16}$$

Let us assume that densities ρ_1 and ρ_2 are the same as above, and additionally let $h_1 = h_2 = h$. Thus, Eq. (6.16) gives:

$$C_i = \sqrt{\frac{\rho_2 - \rho_1}{\rho_2 + \rho_1} gh} = \sqrt{\frac{\rho_2 - \rho_1}{\rho_2 + \rho_1}} C = 0.038C. \qquad (6.17)$$

Again the phase speed of the internal wave is much less than that of a long surface wave of this length.

In coastal regions, near river estuaries, the upper layer is usually very thin, say $h_1 < L_i/20$, over a deep lower layer, where usually $h_2 > L_i/2$. Using these water depth estimates in Eq. (6.13), we obtain:

$$C_i = \sqrt{\frac{\rho_2 - \rho_1}{\rho_1} gh_1}. \qquad (6.18)$$

These waves are non-dispersive (not dependent on wave frequency) and travel slowly because of small density difference. For example, when $h_1 = 5$ m, $\rho_1 = 1\,000$ kg/m^3 (freshwater), and $\rho_2 = 1\,023$ kg/m^3, the phase velocity, C_i, becomes $0.15\sqrt{gh_1} = 1.05$ m/s.

By analogy with surface waves, the orbits of internal waves are circular (short waves) or elliptical (long waves), and orbit dimensions decay away from the interface. We note that internal waves in the ocean produce only very small vertical displacement of the free surface; usually they are smaller by a factor of the order $(\rho_2 - \rho_1)/\rho_1$ (say $\sim 10^{-3}$) than the internal wave displacement.

Suppose now that some object is located at water depth $z = -h_1$ (see Fig. 6.5). This object can be stationary when the balance between its weight and the buoyancy force is maintained. However, due to motion of the internal waves along the pycnocline, the object's buoyancy varies depending on the density of the surrounding fluid. In 1963, the nuclear submarine $U.S.S.$ $Thresher$, with all crew members, was lost in the West Atlantic. There had been no indication of equipment malfunction, or unusual storm weather. As submerged submarines attain neutral buoyancy by flooding or jettisoning sea water from ballast tanks, there are speculations that $U.S.S.$ $Thresher$ was probably cruising along a pycnocline when it encountered a large internal wave and suddenly dropped to a greater depth because of the lower density of the ambient water. Evidently the incident occurred too rapidly and submarine crew was not able to arrest the ship's fall (Pinet, 1992).

The model of internal waves in which a layer of water of uniform lower density is located over a layer of uniform higher density, with a sharp interface is quite realistic for coastal regions. River runoff occupies the upper layer of low density, over the deep layer of much higher salinity, and a sharp gradient of salinity (*i.e.* density) occurs between them. Two-layer model is also applicable when one tries to describe an upper oceanic, well-mixed layer located over deeper water. The thermocline is then fairly abrupt and separates water masses above and below, each of which is almost homogeneous. However, in most cases in the open sea, water density varies continuously with depth in a more complicated manner.

6.4 Internal Waves when the Density Varies Continuously with Depth

6.4.1 Modal Structure of Internal Waves

In general, ocean water is continuously stratified, $i.e.$ the water density varies continuously with the depth, as shown in Fig. 6.2. In such situations, internal waves still occur, but in contrast to the two layer model, the internal waves can also propagate in non-horizontal direction. Thus, wave number should be treated as the vector $\vec{k} = \vec{k}(k_x, k_y, k_z)$ in three dimensional space. This fact is of basic importance for the dynamics of ocean water and biological life in the ocean, as internal waves propagating in the vertical direction transport energy and matter from the surface to the bottom of the ocean and $vice\ versa$.

Let us briefly describe the conditions which are satisfied by internal wave motion when the Brunt-Väisälä frequency, N, changes with depth. Consider an ocean with a constant water depth, h. At the impermeable bottom, the vertical component of the internal wave velocity should be zero, $i.e.$ $w = 0$ at $z = -h$. As we have mentioned above, the internal waves produce only small disturbances of the free surface, with amplitudes $O(\Delta\rho/\rho)$ times the amplitude of internal waves. Because of this, at the free surface we can also assume that $w = 0$ at $z = 0$, as far as the internal waves are concerned. In fact the surface is essentially a rigid surface for the internal waves because stratification is weak. Therefore, the top and bottom conditions, with $w = 0$ at $z = 0$ and $z = -h$, act as a wave guide for internal waves.

For the prescribed Brunt-Väisälä frequency, N, the frequency of internal waves, ω, cannot be arbitrary. At a given depth, this frequency is bounded above by the Brunt-Väisälä frequency, $N(z)$, and below by the inertial frequency $f = 2\omega_E \sin\phi$, which varies from about 1.45×10^{-4} rad s^{-1} at the poles ($T_f \sim 12$ hr) to 0 at the Equator (T_f infinite). The inertial frequency is the minimum frequency that the free internal wave motion can possess. Assuming that $N(z)$ may be as large as 0.01 rad s^{-1} ($T_N \approx 10$ min), we obtain the band of possible internal wave frequencies as shown in Fig. 6.6. At a given depth z, the Brunt-Väisälä frequency, $N(z)$, determines the maximal frequency for the waves which occur at that depth. Since $N(z)$ is greatest in the thermocline where the variation of density with depth is greatest, internal waves of the largest frequency occur in the thermocline.

For a given profile of the Brunt-Väisälä frequency, $N(z)$, the structure of internal waves can be determined only numerically and will not be discussed here. For more details of such solutions, the reader should consult books by Roberts (1975), Phillips (1977), Le Blond and Mysak (1978), Miropolskiy and

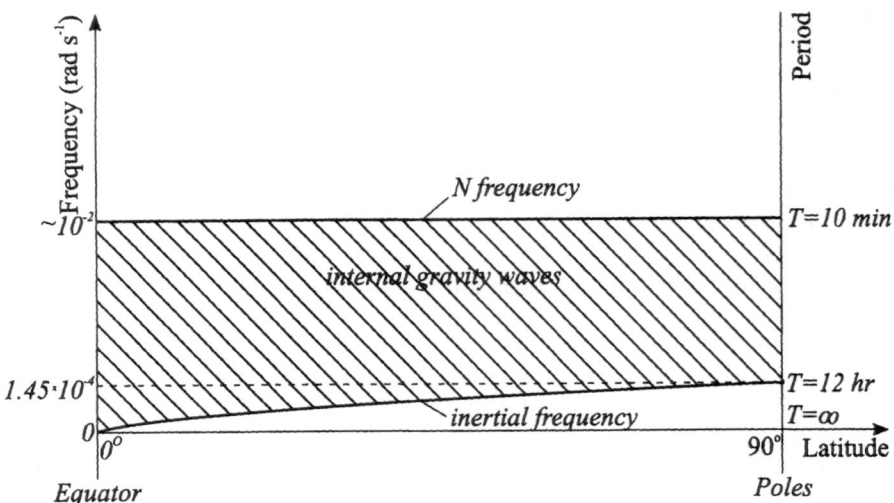

Fig. 6.6: Allowable frequencies and periods for internal gravity waves

Monin (1978), and many articles in professional literature. Some cases of the frequency profile are more tractable, for example, the case when $N_0 =$ constant. This case corresponds to the exponential profile of density (Fig. 6.7):

$$\rho(z) = \rho_0 \exp\left(-N_0^2 z/g\right), \tag{6.19}$$

as from Eq. (6.10) we obtain:

$$N(z) = \left\{-\frac{g}{\rho}\left(-\frac{N_0^2}{g}\right)\rho\right\}^{1/2} = N_0 = \text{Const.} \tag{6.20}$$

As was shown by Krauss (1972) and Phillips (1977), the assumption that $N =$ constant (or when the density profile is given by Eq. 6.19) is an acceptable local approximation to the thermocline.

The fact that the vertical velocity, w, of internal waves at the sea surface and the sea bottom should vanish, suggests the following expression for evolution of velocity, w, in space and time (Pond and Pickard, 1983):

$$w(x, z, t) = a \sin k_z (z + h) \sin(k_x x - \omega t), \tag{6.21}$$

in which h is the water depth, and k_x and k_z are the wave numbers of oscillations in the horizontal, x, and vertical, z, directions. To find k_z, we again use the fact that at $z = 0$, the velocity component $w = 0$. Thus,

$$\sin k_z h = 0, \tag{6.22}$$

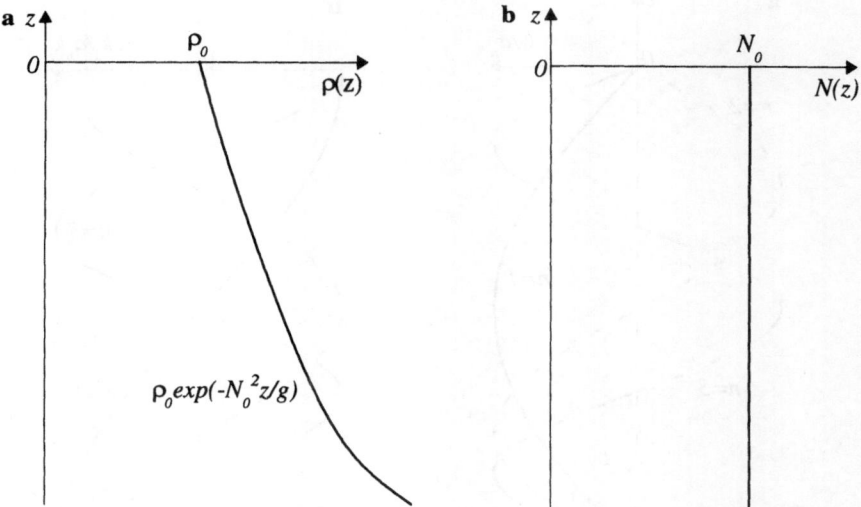

Fig. 6.7: Vertical profile of: **a** water density, **b** the Brunt-Väisälä frequency for the case when N is constant

and

$$k_z h = n\pi,$$ (6.23)

in which $n = 1, 2, 3, ...$ Therefore, internal wave motion is only possible for selected wave numbers, k_z, or selected lengths of the vertical component of the motion, $L_z = 2\pi/k_z$.

Substituting Eq. (6.23) into Eq. (6.21) yields:

$$w(x, z, t) = a_n \sin\left[\frac{n\pi(z+h)}{h}\right] \sin(k_x - \omega t).$$ (6.24)

The wave number k_x should be found from the dispersion relation which takes the form (Roberts, 1975):

$$\left(\frac{N^2 - \omega^2}{\omega^2 - f^2}\right) k_x^2 - \left(\frac{n\pi}{h}\right)^2 = \left(\frac{N^2}{2g}\right)^2.$$ (6.25)

If $\omega^2 \gg f^2$ (waves are of short period), the Earth's rotation may be neglected and Eq. (6.25) becomes:

$$k_x^2 = \left(\frac{N^2}{\omega^2} - 1\right)^{-1} \left[\left(\frac{N^2}{2g}\right)^2 + \left(\frac{n\pi}{h}\right)^2\right].$$ (6.26)

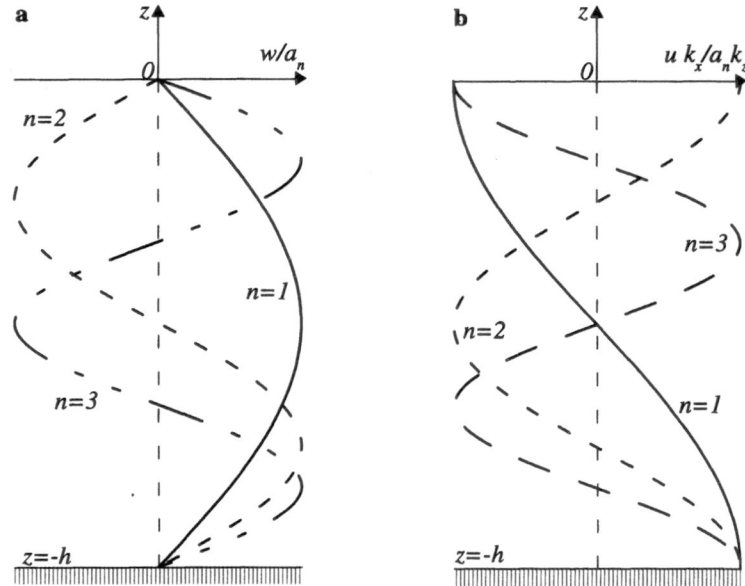

Fig. 6.8: Variations with depth of amplitudes of velocity components for the first three modes of the internal waves, when N is constant: **a** vertical component, **b** horizontal component

Using the mass conservation equation (see, for example Eq. 2.20), we can express the horizontal components of velocity as follows (Pond and Pickard, 1983):

$$u(x, z, t) = \frac{a_n k_z}{k_x} \cos\left[\frac{n\pi(z + h)}{h}\right] \cos(k_x x - \omega t). \tag{6.27}$$

The amplitudes, a_n, depend on the forcing of the internal wave by external mechanisms. Equations (6.24) and (6.27) illustrate that there is only a series of allowable modes and frequencies of internal waves. For an illustration of the modal structure of the vertical and horizontal velocities, the first three modes $(n = 1, 2, 3)$ are shown in Fig. 6.8. Figure 6.8a shows the amplitude variation with depth for the vertical velocity component, while Fig. 6.8b illustrates the variation of the amplitude of the horizontal velocity component. Note that the mode number, n, equals the number of zero crossings for the velocity u or the number of extrema (maxima and minima) for the velocity w.

Figure 6.8a suggests that the amplitude of the first mode of the velocity component, w, is positive for all water levels. Therefore, when the phase $k_x x - \omega t < \pi$, the vertical velocity w is directed upwards. For the second $(n = 2)$ and third mode $(n = 3)$, the velocity w is expected to be positive only for $-h/2 < z < 0$, and for $-2h/3 < z < 0$, respectively. Similar observations can be made for the other wave phases and the u component of velocity. In

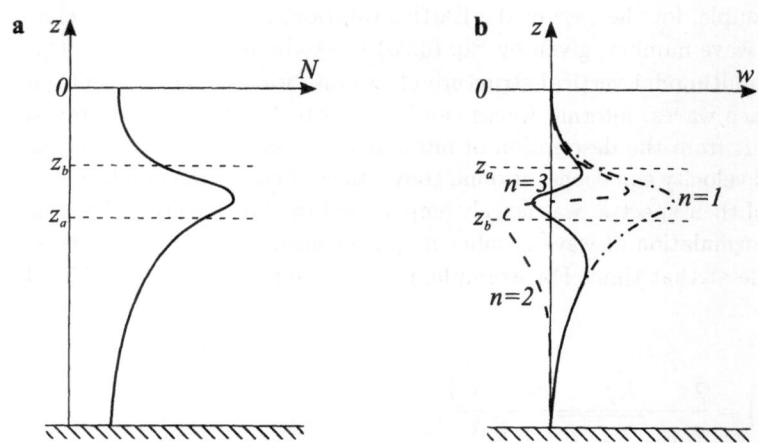

Fig. 6.9: Internal wave trapping in thermocline: **a** vertical profile of frequency $N(z)$, **b** first three modes of the vertical velocity

real situations, internal waves are represented by a summation of many modes. As was shown by Krauss (1972), in the shallow Baltic Sea, most internal waves are describable using only the first three modes for density profile being almost exponential (see Eq. 6.19). However, for deep water internal waves, five or ten modes are needed for a good approximation.

In tropical or subtropical waters, the upper layer of the ocean is usually well mixed by the wind and a sharp thermocline often appears near the ocean surface. The Brunt-Väisälä frequency, N, has a strong maximum at the thermocline, which is schematically shown in Fig. 6.9a. For frequencies approaching N_{max}, the internal waves will be trapped in the narrow depth range of the thermocline. An illustration of the wave trapping by the pycnocline is shown in Fig. 6.9b. The first three modes of the wave motion have an oscillatory behaviour only within the small depth range $z_a < z < z_b$; outside of this range the wave amplitude decays exponentially. Biological consequences of trapping of internal wave energy within a narrow depth range for plankton oscillation and primary production of the ocean will be discussed in Chap. 15.

In general, internal waves can propagate in any direction in three dimensional space. Some of them eventually reach the surface or the bottom and are reflected with the angle of reflection being equal to the angle of incidence. When the bottom is sloping, the internal wave may be reflected backwards from the bottom instead of forward. Such reflection occurs when the bottom slope is greater than the ray slope with respect to the horizontal plane.

The above analysis is valid under the assumption that the frequency of the internal wave is much higher than the inertial frequency associated with the Earth's rotation. However, when the frequencies ω and f are of a similar order of magnitude, the mode structure of the internal waves will be slightly different.

For example, for the case of the Earth's rotation, a more complex relationship for the wave number, given by Eq. (6.25) has to be used in place of Eq. (6.26).

The multimodal vertical structure of internal waves suggests that, similarly to surface waves, internal waves can be described in terms of energy spectra. The shift from the description of internal waves as an analytic solution of the vertical velocity equations, as done above, to the description of internal waves in terms of their spectra, was largely emphasized by the Garrett and Munk (1972, 1975) formulation of wave number frequency spectrum, based on observations available at that time. For example, the spectrum proposed in 1975 takes the form:

$$S\left(\tilde{k},\tilde{\omega}\right) = \frac{2}{\pi} \frac{E\tilde{f}}{\tilde{\omega}\sqrt{\tilde{\omega}^2 - \tilde{f}^2}} \frac{A\left(\tilde{k}/\tilde{k}_*\right)}{\tilde{k}_*}, \qquad (6.28)$$

where $\tilde{k}$ is the dimensionless horizontal wave number, $\tilde{\omega} = \omega/\tilde{N}$ is the dimensionless frequency with $\tilde{N} = 3$ cycle per hour, $\tilde{f} = f/\tilde{N}$ is the dimensionless inertial frequency, $\tilde{k}_x = 6\pi\sqrt{\tilde{\omega}^2 - \tilde{f}^2}$ is the horizontal wave-number scale, $E = 6.3 \times 10^{-5}$, and $A(\lambda) = (t-1)(1+\lambda)^{-1}$, where $t = 2.5$, and $\lambda = k/k_*$. Formula (6.28) is valid under the assumption of horizontal isotropy of the internal wave field.

6.4.2   Topographic Effects

Observations, as well as numerical simulations, help to develop some understanding of the complexity of internal waves. As mentioned in Sect. 6.1, the interaction of semidiurnal tides with shelf topography may be one of the sources of internal waves. They have wavelengths of a tens of kilometres and may propagate out of the generation area. During measurement program in the Andaman Sea near northern Sumatra, large-amplitude, long internal waves with currents as high as 1.8 m/s were observed (Osborne and Burch, 1980). Photographs of the Andaman Sea's surface by LANDSAT satellite, as well as photographs taken during the Apollo-Soyuz mission, showed that the shallow waters around the Andaman Islands and near the southernmost point of Nicobar Islands, or close to northern Sumatra are potential sources of the observed internal waves. A set of current metres and thermistors was placed at approximate depths of 53, 87, 116, 164 and 254 metres, in the ocean with 1100 m water depth. In Fig. 6.10, a typical temperature signal of an internal wave, recorded at the depth of 164 m, is shown. The wave pattern is similar to the solitary shape described in Sect. 4.2.3, suggesting a soliton interpretation of the signal (see also Massel, 1989). The leftmost soliton in Fig. 6.10 is the largest and leads the packet. Similar packets occur, on average, every 12 hour and 26 minutes, which suggests a link between soliton generation and the semidiurnal tide. The average number of observed solitons is six or seven.

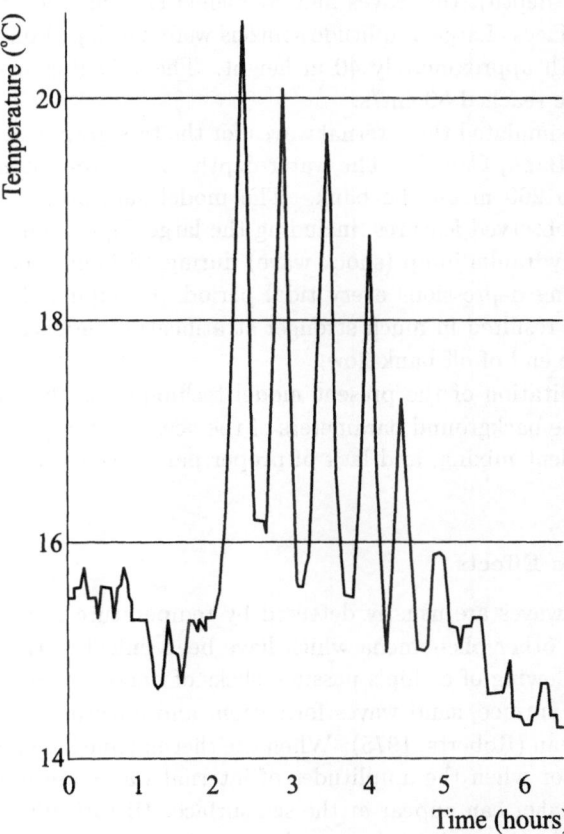

Fig. 6.10: Temperature signal of internal wave at 164 m depth (adapted from Osborne and Burch, 1980)

Usually the generation of internal waves in the ocean is complicated by many factors, including finite amplitude topography, spatially-varying tidal current strength and phase, spatially and temporally varying stratification, boundary-layer and turbulent effects, breaking waves and associated mixing. To develop some understanding of the influence of bottom topography on wave generation in general, and to help interpret the experimental data, idealized numerical experiments were carried out.

Measurements and numerical simulations of internal waves, using the Korteweg-de Vries equation, for the Australian North West Shelf showed the development of the waveform to the formation of shocks and solitons as it propagated shoreward over the continental slope and shelf (Holloway *et al.*, 1997). At the deeper slope mooring ($\sim$ 110 m water depth), relatively smooth internal waves were observed with solitons or short period oscillation waves at the front face of the waves. When the internal tide had propagated to the break mooring

($\sim$ 80 m water depth), the waves had steepened showing shock on both the front and back faces. Large amplitude solitons were developed on the front face of the wave with approximately 40 m height. The velocities associated with the internal tide reached 80 cm/s.

Lamb (1994) simulated the internal waves for the topography at the northern side of George Bank, Canada. The water depth varied from 65 m, on the top of the bank, to 260 m off the bank. The model successfully reproduced a number of the observed features, including the large depression of pycnoclines resulting in a hydraulic jump (shock wave) during off-bank flow, and two on-bank propagating depressions every tidal period. Changing the density field in a way which resulted in much stronger stratification induced internal wave breaking by the end of off bank flow.

The main limitation of the present model techniques is the large temporal variability in the background parameters of the ocean water, such as stratification and turbulent mixing, and lack of proper parameterization of the effects of dissipation.

6.4.3 Surface Effects

While internal waves are usually detected by temperature or salinity fluctuations, there are other phenomena which have been linked to their occurrence. These include slowing of a ship's passing, slicks of darker bands of muddy waters on the sea surface, sand waves formation, and affecting sound-scattering layers in the ocean (Roberts, 1975). When the thermocline lies sufficiently close to the surface, or when the amplitudes of internal waves are unusually large, the internal breaker can appear at the sea surface. Historically, the Andaman Sea was probably the region where such phenomena were first observed. According to LaFond (1966), breaking whitecaps *emitted a low roar as they passed a drifting ship on a calm sea.* More evidence on the surface effects caused by internal waves has been reported by Osborne and Burch (1980). Photographs taken on board a survey vessel show a long band of breaking waves about 1.8 m high. Moreover, LANDSAT images have shown internal waves with crests as long as 150 km, and wavelength as great as 15 km in the same area.

Burrage *et al.* (1996), using the ERS-1 Synthetic Aperture Radar (SAR) imagery detected a presence of internal waves on the Australian North West Shelf (a description of the SAR sensor is given in Chap. 9). These images are consistent with sea-truthing data obtained from a conventional current meter and thermistor chain mooring. These instruments were sampled sufficiently rapidly to resolve internal wave variability. Additional data were also obtained from ships using a conductivity, temperature, and depth (CTD) probe to obtain profiles of temperature, salinity and density at 2 m depth intervals.

Images containing thin, elongated curvilinear features, being almost certainly signatures of bioslicks, were obtained under wind speeds of 2 m/s and less. These sometimes appeared in association with broad or narrow rectilinear alternating light and dark bands which, based on the shelf and band geometry,

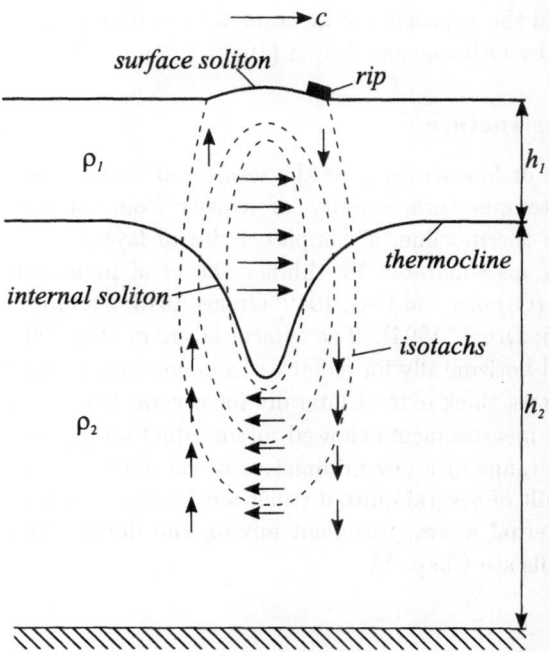

Fig. 6.11: Influence of internal soliton on sea surface (adapted from Osborne and Burch, 1980)

and on *in situ* mooring and hydrographic data, are undoubtedly internal tidal wave signatures. The presence of the internal waves is also manifested by creating regions of short, choppy, breaking waves. Perry and Schimke (1965) observed this phenomena off the north-west coast of Sumatra, and Gargett and Hughes (1972) made similar observations off British Columbia. This surface phenomenon is known as tide rip, current rips, or disturbed water. In Russian it is known as 'suloy', and the information on its occurrence is included in some log-books, for example for the White Sea, where it appears in many places (Monin and Krasitskiy, 1985). According to Perry and Schimke (1965), the distinct zones of whitecaps ranging from 200 m to 800 m in width, stretching to a distance of 30 km, move on an otherwise undisturbed sea. Within these zones, steep randomly oriented waves with heights of about 0.3 to 0.6 m were observed. Phillips (1977) explained the presence of the tide rip as a result of modulation of surface waves induced by internal waves. The strongest modulation results when the phase velocity of the internal wave, C_i, is equal to the group velocity of the surface waves, C_g, which is known as the resonance condition. Usually phase velocity, C_i, is small, *i.e.* $C_i \leq 1$ m/s, and resonance condition is satisfied for the surface waves with wavelengths of the order of 0.1–1.0 m. To illustrate the interaction mechanisms between surface and internal waves, Fig. 6.11 shows an internal soliton in a two-layer fluid of the finite depth and the approximate location of the surface rip, as observed in the Andaman Sea by Osborne and Burch (1980).

6.4.4 Microstructure

Microstructure or fine-structure is the term used to describe the departure of the values of temperature, salinity, or density from the mean value. Occasionally in the thermocline, a complex series of layers of approximately homogeneous regions separated by thinner sheets of high static stability have been observed (Osborn and Cox, 1972; Gregg, 1973; Phillips, 1977; Monin and Ozmidov, 1985; Druet, 1994). The typical layers of 10 to 20 metres thick can often be traced horizontally for as far as 15 kilometres, while the layers of only 20–30 centimetres thick extend laterally for several hundred metres. Simultaneous, vertical measurements showed intense fluctuations of temperature and salinity over a range of a few millimetres or centimetres. The microstructure can be the result of several quite distinct mechanisms, such as the influence of small scale internal waves, turbulent mixing and double-diffusive phenomena (for more details see Chap. 8).

7 Ocean Currents

7.1 Introduction

Water is continually in motion at all depths, even when the sea appears perfectly calm and flat. In Chaps. 3–6 of this book, we discussed a specific type of water motion, namely periodic wave motion. Waves can be as small as ripples on the sea surface, and as large as long tidal waves with a wavelength of thousands of kilometres. Waves can travel on the sea surface, or along interfaces dividing water masses of different densities (internal waves). However, in all these cases the water motion is periodic, or can be represented as a sum of periodic motion.

Both everyday observations and specific measurements provide a lot of evidence of a different type of water motion in the ocean which is not periodic. For example, a sailing boat with sails hanging listlessly from its spars may appear to be motionless, but it is actually drifting with the slow surface currents. This persistent water motion is responsible for slowly transporting large volumes of surface and subsurface water over vast distances. The nature of flow associated with large-scale ocean currents depends on a few dominant driving mechanisms: wind stress, pressure gradients, water density gradients, and the Coriolis effect. For clarity of analysis it will be useful to divide the currents into two fundamental groups of flow: wind-driven surface and near-surface currents, and density-driven subsurface currents. Although the surface and near-surface currents affect only about 10 percent of the ocean's volume, the vast part of all oceanographic studies was and still is devoted to them. Water motion in the ocean depths still requires more efforts from present and future generations of physical oceanographers.

Numerical modelling of ocean circulation, and the coupling of oceanic and atmospheric circulation modelling has grown intensively in recent years. Such models can skilfully mimic observed oceanic features and help in understanding and predicting the global climate system and its impact on life in the oceans. Present models do seem to encompass the major features observed in ocean circulation.

Continental shelf and coastal waters play a specific role for the global ocean ecology, linking the land masses with the ocean waters. Because of interactions with decreasing depths and coasts, the water circulation and associated chemistry and biology are very complex. Therefore it is difficult to generalize about the coastal and shelf environments. At the end of this chapter, we provide only a description of some specific phenomena, such as wave-driven currents and upwelling.

7.2   Wind Patterns on Earth

7.2.1   Forces Driving the Wind

The significance of the wind field over the ocean for generation of surface waves has been discussed in Sec. 3.5. In particular, the link between wave parameters and the local wind speed, V_w, and wind fetch, X, has been documented. Local winds may be generated by local atmospheric systems which may last only a few days. However, ocean currents in meso- and global scales are driven by average conditions, prevailing for months or years. Therefore, prior to discussing the main wind-driven ocean currents, we will briefly characterize the major patterns of wind on the Earth.

The wind field in the vicinity of the Earth's surface is driven by the interaction of pressure gradients and the Coriolis effect. This balance can be represented in the form of two equations (Young and Holland, 1996):

$$\left.\begin{array}{l} \dfrac{V_w^2}{r} + fV_w - \dfrac{1}{\rho_a}\dfrac{\partial p}{\partial n} = 0 \\[2mm] \dfrac{dV_w}{dt} + \dfrac{1}{\rho_a}\dfrac{\partial p}{\partial s} = F \end{array}\right\}, \tag{7.1}$$

in which r and s are the coordinates normal and tangential to the air parcel trajectory, respectively, V_w is the wind speed, f is the Coriolis parameter (see Sect. 5.3), ρ_a is the air density, p is the surface pressure, and F is the frictional dissipation at the Earth's surface. The term $\partial p/\partial n$ is the pressure gradient normal to the direction of motion of the air element, while the term $\partial p/\partial s$ is the pressure gradient along the streamline. The balance of forces for an air element in the vicinity of a low pressure centre in the Northern Hemisphere is shown in Fig. 7.1. Equation (7.1) results in the wind velocity as follows:

$$V_w = \frac{-fr}{2} + \sqrt{\left(\frac{fr}{2}\right)^2 + \frac{r}{\rho_a}\frac{\partial p}{\partial n}}. \tag{7.2}$$

In a low pressure system, or cyclone, the Coriolis and centrifugal forces act in the same direction, and very strong winds and sharp curvature of the trajectories are generated (Fig. 7.1a). For a high pressure system, or anticyclone,

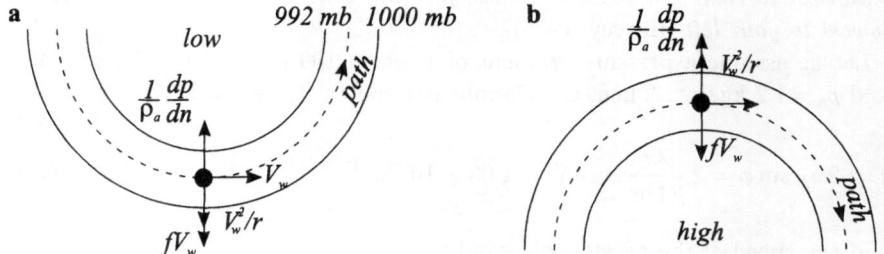

Fig. 7.1: Balance of forces for an air particle in the Northern Hemisphere: **a** low pressure system, **b** high pressure system

the centrifugal and pressure gradient forces oppose the Coriolis force. Therefore, anticyclones generate weak wind with only slightly curved trajectories (Fig. 7.1b).

Near the centre of a tropical cyclone, the Coriolis term is small compared with the centrifugal force, which results in the following expression for the wind speed:

$$V_w = \sqrt{\frac{r}{\rho_a} \frac{\partial p}{\partial n}}. \tag{7.3}$$

At the Earth's surface, friction slows the wind's speed which results in decreasing the centrifugal (V_w^2/r), as well as the Coriolis (fV_w) forces, but the pressure gradient does not change. A new balance of forces is achieved, with air at the surface flowing out of anticyclones and towards cyclones (Young and Holland, 1996). The influence of surface friction decays quickly with height above the ground and is generally negligible at an altitude of about 1 km. Therefore, the lowest atmospheric layer is often known as the boundary, of friction layer and the 1 km level is called the gradient level.

7.2.2 Geostrophic Wind

When the frictional dissipation is neglected and the isobars are straight lines ($F = 0$, $V_w^2/r = 0$), Eq. (7.1) becomes:

$$V_g = \frac{1}{\rho_a f} \frac{\partial p}{\partial n} = \frac{1}{2\rho_a \omega_E \sin \phi} \frac{\partial p}{\partial n}. \tag{7.4}$$

This flow, known as geostrophic flow, is typical for much of the large-scale flow in the atmosphere. It blows parallel to the isobars with the area of high pressure on its right in the Northern Hemisphere. In the Southern Hemisphere the high pressure is on its left. In the 19th century, Buys-Ballot's Law formulated this observation as follows: *In north latitudes, face the wind and the barometer will be lowest to right. In south latitudes, face the wind and the barometer will be lowest to your left* (Harvey, 1985).

Let us assume a pressure gradient of 1 mb in 100 km in latitude, $\phi = 45°$, and $\rho_a = 1.2$ kg/m^3. Then, the Coriolis parameter, f, becomes:

$$f = 2\omega_E \sin \phi = 2 \frac{2\pi}{24 \text{ hr}} \sin 45° = 1.03 \times 10^{-4} \text{s}^{-1}, \tag{7.5}$$

and the speed of the geostrophic wind is:

$$V_g = \frac{1}{1.2 \text{ kg/m}^3 \times 1.03 \times 10^{-4} s^{-1}} \frac{10^2 \text{ kg/m/s}^2}{10^5 \text{ m}} = 8.1 \text{ m/s}. \tag{7.6}$$

On synoptic charts with systems of isobars, a geostrophic wind scale is provided for determining wind speed, V_g, from the spacing of the isobars. Usually, a standard spacing of 4 mb is used. We note that the geostrophic wind assumption is not valid for curved isobars or close to the Equator, where the Coriolis parameter f becomes zero.

7.2.3 Major Surface Wind Patterns

As was shown above, horizontal pressure gradients are the most important forces responsible for the initiation and maintenance of large-scale motions in the atmosphere. Prior to describing the global surface wind patterns, we consider the simplest example of the relationship between wind and pressure gradients, namely the sea breeze and land breeze phenomenon on the coast. During the day the temperature of the land surface rises higher than that of the sea surface, resulting in a horizontal pressure gradient from the sea to the land. This gradient, together with a reverse flow at higher levels and weak rising and sinking air motion, constitutes the **sea breeze** (Fig. 7.2a). During the night, when radiational cooling of the land is rapid, the lower air becomes cooler over the land than over the water, and thus the horizontal pressure

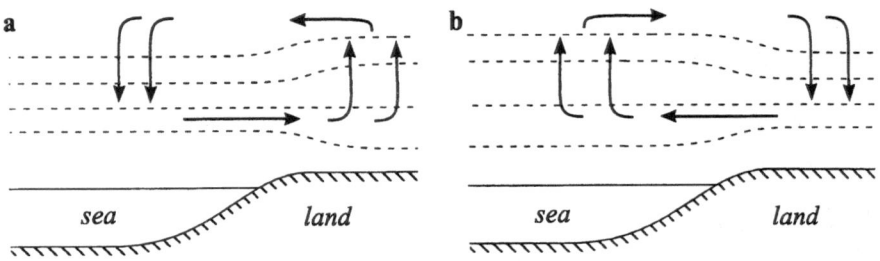

Fig. 7.2: Breezes: a sea breeze, b land breeze

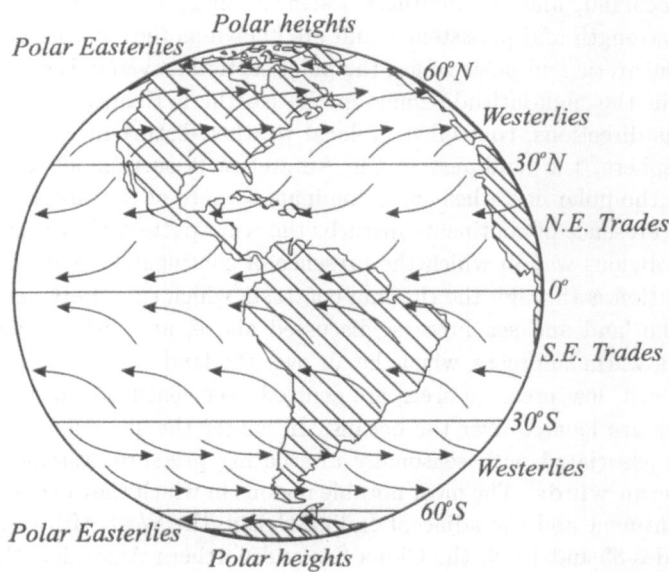

Fig. 7.3: Simplified global wind patterns (adapted from Pinet, 1992)

gradients are reversed. Air now moves from the land to the sea as a **land breeze** (Fig. 7.2b).

Just as sea and land breezes are the result of horizontal pressure gradients, induced by uneven temperature distributions over the land and the sea, the major planetary wind belts are induced by uneven temperature distributions between low and high latitudes, and the Earth's rotation. A schematic representation of the global wind pattern is shown in Fig. 7.3, with the assumption that no land areas exist to modify the belt arrangement of pressure zones.

In two subtropical high-pressure belts, the pressure gradient drives equatorward moving air, from high to low pressure zones. As a result, two belts of **Trade Winds**, the northeastern and southeastern trades, are established. The Trade Winds are highly persistent winds converging near the Equator in a narrow Intertropical Convergence Zone (ITC). The Trade Winds and ITC shift seasonally north and south along with the shifting of pressure belts and isotherms.

In the subtropical high-pressure belt (between 25° and 40°N and S), large and stagnant anticyclones are located with weak winds distributed around a wide range of compass directions. Between latitudes 35° and 60°N and S there is the belt of prevailing westerly winds, or westerlies. These surface winds are shown in Fig. 7.3 as predominantly blowing from the south-west in the Northern Hemisphere and from the north-west in the Southern Hemisphere. In the Southern Hemisphere, between latitudes 40° and 60°, there is an almost empty ocean area, eastward from the South Atlantic Ocean to Australia, Tasmania,

New Zealand, and the Southern Pacific islands, where the westerlies gather great strength and persistence, and are known as the 'roaring forties'.

In the arctic and polar zones, the polar easterlies system is established. However, in the high-latitude zones in the Northern Hemisphere, winds blow in various directions, controlled by local weather disturbances. In the Southern Hemisphere, the landmass of the Antarctica is surrounded by a vast ocean where the polar easterlies are a dominant feature of the circulation (Fig. 7.3).

The presence of continents disturbs the wind pattern shown in Fig. 7.3. The most obvious way in which the presence of continents affects the atmospheric circulation is through the thermal contrasts which exist between land and sea, and the land and sea breezes, discussed above, are their simplest examples. During warm summers, when the air over the land is less dense than that over the ocean, low pressure areas are centred over continents while high pressure centres are located over the oceans. In winter the situation is reversed. The winds associated with seasonally alternating pressure systems are known as **monsoon winds**. The most notable regions in which they occur are the Indian subcontinent and the adjacent Indian Ocean, the West African coast between latitudes 5° and 15°N, the China Sea and northern Australia (Harvey, 1985).

7.2.4   Storms and Cyclones

A large variety of storms are encountered over the oceans, and they can be grouped into three main categories: 1. tropical cyclones, 2. mid-latitude cyclones, and 3. hybrid systems. The latent heat released by condensation of water vapour in clouds and the release of the baroclinic energy stored in horizontal temperature gradients are the main sources of the energy which drives these systems.

Tropical cyclones are the most powerful and destructive type of storm. They develop over oceans in the latitudes 8° to 15°N and S. They cannot develop closer to the equator as the Coriolis effect, required for their formation, is negligibly small. There are some regions, over tropical and subtropical oceans, where tropical cyclones occur (Strahler and Strahler, 1992), namely: West Indies, Gulf of Mexico and Caribbean Sea, western North Pacific, including the Philippine Islands, China Sea and Japanese Islands, Arabian Sea and Bay of Bengali, coastal region off Mexico and Central America, south Indian Ocean, off Madagascar, and western South Pacific, in the region of Samoa and the Fiji Islands, and the east and north coast of Australia.

High sea-surface temperatures, over 26°C, are required to develop tropical cyclones which, once formed, move westward through the Trade Wind belt. However, the movement of tropical cyclones can be complex and difficult to predict, and there are many exception to the above rule. In the western South Pacific, cyclone paths are extremely complicated, with sharp speed and direction changes. As a example, in Fig. 7.4 is shown the path of tropical cyclone 'Justin', which for almost 3 weeks during March 1997 disrupted life along the eastern coast of Australia, in the Coral Sea and over south-east Papua New

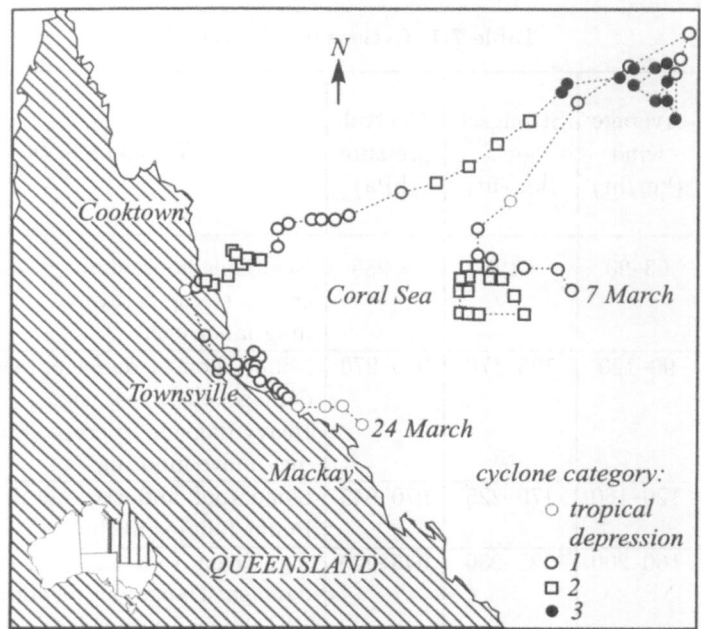

Fig. 7.4: Path of tropical cyclone *Justin* along the east coast of Australia, for the period 7–24 March, 1997 (adapted from Bureau of Meteorology, 1998)

Guinea. On this figure, the categories of cyclone severity at particular locations of the cyclone's eye are given. It should be noted that there are five categories of cyclone, increasing from 1 to 5. Associated central pressure, p_c, wind velocities and typical effects are described in Table 7.1.

A tropical cyclone is an almost circular vortex with a centre of extremely low pressure into which winds are spiralling at high speed (Fig. 7.5, see colour plate p.564). At the centre of a tropical cyclone is the 'eye' in which calm conditions prevail and the pressure reaches its lowest value, p_c, known as the central pressure. The central pressure can be converted to maximum sustained wind by the approximate relationship (Young and Holland, 1996):

$$V_{max} = 1.74(p_0 - p_c)^{0.644}, \tag{7.7}$$

where V_{max} is the speed in m/s, p_0 is the ambient pressure, usually taken as $p_0 = 1010 hPa$, and p_c is the pressure at the cyclone centre. Barometric pressure in the tropical cyclone centre commonly falls to 950 mb or lower.

Wind velocity within the cyclone increases from zero at the cyclone eye, to its maximum value, V_{max}, some distance (20–40 km) from the eye, and then slowly decreases as the distance from the eye increases. The strongest winds are typically found in the front right (left) quadrant for cyclones in the Northern

Table 7.1: Cyclone severity categories

Cat.	Average wind (km/hr)	Strongest gust (km/hr)	Central pressure (hPa)	Typical effects
1	63–90	<125	> 985	Negligible house damage, damage to crops, trees, caravans. Craft may drag moorings.
2	90–120	125–170	985–970	Minor house damage, significant damage to trees, signs, caravans. Heavy damage to crops. Small craft may move moorings.
3	120–160	170–225	970–945	Some roof and structural damage. Power failures likely.
4	160–200	225–280	945–920	Significant roof loss and structural damage. Dangerous airborne debris. Wide spread power failures.
5	> 200	> 280	< 920	Extremely dangerous with widespread destruction.

(Southern) Hemisphere. Tropical cyclone can vary in diameter from 150 to 500 km. Other types of storms are described in detail by Pinet (1992), Strahler and Strahler (1992), and Young and Holland (1996).

7.3 Wind-Driven Surface and Near-Surface Currents

7.3.1 Forces Driving Ocean Currents

As mentioned in the Introduction, prevailing ocean currents that flow steadily for longer period of time (months or years) are the result of combined actions of three main physical mechanisms: wind stress, pressure gradients, and the Coriolis effect. Wind creates friction between moving air and water at the sea surface, which eventually results in the generation of waves and currents. The generation of waves was treated in Sect. 3.5 (see also Massel, 1996a, for a detailed explanation), and here we will concentrate on current generation.

Generally, wind flow over the sea surface is turbulent. However, as shown in Sect. 2.5, in the immediate vicinity of the sea surface, turbulent fluctuations are suppressed by the presence of the surface, and a laminar sublayer develops. Within this sublayer, the stress, τ_0, induced on the sea surface by the wind, takes the form (see Eq. 2.52):

$$\tau_0 = \rho_a u_*^2, \tag{7.8}$$

in which ρ_a is the air density, and u_* is the friction velocity.

In routine meteorological observations, wind speed is measured at the anemometer height $z = 10$ m above mean sea surface, and Eq. (7.8) is usually written as:

$$\tau_0 = C_{10}\rho_a V_{10}^2, \tag{7.9}$$

where V_{10} is the wind velocity at height $z = 10$ m, and C_{10} is the drag coefficient. Experimental data on the drag coefficient, C_{10}, shows a large scatter, typically from 3×10^{-4} to 5×10^{-3}, and functional dependence on wind speed is not well pronounced. Garratt (1977), in his comprehensive review of drag coefficients over oceans and continents, suggested that in the velocity range $4 < V_w < 21$ m/s, C_{10} can be approximated as:

$$C_{10} \approx 0.00075 + 0.000067V_w. \tag{7.10}$$

The actual speed of the ocean current is a small fraction of the wind speed as the transfer of energy from air to water is an inefficient process. Usually a value of 3 or 5 percent of the wind speed is a useful approximation for the ocean current speed.

The pressure-gradient force is another factor which influences the generation of ocean currents. This is a consequence of horizontal variations in the level of the water surface, and the resulting sea surface slope. The ocean surface is not flat, but forms a complicated pattern of water 'hills', formed by convergence currents piling water up, and water 'valleys', being the result of diverging currents causing water to move apart.

In Sect. 5.3 we showed that due to the Earth's rotation, any particle of air or water not attached to the solid earth experiences an apparent change in direction of movement (deflection). This is known as the Coriolis effect. The magnitude of Coriolis deflection is proportional to the speed of the moving particle and its latitudinal position. Therefore, the Coriolis acceleration, a_C, associated with the particle deflection, takes form (see Eq. 5.26):

$$a_C \approx 2\omega_E u \sin\phi = fu, \tag{7.11}$$

in which ω_E is the Earth's rotation frequency, u is the current speed, f is the Coriolis parameter, and ϕ is the latitude of the point under consideration.

The combination of forces induced by wind stress, pressure gradients, and Coriolis effect drives the currents on the Earth's surface. The balance of these forces can be written as a three component equation, with the coordinates x, y

and z, and their respective velocity components u, v and w, being positive in the east, north and upward directions, respectively; the origin of coordinates is located at the sea surface. Thus, we have (Pond and Pickard, 1983):

$$\left. \begin{array}{l} \text{Pressure} \quad \text{Coriolis} \quad \text{Gravity} \quad \text{External forces} \\[4pt] \rho_w \dfrac{du}{dt} = \quad -\dfrac{\partial p}{\partial x} \quad +\rho_w f v \qquad\qquad +F_x \\[10pt] \rho_w \dfrac{dv}{dt} = \quad -\dfrac{\partial p}{\partial y} \quad -\rho_w f u \qquad\qquad +F_y \\[10pt] \rho_w \dfrac{dw}{dt} = \quad -\dfrac{\partial p}{\partial z} \qquad\qquad -\rho_w g \quad +F_z \end{array} \right\}, \qquad (7.12)$$

in which F_x, F_y and F_z are the external body forces. It should be noted that Eq. (7.12) is the Euler equation, discussed in Chap. 2 and Appendix C.3, and expressed here in a co-ordinate system rotating with the Earth. For simplicity, the convective inertia terms have been omitted from Eq. (7.12).

7.3.2 Geostrophic Flow

Let us make a further simplification and assume that the currents are constant $(du/dt = dv/dt = dw/dt = 0)$, and all forces, F_x, F_y, and F_z are zero. Therefore, Eq. (7.12) becomes:

$$\left. \begin{array}{l} \rho_w f v \;=\; \dfrac{\partial p}{\partial x} \\[10pt] -\,\rho_w f u \;=\; \dfrac{\partial p}{\partial y} \\[10pt] -\,\rho_w g \;=\; \dfrac{\partial p}{\partial z} \end{array} \right\}. \qquad (7.13)$$

The third equation in (7.13) is the hydrostatic equation in differential form. It gives the pressure increment dp due to a thin layer dz of fluid of density ρ_w.

Similarly to the geostrophic flow in the atmosphere, the first two equations of (7.13) permit us (at least in principle) to determine the speeds of currents, u and v. Such currents are known as **geostrophic currents**. Some potential techniques of such calculations are described by Pond and Pickard (1983). Recently techniques based on using radar altimetry data from satellites have received much attention. In Chap. 9, we will discuss some results of the TOPEX/POSEIDON mission for determining of the sea surface position and its gradient.

In the Northern Hemisphere, the Westerlies and Trade Winds induce transport which causes water to flow towards the centre of the ocean. This converging flow piles the water in the ocean centre, generating a pressure gradient, directed down-slope. When water begins to flow radially outward, the Coriolis

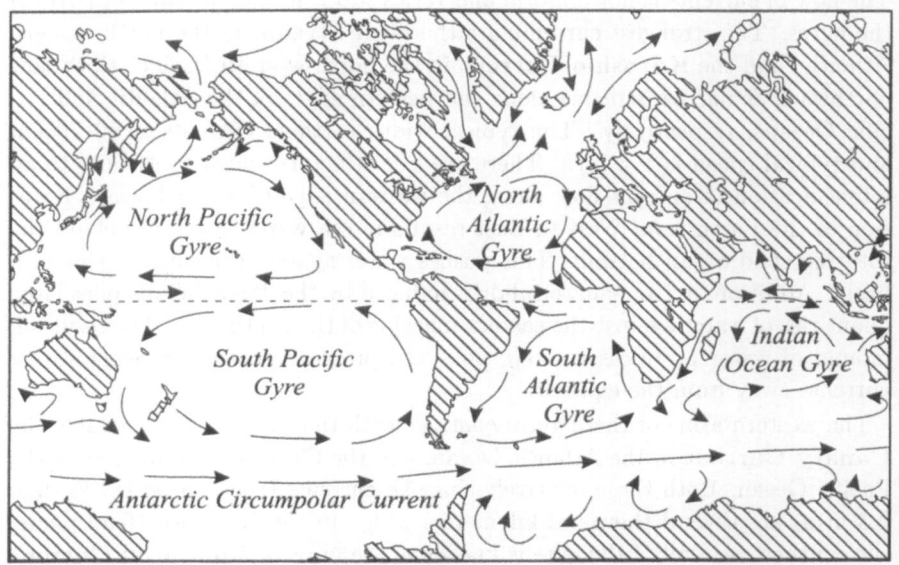

Fig. 7.6: Schematic representation of the global surface current pattern (adapted from Pinet, 1992)

effect bends the current to the right. The continents interrupt the transport and deflect the westerly flow at the Equator poleward, while the easterly flow towards the mid-latitudes is deflected equatorward. The final result in the Northern Hemisphere (the North Pacific and North Atlantic) is the closed ocean gyre with a clockwise rotation. A mirror image of this geostrophic system develops in the Southern Hemisphere (the South Pacific and South Atlantic gyres with anticlockwise rotation). This fact for the North Atlantic was known to Spanish and Portuguese navigators in the early 1500's and was subsequently recognized for other oceans as navigational records accumulated.

The geostrophic model of ocean currents agrees well with observed large-scale ocean circulation. A good example of geostrophic currents is in the western North Atlantic Ocean, known as the Sargasso Sea. This name comes from the brown seaweed *Sargassum* that floats there over much of the waters. The Sargasso Sea consists of a large lens of warm water encircled by a belt of geostrophic current rotating clockwise. The warm, light water is in gravitational equilibrium with colder and denser water to the west of the Gulf Stream (Pinet, 1992).

The geostrophic gyres in the Northern and Southern Hemispheres are shown in Fig. 7.6 as a part of the global surface water current pattern.

7.3.3   General Pattern of Wind-Driven Ocean Circulation

The two major circulation ellipses (gyres) lie between the lines of 60° latitude. The flow of currents is not uniform and varies substantially in different parts of the gyres. The strongest currents are the **Gulf Stream** in the north-western Atlantic, and the **Kuroshio Current** in the north-western Pacific. Both flow northward as narrow streams along the western edges of the Atlantic and Pacific Oceans, respectively. The typical width of these currents is 50–75 km and typical speed is 1–3 m/s. There are a few factors which contribute to an intensification of current flow along the western edge of ocean basins. Rotation of the Earth from West to East results in sea water 'lagging' behind the rotating solid ocean bottom. This pushes water against the western side of a basin. Additionally, an equatorial drift induced by the Trade Winds piles large quantities of water against the eastern margins of the continents. The resulting mound of water produces a steep, poleward pressure gradient causing strong currents away from the equator.

The western arms of these gyres contrast with the eastern arms, namely the **Canary Current** in the Atlantic Ocean, and the **California Current** in the Pacific Ocean. Both these relatively sluggish currents (their speed is less than 0.3 m/s) are several thousand kilometres wide. In the Northern Hemisphere, the southern section of the gyre is known as the **North Equatorial Current** and its northern section is the **North Atlantic Current** (in the Atlantic Ocean) and the **North Pacific Current** (in the Pacific Ocean). The pattern of the northern sections of the circulation gyres are complicated due to the interaction with currents flowing out of the polar seas. In the North Atlantic, these currents are the **Labrador Current** and **East Greenland Current**, west and east of Greenland, respectively. In the Pacific Ocean, the northern section of gyre is joined by the **Oyasho Current** in the Bering Sea.

Similar gyres in the Southern Hemisphere are the **Brazil Current**, at the western edges of the South Atlantic basin, and the **Benguela Current**, at the eastern edge of the basin, with the **South Equatorial Current** at the low latitudes. The **Brazil Current** is weaker than the Gulf Stream because part of the South Equatorial Current is deflected by the South American continent into the Northern Hemisphere.

The South Pacific Gyre includes the **South Equatorial Current**, **East Australian Current** and **Peru Current**. The Indian Ocean Gyre is less stable than the persistent gyres of the North Atlantic and North Pacific; it varies seasonally in response to the reversal of the monsoon winds. The southern section of the gyres in the Southern Hemisphere forms the **Antarctic Circumpolar Current** circulating around the Antarctic Continent.

On both sides of the Equator, the Trade Winds generate the **North Equatorial Current** and **South Equatorial Current** which are deflected poleward by land masses. Due to complexities of the wind between the Trade Wind belts, there is another surface, eastward-flowing current known as the **Equatorial Countercurrent** and a subsurface eastward-flowing current, the **Equatorial**

Undercurrent (known also as the **Cromwell Current** in the Pacific Ocean); both currents are weak and change locations with the seasons.

The global circulation of surface currents, consisting of water moving in large gyres, is geostrophic in character, *i.e.* pressure-gradient forces balance the Coriolis effects. However, it is not a perfectly closed system and the volume of water circulating in the gyres is not constant. A classic example is the water removed from the North Atlantic gyre due to the separation of the Gulf Stream from the main flow. This removed water must be replaced by a flow from another source, presumably by some upwelling mechanisms.

Modern observation techniques, such as satellite monitoring of the sea surface, free-floating and bottom anchored buoys, and sophisticated research vessels have revealed the very complicated structure of the main circulation gyres. In particular, the presence of large mesoscale eddies (with diameters of 200–400 km) and rings (of 100–300 km in diameter and $\leq$ 3 km thick), drifting slowly both with and against main current have been detected. It was shown that mesoscale eddies may contain a substantial portion of the kinetic energy of the oceans. The significance of these structures for the ocean life is enormous and we will discuss this subject in Part III.

The volume of water transported by the large ocean currents usually is measured in Sverdrup units, Sv, named in honour of Harald Sverdrup. One Sv is equal to 10^6 m^3/s. Typical values of volume flow are: Pacific Equatorial Current 10–70 Sv, Gulf Stream 50–150 Sv, and Circumpolar Current up to 290 Sv.

Both wind driving and the effects of density changes are important for the overall circulation, but the former probably dominates in the upper 1000 m in most regions of the ocean. We will consider wind-driven flows in next section, while density effects will be discussed in Sect. 7.4.

7.3.4 Types of Surface Current Flows

Until now we have discussed the spatial patterns of the surface currents, neglecting their vertical structure. As was explained in Chap. 1, once a water particle is set in motion, it exerts frictional drag on the particles beneath. This process continues, slowly transferring the wind's momentum down into the water column. One of the patterns of such a transfer is the Ekman's spiral.

Ekman's Spiral Flow. During the Norwegian North Polar Expedition in 1893–1896, on the famous research vessel *Fram*, Fridtjof Nansen observed that the drift of ice with respect to the wind did not follow the wind direction, but deviated to the right by 20° to 40°, when looking in the direction of the wind. On Nansen's suggestion, the Swedish physicist V. Walfried Ekman (1905) investigated the problem mathematically and laid the foundation for one of the most important theoretical developments in dynamic oceanography. The physical reasoning behind Ekman's solution is as follows. Consider the water column as composed of infinitesimally thin horizontal layers and assume that

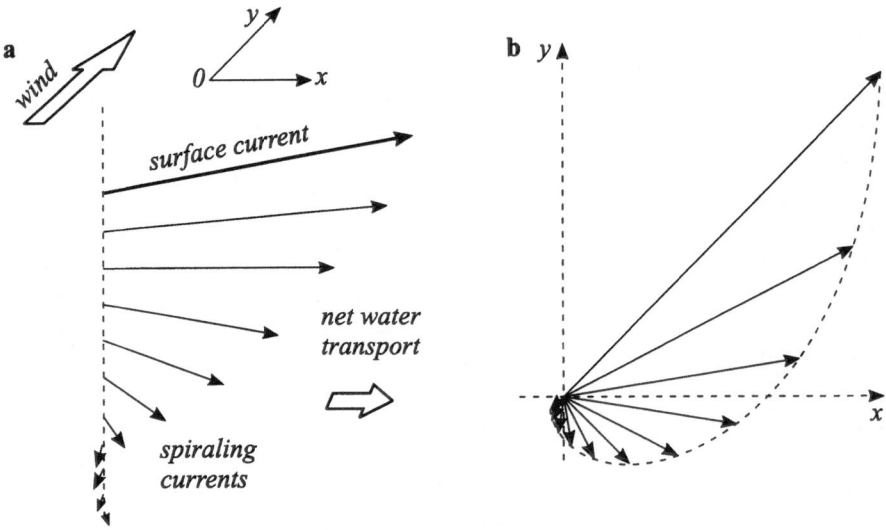

Fig. 7.7: Ekman's transport: **a** perspective view of spiralling current in the Northern Hemisphere, **b** plan view of the Ekman's spiral

each layer has a constant velocity. However, when the wind-driven current deepens into the flow, its speed diminishes from layer to layer.

In the Northern Hemisphere, the Coriolis effect causes the surface current to flow to the right of the generating wind. Each water layer is deflected slightly to the right of the layer above it, producing a spiralling current; this spiral is known as Ekman's spiral (Fig. 7.7). To determine the exact direction of current to the wind, it is necessary to apply a quantitative argument, as Ekman did (1905). He assumed a steady wind, blowing for a long time, so $du/dt = dv/dt = dw/dt = 0$. Moreover, he considered the case of a barotropic condition and no geostrophic flow, when the water is treated as homogeneous, with no slope at the surface, *i.e.* $\partial p/\partial x = \partial p/\partial y = 0$. Thus, Eq. (7.12) becomes (the third equation is simply the hydrostatic equation and can be omitted in further calculations):

$$\left. \begin{array}{l} \rho_w f v + F_x = 0 \\ -\rho_w f u + F_y = 0 \end{array} \right\}, \tag{7.14}$$

in which F_x and F_y are the frictional forces within the water. In Chap. 1 we showed that the friction stress, τ, on the plane parallel to the flow is given by:

$$\tau = \mu \frac{\partial u}{\partial z} = \rho_w \nu \frac{\partial u}{\partial z}, \tag{7.15}$$

in which ν is the coefficient of the kinematic viscosity which is applied only to water in smooth laminar flows with small Reynolds numbers (see Sects. 2.4 and 2.5).

In the ocean, where the motion is generally turbulent, the coefficient of kinematic viscosity is replaced by the coefficients of turbulent viscosity (known also as eddy viscosity coefficients) A_x, A_y, and A_z, for the x, y and z directions, respectively (see also Sect. 1.2). Thus, the eddy friction stress:

$$\tau = \rho_w A_z \left(\frac{\partial u}{\partial z}\right),$$

(7.16)

expresses the force of one layer of fluid on an area of its neighbour above or below. To substitute in Eq. (7.14), we need an expression for the forces F_x and F_y per unit of mass of fluid. It can be shown that these forces can be written as follows:

$$\left. \begin{aligned} F_x &= \frac{\partial \tau_x}{\partial z} = \frac{\partial}{\partial z}\left(\rho_w A_z \frac{\partial u}{\partial z}\right) = \rho_w A_z \frac{\partial^2 u}{\partial z^2} \\ F_y &= \frac{\partial \tau_y}{\partial z} = \rho_w A_z \frac{\partial^2 v}{\partial z^2} \end{aligned} \right\}.$$

(7.17)

As very little information on the variation of A_z with depth is available, this value is assumed to be constant. After substitution of Eq. (7.17) into Eq. (7.14) we have:

$$\left. \begin{aligned} fv + A_z \frac{\partial^2 u}{\partial z^2} &= 0 \\ -fu + A_z \frac{\partial^2 v}{\partial z^2} &= 0 \end{aligned} \right\}.$$

(7.18)

For simplicity let us assume that the wind blows in the y direction (Fig. 7.7). Then, the solution to Ekman's equations is (Pond and Pickard, 1983):

$$\left. \begin{aligned} u(z) &= \pm V_0 \exp\left(\frac{\pi z}{h_E}\right) \cos\left(\frac{\pi z}{h_E} + \frac{\pi}{4}\right) \\ v(z) &= \pm V_0 \exp\left(\frac{\pi z}{h_E}\right) \sin\left(\frac{\pi z}{h_E} + \frac{\pi}{4}\right) \end{aligned} \right\},$$

(7.19)

where V_0 is the total Ekman's surface current. Usually the depth h_E is arbitrarily taken as the effective depth of the wind-driven current and it is known as the **Ekman's layer**. At a level $z = -h_E$ the speed falls to $\exp(-\pi) = 0.04$ of that at the surface.

Strictly speaking, the depth h_E is a function of eddy viscosity coefficient, A_z, and the Coriolis parameter, f, i.e. (Pond and Pickard, 1983):

$$h_E = \pi \left(\frac{2A_z}{|f|}\right)^{1/2}.$$

(7.20)

Very often the depth, h_E, is compared to the depth of the upper mixed layer. However, it should be noted that the mixed layer depth is the result of long-term history of the wind action in a given area. On the other hand, the Ekman's depth is influenced much more by even short periods of strong winds at the time of observation, and it should be much smaller than the mixed-layer depth.

The total Ekman's surface current velocity depends on the wind stress at the surface through the expression:

$$V_0 = \frac{\sqrt{2}\pi\tau_0}{h_E\rho_w |f|},$$ (7.21)

in which τ_0 is given by Eq. (7.9). When $C_{10} \approx 1.4 \times 10^{-3}$ and $V_0/V_{10} \sim (0.013-0.030)$, the Ekman's depth, h_E, is typically between 200 m (for $V_{10} = 20$ m/s and $\phi = 10°$) and 45 m (for $V_{10} = 10$ m/s and $\phi = 80°$).

Equation (7.19) indicates that at the sea surface ($z = 0$), the current speeds are:

$$\left.\begin{aligned} u(0) &= \pm V_0 \cos\left(\frac{\pi}{4}\right) \\ v(0) &= \pm V_0 \sin\left(\frac{\pi}{4}\right) \end{aligned}\right\}.$$ (7.22)

This means that the surface current flows at 45° to the right (left) of the wind direction in the Northern (Southern) Hemisphere.

At the depth $z = -h_E$, which is given by Eq. (7.20), it is seen that:

$$\left.\begin{aligned} u(-h_E) &= \pm V_0 \exp(-\pi) \cos\left(-\frac{3\pi}{4}\right) \\ v(-h_E) &= V_0 \exp(-\pi) \sin\left(-\frac{3\pi}{4}\right) \end{aligned}\right\}.$$ (7.23)

This shows that at $z = -h_E$ the current vector has decreased to the value $\exp(-\pi)$ times the surface speed ($\sim 0.043U_0$), and that the current direction is exactly opposite to the surface current direction.

The Ekman's spiral is illustrated in Fig. 7.7 for the Northern Hemispheres. Figure 7.7a shows the perspective view of the current velocity decreasing and rotating clockwise with increasing depth, and Fig. 7.7b gives the plan view of velocities at equal depth intervals ($\Delta z = h_E/10$) and the Ekman's spiral.

Of special importance is the total vertically integrated horizontal water volume or mass transport in wind-driven currents. The total mass transport is obtained by integrating Eq. (7.19) over the depth between $z = 0$ and $z = -2h_E$, after multiplication by the constant water density, ρ_w. The lower level was chosen deep enough such that the influence of the wind driven current would be essentially zero i.e. $z = -2h_E$ where the speed is $\exp(-2\pi) = 0.002$ of that at the surface. If M_x and M_y are the x and y components of the total mass transport vector, across a section of unit width, we obtain:

$$\left.\begin{array}{rcl} M_x & = & \rho_w \displaystyle\int_{-2h_E}^{0} u(z)dz = \dfrac{\tau_0}{f} = \rho_w \dfrac{V_0 h_E}{\sqrt{2\pi}} \\[3mm] M_y & = & \rho_w \displaystyle\int_{-2h_E}^{0} v(z)dz = 0. \end{array}\right\} \tag{7.24}$$

It should be kept in mind that the wind blows in the y direction. This remarkable result of Eq. (7.24) follows that the total transport in a wind-driven current is directed 90° *cum sole* to the wind direction (see Fig. 7.7a). Therefore, in the Northern Hemisphere the total transport is to the right, and in the Southern Hemisphere it is directed to the left when one faces in the direction toward which the wind is blowing. It is interesting to note that the total transport is proportional to the wind stress and independent of the value of the turbulent viscosity coefficient in the water. Thus, the total transport is correct even if the details of the Ekman's spiral are not, because assumptions on which it is based (a steady wind, an infinite and homogeneous ocean, and no other forces acting) are somewhat unrealistic. However, observations of the surface current away from the land have shown speeds and deviations of the surface current similar to those predicted by Ekman.

Ekman transport converges in some regions and diverges in others, resulting in a vertical flow at the bottom of the surface boundary layer. This flow replaces or removes the converging or diverging water mass. The mechanism of generation of the flow through vertical movement into and out of the surface layer is known as Ekman pumping (Tomczak and Godfrey, 1994).

For Ekman's solution to be true requires that in the Northern Hemisphere there must be an inflow from the left of the wind direction to replace the flow to the right. This requirement is usually satisfied very far from the coastline. However, when the wind blows parallel to the coastline which is on the left of the wind (in the Northern Hemisphere), the Ekman's layer is skimmed away from the coast and flow from below the surface must replace it. This phenomenon is known as **coastal upwelling**. It usually occurs along eastern coasts of the ocean basins. In the Southern Hemisphere, the transport is to the left of the wind, so wind must blow in the northerly direction for upwelling to occur. In other words, we can say that upwelling occurs when the wind blows equatorward along the eastern boundary of an ocean in either Hemisphere.

A **downwelling** phenomenon produces the opposite effect whereby water converging on a coast is forced downward, carrying warm surface water to the ocean's depth. Such a situation occurs when local winds initiate Ekman's transport, which causes water to impinge on the western continental edges of ocean basins in both Hemispheres and the water is forced to sink.

Sverdrup extended Ekman's theory of wind-induced currents by retaining the pressure terms in Eq. (7.12), but he abandoned any attempt to determine the details of the vertical structure of current velocities u and v. Instead, he developed relationships for the total transport in the x and y directions in the whole layer affected by wind action (Kamenkovitch, 1978; Pond and Pickard, 1983; Whitehead, 1995):

$$\beta M_y = \text{curl}_z \tau. \qquad (7.25)$$

Equation (7.25) is known as the Sverdrup equation in which M_y is the total transport in the y direction, τ is the wind stress at the sea surface, $\beta = \partial f / \partial y$, and:

$$\text{curl}_z \tau = \frac{\partial \tau_y}{\partial x} - \frac{\partial \tau_x}{\partial y}, \qquad (7.26)$$

is the vertical component of the curl of the wind stress (for definition of curl, see Appendix C).

Sverdrup's extension of the Ekman theory gains the possibility of having a coastal boundary at one side of the ocean, which is a step towards a more realistic situation compared to Ekman's boundless ocean. The details of Sverdrup's theory are beyond the scope of this book. The interested reader should consult some of Sverdrup's papers (Sverdrup, 1945, 1947) or Pond and Pickard (1983) for an in-depth discussion.

Langmuir Circulation. Persistent winds blowing across the sea surface may induce small-scale vertical water motion. Langmuir (1938) observed that when

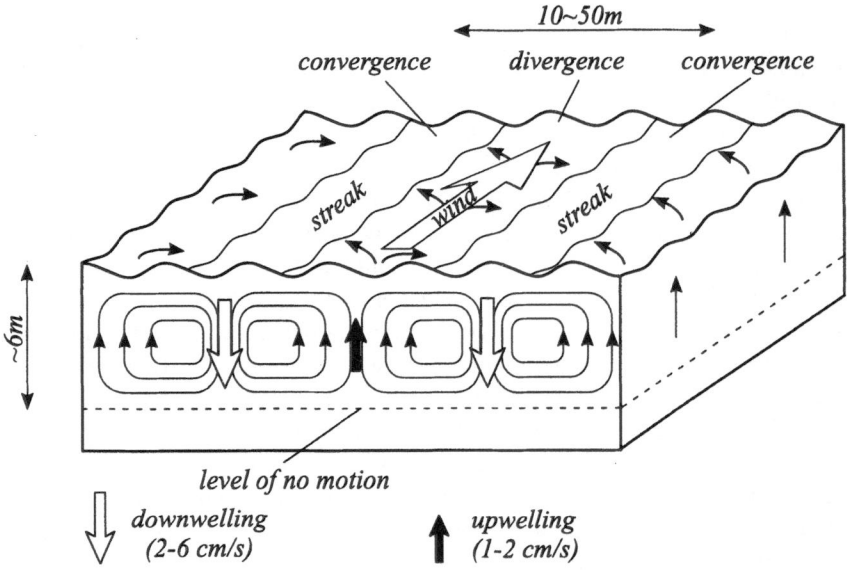

Fig. 7.8: Langmuir circulation (adapted from Pinet, 1992)

wind is stronger than $\sim$ 3.5 m/s, a counterrotating, convective water flow pattern of helical vortices is set up parallel to the wind. The lines are typically spaced about 100 to 200 m apart and sometimes are as much as 500 m long. This wind-driven flow pattern is known as **Langmuir circulation**. Each convection cell is 10–50 m broad and 5–6 m deep (Fig. 7.8).

The formation of the helical vortices is due to thermal instability in the surface water. As the counterrotation of adjacent cells exists, floating material, such as oil-surface films, seaweeds or bubbles, aggregates at zones where currents converge. The Langmuir circulation is a short-term response of the water surface layer to wind stress, and wind action for only a short duration (several minutes or hour) is needed to generate it.

Circulation in Tropical Oceans. In Sect. 7.3.3 we mentioned that the Trade Winds blowing towards the Equator induce the western-flowing North Equatorial Current and South Equatorial Current. Between both Trade Wind systems exists a zone of weak variable wind, known as the **doldrums**. In this region, a vigorous **Equatorial Countercurrent** flows in opposition to the westward North and South Equatorial Currents. The current extends large distances; for example, in the Pacific from the Philippine Islands to Columbia in South America, a distance of almost 15,000 km. Its width varies seasonally and reaches 400–500 km during the northern summer. In the Atlantic Ocean, the countercurrent is even stronger (Pinet, 1992).

Sverdrup's theory, mentioned above, explains the existence of such equatorial current system consisting of two westward currents and an eastward current. In the Pacific, this system is not symmetrical about the Equator but is moved northward. This fact corresponds to a similar displacement of the Trade Winds.

In the Atlantic Ocean, the current distribution is similar. However, in the Indian Ocean the wind pattern changes seasonally causing the seasonal changes of the current. Due to monsoonal activity over the ocean, there is only a two-current system: the wind-driven Southwest Monsoon Current to the east, crossing the Equator, and the South Equatorial Current to the west, well south of the Equator.

In the early 1950s, oceanographic surveys in the tropical Pacific established the existence of an undercurrent beneath the surface, flowing from West to East, known as the **Equatorial Undercurrent**. Subsequent experiments have verified the worldwide existence of this current. Maximum speed of the Equatorial Undercurrent is about 1.5 m/s or more, at depths rising from 200 m in the West to 50 m or less in the East. The dimensions are quite remarkable, with a length of 10,000 km, width of 400 km, and 200–300 m thickness. The annual average transport is estimated at about 40 Sv, with maximum transport up to 70 Sv.

A similar current exists in the Atlantic and there is also evidence for an equatorial undercurrent in the Indian Ocean during the northwest monsoon (Pickard and Emery, 1982). Field experiments and theoretical analysis have shown that the Equatorial Undercurrent is a geostrophic current with the pressure-gradient force balanced by the Coriolis type force. The reader wanting more details can consult papers of Knauss (1961) and Leetmaa *et al.* (1981).

7.3.5 Currents in Non-Homogeneous Ocean

As was shown in Chap. 6, the density of the water in the oceans increases with depth. Only in some exceptional cases, especially near the sea surface, can the density of the water slightly decrease with depth before the usual increase begins. In such stratified ocean, an important relationship exists between the mass distribution and the relative velocity distribution in the vertical direction. For simplicity, let us assume that a geostrophic current flows in the y direction and the density is a function of the z coordinate, $\rho_w = \rho_w(z)$. Thus, the balance of forces (7.12) becomes:

$$\left.\begin{array}{rcl} \rho_w(z)fv & = & \dfrac{\partial p}{\partial x} \\[2mm] -\rho_w(z)g & = & \dfrac{\partial p}{\partial z} \end{array}\right\}. \tag{7.27}$$

After combining the equations above, we obtain:

$$\frac{\partial v}{\partial z} = -\frac{g}{\rho_w f}\frac{\partial \rho_w}{\partial x} - \frac{v}{f}\frac{\partial \rho_w}{\partial z}. \tag{7.28}$$

This equation shows that in a geostrophic current, the vertical velocity gradient depends on the density gradients in lateral and vertical directions. However, in most practical cases, the last term on the right-hand side of the equation is negligible small. Thus:

$$\frac{\partial v}{\partial z} \approx -\frac{g}{\rho_w f}\frac{\partial \rho_w}{\partial x}. \tag{7.29}$$

This simplification expresses the fact that in the Northern Hemisphere, the denser water is found to the left of the current if the current velocity decreases with depth. In the Southern Hemisphere, the directions are reversed. The continuous vertical stratification of the ocean strongly influences sea surface temperature in the upwelling regions. As was shown in Chap. 6, with continuous stratification it is possible for waves to propagate not only horizontally, but also vertically. Subsequently, waves can generate horizontal pressure gradients that drive currents. However, the vertical structure of these currents is different from the vertical structure of wind driven currents.

   At present the prediction of currents in the ocean, with continuously stratified water, is mostly based on computer modelling techniques. Two linear solutions for equations of motion on an equatorial plane for currents, induced by atmospheric forcing and density gradients, were described by Philander (1990). However, to solve the fully nonlinear equation it is necessary to apply the General Circulation Models (see Sect. 7.5).

7.4   Thermohaline Circulation

The wind-induced currents are limited to the uppermost levels of the ocean. Ocean waters lying beneath, although not directly affected by the wind, are in motion as well. The important causes for this motion are the differences in temperature and salinity, resulting from differential heating, evaporation and precipitation at the sea surface. All of these factors eventually affect sea water density. Vertical convection, circulation and mixing provide the mechanisms for distribution of sea water density differences from the sea surface into a deeper ocean. Without deep mixing, the ocean would turn, within a few thousand years, into a stagnant pool of cold salt water with equilibrium maintained locally by near-surface mixing. The horizontal density differences, and consequently horizontal pressure differences, generate subsurface currents known as **thermohaline circulation**. This circulation is a very slow process, and is difficult to study. On the other hand, it should be remembered that the deep ocean, where the thermohaline circulation is dominant, contains about 90 percent of the total volume of ocean water.

From the Equator to approximately 50 degrees latitude there is a surface layer of water, less than 1 km deep, known as the 'thermocline', which has temperatures ranging from 5°C up to 28°C. The salinity in the thermocline also varies, and is highest in the tropics due to high rates of evaporation, and is much lower in temperate zones. In the rest of the ocean, temperature decreases with latitude and depth.

The deep and bottom water in the oceans apparently sinks and spreads away with a time scale of years to centuries, and eventually attains an equilibrium in terms of mass distribution. However, small volumes of sea water maintain their temperature and salinity identity even after they have travelled thousands of kilometres from their point of origin on the sea surface. This property of the deep-ocean water allows oceanographers to trace its lengthy path and determine its area of origin.

In general, sea water is very homogeneous in terms of temperature and salinity; over 75% of all ocean water has a temperature in the range of 0°–5°C and a salinity of 34–35 ppm. This means that deep oceans are filled with cold water originating from the polar latitudes. Usually in the ocean, four classes of water are distinguished: **central waters** which extend below the ocean surface down to the underside of the main thermocline, **intermediate waters** extending down to ~ 2 km, **deep waters** lying further down, and **bottom waters** which are very cold and very dense. At present, thermocline circulation is modelled as being a part of the ocean general circulation (Whitehead, 1995; McWilliams, 1996). Descriptions of water structure in particular ocean basins can be found elsewhere (for example, Pinet, 1992), and the energetics of tidal and wind mixing needed to maintain the global abyssal density distribution was recently discussed by Munk and Wunsch (1998).

7.5   Modelling of Oceanic General Circulation

First we will define the meaning of 'oceanic general circulation'. According to McWilliams (1996), the oceanic general circulation is a system which includes currents at horizontal space and time scales larger than the mesoscale (which is of the order of 100 km and three months), the associated pressure, density, temperature and salinity fields, plus all other elements involved in establishing the dynamical balance for these fields. Sometimes the general circulation also includes the biochemical processes. Various oceanic general circulation models (OGCM) have been developed relatively recently. A detailed description of the OGCM is beyond the scope of this book. Here we will only provide some basic characteristics of these models from the point of view of fluid mechanics and illustrate some model results.

The OGCM's are based on the Navier-Stokes equations (see Appendix C) for the rotating Earth and for sea water which comprises water and dissolved salts. In the vertical direction, the hydrostatic approximation is assumed when the vertical pressure gradient is totally determined by the vertical density profile. Until recently the common practice has been to assume the ocean has a 'rigid lid' at the surface. Thus no fluid is transferred across the surface, although momentum, heat, and chemical components are. Under such an assumption, surface waves are excluded and the dynamics of tides are distorted. The set of the momentum equations and the mass-conservation equation, together with equations for heat and salt conservation, plus equations of state used to determine the velocities u, v, w, the pressure p, etc. are known as the primitive equations (Mellor, 1991).

The most widely used ocean model is that developed in the 1960s, at the Geophysical Fluid Dynamics Laboratory of Princeton University, formerly a part of the Institute for Advanced Study (Bryan, 1969; McWilliams, 1996). Since then, a lot of improvements and modifications have been added to the model. In the high-resolution version, the grid spacing is 1° longitude and 0.3° latitude between 10°N and 10°S, and 10 m in the vertical in the upper 100 m of the ocean. The spacing increases poleward of 10° latitude and below the depth of 100 m. The initial conditions for the model are climatological oceanographic data (Levitus, 1982).

The major currents in the oceans which are at least qualitatively reproduced by the OGCM are (McWilliams, 1996):

- the eastward Antarctic Circumpolar Current,
- the westward, horizontally divergent surface currents in the equatorial Atlantic and Pacific, lying just above equatorial undercurrents,
- the Indonesian throughflow,
- the Somali gyre in the western tropical Indian Ocean (occurring mainly during northern-hemisphere summer),
- the subtropical gyres with western boundary currents in all basins,

- the subpolar gyres in the North Atlantic and Pacific basins, and
- the flow in the Greenland-Norwegian Sea.

The OGCM velocities computed for these currents are too weak and their patterns are too broad compared with reality. However, the total transport of the currents is similar to that observed in the oceans. A discussion on other solution features can be found in a review article by McWilliams (1996).

Ocean models used in simulation of the Earth's climate do not resolve mesoscale eddies because of the computational cost. Therefore in the short term, eddy-resolving global ocean models cannot be used in climate simulations, even on modern supercomputers. On the other hand, energetic mesoscale eddies are important in the transport of heat, salt, and passive tracers such as radiocarbon and freon in the oceans. Their importance has been documented following observations in the Antarctic Circumpolar Current, equatorial Pacific and North Atlantic oceans. Thus, there is a need for parameterization of mesoscale eddies for climate models. Recently, such parameterization of the effects of these eddies, based on an adiabatic down-gradient diffusion of the thickness between neighbouring isopycnal surfaces has been proposed by Danabasoglu *et al.* (1994).

A very important motivation for developing ocean general circulation models is their use in coupled atmosphere-ocean models to study climate and its changes. More than fifty years ago, Sverdrup (1945) wrote: *It is not yet possible to deal with the system atmosphere-ocean as one unit, but it is obvious that, in treating separately the circulation of the atmosphere, a thorough consideration of the interaction between the atmosphere and the oceans is necessary.* The first simulation of climate with a combined ocean-atmosphere model was that discussed by Manabe (1969) and Bryan (1969). A nine-level atmospheric model, known as the GFDL (Geophysical Fluid Dynamics Laboratory) ocean-atmosphere model, was used to calculate values of atmospheric variables on a grid with space dimensions of about 500 km, and a five-level oceanic model (Bryan, 1969). In spite of many simplifying assumptions, the model accurately simulates seasonal patterns of rainfall in the tropics and associated wind fields. The locations of rainbelts and associated disturbances are determined primarily by the distribution of the sea surface temperature.

Another coupled atmosphere and ocean model was developed at the National Center for Atmospheric Research (NCAR), the United States (Washington *et al.*, 1980). This model links separate existing models of the atmosphere, ocean and sea ice. The atmospheric part of the model uses a generalized vertical coordinate with eight layers, each $\sim$ 3 km thick, and 5° horizontal grid spacing over the entire Earth. The ocean model is a modification of the GFDL model mentioned above, developed by Bryan (1969). In the sea ice part of the model, a simplified calculation of heat flux through sea ice is used. As the density of the atmosphere is about 1000 times smaller than that of sea water, the coupled atmosphere-ocean model should deal with two different time scales, for the atmosphere and the ocean environments. In the NCAR model, an

effective technique was used to overcome this problem. The simulated ocean
surface temperature pattern is reasonably similar to the observed pattern and
the major ocean current gyre systems are reproduced in the ocean model.

Since the International Geophysical Year of 1957–1958, numerical simulations
of the behaviour of the atmosphere-ocean system have been supplemented by
international projects aimed at understanding the role of the oceans in climate change. The number of various projects is quite numerous; we note
here some of them, such as GATE (Atlantic Tropical Experiment, 1974), NOR-
PAX (North Pacific Experiment, 1972–82), MODE (Mid-Ocean Dynamics Experiment, 1978), POLYMODE, CUEA (Coastal Upwelling Ecosystems Analysis Program, 1972–77), ERFEN (Estudio Regional del Fenomeno de El Niño,
1974), and TOGA (Tropical Ocean-Global Atmosphere Program, 1985–94).
The reader should consult Perry and Walker (1977), Pond and Pickard (1983),
Glantz (1996), and Anderson *et al.* (1998) for in-depth discussions of these
programs. The last three programs have been designed to provide a better
understanding of the El Niño phenomenon which has attracted an increasing
amount of attention from scientists and the world community as a whole. We
will discuss El Niño in the next section.

7.6   El Niño and Southern Oscillation

7.6.1   Introduction

Initially the term **El Niño**, which in Spanish means child, referring to Jesus as
an infant, described a local current which annually appears off the coast of Peru,
beginning around Christmas time and temporarily replaces the usually cold,
nutrient rich water which wells up to the ocean's euphotic zone. At present,
the term El Niño describes not only the local current off the coast of Peru, but
the infrequent *anomalous* warming of sea surface waters off the coast of Peru
and in the central equatorial Pacific Ocean and the associated changes in the
circulations of the Pacific and the global atmosphere when the Trade Winds
diminish. The term El Niño is also associated with ecological and economic
disasters, correlated with devastating droughts over the western tropical Pacific,
terrential floods over the Eastern Pacific, and unusual weather patterns over
various parts of the world. The term **La Niña** (child girl) is the opposite of
El Niño and is used to describe a time when sea surface temperatures in the
Central and Eastern Pacific are unusually low and when the Trade Winds are
very intense (Philander, 1990; Glantz, 1996).

Actually the El Niño event is linked with the Southern Oscillation and the
acronym ENSO is used to describe the much broader basin-wide events in the
equatorial Pacific. The term Southern Oscillation was first used by Walker in
1924 to describe the inter-annual pressure fluctuations over the Indian Ocean
and eastern tropical Pacific. Walker, who was Director General of the Ob-
servatories in India, observed that *when atmospheric pressure is high in the
Pacific Ocean it tends to be low in the Indian Ocean from Africa to Australia.*

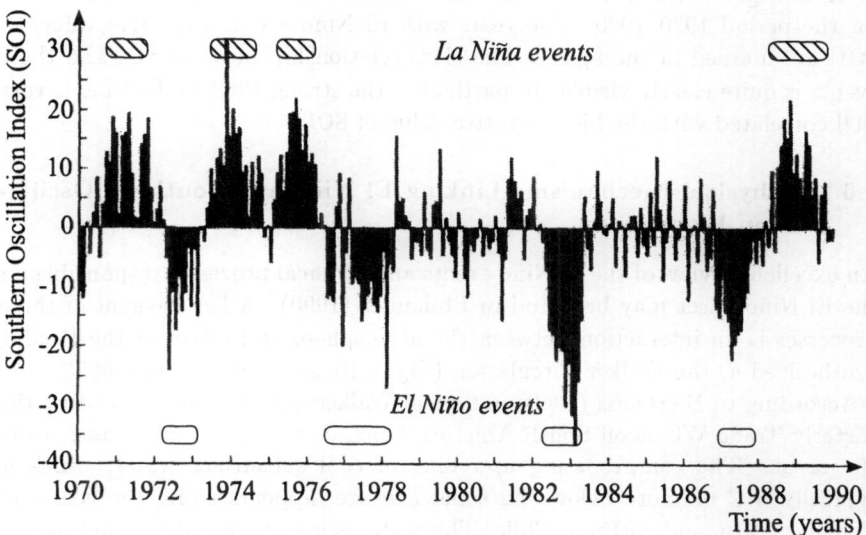

Fig. 7.9: Monthly averages of the Southern Oscillation Index (SOI) and El Niño (La Niña) events for period 1970–1990 (adapted from Glantz, 1996)

His observation was based on the pressure records from locations such as Darwin (Australia), Canton Island in the equatorial Central Pacific, and Santiago (Chile). At present, the Southern Oscillation is monitored using the differences between sea level pressure at Tahiti (French Polynesia) and Darwin (Australia). These differences are converted into an index known as the Southern Oscillations Index (SOI). Thus we have:

$$SOI = (p_a)_{\text{Tahiti}} - (p_a)_{\text{Darwin}}, \tag{7.30}$$

in which $(p_a)_{Tahiti}$ and $(p_a)_{Darwin}$ are the sea level pressures at Tahiti and Darwin, respectively. Usually, there is a low pressure system in the region of Indonesia and northern Australia, around Darwin, providing rains to the region. At the same time, there is a high pressure system in the southeastern Pacific, around Tahiti.

We note that in many publications the term El Niño is used interchangeably with the term El Niño-Southern Oscillation (ENSO), even within the same publication. However, in this book, for consistency only the term El Niño is used, which encompasses both a localized coastal warming of the sea surface in the Eastern Pacific and a much broader basin-wide processes of interaction between the atmosphere and the ocean.

Only in 1969, Bjerkness identified the physical mechanisms that linked El Niño and Southern Oscillation phenomena. He proposed a model, described in the next section, which has confirmed that there is a very strong correlation between observed El Niño events and the SOI. An example of such correlation is

given in Fig. 7.9, in which the time series of monthly averages of SOI are given
for the period 1970–1990. The years with El Niño events (negative values of
SOI) are marked in the figure. The close relationship between SOI and these
events is quite clearly visible. In particular, the strong 1982–85 El Niño is very
well correlated with the high negative value of SOI.

7.6.2 Physical Mechanisms Linking El Niño and Southern Oscillation Phenomena

An excellent review of the El Niño events and physical processes responsible for
the El Niño onset may be found in Philander (1990). A key element of these
processes is an interaction between the atmosphere and ocean in the Pacific,
synthesized as the Walker Circulation (Fig. 7.10, see colour plate p.565).

According to Bjerkness (1969), a typical Walker Circulation is forced by the
easterly Trade Winds off South America which drive surface water away from
the coast. The compensating upwelling of cool subsurface water, which is
typically 5°C or more below the zonal average, appears along the coasts of
Ecuador, Peru, and northern Chile. This water is maintained along the Equator
by the Coriolis force which is to the left in the Southern Hemisphere (coastal
upwelling is described in some detail in Sect. 7.7). Dry air from the Eastern
Pacific flows along the Equator as a part of the Trade Winds. On this journey,
air is warmed and moistened as it moves over the progressively warmer waters.
As the westward Trade Winds drive the warm surface water westward and
expose cold water to the surface in the east, the thermocline becomes deep in
the western and shallow in the eastern regions of the Pacific (see Fig. 7.11a).

The pressure gradient associated with the zonal slope of the thermocline balances the wind stress. At the western end of the equatorial Pacific Ocean
lies the pool of the warmest ocean water on Earth, which provides for sustained upward motion of air and rain clouds. The return flow in the upper
troposphere closes the typical Walker Circulation (Fig. 7.10a, see colour plate
p.565). However, occasionally the Eastern Pacific warms up, the Walker
Circulation weakens and causes the convective zone of heavy rainfall to move
eastward, into the central and eastern tropical Pacific (Fig. 7.10b, see colour
plate p.565), and an El Niño cycle commences.

All El Niño events are different, but historical data provides a composite
picture of the canonical event. This composite is based on the fact that many
aspects of El Niño are closely linked to the annual cycle. The main phases of
the canonical El Niño can be summarized as follows (Cane, 1984):

- *Prelude*: The easterly winds, stronger than average, appear in the western
  equatorial Pacific for at least 18 months before an El Niño event and move
  water toward the west. Sea lever is unusually high in the west and low
  in the east. This results in deepening of the thermocline in the west and
  warming sea surface temperature (SST) above average in the far west.
- *Onset*: In the fall of the year proceeding an El Niño, a warm SST anomaly
  extends across the South Pacific between 15°S and 30°S. In September

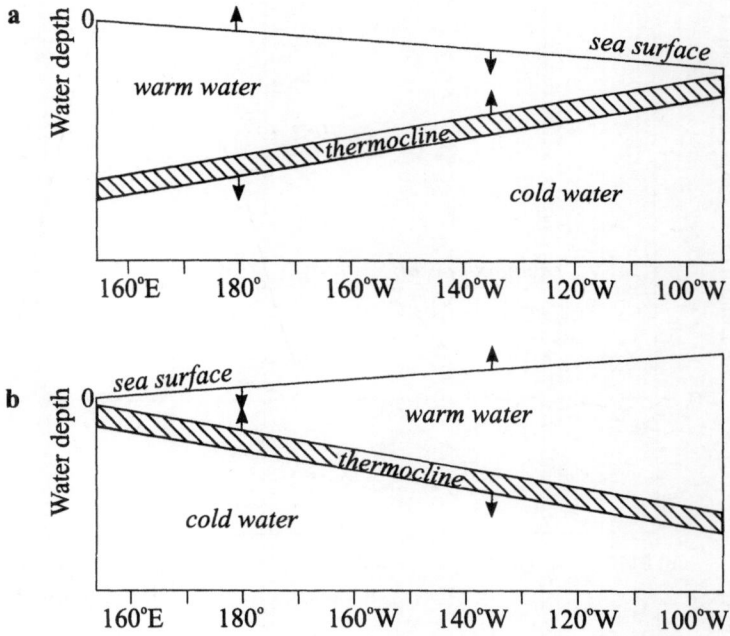

Fig. 7.11: Oscillations of sea level and thermocline: **a** strong Trade Winds (typical sea level slope), **b** weak Trade Winds (El Niño event); adapted from Pinet, 1992

and October, the easterlies begin to diminish along the Equator west of the date line, and the sea level slope along the Equator begins to relax.

- *Event*: Off the coast of South America temperature starts to rise in December or January and continues to build in magnitude till June. However, during the first several months it still is difficult to distinguish between an El Niño and normal seasonal warming. In April–June, the sea level rises in a narrow region along the South America coast and the thermocline becomes deeper (Fig. 7.11b). There is strong southward flow at the coast, and westerly wind anomalies appear along the Equator from 100°W to 170°E. During next 6 months after the peak SST at the coast, the warm anomaly spreads north-westward and then westward along the Equator, until the anomaly is in the Central Pacific.

- *Mature phase*: Another warming at the coast begins about December and reaches its peak early in the following year. However, the coastal SST anomaly drops off sharply and becomes even colder than normal by March. The colder water spreads westward from the coast and reaches the date line late in the year. Winds relax toward their normal pattern and the westward sea level slope is re-established.

Typical sea surface temperature anomalies during a canonical El Niño event are shown in Fig. 7.12a. These mechanisms involve complex ocean and

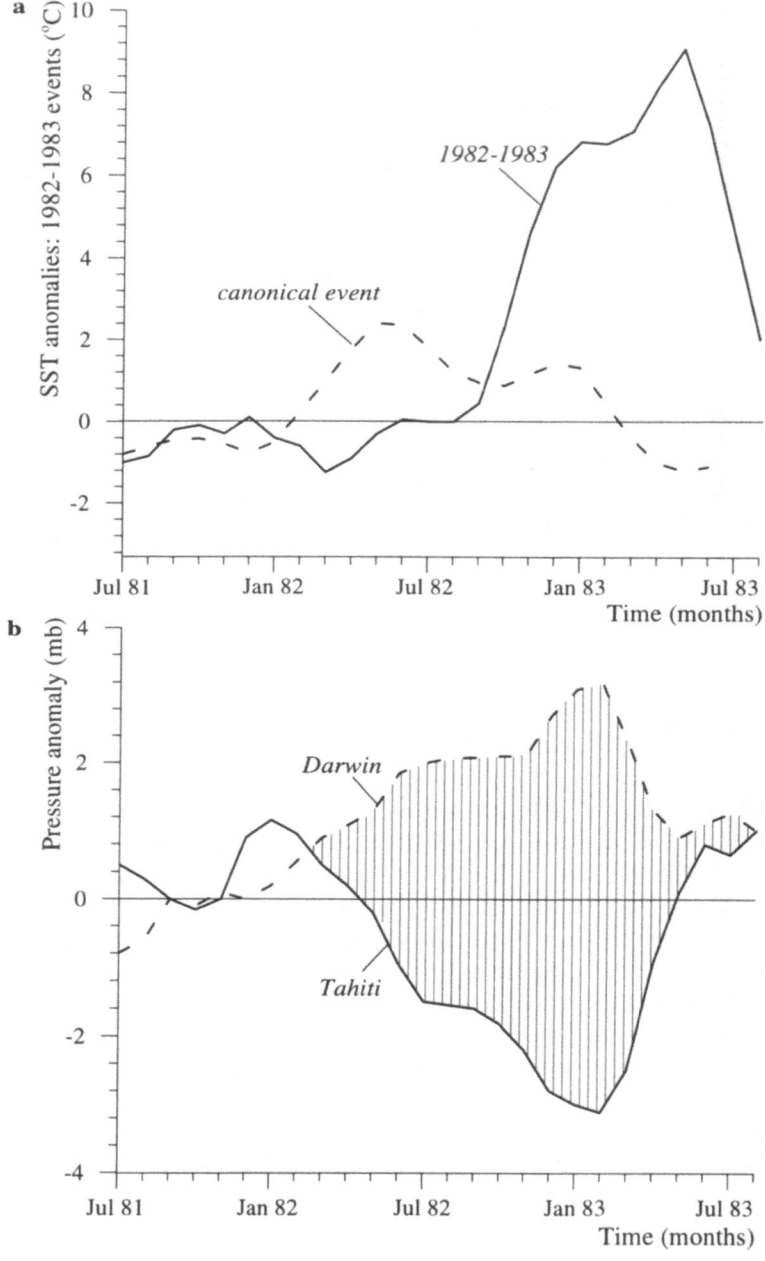

Fig. 7.12: Temperature and pressure anomalies during 1982–83 El Niño event: **a** sea surface temperature (SST) anomaly pattern at Puerto Chicama (7.7°S, 79.3°W) for the 82–83 event and for canonical event, **b** sea-level pressure at Tahiti and Darwin during event. The difference between pressure at Tahiti (solid line) and pressure at Darwin (dashed line) is equal to SOI (adapted from Rasmusson and Wallace, 1984)

atmosphere interactions, some of which are not well understood. However, it is recognized that the sea surface temperature gradients – the cold water off Peru and the warm water in the western tropical Pacific – are necessary for the atmospheric pressure gradients that drive the Walker Circulation.

As the processes in the Pacific Ocean and atmosphere and resulting interactions are quite complicated, there are suggestions to monitor the changes in sea surface temperatures in a specific areas, or the surface winds, or directly focus on the Southern Oscillation as the leading indicators of an El Niño event. In particular, four such regions in the equatorial Pacific have been identified (Glantz, 1996):

- *Region 1* – coastal upwelling area off the coast of Peru and Ecuador. This area is particularly sensitive to variations in air-sea interactions in the central and eastern equatorial Pacific.

- *Region 2* – the Galapagos Islands area is sensitive to seasonal, as well as to El Niño induced changes in the marine environment. These islands are normally dry for half of the year (from June to December) and wet from January to May. However, during El Niño heavy rains fall over the Galapagos. For example, during the 1982–83 event, the rainfall reached 3224 mm for the year, while the yearly average is about 200 mm. This causes flora and fauna changes on the islands.

- *Region 3* – central equatorial Pacific where there is a large El Niño signal in terms of changes in surface winds and warming of sea surface water.

- *Region 4* – western equatorial Pacific with its warm pool. During an El Niño event there is a change in sea surface temperature, with warm waters tending to move toward the Central Pacific.

A particularly comprehensive observing system has been developed as a result of the Tropical Ocean-Global Atmosphere (TOGA) experiment in 1985–1994 (McPhaden *et al.*, 1998). TOGA initiated and supported efforts to provide real-time measurements of the key oceanographic variables such as surface winds, sea surface temperature, subsurface temperature, sea level and ocean current velocity. More details on the TOGA program are given in Sect. 7.6.4 and in a set of review articles edited by Anderson *et al.* (1998).

7.6.3 El Niño of 1982–1983

Having defined an El Niño event, and identified its major oceanic and atmospheric characteristics, the historical records show that similar events were observed in the past, in 1957-1958, 1965, 1969, 1972, 1976, and 1982–1983. In the last decade, two or more El Niño events took place: in 1986–1987, 1991–1995 and 1996–1997. However, no major objective accounting of the physical, social and economic impacts and consequences of these events has yet been undertaken.

The El Niño of 1982–1983 was exceptional because of the very large amplitude it attained and because of the unusual way in which it evolved. None of the

usual precursors of an El Niño event were present. The easterlies in the west had
not been especially strong and there was no tendency for SST to be unusually
low in the east and high in the west, and the thermocline was not unusually
deep there. Anomalously warm sea surface temperature first appeared in the
Central Pacific, instead of off the coast of Peru. Subsequently, this warm sea
surface temperature migrated gradually eastward during most of 1982. These
conditions reached the coast of South America in late 1982, and westerly winds
prevailed over much of the western and central equatorial Pacific. Anomalies
reached their peak amplitude in February 1983, and started to attenuate when
sea level pressure at Darwin began to decline, and the westerly surface winds
in the western tropical Pacific disappeared.

In Fig. 7.12a, the sea surface temperature (SST) anomaly pattern for the
1982–1983 El Niño is shown. This pattern is different from the canonical
pattern based on previous El Niño events, observed between 1950 and 1975
(Rasmusson and Wallace, 1984), in terms of the time of El Niño onset and
the distribution of temperature anomalies. Figure 7.12b shows the corresponding sea-level pressure anomalies at Tahiti and Darwin during the 1982–1983
episode. The difference between two pressures, indicated by shading, represents the Southern Oscillation Index (SOI) – see Eq. (7.30). Because of its
unusual pattern, different from the canonical event, El Niño 1982–1983 was
unexpected by most scientists.

Most of the weather anomalies occurring in 1982 and 1983 around the world
were linked with the occurrence of El Niño. Compilation of the damage hypo-
thetically associated with the 1982–1983 El Niño has been published by the National Oceanic and Atmospheric Administration of the United States (Glantz,
1996). They include severe droughts in Indonesia, Australia, and north central
states of the United States, devastating coastal storms and mud slides along the
southern California coast, sharp decline of salmon harvests along the northwest
coast of the Pacific, severe infrastructure damage in Peru and Ecuador as the
result of heavy flooding, and a major drought in the West African Sahel.

Similar connections have also been observed in the past for other El Niño
events. A special word, 'teleconnections', is used to describe the linkages between these climate anomalies some distance from each other. It was apparently
first used in 1935 by a Swedish meteorologist, Anders Angström, in his article
on the North Atlantic region. Later, Flohn and Fleer (1975) published a chart
illustrating alleged teleconnections, suggesting regional climate connections,
sometimes at very great distances. In particular, they showed the relations between El Niño and drought in North-East Brazil, water level changes (decline)
of Lake Chad, runoff of the Nile River, and drought in India.

Although in Australia droughts have occurred in the past, and their link
with variations in sea level pressure across the Pacific have been known for
many years, the impact of 1982–1983 El Niño was devastating: bush fires, dust
storms, agricultural and livestock losses. Droughts turned apocalyptic when
bush fires erupted across southeastern Australia in February 1983. Fireballs

roared through woodlands near Melbourne, and volatile eucalyptus trees ignited explosively. In all, 75 people lost their lives, 8000 became homeless, and a million acres of forest and farmland were destroyed. By the time rains came again in March 1983, the Australian agricultural economy had shriveled. Farm income fell by nearly half, and farm related losses totaled two and a half billion dollars (Canby, 1984).

Since 1983, Australian meteorologists have been active in searching for possible linkages between El Niño, especially the SOI, and changes in the frequency, location, and intensity of tropical cyclone activity in the eastern Australian region, droughts, low river flow in the Darling River, outbreaks of mosquito-borne Murray Valley encephalitis in southern Australia, and other phenomena (Nicholls, 1986, 1991). Other aspects of teleconnections in large-scale features of atmospheric circulation, especially those related to El Niño, have recently been discussed by Trenberth *et al.* (1998) using the TOGA experiment evidence.

In the community's perception, an El Niño event is a 'bad time' and its impact on human activities is adverse. However, as was pointed out by Philander (1990), the years with El Niño were known in Peru as *años de abundancia* (the years of abundance), when *the sea is full of wonders, the land even more so. First of all desert becomes a garden... The soil is soaked by the heavy downpour, and within a few weeks the whole country is covered by abundant pasture. The natural increase of flocks is practically doubled and cotton can be grown in places where in other years vegetation seems impossible* (Philander, 1990). At that time, the shallow waters off Peru are full of yellow and black water snakes. However, the birds and marine life that usually are abundant temporarily disappear. Arntz (1984) also provided some examples of positive changes associated with the 1982–83 El Niño.

7.6.4   Observing and Forecasting El Niño

As was showed above, observing El Niño has a long history, starting from individual local observations to more coordinated field work in the 1970s and early 1980s. Apart from these achievements, an understanding of the physical processes responsible for the El Niño cycle is still limited. The real breakthrough was initiated by the recently completed TOGA experiment. A major accomplishment of TOGA was the development of an ocean observing system to support seasonal-to-inter annual climate studies. The main TOGA observing system components included *in situ* oceanographic measurements (an island and coastal tide gauge network, drifting buoys arrays, moored buoys to provide surface wind, SST and current measurements, volunteer observing ship and expendable bathythermograph program for upper ocean temperature profiles); satellite measurements (SST, sea level and wind measurements), and *in situ* meteorological measurements from an expanded World Weather Watch network (McPhaden *et al.*, 1998). For each type of measurement, specific resolution and accuracy requirements were established and by the end of TOGA it was found that the observing system met many of the data requirements in the equatorial Pacific Ocean between 8°N and 8°S. A detailed description of all instrumentation used during TOGA was given by McPhaden et al. (1998). Fortunately, TOGA spanned a decade in which there was both a large swing from El Niño to La Niña conditions (1991–1995). Thus, the observing system contributed to scientific progress in modelling studies of short-term climate variability during these TOGA phases.

An excellent overview of the process of forecasting El Niño, not only from the scientific perspective, can be found in Glantz (1996) and Anderson et al. (1998). Here, we only concentrate on the possibility of predicting El Niño from the geophysical point of view. Studies with coupled atmosphere-ocean models indicate that the interactions between atmosphere and ocean manifest themselves in terms of modes which are inter-annual fluctuations between warm El Niño and cold La Niña, and hence correspond to possible Southern Oscillations. These periodic oscillations are predictable, provided there are no high-frequency ('weather' type) disturbances which are not correlated with sea surface temperature changes. The impact of the high frequency components results in irregular, inter-annual Southern Oscillations.

As the Southern Oscillation is a low frequency event, it will therefore be possible to make predictions by extrapolating this low-frequency trend. The success of such extrapolation depends on the correlation between various Southern Oscillation indices. For example, it was discovered by Walker that sea surface temperature in the central equatorial Pacific during the northern summer, and rainfall variations over India during those months, are highly correlated and can be used as predictors for subsequent developments (Philander, 1990). Another example of the correlation of various events is the seasonal pressure anomaly at Darwin and rainfall in India. If during March, April and May, the anomaly is above normal and is increasing, then heavy rains over India in June, July and August are unlikely. Also, when the pressure is below normal and is falling, droughts are unlikely. Advanced statistical techniques have recently been used to identify the principal modes of climate variability on different space and time scales (Latif et al. 1998).

Barnett et al. (1988), applying canonical correlation analysis developed a method to predict the El Niño event of 1986 at lead times of 3 to 9 months. Long-term predictions for periods of a few years indicate that if an El Niño occurs in a certain year, then it is highly likely that it will appear again 3 to 4 years later. However, only the TOGA experiment has provided more reliable experimental data on the atmosphere and ocean circulations, and inspired the development of a hierarchy of various El Niño prediction schemes. Besides the different explanation of nature of the El Niño, offered by those schemes, it is now commonly believed that El Niño is a fundamental oscillatory mode of the coupled atmosphere-ocean system and the memory of the coupled system resides in the ocean thermocline state (Latif et al., 1998). According to the 'delayed action oscillator' scenario, the easterly winds over the Western Pacific, prevailing during the cold phase of the El Niño, force an upwelling which, in the form of a Kelvin wave packet, propagates eastward along the Equator. This causes cooling at the sea surface in the Eastern Pacific and shallowing of the thermocline. At the same time, the ocean response in the west to the easterly winds induces a downwelling Rossby wave packet which propagates westward (for a description of Rossby waves see Sect. 3.9). The Rossby wave is reflected at the western boundary and propagates eastward in the form of downwelling Kelvin waves, which affects the SST. The delay of propagation of the Rossby waves from the wind stress region to the western boundary, and the return as a reflected Kelvin waves to the eastern basin, provide the memory for the oscillation. A positive SST anomaly develops, growing due to unstable air-sea interactions in the warm phase of El Niño. Thereafter the sequence of events repeats itself but with a reversed sign, every 3 to 5 years. Thus, in this sense El Niño is a low-frequency basin-wide mode of oscillation (Neelin *et al.* 1998).

The hierarchy of El Niño prediction models also includes more sophisticated coupled ocean-atmosphere models. The simplest coupled models designed for predictions of an El Niño event are the limited area models. Zebiak and Cane (1987) developed a coupled atmosphere-ocean model to reproduce certain key features of the observed El Niño phenomenon. The model calculates perturbations about the climatological mean state that are specified from observations. In particular, the model produces recurring warm events that are irregular in both amplitude and spacing, but favour a 3–4 year period. The predicted warm events, which include equatorial westerly wind anomalies in the Central Pacific and large SST anomalies in the Eastern Pacific, are in general agreement with observations. Zebiak and Cane used their model to successfully forecast the 1986–87 El Niño.

Another class of ocean-atmosphere models consists of physical ocean models coupled to empirically derived atmosphere models. They are known as hybrid coupled models. The hybrid coupled models generally yield useful El Niño predictions for lead times up to about 1 year. Thus, the anomaly correlation coefficient between observed and predicted SST anomalies, averaged over the El Niño 3 region (see above) yields values above 0.5 at lead times of 1 year.

The most complex models applied to El Niño forecasting are coupled ocean-atmosphere general circulation models. Their detailed description is beyond the scope of this book and the reader should consult Anderson *et al.* (1998) for an in-depth review. All these models are still idealized and have large numbers of adjustable parameters. The efforts to improve the models and collect necessary experimental evidence are continuing (see for example Godfrey *et al.* 1998) and a Coupled General Circulation Model (CGCM) of the ocean and atmosphere will probably be available before long, capable to anticipate future El Niño events.

However, at present we are in a better position to forecast El Niño event some time in advance. This was the case for the 1997 El Niño. In December 1996, normally westward blowing Trade Winds briefly reversed direction in the far Western Pacific. This change generated a deepening of the equatorial warm-water layer and caused the layer to spread eastward to South America, where it arrived by March 1997. The combination of deepening warm water and weakening Trade Winds resulted in a warming of the far eastern tropical Pacific, which was first noticeable in May, and by the year's end the warming episode had spread westward with temperatures of several degrees Celsius above normal in the central and eastern tropical Pacific. This caused an unusual drought with incidents of prolonged forest fires in the far Western Pacific.

El Niño has been a key player in the earth's climatic variability for a very long time, and has only recently become appreciated in scientific, economic, and social terms. Because this is a book on fluid mechanics and its ecological implications, I am limiting the discussion of the El Niño consequences to the biological ones only. This discussion is given in Chap. 15.

7.7   Continental Shelf Flow

7.7.1   Overview of Continental Shelf Waters

A continental shelf is an almost flat plain or terrace that borders a continent and slopes gently towards the ocean basin. The offshore boundary of the continental shelf is the **shelf break**, usually at a water depth of about 130–200 m. Continental shelf widths range from a few kilometres, along the Pacific coast of the North and South America, to more than 1000 km in the Arctic Ocean.

Seaward of the continental shelf break is a steeper **continental slope** of about 4°, continuing to a water depth of 2–3 km. At the toe of the continental slope, the ocean floor flattens with average gradient of 1°. This broad underwater plain of sediment is known as the **continental rise**. Some of the continental rises extend seaward more than 500 km to water depths approaching 4000 m.

As the continental shelves are situated between land and open ocean, they are influenced by both. Rivers transport freshwater and dissolved chemical components, including nutrients and mud, into shelf waters. These land-origin substances are subsequently mixed, dispersed by winds, waves, currents and tides. In general, there is a large variety of currents with time scales ranging from seconds (turbulence, wind-generated waves), to months and years (climatic variations).

Due to intensive mixing induced by waves and currents, there is efficient recycling of nutrients derived from microbiological decomposition of organic matter near the ocean floor. This nutrient regeneration is essential for plant growth and to provide abundant organic matter to support large fish stocks. In addition, cold nutrient-rich oceanic waters and associated materials intrude onto the shelf, promoting conditions favourable for higher productivity on the continental shelf than in the open ocean. Although the waters of the continental shelves compose only 10 per cent of the ocean area, they provide 90 per cent of the world's annual fish catch (Pinet, 1992).

The continental shelves are also used as a final destination for various types of anthropogenic materials, and all types of pollutants, such as sewage, heavy metals, petroleum compounds and toxic pesticides. The efficiency of removing these waste contaminants from their source is dependent on the mixing and dispersion due to wave and tidal action. These processes are discussed in detail in the next chapter. However, it should be noted that the capacity of the ocean to disperse anthropogenic contaminants is not boundless. There are cases, where due to alterations in water quality, the food web has been interrupted and local fisheries have collapsed.

At present, the tendency to consider the continental shelves as systems with interactive living and non-living elements prevails. There are various attempts to categorize continental shelves, depending on the factors which shape them. As far as water flows on the continental shelves are concerned, the most important factors are the location of the shelf on the ocean boundary (eastern or western boundary), presence or absence of large rivers, and shelf width. In the next section we illustrate the influence of these factors by some specific examples. We start with flows at western boundaries.

7.7.2 Western Boundary Currents

As was shown in Fig. 7.6, western boundary currents are part of the main ocean gyres in the Pacific, Atlantic and Indian Ocean. In the Northern Hemisphere, these currents flow north along the western boundaries of the ocean basins, while in the Southern Hemisphere they flow south. All western boundary currents have a number of common features. They form narrow streams along the western continental rise of ocean basins and they extend to great depths, well below the thermocline. Usually at some point they separate from the basin boundary and continue to flow into the open ocean as narrow jets. This jet-like flows create many instabilities along their paths. However, due to variations in bottom topography, each current possesses its own individual characteristics. The magnitude of some western boundary currents are: 50 Sv for the Kuroshio, 30 Sv for the Gulf Stream and the Brazil current, 25 Sv for the East Australian Current, and 70 Sv for the Agulhas Current. Let us note that, for example, a discharge of 50 Sv means that 4000 cubic kilometres of water is transported per day (Tomczak and Godfrey, 1994).

Probably the most studied western boundary currents are the Gulf Stream in the Atlantic Ocean and the Kuroshio in the Pacific. There is extended professional literature on these currents, for example Stommel (1965), Pickard and Emery (1982), Tomczak and Godfrey (1994), Stommel and Yoshida (1972), Mizuno and White (1983), and others. Therefore, in this section we describe two other western boundary currents flowing in the Southern Hemisphere, *i.e.* the East Australian Current in the Pacific, and Agulhas Current in the Indian Ocean.

East Australian Current. This current is one of the weakest western boundary currents. Its low transport volume results from the fact that part of the

South Equatorial Current is directed to the Indian Ocean through the Indonesian throughflow. The influx of the South Equatorial Current splits at a bifurcation point, located between latitudes of 14° and 18°S. One part of the current continues to flow southward as the East Australian Current and the other as the Coral Sea Coastal Current, which joins with the Indonesian throughflow through the Torres Strait. Tomczak and Godfrey (1994) note that if the Indonesian passage were closed, the transport of the East Australian Current would be doubled. Near latitude 34°S, the current separates from the Australian continent, and its path from Australia to New Zealand is known as the Tasman Front, at which the warm waters of the Coral Sea meet the colder waters of the Tasman Sea. The continuation of the East Australian Current east of New Zealand is the **East Auckland Current**.

The East Australian Current is associated with strong instabilities which result in meanders and eddies. Figure 7.13 (see colour plate p.565) shows sea surface temperature distributions as observed by the satellite NOAA9 on September 19, 1988. Red colour illustrates a formation of warm mesoscale eddies of the current. The East Australian Current generates about three eddies per year and at any particular time about 4–8 eddies may co-exist. When eddies separate themselves from the main current, they maintain their speed (1.5–2.0 m/s) for long periods of time (months). The speed of the East Australian Current varies seasonally and reaches about 1.3 m/s near 30°S during summer. During winter its speed is about 0.7 m/s at the same location (Tomczak and Godfrey, 1994). At the shelf break, in 120 m water depth, just offshore from the Great Barrier Reef (GBR) matrix and close to Myrmidon Reef (18.3°S), measurements at water depths of 27 m and 100 m confirm a southward current speed of 0.55 m/s and 0.2 m/s, respectively (Wolanski and Pickard, 1983).

A very important feature of the east Australian shelf is the Great Barrier Reef, which is not a continuous barrier but a matrix of 2500 individual reefs extending along the shelf rim between latitudes of 9.2°S to 25°S. Water depths on the GBR shelf increase from 30 m to 140 m in the northern section (south of 20°). The lagoon between the GBR and the coast is rather shallow ($\leq$ 60 m) and connected to the Coral Sea by numerous channels. A comprehensive overview of the physical oceanographic processes of the east Australian shelf, and especially the Great Barrier Reef region, can be found in Wolanski (1994) and biological characteristics of the GBR were recently summarized by Alongi (1998). Some aspects of the interactions between water movement on the shelf and biological life in the GBR are discussed in Part III of this book.

Agulhas Current. Currents along the western side of the northern Indian Ocean have important implications for biological productivity. The South Equatorial Current, flowing westward, converges on Africa and Madagascar during the spring and summer, and splits into the northward Somali Current (with transport of 30 Sv) and the southward Agulhas Current. Due to strong upwelling at the region of divergence, the primary productivity rate is very high.

The Agulhas Current is one of the strongest currents of the world ocean with little seasonal variation. Its speed, of about 1.6 m/s (peak speed is about 2.5 m/s), is almost constantly maintained throughout the year; the width of the current is about 90–170 km. Near 31°S, the current transport is estimated as 70 Sv; this transport increases by 6 Sv every 100 km, and near 35°S it carries 95–135 Sv. Further south the current encounters the Antarctic Circumpolar Current and most of its transport turns back to the Indian Ocean Gyre (see Fig. 7.6).

The nature and kinematics of the Agulhas Current termination are very complicated. This current exhibits a unique circulation feature that consists of a turnabout at which its direction is more or less reversed, which is known as the Agulhas retroflection (Lutjeharms and van Ballegooyen, 1988). The area of retroflection lies between 16°E and 20°E and has a diameter of 340 km. The Agulhas retroflection area may move in western, southern or northern directions, together with a range of rings and eddies in various stages of decay.

Another unique aspect of the Agulhas Current is that it runs against the prevailing wind direction during the Southwest Monsoon season, which results in a steepening of wind waves in the region of the current. Further south, the current runs into the swell approaching from the Southern Ocean. Close to the African coastline, between Durban and Port Elizabeth, the steepening of these waves produces giant waves with enormous heights of 15–18 m (see Sect. 3.5.4). When waves propagate into the current, their phase velocity changes according the formula (Massel, 1996a):

$$\frac{C}{C_0} = \frac{1}{2} + \frac{1}{2}\left(1 + \frac{4U}{C_0}\right)^{1/2}, \tag{7.31}$$

in which C is the phase velocity of waves in the presence of current, C_0 is the phase velocity without the current, and U is the current velocity. When waves move against the current, wave phase velocity, as well as wave group velocity, will be reduced. However, as the wave energy flux (see Eq. 3.13) should be conserved, wave height has to rise. It can be shown (Massel, 1996a) that if the opposing current velocity is equal to only 22 percent of the wave speed, C_0, the amplification of wave height is equal to ~ 2.5.

7.7.3 Eastern Boundary Currents and Coastal Upwelling

As we showed in Sect. 7.2, the differential heating of the ocean surface and the rotation of the Earth produce the easterly Trade Winds of the subtropics. However, along the eastern ocean coastlines, the air is dry and hot in summer because air moisture is reduced by rains over the land further east. The resulting pressure difference between land and ocean induces equatorial winds along east coasts. The Ekman transport produced by equatorial winds is directed offshore, resulting in a lower sea surface at the coast. The water

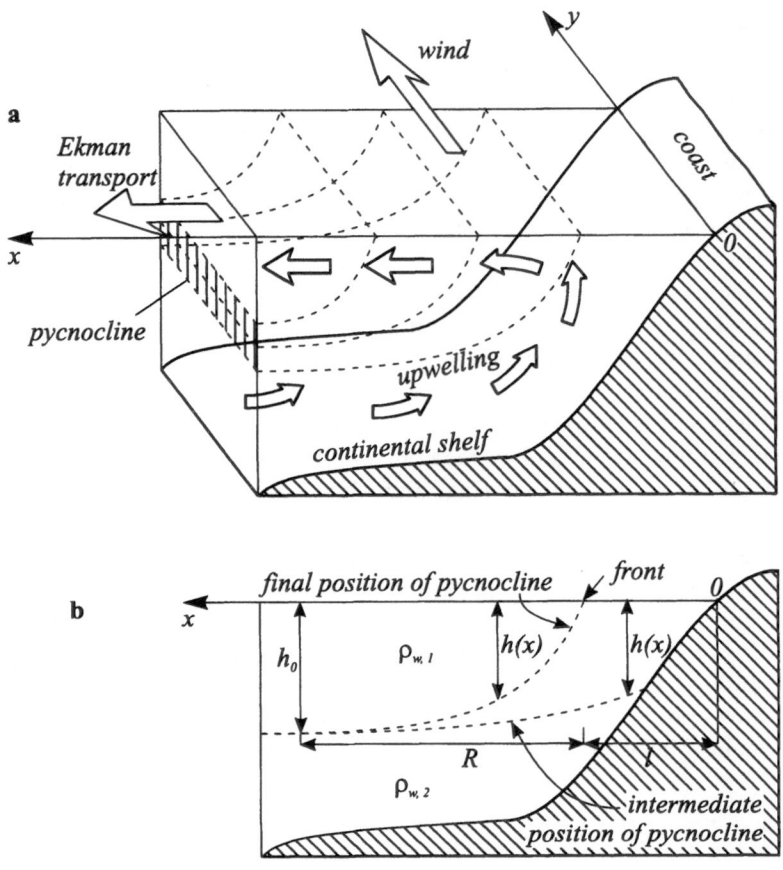

Fig. 7.14: Coastal upwelling along east coast in the Southern Hemisphere

that is removed by the offshore surface drift has to be replaced to satisfy continuity. Because of the presence of the coastal boundary, this replacement cannot be due to horizontal flow. Instead, water is upwelled from the deeper layers (Fig. 7.14). These deeper waters are generally colder than the surface waters, richer in nutrients such as nitrate, and more depleted of phytoplankton populations.

In the simplest two-dimensional model, the upwelled water must come from far offshore, and the onshore flow averaged over all depths must be zero at any location. Thus, consider coastal water bounded by a coast and subjected to a surface stress acting with the coast on its right (Fig. 7.14a). The upper layer supports an offshore drift current and the lower layer is assumed to be infinitely deep and motionless. The position of the pycnocline, which separates waters of densities $\rho_{w,1}$ and $\rho_{w,2}$, depends on the intensity of the wind stress. After a weak or brief wind event, the interface rises but not to the point of reaching the surface. A prolonged wind event may cause the interface to reach the surface and result in the formation of a front (Fig. 7.14b). This front is displaced offshore and cold waters from below are exposed to the surface, which enhances the primary production.

A simple model of coastal upwelling for the case shown in Fig. 7.14, can be expressed by the following system of equations (Cushman-Roisin, 1994):

$$\left.\begin{aligned}
\frac{\partial u}{\partial t} - fv &= g'\frac{\partial \zeta}{\partial x} \\
\frac{\partial v}{\partial t} + fu &= \frac{\tau}{\rho_{w,1}h_0} \\
-\frac{\partial \zeta}{\partial t} + h_0\frac{\partial u}{\partial x} &= 0
\end{aligned}\right\}, \tag{7.32}$$

in which u and v are current components in the x and y directions, respectively, ζ is the small upwards displacement of the interface, f is the Coriolis parameter $f = 2\omega_E \sin\phi$, h_0 is the thickness of the upper layer far offshore, τ is the longshore stress, and g' is the reduced gravity:

$$g' = g\frac{\rho_{w,1} - \rho_{w,2}}{\rho_{w,1}}. \tag{7.33}$$

It should be noted that the first two equations of (7.32) system are the linearized Euler equations with forcing provided by wind stress at the surface and a baroclinic gradient. The third equation simply is a mass conservation equation. At the coast ($x = 0$), the velocity, u, vanishes, while displacement, ζ, vanishes far offshore ($x \to \infty$). Due to the fluctuating nature of winds, we assume that the wind stress, τ, varies as follows:

$$\tau = \tau_0 \sin(\omega t), \tag{7.34}$$

where τ_0 is the reference stress. A solution of Eq. (7.32) satisfying assumed boundary conditions takes the form (Cushman-Roisin, 1994):

$$\left.\begin{aligned}
u(x,t) &= \frac{f\tau_0}{\rho_{w,1}h_0(f^2 - \omega^2)}\left[1 - \exp\left(-\frac{x}{R_\omega}\right)\right]\sin(\omega t) \\
v(x,t) &= \frac{\omega\tau_0}{\rho_{w,1}h_0(f^2 - \omega^2)}\left[1 - \frac{f^2}{\omega^2}\exp\left(-\frac{x}{R_\omega}\right)\right]\cos(\omega t) \\
\zeta(x,t) &= \frac{-fR_\omega\tau_0}{\rho_{w,1}g'h_0\omega}\exp\left(-\frac{x}{R_\omega}\right)\cos(\omega t)
\end{aligned}\right\}, \tag{7.35}$$

in which R_w is the modified deformation radius (Pedlosky, 1979; Smith, 1981):

$$R_w = \sqrt{\frac{g'h_0}{f^2 - \omega^2}}. \tag{7.36}$$

Thus, the upwelling is trapped along the coast within a distance of the order of R_w. For long periods (weeks or months) when $f \gg \omega$, the distance R_w becomes the Rossby radius of deformation:

$$R = \frac{\sqrt{g'h_0}}{f}.$$
(7.37)

It should be noted that the solution (7.35) is applicable for the intermediate stage of upwelling when the colder water is not yet exposed to the surface. At the final stage, after prolonged wind action, the front forms and particles which were initially against the coast (zero), are now at some distance, l, from the coast, *i.e.*:

$$l = \frac{I}{f} - R,$$
(7.38)

in which I is the wind impulse resulting from the integration of the wind-stress term, $\tau/\rho_{w1}h$, over time as follows:

$$I \approx \frac{1}{\rho_{w,1}h_0} \int \tau dt.$$
(7.39)

For the final stage of upwelling, the depth of interface $h(x)$ and longshore velocity $v(x)$ can be expressed as follows (Csanady, 1977):

$$\left. \begin{array}{rcl} h(x) & = & h_0 \left[1 - \exp\left(\dfrac{l-x}{R}\right) \right] \\[2mm] v(x) & = & \sqrt{g'h_0} \exp\left(\dfrac{l-x}{R}\right) \end{array} \right\}.$$
(7.40)

Thus, the interface between the layers rises from the undisturbed depth h_0 to the sea surface in the distance R, and at the sea surface the interface lies a distance l from upwelling regions to the shore (Fig. 7.14b). Comparison of three different upwelling regions, *i.e.* Oregon coast (near 45°N), northwest Africa (near 22°W) and Peru (near 15°S) shows that the radius of deformation, R, equals 14 km, 10 km and 20 km, respectively.

The most impressive coastal upwelling system of the world ocean is found along the coast of Peru. The cold, northerly flowing Peru Current, in conjunction with the coastal upwelling, lowers the sea-surface temperatures (SST) offshore from Chile and Peru (Fig. 7.15a). The thermocline in this region becomes quite shallow (as was schematically illustrated in Fig. 7.11), and nutrient rich waters and abundant anchovies prevail offshore from Peru at these times.

The dynamics and mass balance of the Peruvian coastal upwelling system are strongly three-dimensional and time-dependant. Time scales of the phenomena

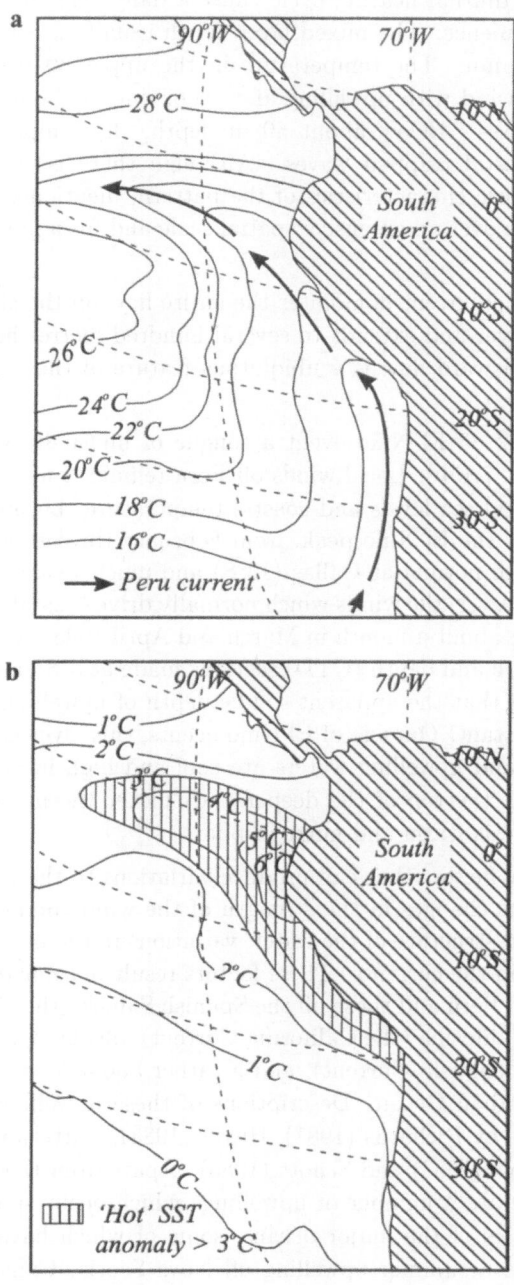

Fig. 7.15: Seasonal sea-surface temperature off the Peruvian coast: **a** typical mean SST of the Eastern Pacific, **b** SST anomaly during El Niño 1982–1983 (adapted from Pinet, 1992)

range from a day, for a sea breeze, to quasi-decadal El Niño. Near 15°S, a strong sea breeze and a diurnal heating cycle cause a daily cycle of the evolution of near surface turbulence. The mixed layer depth usually is about 20 m during summer and autumn. The temperature in the upper 30 m or so is largely governed by the wind with upwelling and deepening of the mixed layer being important processes. Below about 50 m depth, the temperature is largely controlled by coastal trapped waves. Although there is substantial spatial variability, the basic characteristics of the pattern, mentioned above, seem to persist alongshore over an extensive spatial scale and for a long time (Brink *et al.*, 1983).

A poleward undercurrent dominates the entire flow on the shelf beneath the surface Ekman layer and extends to several hundred metres in depth over the slope. Such a poleward flow is a ubiquitous feature of the major coastal upwelling regions.

During the 1982–83 El Niño event a tongue of unusually warm water was observed (see Fig. 7.15b). Local winds off Peru remained normal for at least a month after coastal sea levels and coastal temperatures began to rise in early October 1982. At the El Niño peak, from February through May 1983, winds were stronger than normal at Callas (12°S) and much weaker than normal at Talava (5°S). However, the winds which normally drive coastal upwelling in this area collapsed for about a month in March and April 1983. A series of conductivity, temperature and depth (CTD) sections made at 5°S and 10°S (Huyer *et al.*, 1987) showed that the apparent source depth of upwelling waters (50–100 m) remained constant. Outside of El Niño events, this layer is normally below the thermocline and upwelling waters are cool and high in nutrients. During El Niño it was at the top of the deeper and thicker thermocline. Therefore, ascending water was warm and nutrient-poor.

There are many factors that can produce variations to the typical pattern of upwelling, such as changes in the strength of the wind component parallel to the shore, vertical structure of the water, variations in the bottom bathymetry, and instabilities in the currents. These factors result in some differences in the upwelling sites off Peru and those off the Spanish Sahara (the Canary Current), off Oregon and California (the California Current), off the west coast of Southern Africa (the Benguela Current), and a rather anomalous Somali upwelling in the western Indian Ocean. Descriptions of these upwelling systems are in a series of papers by Richards (1981), Huyer (1983), Mittelstaedt (1983), Nelson and Hutchings (1983) and Schott (1983). Apart from these major coastal upwellings, there are a number of upwellings which occur sporadically at the western boundaries of the major oceans, some of which have been well documented, such as a summer upwelling off Nova Scotia, Canada, and that of the northwest corner of Spain, near Cape Finistere. The physical environment and biological production of some of these regions is reviewed by Tomczak and Godfrey (1994) and Mann and Lazier (1996). We will return to the biological consequences of coastal upwelling in Chap. 15.

7.7.4   Other Examples of Upwelling

Apart from coastal upwelling, oceanographers recognize two other types of upwelling: polar and equatorial upwellings. These upwellings are also highly productive, but the mechanisms that promote the upward motion of water in each case is different. We start with the polar upwelling.

Polar Upwelling. In the vast, harsh, desolate and inhospitable habitat of waters surrounding Antarctica, which are informally known as the Southern Ocean, exists the **polar upwelling** of cold water, rich in nutrients, which supports a large community of marine animals. Due to the absence of land barriers, circulation around the Antarctica is circumpolar, and consists of two major wind-generated currents that rotate in opposite directions. Closer to the Antarctic continent, the polar easterlies blow water in a counterclockwise direction forming the **East Wind Drift**. At the Palmer Peninsula, this flow is deflected northward into the **West Wind Drift**, rotating in a clockwise direction. Both currents are located inside the **Antarctic Convergence**, a broad zone about 150 km wide. In the Antarctic Convergence area, the colder water of the West Wind Drift plunges beneath warmer, less dense water flowing southward out of the Atlantic, Pacific and Indian Ocean. Complicated interactions of these water masses within the Antarctic Convergence causes a variety of turbulent flows, vortices and meandering loops. However, the zone remains stationary, located at about 50°S latitude (Pinet, 1992).

Close to the Antarctic continent, during southern winter, the surface water temperature drops as low as minus 1.9°C and the salinity of the remaining water increases, forming the heaviest ocean water on earth, known as the **Antarctic**

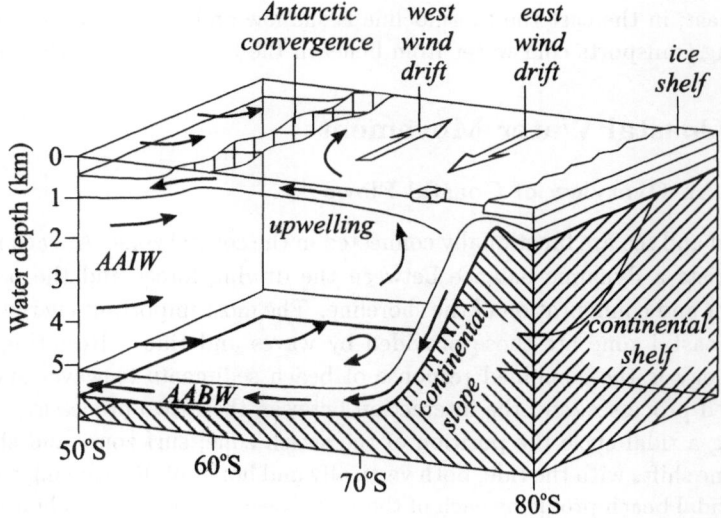

Fig. 7.16: Schematic three-dimensional pattern of circulation in the Southern Ocean

Bottom Water (AABW). Due to its high density, this water sinks and slowly spreads northward along the sea bottom. Part of this water, warmed slightly by mixing with the East and West Drifts water, remains near the surface. When it encounters less dense water along the Antarctic Convergence, the water sinks forming the **Antarctic Intermediate Water** (AAIW).

The continuity of mass requires a compensation for the loss due to down-welling of the Antarctic Bottom Water and the Antarctic Intermediate Water. This compensation is provided by the **Circumpolar Deep Water**, upwelling to the surface, and transporting dissolved nutrients which are the basis for the food web of this area. A schematic three-dimensional representation in the Southern Ocean is given in Fig. 7.16. Biological consequences of upwelling along the Antarctic Convergence are discussed in Chap. 15.

Equatorial Upwelling. As was shown in Sect. 7.3.4, the surface circula-tion of equatorial water is very complex, consisting of the wind-driven North and South Equatorial Currents, Equatorial Countercurrent, and Equatorial Undercurrent. Superimposed on this horizontal transport is the slow verti-cal movement of water, known as **equatorial upwelling**. Coriolis deflection, which changes its sign on both sides of the Equator, induces divergence of the equatorial water, inducing the upwelling of cold water.

This upwelling is clearly indicated by the presence of a band of water around the Equator, of 2°–9°C colder than the seawater on either side. Upwelling is stronger in summer than in winter causing high concentration of nitrate during that season. Within the equatorial band of colder water in the Atlantic and Pacific Ocean, temperature also varies spatially. The water in the eastern part of this colder band is cooler than that in the western part. However, due to tilting of the thermocline to the West (see Fig. 7.11), Ekman transport causes warm water above the thermocline to ascend to the sea surface in the west. In contrast, in the east the thermocline is shallow and weak, and even weaker upwelling transports cold water from beneath the thermocline to the surface.

7.8 Coastal Water Movement

7.8.1 An Overview of Coastal Flows

The land and sea are functionally connected in the coastal zone. At each instant there exists a dynamic balance between the driving forces and the position, character and configuration of the shoreline. The most important energy inputs in the coastal zone are those provided by waves and tides. Resulting beach morphology is the combined response of beach sediments to wave- and tide-generated processes, and the interaction between these two processes.

During a tidal cycle the position of the swash zone, surf zone, and shoaling wave zone shifts with the tide, both vertically and horizontally, causing a change of intertidal beach profile by each of these processes about every 12 hours. The rate of migration of the tide across the profile is a function of the tidal range,

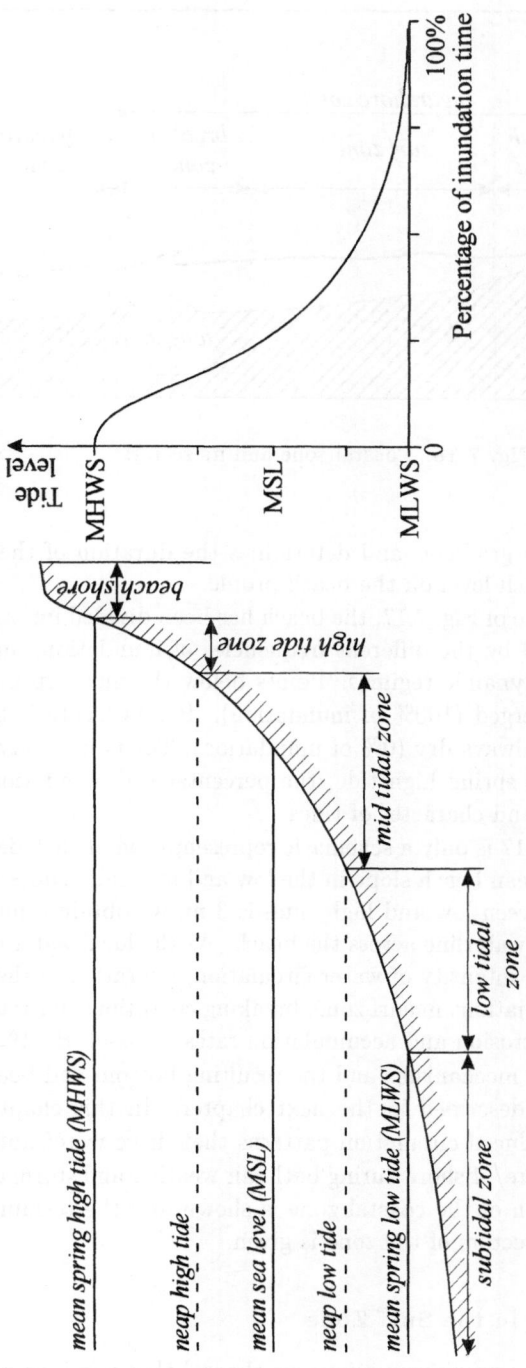

Fig. 7.17: Dependence of percentage of inundation time on tide level (adapted from Massel, B., 1998)

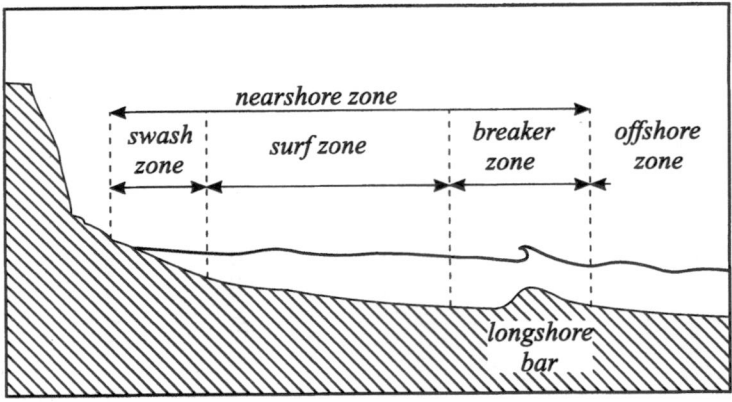

Fig. 7.18: Coastal zone and its sectors

tidal stage and beach gradient, and determines the duration of the hydrodynamic processes at each level on the beach profile.

On the left-hand side of Fig. 7.17, the beach has been divided into zones, each
of them distinguished by the different frequencies of inundation, morphologic
characteristics, and dynamic regimes. Points below the mean spring low tide
are constantly submerged (100% of inundation). Points located above mean
spring high tide are always dry (0% of inundation). For points between mean
spring low and mean spring high tide, the percentage of inundation depends
on the beach profile and character of tides.

The curve in Fig. 7.17 is only a schematic representation of such dependence.
Assuming that the mean beach slope in the low and mid tidal zones is 1%, and
vertical distance between low and high tides is 3 m, we obtain a migration as
large as 300 m of the waterline across the beach. As the local water depth controls wave height and intensity of water circulation, migration of the waterline
causes continuous variations in surf zone, breaking conditions, current patterns
and velocity, and in erosion and accumulation rates (Massel, B., 1998).

Sediment transport mechanisms and the resulting bottom and beach erosion
or accretion will be described in the next chapter. In this chapter we will
examine the wave-induced circulation patterns that drive water and sediment
alongshore and onshore/offshore during both fair weather and storm conditions.
In Fig. 7.18 a division of the coastal zone is shown and the terminology that
describes particular sectors of the zone is given.

7.8.2 Circulation in the Surf Zone

Introduction. In the previous sections we showed that wind-driven currents
constitute a very important component of the oceanic circulation. In the proximity of a coastline, the water movement becomes restricted, leading to a rising

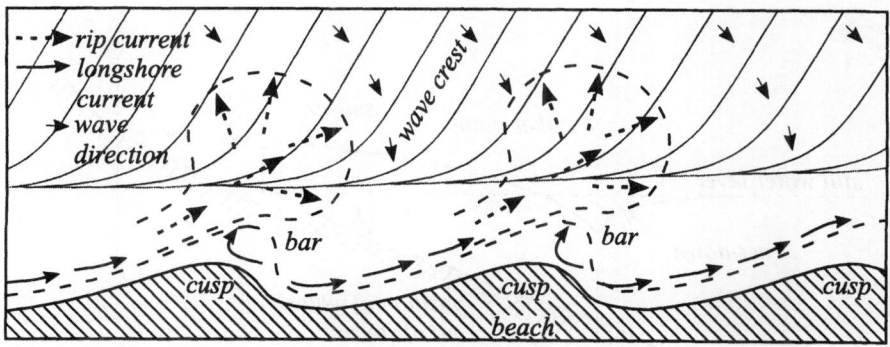

Fig. 7.19: Nearshore circulation

of the water level and the development of surface slopes which may extend away from the coast. The surface slopes give rise to horizontal pressure gradients, and these in turn generate currents known as **gradient currents**. Moreover, when waves approach the shore at some angle, the longshore component of wave energy creates a **longshore current** that flows alongshore, confined between the first breaker and the shoreline. Typical values of the longshore current velocity are well under one metre per second and the highest velocity is observed at the breaker line. Field experiments have generally recognized the variability of longshore currents across the surf zone and down the coast. It should be noted that the longshore current plays an important role in the generation of sediment transport (see Chap. 8).

Longshore variations in breaker heights create gradients of the mean water level within the surf zone, which generates currents flowing from positions of the highest breaker to the lowest breakers. Here, the longshore currents converge and turn seaward as **rip currents**. Rip currents are narrow strong return flows directed through the surf to sea. Further offshore, the flow in the rip current weakens and radiates to form a bulbous head from which its slack water eventually mixes with the surrounding nearshore water. Together with longshore currents, the rip current can create a two-dimensional coastal current system within and beyond the surf zone, known as **nearshore circulation** (see Fig. 7.19). Hence, the system of water flow forms a series of coastal circulation cells in which offshore water enters, flows through, and then exits the surf zone. The longshore dimension of circulation cells varies from 30 to 400 m. In the case of tides in coastal seas, there are currents associated with the rising or falling of the sea surface, known as **tidal currents**.

There are three main processes influencing the mean water level close to the coastline, namely tidal oscillations, storm surges and wave-induced se-up. The first two mechanisms have been described in Chaps 5 and 3, respectively. Wave set-up is a local process strongly related to the local wave field. When waves

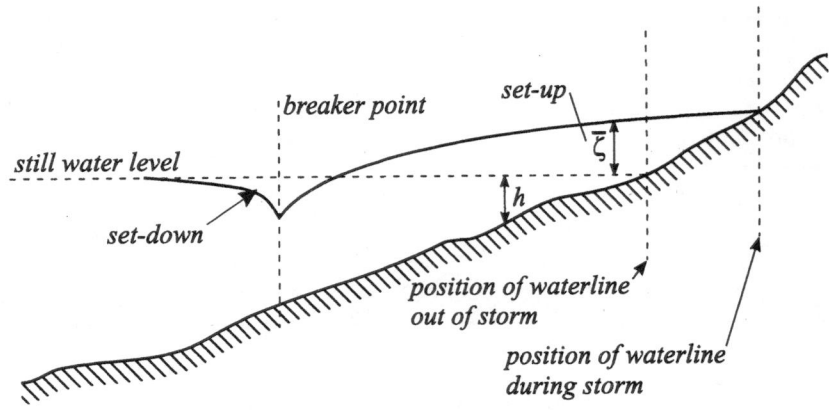

Fig. 7.20: Schematic representation of mean water level changes due to wave breaking (adapted from Massel, B., 1998)

approaching the shore start to break, wave height decreases and part of the wave energy is used to raise mean water level (see Fig. 7.20).

On the contrary, offshore of the breaking point, wave height increases and mean water level decreases (set-down), and the lowest mean water level is observed at the breaking point. As shown in Fig. 7.20, set-up causes the movement of the water line further onshore. The change of mean water level, $\bar{\zeta}$, with respect to still water level is given by the equation developed by Longuet-Higgins and Stewart (1964):

$$\rho_w g \left(h + \bar{\zeta} \right) \frac{d\bar{\zeta}}{dx} + \frac{dS_{xx}}{dx} = 0, \tag{7.41}$$

in which S_{xx} is the cross-shore component of so called radiation stress tensor:

$$S_{xx} = \frac{1}{8} \rho g H^2 \left[m \left(\cos^2 \theta + 1 \right) - \frac{1}{2} \right], \tag{7.42}$$

where m is given by Eq. (4.22), H is the wave height and θ is the angle between wave ray and normal to the beach.

For a mildly sloping bottom, the set-down at the breaking point is about 5% of the breaking depth, while the maximum set-up is usually about 15% of the breaker depth or about 20% of the breaking wave height.

Longshore Current. When waves approach a straight coastline at an oblique angle, a longshore current is established, flowing parallel to the coastline in the nearshore zone. Maximum longshore current velocity is located close and shoreward of the breaking line. The velocity of the current quickly decreases to zero outside the nearshore zone, so it is clearly wave-induced and cannot be at-

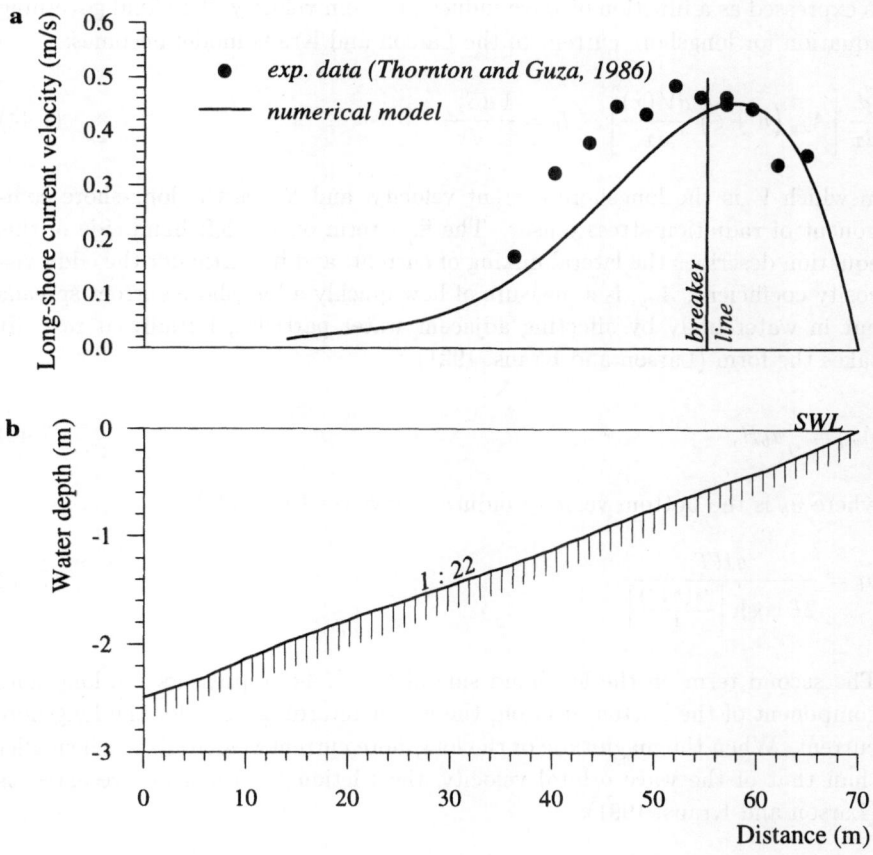

Fig. 7.21: Comparison of measured and calculated long-shore current velocities at Leadbetter Beach, California: **a** long-shore current velocity, **b** water depth (adapted from Massel, B., 1998)

tributed to ocean currents or tides. In Fig. 7.21, a comparison of observed and predicted longshore velocity at Leadbetter Beach (California) is shown. Experiments were conducted as part of the Nearshore Sediment Transport Study (Seymour, 1989). Waves with initial significant height $H_s = 0.79$ m and period $T = 14.2$ s approached the coast at an angle of 9°.

The method for prediction of the longshore current velocity is based on the following balance of forces acting on a water element, *i.e.*:

$$\text{driving force} + \text{bottom friction} + \text{lateral friction} = 0. \tag{7.43}$$

Larson and Kraus (1991) demonstrated that the driving force is proportional to the gradient of wave energy in the surf zone, while lateral and bottom friction is expressed as a function of wave-induced bottom velocity. The final governing equation for longshore current in the Larson and Kraus model becomes:

$$\frac{d}{dx}\left[A_{xy}\left(h+\bar{\zeta}\right)\frac{dV(x)}{dx}\right] - f_b = \frac{1}{\rho}\frac{dS_{xy}}{dx}, \tag{7.44}$$

in which V is the longshore current velocity, and S_{xy} is the long-shore component of radiation stress tensor. The first term on the left-hand side of this equation describes the lateral mixing of current, and in particular the eddy viscosity coefficient, A_{xy}, is a measure of how quickly a longshore current spreads out in water body by affecting adjacent water particles, initially at rest. It takes the form (Larson and Kraus, 1991):

$$A_{xy} = \frac{1}{2}u_b H, \tag{7.45}$$

where u_b is the bottom velocity induced by waves (Eq. 4.25):

$$u_b = \frac{gHT}{2L\cosh\left[\frac{2\pi\left(h+\bar{\zeta}\right)}{L}\right]}. \tag{7.46}$$

The second term on the left-hand side of Eq. (7.44) represents the longshore component of the bottom friction, the major retarding force for the longshore current. When the magnitude of the longshore current velocity is much smaller than that of the wave orbital velocity, the friction term can be presented as (Larson and Kraus, 1991):

$$f_b = \frac{2}{\pi}c_f u_b V(x)\left(1 + \sin^2\theta\right), \tag{7.47}$$

in which c_f is the bottom friction coefficient (usually of order of 0.005–0.01). The term on the right-hand side of Eq. (7.44) is responsible for the wave-induced force necessary for the generation of longshore current. It can be presented in the form:

$$S_{xy} = \frac{1}{2}mE\sin(2\theta) = \frac{\rho gH^2}{16}m\sin(2\theta), \tag{7.48}$$

where m is given by Eq. (4.22). When the angle θ is equal to zero (waves approach perpendicularly to the beach), longshore current ceases. Intensity of the longshore transport increases with increased angles of incidence. An example of a comparison of predicted longshore current with field measurements is shown in Fig. 7.21.

Rip Currents and Edge Waves. There are several mechanisms which are responsible for the generation of rip currents. The regular spacing of rip currents can be induced by the presence of regularly varying bottom topography

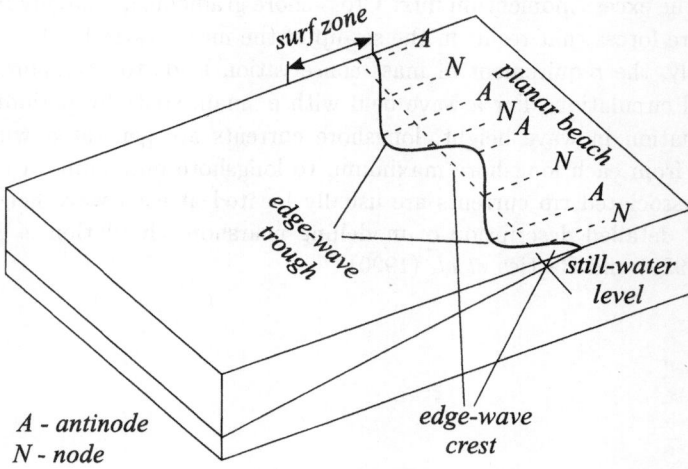

Fig. 7.22: Nodes and antinodes

as shown by Mei and Liu (1977). Coastlines with large offshore bars present an extreme example of such situations. Water is setup between the shore-line and the bar and then flows seaward through rip channels. Also, there is some evidence that synchronous trains of incident waves can interact to create rip currents which are spaced depending on the deep water wave-length and directions of the waves.

Another possible mechanism of current generation involves the concept of synchronous edge wave-incident wave interaction. **Edge waves** appear to be low-amplitude, standing waves with crests that are aligned at right angles to the shoreline with amplitudes decreasing from shore (Guza and Inman, 1975; Massel, 1989). A fraction of the energy from the incoming swell is reflected from the beach and trapped in the nearshore zone. Like all standing waves, edge waves possess nodes and antinodes (Fig. 7.22). The crests and troughs of edge waves are oriented normally to the beach, and the water level does not fluctuate with time at the nodes. The sinusoidal appearance of the water line along a planar beach may be an indication of edge wave action. Such configuration of the beach may help generate rip currents (Bowen, 1969).

Nearshore Circulation. Circulation in the coastal zone is mostly dominated by wave-induced forces associated with the processes of refraction, shoaling and breaking. In the breaker zone, a considerable part of the kinetic energy is transferred into breaking-induced turbulence. Incident wave periods at exposed coasts are typically of the order of 5–10 s. However, measurements also identify motions with periods of more than 20 s.

The temporal and spatial distribution of wave energy and its dissipation de-termines the nearshore circulation. Present-day models are capable of realistic predictions of the depth-integrated, time-average dissipation rate. The driving force for horizontal circulation in the nearshore zone is provided by spatial gradients of the excess momentum flux. Cross-shore gradients are largely balanced by pressure forces that result in the set-up of the mean water level.

Generally, the requirement of mass conservation leads to nearshore cells of horizontal circulation. For a wave field with a small, spatially periodic, longshore variation in wave height, longshore currents are generated within the surf zone from each longshore maximum, to longshore minimum, of the wave height. Associated rip currents are usually located at each wave height minimum. A detailed description of modelling nearshore circulation is given by Massel (1989) and Battjes *et al.* (1990).

8 Transport in the Oceans and Coastal Zone

8.1 Introduction

In previous chapters in Part II of this book, we have discussed various motions of ocean water. The scales of these motions have varied from seconds, for wave motion, to years, for planetary currents. Ocean water contains, or carries with it various types of matter such as nutrients, gases, organic and non-organic particles, and dissolved chemical substances which move with the ocean water. Here we assume that this matter is passive, which means that it has no ability to move by its own means with respect to the fluid. Therefore, for example, the locomotion of marine organisms is excluded from the discussion, and this problem will be taken up in Chap. 11.

An important and common outcome of the exchange between two volumes of passive matter which satisfy a conservation law is that a change in the amount of one element will cause a change in the amount of the other element, according to the conservation law. For example, an increase in the amount of element A will cause a decrease in the amount of element B. All variety of such exchanges constitutes what are known as **transport phenomena**. In general, transport phenomena correspond to the transfer of matter, energy and momentum. Transport of energy and momentum determines the nature of water motion itself, and in the previous chapters we have shown how the balance of energy and momentum is included in the governing equations of water motion.

There are various mechanisms by which matter is transported through the fluid environment. Probably one of the most important is **diffusion**. The concept of diffusion comes from the irregular motion of a group of particles, which are initially concentrated near some point of space, and after time spread out, gradually occupying more volume around the initial point. Diffusion is in fact a random process.

In laminar flow, the transfer of a substance between laminar layers of fluid is only due to molecular diffusion, which is an extremely inefficient transport mechanism. However, laminar flow is seldom found in nature. When small particles are placed in a fluid influenced by turbulence, the associated diffusion due to turbulence is known as turbulent diffusion, in distinguishing it from the molecular diffusion. Turbulent diffusion is different from molecular diffusion due to random molecular motion.

When particles are subjected to molecular or turbulent diffusion, their initial concentration will change in space and time. The determination of particle concentration at a given point, and at a given instance of time, is of fundamental importance for biological life in the ocean. The concentration of organic and non-organic particles or chemical substances strongly influences many elements of biological life in the ocean, such as metabolic processes. On the other hand, concentration of pollutants determines the degree of threat to marine plants and organisms. We will resume this discussion in Chap. 13.

Molecular and turbulent diffusions are not the only means of transport of matter in the ocean environment. Particles and substances can also be transported by **convection**, where convection includes transport by either the bulk movement of a fluid relative to marine organisms or by movement of an organism relative to the fluid. In the next section we will examine these physical mechanisms of transport. We will start with the basic information on diffusion.

8.2 Basics of Diffusion

8.2.1 Randomness of Diffusion

Consider a fluid in which a number of particles are suspended. If the particles move around at random, transfer of kinetic energy of particles will occur. This is a consequence of the interaction of neighbouring particles through the occurrence of occasional collisions. The direction taken by a particle after a collision is random. If it hits another particle, it may return back to where it came from. Therefore, subsequent positions of the particle are random, and as such it can be treated as a random variable which satisfies some laws of probability. Because of the randomness of this mechanism, Okubo (1980) defined diffusion to be *a basically irreversible phenomenon by which matter, particle groups, population, etc. spreads out within a given space according to individual random motion.* The diffusion is irreversible in the sense that the probability of returning to the original state decreases, when the number of diffusing particles increases.

It should be noted that diffusion was defined as the dispersion of some property of the fluid, for example salinity, without any net mass transfer of the fluid itself. In the oceans we distinguish diffusive transfer due to molecular motion of the water particles (molecular diffusion) and due to turbulence of the water mass (turbulent diffusion).

8.2.2   Random Walks

As we mentioned above, particles which undergo diffusion spread in space in a very irregular manner. This is known as a **random walk** through space. In 1826, a botanist, Robert Brown observed with the aid of a microscope, an irregular motion performed by small particles of colloidal size immersed in a fluid which can serve as an example of random walk. The nature of these motions remained a puzzle until the beginning of our century when Albert Einstein showed that this 'Brownian motion' is maintained by collisions with the molecules of the surrounding fluid. During each collision, a randomly oriented impulse on the particle results in their spreading. Statistical analysis of Brownian motion is a starting point for an attack on the more difficult problems of diffusion. Therefore, for better understanding of the random walk, following Okubo (1980), we will consider a simple case of particles which can spread from their initial position at point O in the one-dimensional system of cells containing particles (Fig. 8.1).

Cells are separated by a small distance λ. We assume that particles can move a distance λ to the right or left during each short time step τ. Because the motion is considered completely random, the probability of moving either to the left or right is $1/2$. This means that after time τ, a given population will spread, and one half will occupy a cell at distance λ to the right of the origin and one half will occupy a cell the same distance to the left. However, we do not know which individual particles belong to which cell. During the next time step τ, each particle has an equal chance to again move a distance λ to the adjacent cell, right or left, independently of the previous motion. Therefore, after time 2τ, one fourth of the population will occupy a cell at a distance 2λ to the right of the origin, one fourth will occupy the cell the same distance to the left of the origin, and half of the population will return to the origin cell. What will be the spatial distribution after time $n\tau$, when n becomes large?

After time $n\tau$, particles can occupy cells at $-n\lambda$, $-(n-1)\lambda$, ..., $(n-1)\lambda$ and $n\lambda$ distances from the origin. However, the spatial distribution of the particles is not uniform. For example, the probability that a particle occupies the cells farthest to the right (or left) is very small, as such a situation requires n successive movements of the particle in only the right (or left) direction. In

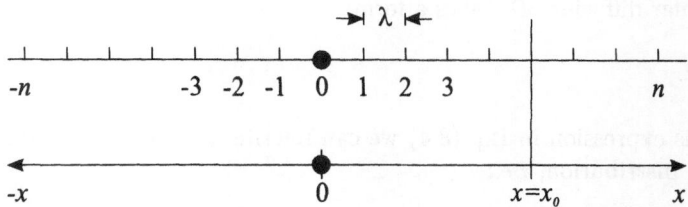

Fig. 8.1: Spreading of particles along one-dimensional cells

general, assuming that the variable $(n \pm m)$ is even, the probability that a particle reaches the cell situated a distance $m\lambda$ to the right of its initial position after time $n\tau$ can be expressed as follows (Ochi, 1990):

$$p(m, n) = \frac{n!}{2!} \frac{1}{[(n+m)/2]! \, [(n-m)/2]!}. \tag{8.1}$$

The symbol $n!$ denotes the factorial which is the product of the first n positive integers, i.e.:

$$n! = 1 \times 2 \times 3 \ldots (n-1) \times n. \tag{8.2}$$

When n becomes very large (or λ and τ become very small), the probability distribution of Eq. (8.1) converges to the Gaussian probability density function with a mean of zero, and variance n (see Eq. 4.80), i.e.:

$$\lim_{n \to \infty} p(m, n) = f(m, k) = \frac{1}{\sqrt{2\pi n}} \exp\left(-\frac{m^2}{2n}\right). \tag{8.3}$$

It can be found that the distribution in Eq. (8.1) can be approximated by the Gaussian distribution when $n > 6$.

As the particle moves a distance λ in each step, we may write $x = m\lambda$ and $n\tau = t$. Then Eq. (8.3) can be rewritten as:

$$f(x, t) = f(m, k) = \frac{1}{2\sqrt{\pi D t}} \exp\left(-\frac{x^2}{4Dt}\right), \tag{8.4}$$

in which:

$$D = \lim_{\lambda \to 0; \tau \to 0} \frac{\lambda^2}{2\tau}, \tag{8.5}$$

where D is known as the **coefficient of molecular diffusion**, or simply diffusivity. The limit in Eq. (8.5) is taken in such a way that the ratio of λ^2 and τ becomes constant in the limit. Coefficient D has dimension of $(\text{length})^2/\text{time}$.

Equation (8.4) has the form of a Gaussian (or normal) distribution. In particular, the relationship between variance, σ_x^2, of the distribution and coefficient of molecular diffusion, D, takes a form:

$$\sigma_x^2 = 2Dt. \tag{8.6}$$

Using this expression in Eq. (8.4) we can rewrite it in the standard form of a Gaussian distribution, i.e.:

$$f(x, t) = \frac{1}{\sqrt{2\pi}\sigma_x} \exp\left(-\frac{x^2}{2\sigma_x^2}\right). \tag{8.7}$$

It should be noted that the coefficients ν (see Sect. 1.2), A_x, A_y and A_z (see Sect. 1.2 and 7.3), and D have the same dimensions. However, their meanings are different: ν is the coefficient of kinematic viscosity and represents the diffusivity for momentum in laminar flow; A_x, A_y and A_z are the coefficients of turbulent viscosity, and they are measures of the diffusivity for momentum in turbulent flow; D is a measure of the diffusivity for matter.

In Sect. 8.3.4 we will exploit and expand the random walk model to determine the concentration of particles released under various environmental conditions.

8.2.3   Fick's Equation of Diffusion

Moving particles transport matter. However, determination of the transport by monitoring individual particles on a microscopic level is very impractical. It is much easier to calculate transport when diffusion is examined from a macroscopic point of view.

On a macroscopic level, a suitable measure is the flux of matter between areas of different concentrations. Let us consider a plane with area A, located at $x = x_0$ and separating two neighbouring cells (see Fig. 8.1). We assume that after some n steps, or after time $t = n\tau$, there will be $N(x_0)$ particles located in a cell to the left of the plane, and similarly there will be $N(x_0 + \lambda)$ particles located in the cell to the right of the plane. All particles undergo the same type of motion as described in the previous section. So, at time $t + \tau$, half the particles at $x = x_0$ can move to the right through the plane, to a cell located at $x = x_0 + \lambda$. Similarly, half the particles from the cell at $x = x_0 + \lambda$ can move to the left.

Therefore, the net number of particles that cross the plane in the positive x direction in time t is:

$$\text{net number of crossing particles} = \frac{N(x_0)}{2} - \frac{N(x_0 + \lambda)}{2}. \tag{8.8}$$

The number of crossing particles usually is presented in a form of a flux density I_x which is the net number of particles crossing a plane per time and per area. So, we have:

$$I_x = \frac{-[N(x_0 + \lambda) - N(x_0)]}{2A\tau}, \tag{8.9}$$

or:

$$I_x = -\frac{D}{\lambda}\left[\frac{N(x_0 + \lambda)}{A\lambda} - \frac{N(x_0)}{A\lambda}\right], \tag{8.10}$$

where the definition (8.5) of the coefficient of diffusion, D, has been used. For ocean water, the coefficient D is of the order of 10^{-9} m^2/s (Ozmidov, 1986), and is about 10,000 times smaller than the diffusion coefficient for air. As a result, terrestrial organisms that rely on the diffusive delivery of oxygen and carbon dioxide can metabolize faster than their aquatic counterpartners (Denny, 1993).

We note that the product $A\lambda$ is the cell volume, so that the terms in brackets each represent the number of particles per volume of a given cell. In other words, these terms are measures of particle concentration, c; therefore we have:

$$I_x = -D\frac{c(x_0 + \lambda) - c(x_0)}{\lambda}. \tag{8.11}$$

The concentration is also measured as the volume of the substance, or as substance units per unit volume of fluid. Thus, the units of concentration can be expressed, for example, in ppm, mg per litre, or the number of *Escherichia coli* per litre. If we assume that $\lambda \to 0$, Eq. (8.11) becomes:

$$I_x = -D\frac{\partial c}{\partial x}, \tag{8.12}$$

showing that the flux of particles in the x direction is proportional to the product of the gradient in concentration c and the molecular diffusion coefficient D. The negative sign indicates that transport proceeds from a position of higher to lower concentration. Equation (8.12) is known as Fick's equation of diffusion.

In 1855 a German physiologist, Adolph Fick, argued that Fourier's law of heat flow should be applied for the diffusion of salt in its solvent, and the governing equations should be similar. In fact, there is a direct and complete analogy between heat flow and molecular diffusion and, as far as the mathematical description is concerned, the processes are identical.

Let us now assume that particles are released from some single point (say from a reference origin) into three-dimensional space. Then the radial flux of matter will be given by the following spherical version of the Fick's equation:

$$I_r = -D\frac{\partial c}{\partial r}, \tag{8.13}$$

in which r is the radial distance from the point of particle release. Equation (8.13) is valid assuming that particle release is isotropic, *i.e.* particles have the same chance to be spread in any direction. Finally, we note that concentration, c, usually is expressed as mass per volume (kg/m^3) and therefore flux density has the units of $kg\ m^{-2}s^{-1}$.

Typical time and spatial scales of molecular motion are 10^{-12} s and 10^{-7} cm, respectively. These values are considerably smaller than typical time and space scales of observation, thus molecular diffusion obeys Fick's law. On the other hand, in a turbulent environment, the scales of motion are much larger than the distances of molecular motion. These scales are of the same order of magnitude as typical time and length scales of observations (Batchelor, 1967). Moreover, the turbulent motion of particles is continuous, and oscillations separated by small time intervals are correlated with each other. Therefore, a random walk model which assumes that all particle movements are uncorrelated can not be applied in a straightforward manner for turbulent environment. We will return to modelling turbulent diffusion later in this chapter.

8.3 Concentration of Matter for Molecular and Turbulent Diffusion

8.3.1 Concentration of Matter for Molecular Diffusion

Fick's Eq. (8.12) describes the relationship between the flux of matter, I_x, and the gradient of concentration, c. Let us extend Fick's equation to a three-dimensional volume of fluid as shown in Fig. 8.2, assuming that the coefficient of molecular diffusion, D, remains constant. Thus, during a small time increment, Δt, the difference between the rates of particle influx into the volume (through section $ABFE$) and particle outflow from the volume (through section $CDHG$), normal to the z axis, is:

$$\Delta m_z = D \left(\frac{\partial c}{\partial z}\Big|_{z=\Delta z} - \frac{\partial c}{\partial z}\Big|_{z=0} \right) \Delta x \Delta y \Delta t \approx D \frac{\partial}{\partial z} \left(\frac{\partial c}{\partial z} \right) \Delta x \Delta y \Delta z \Delta t =$$

$$= D \frac{\partial^2 c}{\partial z^2} \Delta x \Delta y \Delta z \Delta t. \tag{8.14}$$

In a similar way we can determine fluxes through other sections of the control volume of fluid, and the final change of mass in the volume becomes:

$$\Delta m = D \left[\frac{\partial^2 c}{\partial x^2} + \frac{\partial^2 c}{\partial y^2} + \frac{\partial^2 c}{\partial z^2} \right] \Delta x \Delta y \Delta z \Delta t. \tag{8.15}$$

Dividing both sides of Eq. (8.15) by the volume of the element $\Delta x \Delta y \Delta z$ and time increment, Δt, and assuming that $\Delta x \to 0$, $\Delta y \to 0$, $\Delta z \to 0$, and $\Delta t \to 0$, we obtain:

$$\frac{\partial c}{\partial t} = D \left(\frac{\partial^2 c}{\partial x^2} + \frac{\partial^2 c}{\partial y^2} + \frac{\partial^2 c}{\partial z^2} \right) = D \nabla^2 c, \tag{8.16}$$

in which ∇^2 is the Laplacian given by Eq. (C.14). Equation (8.16) is the equation of molecular diffusion for non-moving fluid. If the fluid is moving with some velocity, $\mathbf{u}$, additional terms should be added to the diffusion equation to describe the influence of the advection on the concentration c, *i.e.*:

$$\frac{\partial c}{\partial t} + \frac{\partial (cu)}{\partial x} + \frac{\partial (cv)}{\partial y} + \frac{\partial (cw)}{\partial z} = D \left(\frac{\partial^2 c}{\partial x^2} + \frac{\partial^2 c}{\partial y^2} + \frac{\partial^2 c}{\partial z^2} \right), \tag{8.17}$$

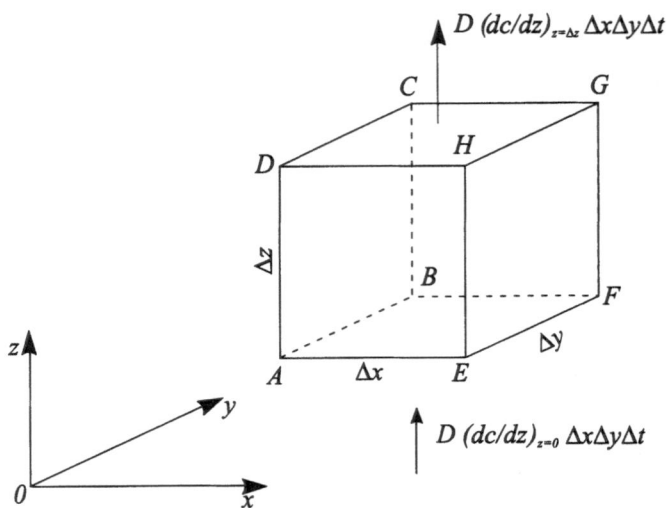

Fig. 8.2: Fick's equation and three-dimensional volume of fluid

or:

$$\frac{\partial c}{\partial t} + \nabla(\mathbf{u}c) = D\nabla^2 c, \tag{8.18}$$

in which ∇ is the nabla operator defined by Eq. (C.10). In particular for diffusion of particles in an (x, y) plane with uniform water flow along the x axis, Eq. (8.17) becomes:

$$\frac{\partial c}{\partial t} + u\frac{\partial c}{\partial x} = D\left(\frac{\partial^2 c}{\partial x^2} + \frac{\partial^2 c}{\partial y^2}\right). \tag{8.19}$$

The terms on the left side of this equation describe the rate of change due to advective mechanisms which carry particles with the current, while the terms on the right side describe processes due to the molecular nature of the diffusion which spreads the particles. These two processes are shown schematically in Fig. 8.3. At time $t = 0$, some initial distribution of particles is given (stage a). Due to current with velocity u, this distribution is advected along the x axis (stage b). At the same time molecular diffusion is acting and particles are spread further (stage c). A combination of advection and diffusion results in 'cloud' of particles being widely distributed and at the same time, shifted along the x axis (stage d).

The proportion by which the two mechanisms of advection and diffusion contribute to the final particle transport can be measured by the ratio of advection

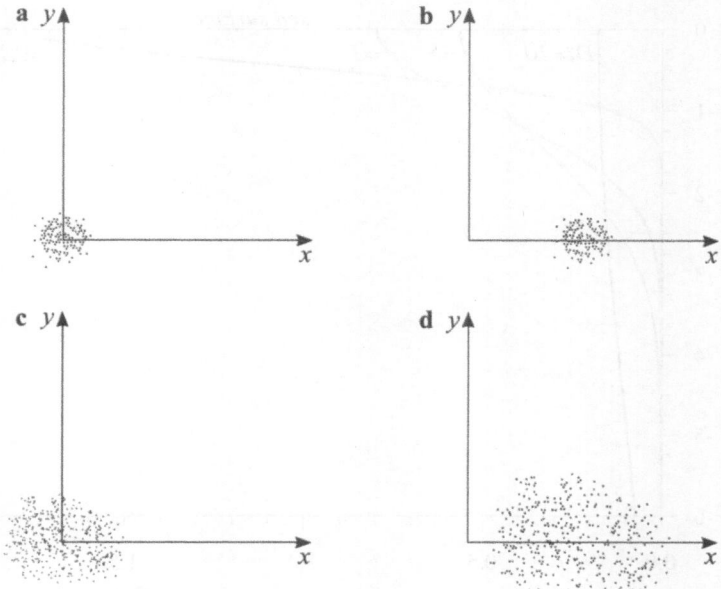

Fig. 8.3: Advective and diffusive processes involved in the transport of particles on plane (x, y): **a** initial stage, **b** advection by velocity U only, **c** diffusion only, **d** advection and diffusion

and diffusion terms, *i.e.*::

$$\frac{u\dfrac{\partial c}{\partial x}}{D\nabla^2 c} \approx \frac{U\dfrac{c}{L}}{D\dfrac{c}{L^2}} = \frac{UL}{D}, \tag{8.20}$$

where U is the characteristic velocity of the transporting fluid, L is the characteristic length of the flow domain and D is the diffusion coefficient. The ratio

$$P_e = \frac{UL}{D}, \tag{8.21}$$

is known as the Péclet number Pe. Values of Pe greater than one demonstrate that advection dominates over diffusion. In the case of molecular diffusion, when coefficient D is very small, the Péclet number is much greater than one and advection dominates over diffusion.

To solve Eq. (8.16), boundary conditions have to be specified. Firstly, the concentration at a given time instant, $t = t_0$, has to be known; usually $t_0 = 0$

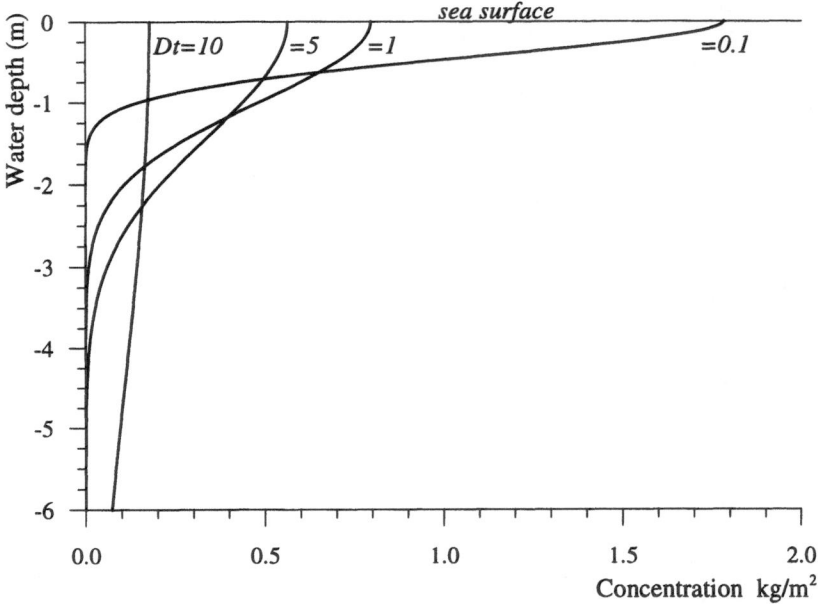

Fig. 8.4: Vertical distribution of concentration for substance deposited instantaneously

is used. In terms of space boundary conditions, the concentration, c, or flux of
particles at the boundaries, should be known.

To illustrate an example of solution of Eq. (8.16), let us assume that horizontal
transport of particles can be neglected. This assumption corresponds to the
situation when some substance from the atmosphere is falling on a large ocean
area. Therefore, the horizontal concentration gradients of this substance will
be very small and Eq. (8.16) simplifies to the one-dimensional equation for
vertical diffusion:

$$\frac{\partial c}{\partial t} = D \frac{\partial^2 c}{\partial z^2}. \tag{8.22}$$

Let us now assume that the total amount of substance at the surface $z = 0$
at time $t = 0$ is equal to M. Then, the solution of Eq. (8.22) becomes (Crank,
1975):

$$c(z, t) = \frac{A}{\sqrt{t}} \exp\left(-\frac{z^2}{4Dt}\right), \tag{8.23}$$

where A is a constant which should be defined. As the total amount of deposited
substance, M, is known, the following equation has to be true:

$$M = \int_{-\infty}^{0} c(z) dz. \tag{8.24}$$

After substituting Eq. (8.23) into Eq. (8.24) and integrating we obtain:

$$M = A\sqrt{\pi D}. \tag{8.25}$$

Thus, Eq. (8.23) becomes:

$$c(z,t) = \frac{M}{\sqrt{\pi Dt}} \exp\left(-\frac{z^2}{4Dt}\right). \tag{8.26}$$

Equation (8.24) indicates that the amount of diffusing substance remains constant and equal to the amount originally deposited at the surface, $z = 0$. Concentration, c, tends to zero, as the distance from the surface increases. For $t = 0$ it vanishes everywhere except at $z = 0$, where it becomes infinite.

In Fig. 8.4, vertical distributions of concentration, c, are shown for some time instants after t_0. It is assumed that the diffusion coefficient $D = 2 \times 10^{-9}$ m^2/s. This is a typical value for the diffusion coefficient for oxygen in water of temperature 20°C (Denny, 1993). Therefore, the values of Dt equal to 0.1, 1, 5 and 10 correspond to times of 5×10^7 s ($\sim$ 579 days), 2.5×10^8 s (2893 days), 0.5×10^9 s (5787 days), and 5×10^4 s (57870 days). It takes almost 8 years for particles to descend 3.2 m below the water surface. Thus, molecular diffusion is an extremely inefficient mechanism of transport. However, as we showed in the previous section, flows in the ocean are generally turbulent, and we need to define a turbulent equivalent of the molecular diffusion coefficient, and find methods for the calculation of substance concentration in turbulent flows.

8.3.2 Concentration of Matter for Turbulent Diffusion

Governing Equations. Turbulent flow is a random motion characterized by fluctuations of water velocity as well as fluctuations of concentration of matter suspended in the fluid. Therefore, we are not able to determine the instantaneous values of concentration at a given point and at a given time. However, in defining the turbulence equivalent of a concentration equation we can exploit the analogy between the transport of mass and the transport of momentum, a process we examined in Sect. 2.4. Thus, we represent the instantaneous concentration and instantaneous velocity components as a summation of their mean values and fluctuation components, *i.e.*:

$$\left.\begin{array}{l} c = \bar{c} + c' \\ u = \bar{u} + u', \quad v = \bar{v} + v', \quad w = \bar{w} + w' \end{array}\right\}. \tag{8.27}$$

After substituting the representation (8.27) into Eq. (8.17) we obtain:

$$\frac{\partial \bar{c}}{\partial t} + \bar{u}\frac{\partial \bar{c}}{\partial x} + \bar{v}\frac{\partial \bar{c}}{\partial y} + \bar{w}\frac{\partial \bar{c}}{\partial z} = D\nabla^2 \bar{c} -$$

$$- \frac{\partial}{\partial x}\left(\overline{u'c'}\right) - \frac{\partial}{\partial y}\left(\overline{v'c'}\right) - \frac{\partial}{\partial z}\left(\overline{w'c'}\right), \tag{8.28}$$

in which overbars denote averaging in a stochastic sense. The last three terms
on the right side of this equation result from the turbulent character of water motion and they describe the input of the turbulent fluctuations into the
balance of mass transport. These terms are unknown *a priori* and have to be
defined through other variables, for example, the mean concentration $\bar{c}$. Using
the so called Boussinesq approximation, we can write:

$$\left.\begin{aligned}
\overline{u'c'} &= -K_x\frac{\partial \bar{c}}{\partial x} \\
\overline{v'c'} &= -K_y\frac{\partial \bar{c}}{\partial y} \\
\overline{w'c'} &= -K_z\frac{\partial \bar{c}}{\partial z}
\end{aligned}\right\}, \tag{8.29}$$

in which K_x, K_y, and K_z are **coefficients of turbulent diffusion** in the x, y
and z direction, respectively. They are also known as eddy diffusion coefficients
or eddy diffusivities. In contrast to the coefficient of molecular diffusion D, the
coefficients K_x, K_y and K_z are not physical properties of fluid, but rather they
describe a stage and scale of motion.

As will be shown later, in most cases turbulent transport is a few orders
of magnitude larger than the molecular transport and molecular term $D\nabla^2\bar{c}$
is usually neglected. Using this fact and the representation of Eq. (8.29) in
Eq. (8.28) gives:

$$\frac{\partial \bar{c}}{\partial t} + \bar{u}\frac{\partial \bar{c}}{\partial x} + \bar{v}\frac{\partial \bar{c}}{\partial y} + \bar{w}\frac{\partial \bar{c}}{\partial z} = \frac{\partial}{\partial x}\left(K_x\frac{\partial \bar{c}}{\partial x}\right) + \frac{\partial}{\partial y}\left(K_y\frac{\partial \bar{c}}{\partial y}\right) + \frac{\partial}{\partial z}\left(K_z\frac{\partial \bar{c}}{\partial z}\right). \tag{8.30}$$

In general, coefficients of turbulent diffusion are functions of coordinates. If
we assume for a moment that these coefficients are constant, then:

$$\text{right side of Eq. (8.30)} = K_x\frac{\partial^2 \bar{c}}{\partial x^2} + K_y\frac{\partial^2 \bar{c}}{\partial y^2} + K_z\frac{\partial^2 \bar{c}}{\partial z^2}. \tag{8.31}$$

Coefficients K in this advection-diffusion equation represent the efficiency of
the turbulent diffusion of any substance or particles suspended in ocean water. Equation (8.30) is the basis for many approximations used in practical
calculations. Some of such solutions will be discussed in the next section.

A similar consideration leads to the turbulent heat exchange and a corresponding advection-diffusion equation takes the form:

$$\frac{\partial \bar{T}}{\partial t} + \bar{u}\frac{\partial \bar{T}}{\partial x} + \bar{v}\frac{\partial \bar{T}}{\partial y} + \bar{w}\frac{\partial \bar{T}}{\partial z} =$$

$$= \frac{\partial}{\partial x}\left(K_x^{(T)}\frac{\partial \bar{T}}{\partial x}\right) + \frac{\partial}{\partial y}\left(K_y^{(T)}\frac{\partial \bar{T}}{\partial y}\right) + \frac{\partial}{\partial z}\left(K_z^{(T)}\frac{\partial \bar{T}}{\partial z}\right), \tag{8.32}$$

in which $\bar{T}$ is the mean temperature and $K_x^{(T)}$, $K_y^{(T)}$ and $K_z^{(T)}$ are the coefficients of turbulent heat exchange. Equation (8.32) is valid assuming that the water density ρ_w is constant and the specific heat $c_p t$ is constant.

Magnitude of the Coefficients of Turbulent Diffusion. The first estimation of the coefficient of vertical turbulent diffusion, K_z, for salt in Kattegat Strait by Jacobsen gave a value of the order of 10^{-3} m^2/s, which is a few orders of magnitude higher than the coefficient D ($\sim 10^{-9}$ m^2/s). Ozmidov (1986) found that the coefficient, K_z, for ^{90}Sr in Atlantic waters was about 3×10^{-3} m^2/s, and in the Sargasso Sea, K_z changes between 2.3×10^{-4} and 9.5×10^{-4} m^2/s for the upper ocean layer up to 50 m water depth, while for water 50–100 m deep, it varies between 1.4×10^{-4} and 9×10^{-4} m^2/s. For deeper water between 100 m and 200 m, coefficient, K_z, changes between 1.4×10^{-3} m^2/s and 2.2×10^{-3} m^2/s.

Therefore, in general the coefficient of vertical turbulent diffusion is in fact a function of vertical coordinate. Dependence of the coefficient, K_z, on z is strongly related to the stratification of ocean waters.

The coefficients of horizontal turbulent diffusion, K_x and K_y, are much higher than the coefficient K_z. Initial estimations of the horizontal coefficients by Montgomery and Sverdrup (Kamenkovitch, 1978) in late thirties showed that they are of the order of 10^3–10^4 m^2/s. On the other hand, Munk et al. (1949) obtained values of K_x and K_y of the order of only 15 m^2/s for diffusion of contaminants in the lagoon of Bikini.

However, it soon became clear that the value of coefficients, K_x and K_y strongly depends on the spatial scale, L, of a given phenomenon. As L becomes larger, more and larger eddies participate in diffusion, and the diffusivity increases. Let us add that the coefficient of vertical turbulent diffusion of heat is of the order of 10^{-5} m^2/s, and the coefficient of vertical turbulent diffusion of salt is of the same order of magnitude.

There are two methods commonly used to quantify turbulent diffusivity. The first one is based on the observation of the rate at which particles spread after being released from some point. Initially, concentration is very high, but with passing time the cloud of particles spreads out. The spreading of the particle cloud is faster when the rate of turbulent diffusion is higher.

The simplest measure of the rate of distribution of particles around the cloud centre is the variance, σ_r^2, of the distances of the particles from the point of release (we assume that release point is at the origin), i.e.:

$$\sigma_r^2 = \frac{1}{N} \sum_{i=1}^{N} r_i^2, \qquad (8.33)$$

in which r_i is the distance of the ith particle from the origin. The directions of diffusivity are directly related to the rate of change of the corresponding variances (Okubo, 1980):

$$
\left.
\begin{aligned}
K_x &= \frac{1}{2}\frac{d\sigma_x^2}{dt} \\
K_y &= \frac{1}{2}\frac{d\sigma_y^2}{dt} \\
K_z &= \frac{1}{2}\frac{d\sigma_z^2}{dt}
\end{aligned}
\right\},
\tag{8.34}
$$

in which σ_x^2, σ_y^2, σ_z^2 are the variances of particle distances in the x, y and z direction, respectively.

Although simple in theory, this approach to quantify K_x, K_y and K_z values can be difficult in practice, as it requires an experimental determination of variances of particle distances in all directions.

The differences in observed values of coefficients of horizontal turbulent diffusion indicate that coefficients K_x and K_y are related to the scale of a given diffusion phenomenon. In the cloud of particles of dimension l, there are active turbulent eddies of dimension l (and smaller). Eddies larger than l simply transport the cloud as a one body, not participating in its diffusion. Okubo (1980) and Ozmidov (1986) provided experimental evidence for the relationship between coefficient K_x (K_y) and the scale, l, in the form:

$$
K \approx k f\left(\frac{l}{L}\right) l^{4/3},
\tag{8.35}
$$

in which l is the scale of the turbulent flow (for example, eddies), k is a constant coefficient, and the function $f(l/L)$ is a certain universal function in which L is the characteristic scale of the main turbulent flow. For example, for water motion with a scale of 10^6 m, the coefficient of horizontal turbulent diffusion can be of the order of 10^4 m^2/s, while for motion with a scale of 10^2 m, it can be of the order of 10^{-1} m^2/s. For an in-depth discussion on scaling and Eq. (8.35), the reader should consult Csanady (1973), Okubo (1980) or Ozmidov (1986).

The relationship (8.35) applies only to horizontal diffusivity. For the diffusion of particles in the vertical direction, water stratification has to be taken into account. In an ocean with stable stratification, any vertical turbulent fluctuations of particles will be suppressed as more energy is needed to transfer more dense water to a higher level, or *vice versa*. The suppressing action of stable stratification may preclude the generation of turbulence and turbulent diffusion. The criteria for turbulent hydrodynamic instability and appearance of turbulence is known as the Richardson number Ri which, for stationary and horizontally uniform flow with stable stratification, takes the form:

$$
Ri = \frac{g}{\rho_w}\frac{\dfrac{\partial \rho_w}{\partial z}}{\left(\dfrac{\partial \bar{u}}{\partial z}\right)^2 + \left(\dfrac{\partial \bar{v}}{\partial z}\right)^2} = N\left[\left(\frac{\partial \bar{u}}{\partial z}\right)^2 + \left(\frac{\partial \bar{v}}{\partial z}\right)^2\right]^{-1},
\tag{8.36}
$$

in which ρ_w is the water density, $\bar{u}$ and $\bar{v}$ are mean flow velocities in the x and y directions, and N is the Brunt-Väisäla frequency (see Eq. 6.10).

Hydrodynamic instability and turbulence can appear when (Monin and Ozmidov, 1985):

$$Ri < \frac{1}{4}. \tag{8.37}$$

If the Richardson number is known, the coefficient of turbulent viscosity A_z can be approximated as (Peters $et\ al.$, 1988):

$$A_z \approx 5 \times 10^{-4}(1 + Ri)^{-3/2} + 2 \times 10^{-5}. \tag{8.38}$$

There is a convenient criterion to distinguish laminar and turbulent motion of ocean water, known as the Cox number Cx:

$$Cx = \left(\frac{\partial \vartheta'}{\partial z}\right)^2 \left(\frac{\partial \bar{\vartheta}}{\partial z}\right)^{-2}, \tag{8.39}$$

in which, for example, $\vartheta = T$ (temperature) or $\vartheta = S$ (salt). For $Cx \gg 1$, the motion is turbulent and for $Cx < 1$ the motion is laminar.

We note that turbulence can also be generated in wind-induced currents due to a strong shearing velocity. However, in this case the stability criterion is expressed as a function of the Reynolds number.

8.3.3 Shear Flow Dispersion

Brief Orientation. Hitherto in the discussion of mixing phenomena, we neglected the presence of any boundaries, such as the sea surface or sea bottom, channel or pipe walls. What is the effect of these boundaries on the spreading of contaminants? It is a well known fact that spreading in rivers and estuaries in the direction of flow is primarily caused by the velocity profile across the river. As this mechanism of spreading is associated with shear flows, it is known as shear effect.

If we consider two molecules being carried in a river flow, one well above the bottom and one near the bottom, the rate of separation caused by the difference in advection velocity will exceed that caused by molecular motion. At each instant of time, the velocity of any single molecule is essentially that of the stream velocity at a given cross-sectional position plus the velocity of the random movement across the cross-section due to molecular diffusion. When we adopt a coordinate system moving at the mean velocity, these random steps (with respect to this coordinate system) will have the same probability of being either backward or forward.

After a sufficiently long time, the motion becomes similar to the random walk, discussed in Sect. 8.2.2, and we should expect that Fick's equation (8.22) describes the particles spreading in the river. However, to include both diffusion effects and velocity shear, or possible circulation effects, we use the term **dispersion** rather than diffusion. Also, the value of coefficient K, which we rename as the **dispersion coefficient**, will be different as it results from the process of dispersion due to shearing.

Theory of shear flow dispersion was initiated by Sir Geoffrey Taylor, who, in his paper published in 1953, described the spreading of dissolved contaminants in laminar flow through a pipe (Taylor, 1953). His method was later extended to the flows between two plates, in open channels and estuaries. In the next subsections, following Fisher *et al.* (1979), we will describe Taylor's method. In Chap. 13, we will apply these results for shear flows in coastal zones and estuaries.

Dispersion of Contaminants in Laminar Flow Between Parallel Walls.
Let us consider flow between two parallel walls, separated by a distance h. The velocity distribution is given by $u(z)$ (see Fig. 8.5). The mean velocity of flow, $\bar{u}$, is defined by the integration, *i.e.*:

$$\bar{u} = \frac{1}{h} \int_{-h/2}^{h/2} u(z)dz. \tag{8.40}$$

We assume that the flow carries a contaminant with concentration $c(y)$ and molecular diffusion coefficient D. Similarly to Eq. (8.40), we define the mean concentration at any cross-section of conduit as:

$$\bar{c} = \frac{1}{h} \int_{-h/2}^{h/2} c(z)dz. \tag{8.41}$$

Subtracting the mean values $\bar{u}$ and $\bar{c}$ from $u(x, z)$ and $c(x, z)$, we obtain the deviations of the velocity and concentration, respectively:

$$\left. \begin{array}{rcl} u'(z) & = & u(x, z) - \bar{u} \\ c'(z) & = & c(x, z) - \bar{c} \end{array} \right\}. \tag{8.42}$$

Hence, the advection/diffusion equation (8.19) takes the form:

$$\frac{\partial(\bar{c} + c')}{\partial t} + (\bar{u} + u')\frac{\partial(\bar{c} + c')}{\partial x} = D\left[\frac{\partial^2(\bar{c} + c')}{\partial x^2} + \frac{\partial^2 c'}{\partial z^2}\right]. \tag{8.43}$$

The terms $\partial\bar{c}/\partial z = \partial^2\bar{c}/\partial z^2 = 0$, as the flow is only in the x direction.

To simplify Eq. (8.43), a transformation of the coordinate system is introduced, such as (see Fig. 8.5):

$$x_1 = x - \bar{u}t, \quad \tau = t, \quad z = z. \tag{8.44}$$

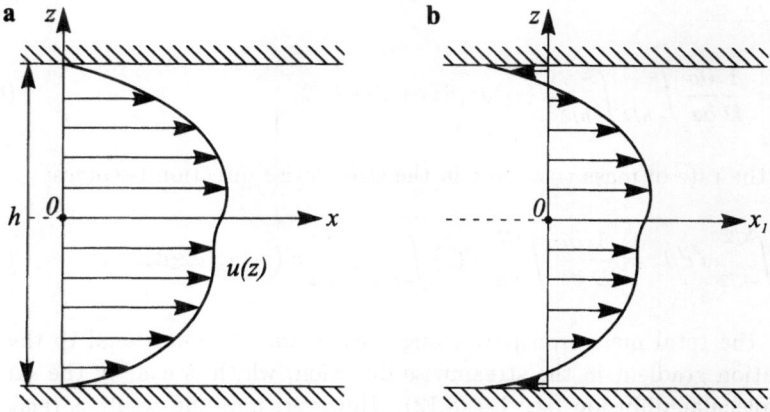

Fig. 8.5: Velocity distribution between parallel plates: **a** velocity distribution in a fixed coordinate system, **b** velocity distribution in a coordinate system moving at the mean velocity

After substituting Eq. (8.44) into Eq. (8.43) and using the 'chain rule' of differentiation we obtain:

$$\frac{\partial(\bar{c}+c')}{\partial\tau} + u'\frac{\partial(\bar{c}+c')}{\partial x_1} = D\left[\frac{\partial^2(\bar{c}+c')}{\partial x_1^2} + \frac{\partial^2 c'}{\partial z^2}\right]. \tag{8.45}$$

As we mentioned in the previous section, the spreading along the direction of flow due to shear effect dominates over molecular diffusion. Therefore, the longitudinal diffusion term in Eq. (8.45) can be neglected, *i.e.*:

$$\frac{\partial\bar{c}}{\partial\tau} + \frac{\partial c'}{\partial\tau} + u'\frac{\partial\bar{c}}{\partial x_1} + u'\frac{\partial c'}{\partial x_1} = D\frac{\partial^2 c'}{\partial z^2}. \tag{8.46}$$

As the general solution of Eq. (8.46) can not be found, Taylor (1953) examined the orders of magnitude of the particular terms in this equation and found that Eq. (8.46) can be substantially simplified as follows:

$$u'\frac{\partial\bar{c}}{\partial x_1} = D\frac{\partial^2 c'}{\partial z^2}. \tag{8.47}$$

It should be noted that the term $\partial\bar{c}/\partial\tau$, which we are interested in, has also been discarded. For a more detailed justification of the simplification of Eq. (8.46), the reader should consult Taylor's (1953) paper, or Fisher *et al.* (1979) book.

The solution of Eq. (8.47), under the condition $\partial c'/\partial z = 0$ at $z = -h/2$ and $z = h/2$, is:

$$c'(z) = \frac{1}{D}\frac{\partial \bar{c}}{\partial x}\int_{-h/2}^{z}\int_{-h/2}^{z_2} u'(z_1)dz_1 dz_2 + c'(-h/2). \tag{8.48}$$

Now, the rate of mass transport in the streamwise direction becomes:

$$Q = \int_{-h/2}^{h/2} u'c'dz = \frac{1}{D}\frac{\partial \bar{c}}{\partial x}\int_{-h/2}^{h/2} u'(z)\int_{-h/2}^{z}\int_{-h/2}^{z_2} u'(z_1)dz_1 dz_2 dz. \tag{8.49}$$

Thus, the total mass transport along the stream is proportional to the concentration gradient in the streamwise direction, which is exactly the same as for molecular diffusion (see Eq. 8.12). However, now the result is related to the flow integrated over the whole cross-section. Using this similarity, we can rewrite Eq. (8.49) in the form:

$$Q = -hK_x\frac{\partial \bar{c}}{\partial x}. \tag{8.50}$$

The coefficient K_x, although similar to the diffusion coefficient, expresses the diffusive property of the velocity distribution; it is known as the longitudinal dispersion coefficient. So, from Eqs. (8.49) and (8.50) we have:

$$K_x = \frac{-1}{hD}\int_{-h/2}^{h/2} u'(z)\int_{-h/2}^{z}\int_{-h/2}^{z_2} u'(z_1)dz_1 dz_2 dz. \tag{8.51}$$

The longitudinal dispersion coefficient, K_x, is a measure of the diffusive property of the velocity on the macroscopic scale of all sections, while the diffusion coefficient, D, is a measure of the molecular diffusivity of fluid on a microscopic scale. Using Eq. (8.50), we can stipulate that the one-dimensional dispersion equation has a form similar to Eq. (8.19):

$$\frac{\partial \bar{c}}{\partial t} + \bar{u}\frac{\partial \bar{c}}{\partial x} = K_x\frac{\partial^2 \bar{c}}{\partial x^2}. \tag{8.52}$$

It is straightforward to extend similar analysis to turbulent shear flow. The only differences are a somewhat different velocity profile and the possibility that the turbulent dispersion coefficient, K, may be a function of cross-sectional position, y. For uniform flow and in the absence of density gradients, the dispersion coefficient is constant and equal to the turbulent diffusion coefficient. In non-uniform flow, when the transported constituent induces density gradients, the dispersion coefficient depends on the distance.

Extension of Taylor's method for unsteady shear flow and dispersion in two dimensions are treated in the books by Csanady (1973) and Fisher et al. (1979).

8.3.4   Some Workable Solutions of the Diffusion Equations

The diffusion equation (8.30) is widely used for the determination of concentration of particles of various types of contaminants in ocean waters. The general form of Eq. (8.30) can only be solved using numerical techniques (Fisher, 1981; Ozmidov, 1986; Noye, 1987; Koutitas, 1988; Partheniades, 1992). However, in some simpler cases, this equation can be solved analytically. Although simplifications are introduced into the equation, analytical solutions remain very useful for a quick estimation of the required concentration, as well as serving as test references for numerical schemes. In this chapter we will discuss a few examples of such simplified solutions of the diffusion equation. Computer programs, ready for use, are given on the CD (see Appendix D).

Instantaneous Release of Substances in Steady Uniform Flow. Let us assume a uniform one-dimensional current with mean velocity, $\bar{u}$, flowing in the x direction. At time $t = 0$, a mass M of some contaminant is instantaneously released at the ocean surface, in the vicinity of point $x = 0$. The distribution of the initial concentration $\bar{c}(x, 0)$ takes the form of the Gaussian curve, *i.e.*:

$$\bar{c}(x,0) = \frac{M}{\sqrt{2\pi}\sigma} \exp\left(-\frac{x^2}{2\sigma^2}\right), \tag{8.53}$$

in which M (kg/m) is the mass of contaminant per unit width, and σ (m) is the parameter characterizing the spreading of the initial contaminant distribution. The contaminant has the same specific gravity as the ambient water, and the coefficient of turbulent diffusion for the contaminant, K_x, is constant. Under these assumptions, a governing equation for $\bar{c}(x, t)$, resulting from Eq. (8.30), becomes:

$$\frac{\partial \bar{c}}{\partial t} + \bar{u}\frac{\partial \bar{c}}{\partial x} = K_x \frac{\partial^2 \bar{c}}{\partial x^2}. \tag{8.54}$$

The solution of this equation, with initial condition (8.53), takes the form (Noye, 1987):

$$\bar{c} = \frac{M}{\sqrt{2\pi(\sigma^2 + 2K_x t)}} \exp\left[-\frac{(x - \bar{u}t)^2}{2(\sigma^2 + 2K_x t)}\right]. \tag{8.55}$$

Solution (8.55) is illustrated in Fig. 8.6. The mass of contaminant (for example, such as solution of fluoresceine dye) released is 1 kg/m and coefficient σ, controlling the initial distribution, is assumed to be equal to 0.125 m. The coefficient of turbulent diffusion, K_x, is set as $K_x = 500$ cm^2/s $= 0.05$ m^2/s. This value is very close to that observed during experiments with fluoresceine in the Black Sea (Ozmidov, 1986). It was also assumed that current velocity $\bar{u}$ is constant and equal to 1 m/s. In Fig. 8.6, three particular time instants

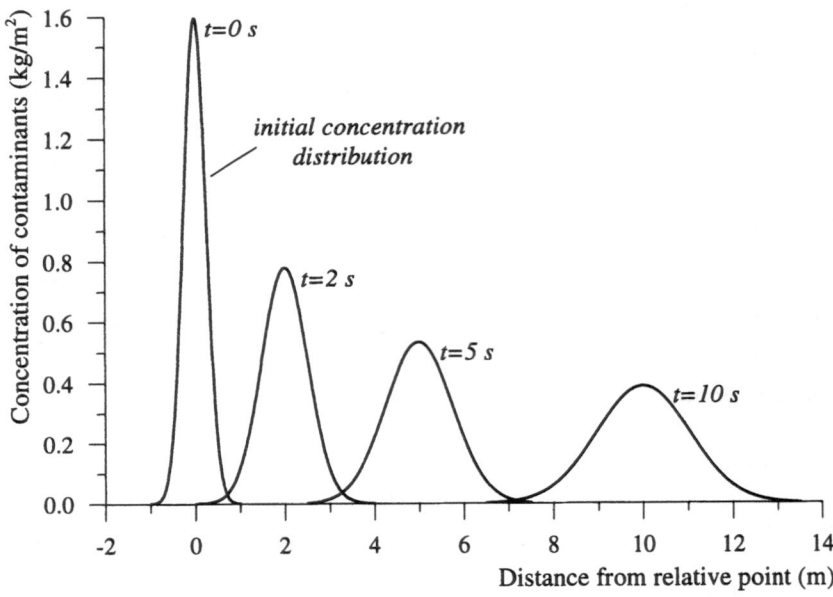

Fig. 8.6: Spreading of contaminant in time and space in a steady uniform flow

($t = 2$ s, 5 s, and 10 s) have been selected. As time increases, the contaminant moves along the x axis and concentration becomes more widely spread. Peak concentration (moving at speed $\bar{u}$) decreases in the positive x direction. As $t \to \infty$, $\bar{c} = 0$ for all x. As there are no sources of sinks of mass, the amount of contaminant at each time t is constant and equal to M.

If the mass of the tracer material is introduced initially at time $t = 0$, in a very small area ($\sigma \to 0$ in Eq. 8.53) around point $x = 0$, the concentration approaches infinity in such a way that the area under the initial curve (which represents the total mass of the substance) is constant and equal to M. This property can be represented by a function known as the delta function $\delta(x)$ (or Dirac delta function). The delta function has two main properties:

$$\delta(x - x_0) = \begin{cases} \infty & \text{when } x = x_0 \\ 0 & \text{when } x \neq x_0, \end{cases} \tag{8.56}$$

and

$$\int_{-\infty}^{\infty} \delta(x - x_0)dx = 1. \tag{8.57}$$

As in our case $x_0 = 0$, the initial condition (8.53) becomes (with $\delta \to 0$):

$$\bar{c}(x,0) = M\delta(x), \tag{8.58}$$

and evolution of the spreading substance can be found from Eq. (8.55) for $\sigma = 0$ as:

$$\bar{c} = \frac{M}{\sqrt{4\pi K_x t}} \exp\left[-\frac{(x - \bar{u}t)^2}{4K_x t}\right].$$

(8.59)

Similarly as above, the tracer material moves downstream and spreads out as a symmetrical Gaussian curve.

Advance Front of Substance in a Steady Uniform Flow. Instead of a finite quantity introduced instantaneously at $x = 0$ and $t = 0$, it is assumed that there is continuous input of substance from time $t = 0$, at a rate of $Q_i = dM/d\tau$ in each increment of time $d\tau$. The concentration increment, $d\bar{c}$, due to an input dM, at time τ can be found from Eq. (8.59) as (Ippen, 1966):

$$\bar{c}(x, t) = \frac{Q_i}{\sqrt{4\pi K_x}} \int_0^t \frac{1}{\sqrt{t - \tau}} \exp\left\{-\frac{[x - \bar{u}(t - \tau)]^2}{4K_x(t - \tau)}\right\} d\tau.$$

(8.60)

This is not a trivial integral and details of integration are given in Appendix C.7. Using Eqs. (C.172) and (C.173) we finally obtain:

$$\bar{c}(x, t) = \frac{2Q_i}{\sqrt{\pi \bar{u}}} \exp(2a) \int_0^c \exp\left[-\left(\frac{a}{x}\right)^2 - x^2\right] dx,$$

(8.61)

or:

$$\bar{c}(x, t) = \frac{Q_i}{2\bar{u}} \left\{\exp(4a)\left[\mathrm{erf}\left(c + \frac{a}{c}\right) - 1\right] + \left[\mathrm{erf}\left(c - \frac{a}{c}\right) + 1\right]\right\},$$

(8.62)

where:

$$a = \frac{\bar{u}x}{4K_x}, \qquad c = \frac{\bar{u}}{2}\sqrt{\frac{t}{K_x}}.$$

(8.63)

Function $\mathrm{erf}(x)$ is known as the error function and its definition is given by Eq. (C.174).

In Fig. 8.7 the evolution in time of the concentration of a substance at given distances is shown. The rate of discharge of substance is $Q_i = 1$ kg/s, the velocity of current is $\bar{u} = 0.2$ m/s, and the coefficient of turbulent diffusion K_x is 0.075 m²/s. At distance $x = 10$ m from the release point, the front of the substance gradually advances after about 20 s following release, and full 'saturation' is achieved after about 90 s. For distances $x = 20$ m and $x = 30$ m, the times of 'saturation' are about 180 s and 250 s, respectively. The concentration at full saturation can be calculated from Eq. (8.62) as follows:

$$\lim_{t \to \infty, b \to \infty} \bar{c}(x, t) = \frac{Q_i}{\bar{u}}.$$

(8.64)

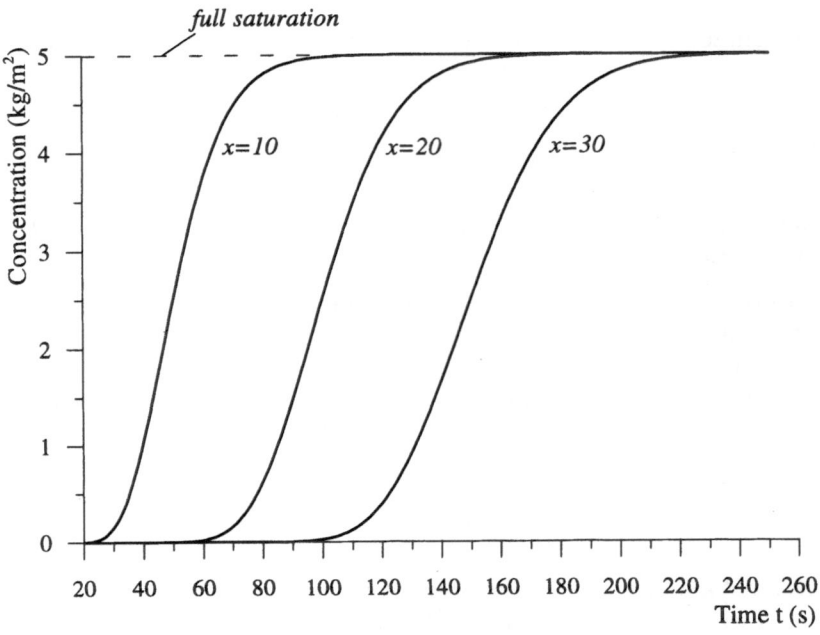

Fig. 8.7: Evolution in time of concentration of substance at particular distances from release point

For the values used in Fig. 8.7, the concentration of full saturation equals 5 kg/m.

Steady Release of a Substance on a Plane in a Steady Uniform Flow. We now consider the case when a contaminant is introduced steadily at one point on the sea surface. The plume of contaminant moves downstream (current velocity $\bar{u}$ is constant) and spreads out at the sea surface as well as below. Ozmidov (1986) suggests using the following variant of the basic advection-diffusion equation (8.30):

$$\bar{u}\frac{\partial \bar{c}}{\partial x} = K_y \frac{\partial^2 \bar{c}}{\partial x^2} + K_z \frac{\partial^2 \bar{c}}{\partial z^2}. \tag{8.65}$$

The contaminant is introduced at point $x = y = z = 0$ at the rate of Q (kg/s). A solution of Eq. (8.65) becomes (Ozmidov, 1986):

$$\bar{c}(x, y, z) = \frac{2Q}{4\pi x \sqrt{K_y K_z}} \exp\left(-\frac{\bar{u}y^2}{4K_y x} - \frac{\bar{u}z^2}{4K_z x}\right). \tag{8.66}$$

In particular at the sea surface ($z = 0$), Eq. (8.66) becomes:

$$\bar{c}(x, y, 0) = \frac{2Q}{4\pi x \sqrt{K_y K_z}} \exp\left(-\frac{\bar{u}y^2}{4K_y x}\right). \tag{8.67}$$

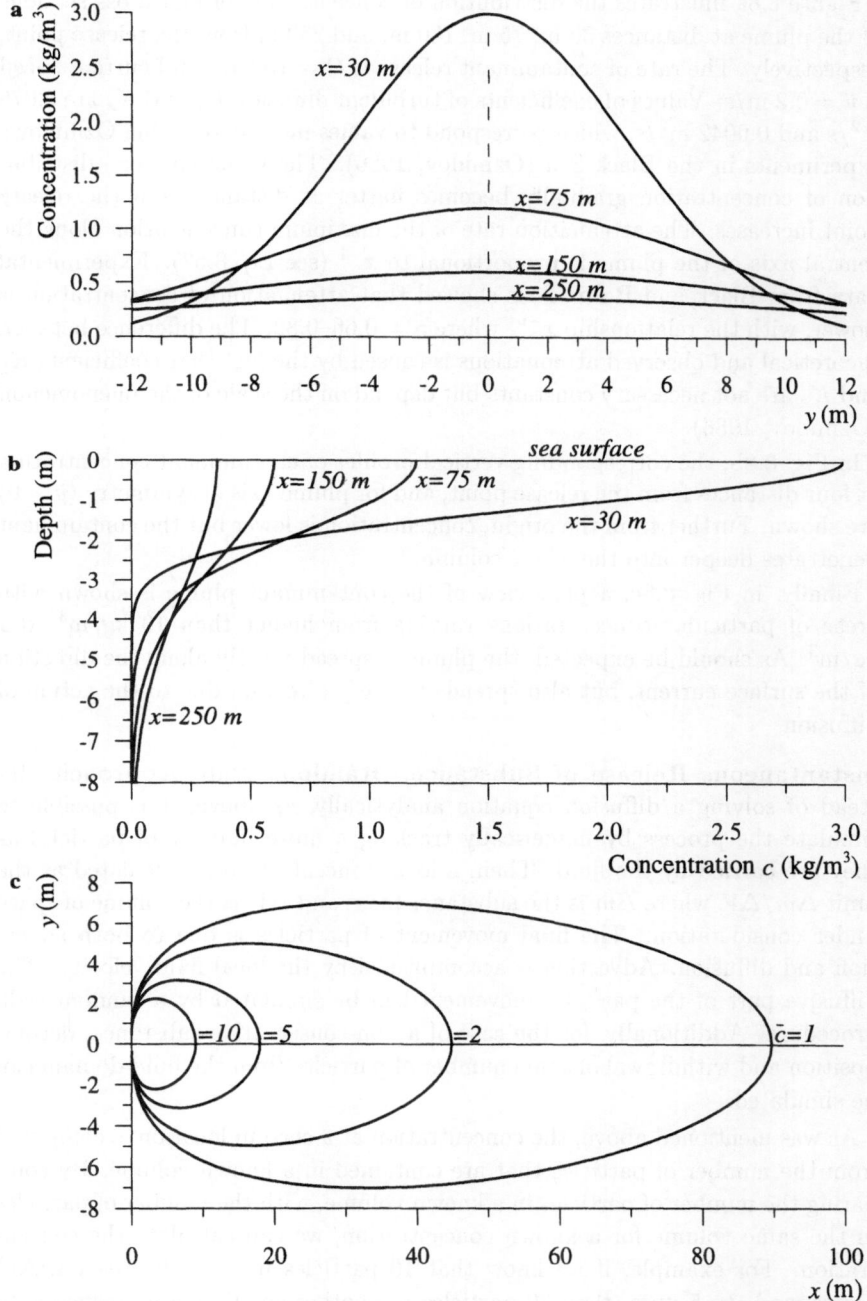

Fig. 8.8: Concentration of plume: **a** at a few cross-sections of the surface plume, **b** vertical profiles of concentration at four distances from the point of release, **c** concentration plume distribution (top view)

Figure 8.8a illustrates the distribution of concentration of four cross-sections of the plume at distances 30 m, 75 m, 150 m, and 250 m from the release point, respectively. The rate of contaminant release is $Q = 10$ kg/s and current speed is $\bar{u} = 0.2$ m/s. Values of coefficients of turbulent diffusion K_y and K_z are 0.075 m^2/s and 0.0042 m^2/s, which correspond to values measured during Ozmidov's experiments in the Black Sea (Ozmidov, 1986). The Gaussian type distribution of concentration gradually becomes flatter as distance from the release point increases. The attenuation rate of the maximum concentration along the central axis of the plume is proportional to x^{-1} (see Eq. 8.67). Experimental data from Black and Baltic Seas showed that attenuation of concentration is slower, with the relationship x^{-n}, where $n = 0.66$–0.82. The difference between theoretical and observed attenuations is caused by the fact that coefficients K_y and K_z are not necessary constants but depend on the scale of the phenomenon (Ozmidov, 1986).

In Fig. 8.8b, the corresponding vertical profiles of contaminant concentration at four distances from the release point, and for plume axis of symmetry ($y = 0$) are shown. Further from the origin, concentration is lower but the contaminant penetrates deeper into the water column.

Finally, in Fig. 8.8c, a plan view of the contaminant plume is shown with areas of particular concentrations varying from higher than 10 kg/m^3 to 1 kg/m^3. As should be expected, the plume is spread mostly along the direction of the surface current, but also spreads in the y direction due to the action of diffusion.

Instantaneous Release of Substance. Random Walk Approach. Instead of solving a diffusion equation analytically, as above, it is possible to simulate the process by numerically tracking a finite number of particles as they are carried by the fluid. Then, a local concentration is calculated as the limit $\Delta m / \Delta V$ where Δm is the substance mass and ΔV is the volume of space under consideration. The final movement of particles is due to both advection and diffusion. Advection is accomplished by the local fluid velocity. The diffusive part of the particles movement can be simulated by a random walk procedure. Additionally, for the case of a non-conservative substance, decomposition and withdrawal of some number of particles from the fluid domain can be simulated.

As was mentioned above, the concentration at a certain location is computed from the number of particles that are contained in a known volume. By comparing the number of particles in a known volume, with the number of particles in the same volume for a known concentration, we can calculate the concentration. For example, if we know that 10 particles in a small area ($\Delta x \Delta y$) corresponds to 5 ppm, then 30 particles in another small area corresponds to a concentration of 15 ppm.

A determination of the individual particle positions in the Lagrangian description is usually based on the Monte Carlo simulation technique. Let us assume that there are n elements of substance at the sea surface where the coordinates of ith element are $x_i^{(0)}$ and $y_i^{(0)}$ at a given time. After a time step Δt, due to advection the coordinate of the ith particle becomes:

$$\left.\begin{array}{rcl} x_i & = & x_i^{(0)} + u\Delta t \\ y_i & = & y_i^{(0)} + v\Delta t \end{array}\right\}. \tag{8.68}$$

At the same time, the particle diffuses in the medium. As was shown in Sect. 8.2.2, on average the position of a particle along each axis follows a normal distribution with variances:

$$\sigma_x^2 = 2K_x t \quad \text{and} \quad \sigma_y^2 = 2K_y t. \tag{8.69}$$

In the following we assume that $K_x = K_y = K_h$; thus, the root-mean-square distance of a particle travelling in time Δt due to diffusion is:

$$l_{rms} = \sqrt{\sigma_x^2 + \sigma_y^2} = \sqrt{4K_h\Delta t}. \tag{8.70}$$

The position of any individual particle due to diffusion in time Δt is generated randomly using one of the available random number generator procedures for the interval $[0, s]$. Thus, the horizontal diffusion step size, l, becomes (Al-Rabek and Gunay, 1992):

$$l = [R]_0^s, \tag{8.71}$$

in which $[R]_0^s$ is the standardized random number in the interval 0 to s. The value of s should be chosen so that l_{rms} is equal to the mean square of all values of l. Usually the random number generators return values in the interval $[0, 1]$. Therefore, the standard deviation of the set of all numbers generated by the random number generator is:

$$\left\{\int_0^1 l^2 dl\right\}^{1/2} = \left\{\int_0^1 \left([R]_0^1\right)^2 d[R]_0^1\right\}^{1/2} = \frac{1}{\sqrt{3}}. \tag{8.72}$$

Thus, the distance that any particle travels by diffusion becomes:

$$l = [R]_0^1 \sqrt{12K_h\Delta t}. \tag{8.73}$$

The value l from Eq. (8.73) guarantees that the root-mean-square distance of a particle travelling during time Δt satisfies Eq. (8.70).

In order to find the new position of a particle after time increment Δt, the diffusion component must be added to the particle travel due to advection:

$$\left.\begin{array}{rcl} x_i & = & x_i^{(0)} + u\Delta t + l\cos\theta \\ y_i & = & y_i^{(0)} + v\Delta t + l\sin\theta \end{array}\right\}, \tag{8.74}$$

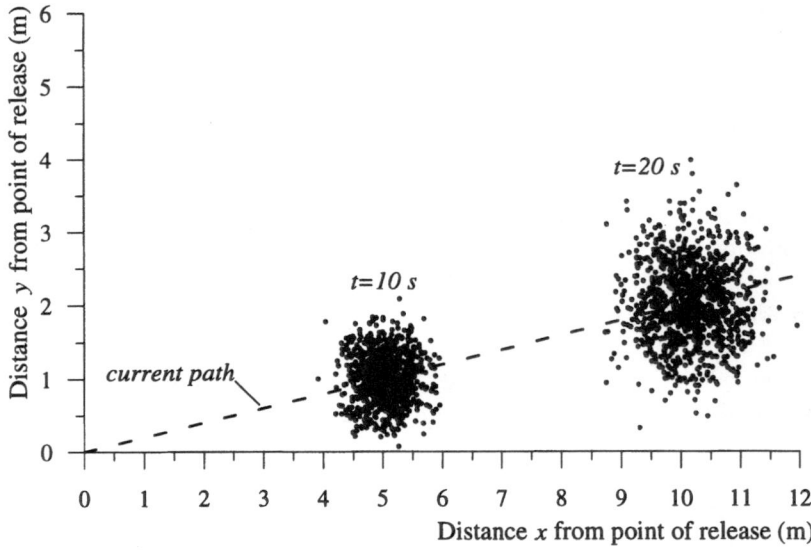

Fig. 8.9: Particles spreading

where the angle θ is assumed to be uniformly distributed in the range between 0 and 2π; this angle can be found from the random number generator as follows:

$$\theta = 2\pi \, [R]_0^1. \tag{8.75}$$

Values of the diffusion coefficient should ideally be determined from field studies, and in order to obtain a sufficiently smooth distribution, a very large number of elements must be used. As was pointed out by Hunter (1987), the random walk type solution is significantly more efficient in cases of higher dimension and also where the substance resides in a patch that occupies only a portion of the total model area. However, in each case, the solution of the diffusion equation provides an asymptotic solution to random walk problems and conversely, an approximate solution to the diffusion equation with complex boundary conditions may be obtained from the Monte Carlo simulation method.

In Fig. 8.9 an example of a simulation of discrete particle spreading, released at the origin of reference at the sea surface is shown. The current velocity components are: $u = 0.5$ m/s and $v = 0.1$ m/s, while the coefficient of diffusion $K_h = 0.01$ m^2/s. The particle spreading is shown for time $t = 10$ s and $t = 20$ s from release, with 1000 particles used for this simulation. The dashed line indicates the current path. 'Density' of the plume significantly decreases as time increases. It should be noted that both plumes are almost circular due to the fact that the coefficients of turbulent diffusion, K_x and K_y, have been assumed equal to each other.

General Cases of Diffusion. In the scenarios discussed above, relatively simple environmental conditions have been applied. In particular, coefficients of turbulent diffusion were assumed to be constant and water depth to be infinite. However, in many coastal and estuary problems, as well as for more detailed descriptions of diffusion and mixing phenomena in the deep ocean, a solution of the full Eq. (8.30) is required. Such a solution can only be achieved by numerical solution.

In general, two main methods are used. In the first method, the basic system of equations (for water circulation, and diffusion equation) is solved numerically, using finite difference or finite element techniques in the Eulerian frame of reference. Examples of applications of such techniques can be found in Fischer (1981) and Ozmidov (1986). We will return to these techniques in Sect. 8.5, where mixing processes in estuaries are discussed. The second group are Lagrangian tracking methods, such as the one described in the previous case in this section. These methods are very flexible and can be applied to three-dimensional space as well as allowing for various types of boundary conditions, such as 'perfectly absorbing surfaces' and reflecting barriers (Csanady, 1973; Fischer et al., 1979).

Dispersion in Shear Flow Between Parallel Plates and in a Tube. Let us consider laminar flow between a pair of flat plates separated by a distance h (Fig. 8.5), with the velocity profile given by Eq. (2.107), i.e:

$$u(z) = \frac{\Delta p}{2\mu l}\left[\left(\frac{h}{2}\right)^2 - z^2\right] = \frac{\Delta p h^2}{8\mu l}\left(1 - \frac{4z^2}{h^2}\right),$$

(8.76)

where h is the distance between plates, Δp is the pressure drop over the plate length l (l is assumed to be large), and μ is the coefficient of dynamic viscosity. Suppose now that a tracer material is injected between the plates, and the time elapsed since the injection is sufficient for complete mixing of the tracer. The diffusion coefficient of the tracer is equal to D. From Eq. (2.109) it follows that the mean velocity, $\bar{u}$, is:

$$\bar{u} = \frac{\Delta p h^2}{12\mu l};$$

(8.77)

thus, the deviation velocity, u', becomes:

$$u'(z) = \frac{\Delta p}{2\mu l}\left(\frac{h^2}{12} - z^2\right).$$

(8.78)

Using Eq. (8.48) we determine the concentration deviation $c'(z)$ as:

$$c'(z) = \frac{1}{D}\frac{\partial \bar{c}}{\partial x}\int_{-h/2}^{z}\int_{-h/2}^{z_2} u'(z_1)dz_1dz_2 + \text{Const} =$$

$$= \frac{1}{D}\left(\frac{\Delta p}{2\mu l}\right)\frac{\partial \bar{c}}{\partial x}\left(-\frac{h^4}{192}+\frac{h^2}{24}z^2-\frac{z^4}{12}\right) + c'\left(-\frac{h}{2}\right). \qquad (8.79)$$

The concentration, c', at $z = -h/2$ can be found from the condition that the average value of c' over the cross-section must be zero. Thus, we obtain:

$$c'\left(\frac{h}{2}\right) = \frac{1}{D}\frac{\partial \bar{c}}{\partial x}\left(\frac{\Delta p}{2\mu l}\right)\frac{h^4}{360}. \qquad (8.80)$$

After substitution Eq. (8.80) into Eq. (8.79) we get:

$$c'(z) = \frac{1}{D}\left(\frac{\Delta p}{2\mu l}\right)\frac{\partial \bar{c}}{\partial x}\left(\frac{-7h^4}{2880}+\frac{h^2}{24}z^2-\frac{z^4}{12}\right). \qquad (8.81)$$

From Eq. (8.51), we obtain the dispersion coefficient K_x in the form:

$$K_x = -\frac{1}{h\frac{\partial \bar{c}}{\partial x}}\int_{-h/2}^{h/2} u'(z)c'(z)dz, \qquad (8.82)$$

and:

$$K_x = \frac{1}{7560}\frac{h^6}{D}\left(\frac{\Delta p}{2\mu l}\right)^2, \qquad (8.83)$$

or:

$$K_x = \frac{2}{945}\frac{h^2}{D}u_{\max}^2, \qquad (8.84)$$

when Eq. (2.109) is used. Note that the longitudinal dispersion coefficient is inversely proportional to the molecular diffusion coefficient D. Let us assume that the tracer material injected between plates of $h = 0.005$ m is salt with $D \approx 10^{-9}$ m^2/s, and let the velocity $u_{\max} = 0.01$ m/s. After substituting these values into Eq. (8.84) we obtain the longitudinal dispersion coefficient $K_x \approx 5\times 10^{-3}$ m^2/s, which is about a million times the magnitude of molecular diffusion coefficient D.

In a similar way we can find the longitudinal dispersion coefficient for laminar flow in a tube (Fischer et al., 1979):

$$K_x = \frac{1}{192}\frac{a^2}{D}u_{\max}^2, \qquad (8.85)$$

in which a is the tube radius.

Other examples of the determination of dispersion coefficients will be given in Sect. 8.5 and Chap. 13.

8.4 Diffusion and Mixing in the Ocean

8.4.1 Introduction

In the open ocean, the vertical structure of the water mass is characterized by the presence of one or more thermoclines, which are regions of large temperature gradient within the upper few hundred meters. The thermocline structure is strongest in equatorial waters where there is a great contrast between the warmed surface layer and the cold deeper water. This contrast weakens at increasing latitudes.

In the upper ocean, from the thermocline to the surface, there is a large variety of water motion such as currents, turbulence, internal and surface waves. All of these motions are responsible for the exchange of matter, momentum and heat between atmosphere and the underlying deep ocean water. Some of these mechanisms have been discussed in previous chapters; in this section we will concentrate on diffusion processes and the resulting mixing in stratified ocean waters.

From the perspective of this book, one of the most important transport processes is the diffusion of gases in the water column and the dispersion of various biogenic elements. For consistency, these problems are left to Part III, dedicated to the application of various physical results, reported in Parts I and II, to marine ecology.

The upper ocean layer is the subject of constant bombardment by a large variety of particles coming from the atmosphere. Matter of atmospheric origin penetrates the ocean in different ways. In order to track the intensity of substance penetration, tracers which are easy to detect are required. One type of such tracers are radioactive isotopes. Generally, in the ocean we observe three types of isotopes: terrigenic isotopes, which are part of the Earth's core, cosmic isotopes coming from the cosmos and surrounding atmosphere, and isotopes resulting from human activity. The first two types of isotopes are supplied to the ocean surface at an almost constant rate, however, some seasonal variation is observed for cosmic isotopes.

Radioactive isotopes are not the only tracers used to study mixing in the ocean. In the previous section, we described the use of fluoresceine dye for mixing in experiments in the Black Sea. During the North Atlantic Tracer Release Experiment (NATRE) in 1992, sulfur hexafluoride was released within a few metres of the $\sigma_t = 28.036$ isopycnal surface, near 300 m depth, in the Canary Basin in the Atlantic Ocean (Ruddick and Walsh, 1997). The lateral dispersion and vertical spreading were observed during the following year.

In the next section we describe the main features of the mixing processes in the upper ocean using the results of tracer releases. We start with radioactive isotope experiments.

8.4.2 Diffusion of Tracers in the Ocean

Radioactive Isotopes. Atmospheric nuclear experiments in the 1950s and 1960s resulted in a large number of radioactive elements which finally settled at the ocean surface and penetrated deep in the ocean. After nuclear explosions in the Bikini and Enewetak atolls in the Pacific Ocean, it was found that radioactive elements had been transported a distance of 2000 km during four months and a distance of 7000 km during eight months (Ozmidov, 1986). Among radioactive elements which pose a large danger for human and ocean organisms are isotopes of Caesium ^{137}Cs and Strontium ^{90}Sr with half-lives of about 30 years, and Cerium ^{144}Ce with a half-life of about one year. These isotopes are almost totally diluted in ocean water and are therefore very useful tracers for studying diffusion.

In the upper ocean layer, above the thermocline, water mixing is intensive and radioactive elements reach the thermocline rather quickly in comparison with the characteristic time for diffusion in the deeper ocean. For isotopes with a long half-life, the coefficient of vertical diffusion, K_z, is usually treated as a constant value. Additionally, at the sea surface, the flux of radioactive elements is assumed to be known, *i.e.*:

$$K_z \left. \frac{\partial \bar{c}}{\partial z} \right|_{z=0} = \begin{cases} 0 & \text{for} \quad t < 0 \\ Q & \text{for} \quad t \geq 0. \end{cases} \tag{8.86}$$

For large areas of the ocean surface, the horizontal gradients of radioactive element concentrations are negligible. The vertical advection velocity, w, is usually also small. Thus, the concentration of radioactive elements can be found from the solution of an equation similar to Eq. (8.22), but with a different boundary condition at the sea surface (Eq. 8.86), and with a turbulent diffusion coefficient, K_z, instead of the molecular diffusion coefficient D, *i.e.*:

$$\frac{\partial \bar{c}}{\partial t} = K_z \frac{\partial^2 \bar{c}}{\partial z^2}. \tag{8.87}$$

The solution of Eq. (8.87) becomes:

$$\bar{c}(z,t) = \frac{2Q\sqrt{t}}{\sqrt{\pi K_z}} \left\{ \exp\left(-\zeta^2\right) - \sqrt{\pi}\zeta \left[1 - \text{erf}(\zeta)\right] \right\}, \tag{8.88}$$

in which $\zeta = z/\sqrt{K_z t}$. Error function $\text{erf}(x)$ is defined by Eq. (C.174).

Experiments have shown that the coefficients of turbulent diffusion of the isotope ^{90}Sr in the Atlantic and Pacific Oceans are about 17.2 cm^2/s and 6 cm^2/s, respectively. In Fig. 8.10, the vertical distribution of concentration for Atlantic conditions is shown. The flux of ^{90}Sr from the atmosphere was determined as 7.4×10^{-6} Bq/m^2/s using Ozmidov (1986) data. The unit of radioactivity, becquerel, (Bq), named after the discoverer Henri Becquerel (1852–1908), is defined as one decay per second. Fig. 8.10 shows that it takes almost 3 years for the ^{90}Sr to reach the water depth of 750 m in the Atlantic Ocean.

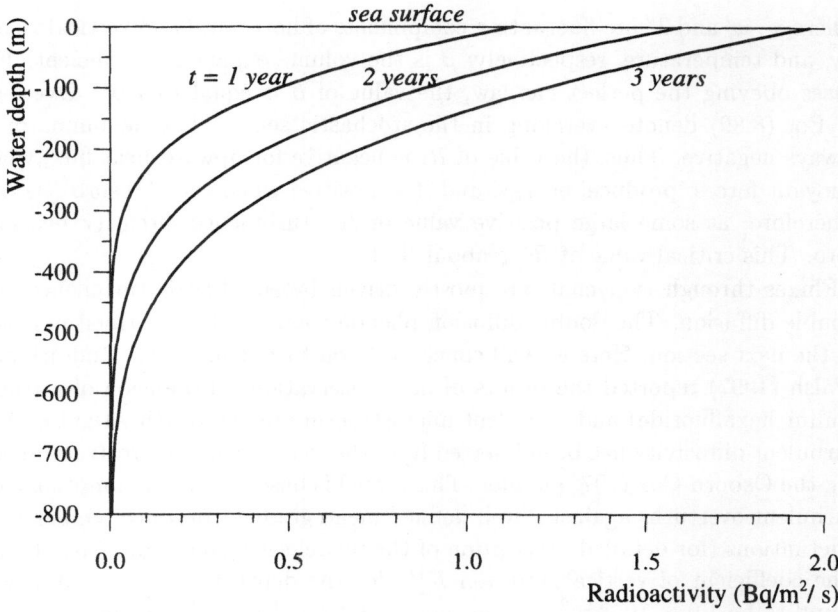

Fig. 8.10: Vertical distribution of concentration for the Atlantic Ocean

8.4.3 Diffusion in a Continuously Stratified Ocean

In previous examples of diffusion, ocean water was treated as homogeneous, with constant density ρ_w. In fact, the ocean is continuously stratified due to heating in its upper layer by the Sun and due to the contents of a large variety of suspended or diluted substances. A particle warmer than the surrounding environment becomes lighter and experiences a positive buoyancy force, while a cooler particle becomes heavier and negatively buoyant. Buoyancy forces can affect diffusion by producing or absorbing turbulent energy and therefore determining the intensity of ocean turbulence.

Particles moving due to extra buoyancy (positive or negative) will enhance the diffusion of surrounding particles. Turbulent energy is continuously depleted by work against the buoyancy forces and the intensity of turbulence diminishes and can even disappear. In such a situation, any diffusion is only due to molecular movement which is a very inefficient mechanism when compared with turbulent diffusion.

As shown above, a convenient parameter for measuring the relative importance of buoyancy effects is the 'Richardson number', which can also be presented as the ratio of thermal to mechanical energy production (Csanady, 1973):

$$Ri = \frac{\beta g \overline{w'T'}}{\overline{u'w'} \dfrac{\partial u'}{\partial z}}, \tag{8.89}$$

where u', w' and T' are fluctuating components of horizontal and vertical velocity, and temperature, respectively; β is the volume expansion coefficient. For
gases obeying the perfect gas law, the value of β is equal to $1/T$. Overbars
in Eq. (8.89) denote averaging in the stochastic sense. The denominator is
always negative. Thus, the value of Ri is negative for upward heat flux, when
buoyant forces 'produce' energy, and it is positive when they 'absorb' energy.
Therefore, at some large positive value of Ri, turbulence intensity becomes
zero. This critical value of Ri is about 0.21.

Fluxes through isopycnals are mostly driven by small scale turbulence and
double diffusion. The double diffusion phenomenon will be described in detail
in the next section. Here we will concentrate on turbulence only. Ruddick and
Walsh (1997) reported the results of field observations of diffusion of a tracer
(sulfur hexafluoride) and turbulent microstructure in the North Atlantic. The
turbulent diffusivity has been inferred from the temperature microstructure using the Osborn-Cox (1972) model. This model is based on the assumption that
turbulent overturns against a well-defined mean gradient produce temperature
fluctuations (for detailed description of the model see Osborn and Cox (1972).
The coefficient of vertical diffusion $K_z^{(T)}$ for the depth range 240–340 m was
found to be 1.0×10^{-5} m^2/s in autumn and 2.2×10^{-5} m^2/s in spring. In the
experimental area, a mixed layer occupied the upper 60 m of the water column
during autumn, whereas during the following spring, the surface mixed layer
thickness increased to about 90 m.

8.4.4 Double Diffusion and Salt Fingers

In turbulent motion, heat and dissolved substance, such as salt, are exchanged
at the same rate of mixing. However, this is not the case at the level of
molecular diffusion. Molecular diffusion of salt is about two orders of magnitude
smaller than the molecular diffusion of heat. Thus, more energy is needed to
exchange salt on the molecular scale than it is to exchange heat. This difference
between diffusivity can cause specific instabilities in stratification (Tomczak
and Godfrey, 1994).

Let us consider a typical ocean thermocline. Temperature and salinity decrease with depth. Thus, the vertical salinity gradient itself would result in
an unstable density stratification. However, this potential instability is more
than compensated by the stabilizing vertical temperature gradient, because
this gradient is reduced much faster than the salinity gradient. If the temperature gradient is reduced, then it is no longer sufficient to compensate for
the salinity-induced instability. As a result, convection appears in the form
of narrow vertical tubes of rising low salinity water between narrow tubes of
sinking salty water, known as **salt fingers** (see Fig. 8.11), and the process is
known as **double diffusion**.

Evidence of the double diffusion mechanism first came from laboratory experiments (Turner, 1967; Taylor and Bucens, 1989). In such experiments, cool

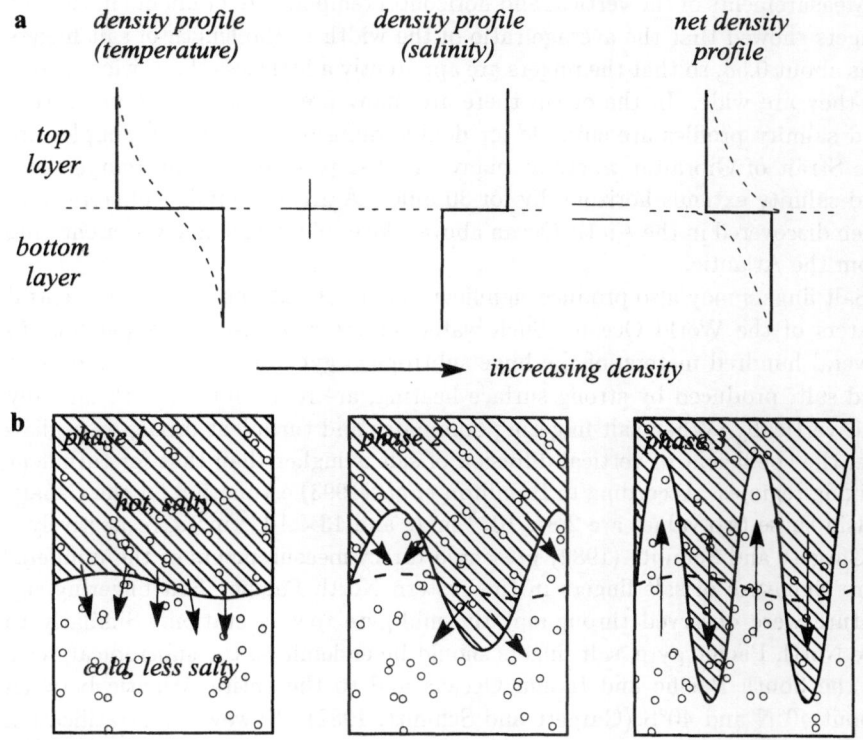

Fig. 8.11: Salt fingers dynamics: **a** initial profiles of density due to temperature and salinity, **b** salt fingers development (adapted from Gregg, 1973)

freshwater is usually introduced below a layer of warm salty water, as is shown in the density profiles in Fig. 8.11a. The solid lines show the initial density profile, whereas the broken lines show the situation a few minutes later. The rapid diffusion of heat has produced an unstable interface pattern (Fig. 8.11b). The open circles and arrows in Fig. 8.11 indicate salt and salt diffusivity, hatched areas denote higher density, while white areas denote lower density. A downward bulge cools the water above the interface very rapidly. This rapid loss of heat makes the water above the bulge heavier than its surroundings, so the water tends to sink (see downward arrow in the third box in Fig. 8.11b). On the other hand, the water below an upward bulge is warmer than its surroundings and tends to rise (see upward arrows the third box in Fig. 8.11b). The result is a convection pattern, in which alternate cells of salty water sink, while adjacent cells of less salty water rise. The salt-fingering pattern of convection persists only for a limited depth below the interface. Negative-buoyancy flux due to the descending saline water induces overturning turbulent motion which creates a well-mixed layer.

Measurements of the vertical and horizontal temperature gradients in the salt fingers showed that the average ratio of the width to the length of salt fingers was about 0.58, so that the fingers are apparently a little less than twice as long as they are wide. In the ocean there are many areas where the temperature and salinity profiles are suitable for double diffusion to occur. Examples are the Strait of Gibraltar where as many as 20 steps of decreasing temperature and salinity extends horizontally for 30 miles. A greater number of steps have been discovered in the Arctic Ocean above a layer of warm saline water entering from the Atlantic.

Salt fingers may also produce significant fluxes of salt and heat in the central waters of the World Ocean. Such water masses comprise the upper few, to several hundred meters, of the huge subtropical gyres in which excess of heat and salt, produced by strong surface heating, are redistributed both laterally and vertically. When salt fingering dominates and turbulent mixing is negligible, the coefficient of vertical diffusion of salt is higher than that for turbulent mixing regimes. According to Hamilton et al. (1993) estimation for the Canary Basin these two values are 2.8×10^{-5} m^2/s and 13×10^{-5} m^2/s, respectively.

Gargett and Schmitt (1982) presented direct measurements of the temperature signature of salt fingers in the eastern North Pacific. Salt-fingering signature were observed throughout the mid-gyre towing station. Similarly to the North Pacific gyre, salt fingers should be endemic to the subtropical gyres of the South Pacific and Indian Oceans and to the entire Atlantic between about 40°N and 40°S (Gargett and Schmitt, 1982). However, a stratification characterized by the presence of double diffusion of heat and salt, when the temperature and salinity gradients make opposing contribution to the density gradient, are not typical in the ocean. Usually a temperature decreases with depth while salinity increases, and stratification becomes stable. The reader should consult Monin and Ozmidov (1985) for an in-depth discussion.

8.5 Diffusion and Mixing in Estuaries

8.5.1 Introduction and Classification

The term 'estuary' is derived from the Latin word *estuarium* which means tidal. A typical estuary can be defined as a semi-enclosed coastal basin that receives an inflow of both freshwater and salt water, and is affected by tidal oscillations. There are various classifications of estuaries depending on the phenomena which are taken into account. In this section we will mostly concentrate on water dynamics, stratification and mixing in estuaries.

Under this assumption, estuaries can be assigned to one of three general types: *the salt-wedge estuary, the partly mixed estuary, and the well-mixed estuary* (Fig. 8.12). The relative contributions of river flow to tidal currents are largely responsible for the chemical, circulatory and sedimentologic characteristics of all these types of estuaries (Pinet, 1992).

Salt-wedge estuaries (Fig. 8.12a) are usually shallow and narrow, with respect to their water depth, and dominated by high river discharge. Tidal currents are weak due to a small tidal range. Fresh and salt water masses do not mix across the sharp halocline, resulting in a distinctive salt-wedge. At the interface, the river currents exert strong shear stresses which generate some entrainment of salt water into the fresh surface water. Continuity of mass requires a compensation of this salt water mixed upward into the freshwater by a net inflow of bottom water through the estuary mouth. This weak flow is known as residual current. In salt-wedge estuaries a significant amount of sediment is transported by river currents in contrast to a small amount of sediment supplied by tidal currents.

A second group of estuaries are those known as *partly mixed estuaries* (Fig. 8.12b). In partly mixed estuaries, stronger tidal currents induce mixing, and stratification is weaker, with salinity varying by no more than a few parts per thousand from the surface to the bottom. During a tidal cycle, the entire

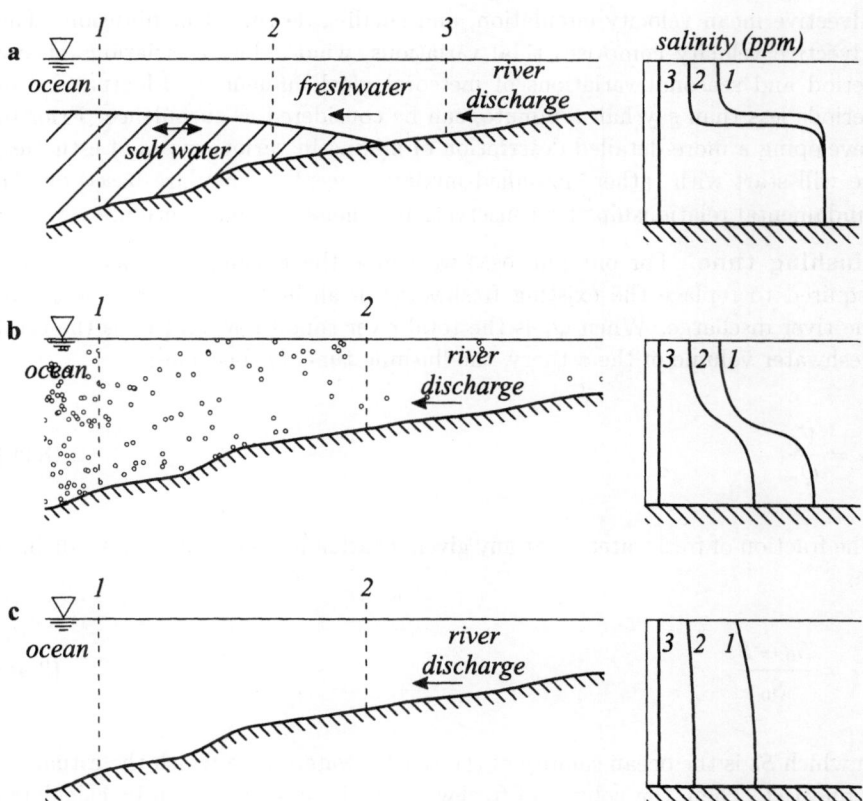

Fig. 8.12: Types of estuaries: **a** salt-wedge estuaries, **b** partly mixed estuaries, **c** well-mixed estuaries

water mass in the estuary moves landward and seaward. Partly mixed water moves upward and downward through the halocline, diluting salt water with the freshwater and *vice versa*. Due to turbulent mixing, the residual circulation, consisting of seaward flow above and landward flow below the halocline, is better developed than in the salt-wedge estuary. The combination of both residual and flood tidal currents suspend a lot of sediment and transport it landward into the estuary.

In *well-mixed estuaries* where tidal energy dominates over river input, the water column is completely mixed by turbulence (Fig. 8.12c). Because of this mixing, there are no residual currents in the vertical plane. Well-mixed estuaries are usually very wide when compared with their depth, and strong currents transport a great amount of marine sediments.

8.5.2   Some Simple Mixing Concepts

Similarly to ocean waters, mixing in estuaries results from a combination of advective mean velocity circulation, and small-scale turbulent diffusion. The advective velocity comprises tidal variations, wind-induced variations of any period and seasonal variations of meteorological influences. Fluctuations of periods less than say half a minute, can be considered as turbulence. Prior to developing a more detailed description of mixing in various types of estuaries, we will start with rather simplified mixing concepts. They are based on the fundamental relationships for conservation of mass and momentum.

Flushing time. For our purposes we define the flushing time as the time required to replace the existing freshwater in an estuary at a rate equal to the river discharge. When Q_r is the total river runoff flow and V_f is the total freshwater volume of the estuary, the flushing time, t_f, becomes:

$$t_f = \frac{V_f}{Q_r}. \tag{8.90}$$

The fraction of freshwater α_f at any given location in an estuary can be defined as:

$$\alpha_f = \frac{S_0 - S}{S_0}, \tag{8.91}$$

in which S_0 is the ocean salinity of the coastal waters with which the estuary is connected. Then, the volume of freshwater V_f in an estuary can be calculated by simple integration:

$$V_f = \int_{V_e} \alpha_f dV = \bar{\alpha}_f V_e, \tag{8.92}$$

where $\bar{\alpha}_f$ is the average freshwater fraction and V_e is the entire volume of the estuary. Using Eq. (8.90) and (8.92) we obtain the flushing time as:

$$t_f = \frac{\bar{\alpha}_f V_e}{Q_r}. \tag{8.93}$$

Tidal Prism Method. Tidal prism models have been used for estuary studies for a considerable period of time. In tidal models, an estuary is represented by a single box (Fig. 8.13), where the volume of sea water, V_p, entering the estuary on the flood tide, is entirely of oceanic salinity S_0. It is assumed that this water is completely mixed with the corresponding volume of freshwater, V_r, over the entire tidal cycle. During ebb tide, the entire quantity of mixed water is totally removed from the estuary. During the following flood tide, sea water of oceanic salinity again enters the estuary. Thus, the average salinity, $\bar{S}$, at high tide is:

$$\bar{S} = \frac{V_p}{V_p + V_r} S_0, \tag{8.94}$$

and the average fraction of freshwater, $\bar{\alpha}_f$ becomes:

$$\bar{\alpha}_f = \frac{V_r}{V_p + V_r}. \tag{8.95}$$

Thus, the tidal prism flushing time, t_f, can be given from Eq. (8.93) as:

$$t_f = \frac{\bar{\alpha}_f V_e}{Q_r} = \frac{V_e T}{V_p + V_r} = \frac{V_e T}{P}, \tag{8.96}$$

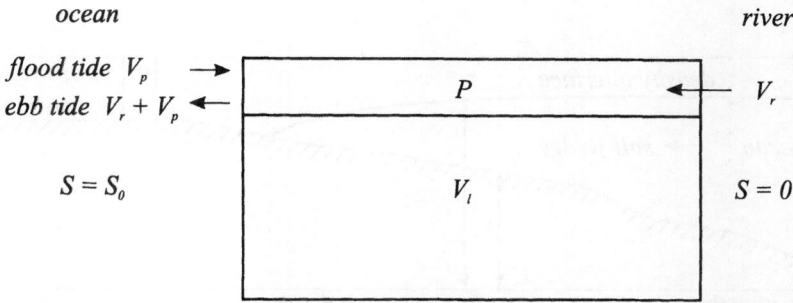

Fig. 8.13: A box model of an estuary; the outflow and inflow volumes are shown for the ebb and flood tides

where $P = V_p + V_r$ is the tidal prism and T is the period of the semidiurnal or diurnal tide.

In general, we do not expect complete mixing during each tidal cycle for an estuary, and some of the mixed water may return on each successive flood tide. As a result, the calculated flushing time may be less than observed. More rigorous tidal prism models have recently been reported by Luketina (1998).

8.5.3 Salt-Wedge Estuary

As we described in the Introduction (Sect. 8.5.1), salt-wedge estuaries are characterized by a small tidal range and by the absence of strong wind shear. Ocean salt water intrudes upstream against the direction of the overlying freshwater flow and a sharp interface separates the fresh and salt water layers. When the river flow, water depth and salinity of ocean water remain constant, the advancing motion of the wedge eventually ceases. This ideal steady-state form is known as an arrested saline wedge, when an equilibrium is maintained between internal buoyant pressure gradients, interfacial shear stress and convective accelerations (Jirka, 1990; Partheniades, 1992).

The overall flow in a salt-wedge estuary is controlled by two non-dimensional parameters: the channel densimetric Froude number $F_0 = Q_f/\sqrt{g_0' h^3}$ and the channel Reynolds number $Re = Q_f/\nu$, where Q_f is the freshwater flow per unit width, $g_0' = g\Delta\rho/\rho$ ($\Delta\rho$ is the density difference) is the reduced gravity acceleration, h is the channel depth, and ν is the coefficient of kinematic viscosity. A salt-wedge estuary can exist when $F_0 < 1$. In the absence of comprehensive field data, most of the available information on the flow and density distribution is based on laboratory experiments. Using detailed laboratory experiments by Sargent and Jirka (1987), a two-layer model for salt-wedge dynamics has been developed by Arita and Jirka (1987). The entire salt-wedge is divided into three regions (Fig. 8.14). In the tip region, the ambient flow bottom boundary layer undergoes a transformation into a mixing layer. This region is,

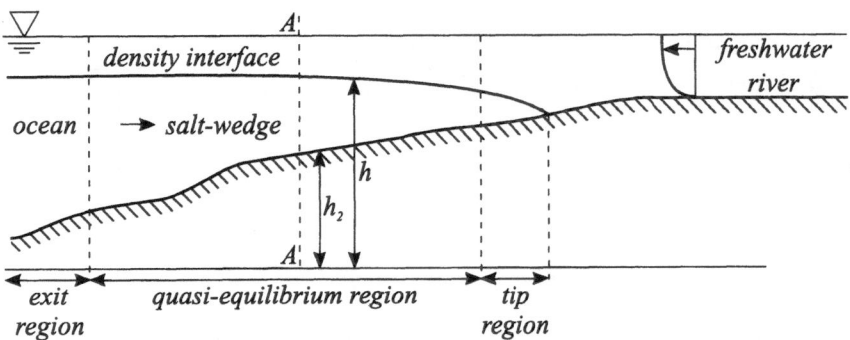

Fig. 8.14: Schematic representation of salt-wedge estuary with three major regions

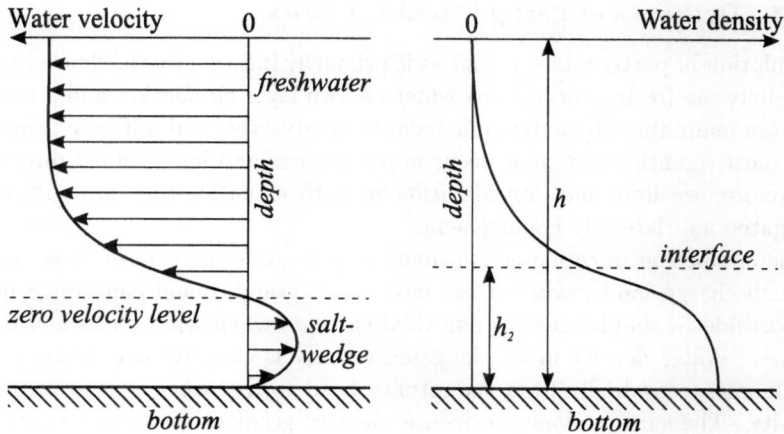

Fig. 8.15: Schematic representation of velocity and density profiles for the quasi-equilibrium region

at most, several water depths long. The exit region is also a non-equilibrium region with non-hydrostatic pressure distribution.

The central region of the salt-wedge estuary is the quasi-equilibrium region where convective inertial forces, buoyant pressure forces and shear forces are in equilibrium; they only slightly depend on longitudinal distance. Therefore, velocity and density distributions become similar. An example of such distributions for the cross-section A–A in Fig. 8.14 is shown schematically in Fig. 8.15. The interface is located at a h_2 distance above the bottom, where the density is 50% of the maximum density. The height of the zero velocity line is about 60% of the interface height h_2. The turbulent freshwater exerts a high shear over the underlying salt layer. This shear is counteracted by a very strong density gradient. As a result, viscous and turbulent shear transfer is taking place at the interface with some net entrainment into the upper layer. These processes provide energy for a weakly turbulent middle layer located between the density interface and the zero velocity line. Generation of turbulence in the layer between density interface and zero velocity is confirmed by a high value of the Richardson number:

$$Ri = -\frac{g}{\rho_w}\frac{\dfrac{\partial \rho_w}{\partial z}}{\left(\dfrac{\partial u}{\partial z}\right)^2}. \tag{8.97}$$

Sargent and Jirka's (1987) experiments showed that Ri reaches its maximum value ≈ 2 at the density interface and drops off very strongly on both sides of the middle layer. A two-layer model by Arita and Jirka (1987) also provides prediction of several global wedge properties, such as wedge length and fluxes of fresh and salt waters.

8.5.4 Dynamics of Partly Mixed Estuaries

Circulation in partly mixed estuaries is primarily induced by the density difference between freshwater and sea water. A two layer circulation and stratification are maintained by a dynamic balance of advective and diffusive processes.
The most essential variations occur in the vertical and longitudinal directions.
Therefore, we limit our consideration here to estuaries that are sufficiently
elongated and laterally homogeneous.

The circulation in estuaries is usually expressed as the sum of three modes:
river discharge mode, wind stress mode, and gravitational convective mode.
We consider a simpler case of non-tidal currents when wind stress can be neglected. Water density in an elongated estuary is constant over water depth,
but increases gradually from the estuary head to the ocean due to increasing
salinity. The surface slope, horizontal density gradient and consequently the
flow velocity are constant as functions of longitudinal distance. We are interested in describing the vertical variation in current velocity rather than the
horizontal variation.

The Navier-Stokes equation (see Eq. C.32) describing the dynamic water
balance in an estuary can be simplified as follows:

$$\frac{\partial p}{\partial x} = \rho_w A_z \frac{d^2 u}{dz^2}, \tag{8.98}$$

in which A_z is the coefficient of turbulent viscosity assumed to be constant.
The pressure in the water column is hydrostatic and can be written as follows:

$$p = \rho_w g (\zeta - z), \tag{8.99}$$

in which ζ is the sea surface elevation. Thus, longitudinal pressure gradient
becomes:

$$\frac{\partial p}{\partial x} = \rho_w g \frac{\partial \zeta}{\partial x} - gz \frac{\partial \rho_w}{\partial x} = \rho_w g i - g \lambda z, \tag{8.100}$$

where i and λ are the longitudinal surface and density gradients, respectively:

$$i = \frac{\partial \zeta}{\partial x}, \quad \lambda = \frac{\partial \rho_w}{\partial x}. \tag{8.101}$$

After substituting relationships (8.101) into Eq. (8.98) we obtain:

$$\frac{d^2 u}{dz^2} = \frac{gi}{A_z} - \frac{g\lambda}{\rho_w A_z} z. \tag{8.102}$$

We assume that there is no wind stress at the sea surface, so:

$$\frac{du}{dz} = 0, \quad \text{at} \quad z = 0, \tag{8.103}$$

and at the bottom $(z = -h)$, velocity u vanishes:

$$u = 0, \quad \text{at} \quad z = -h.$$ (8.104)

The solution on Eq. (8.102) with boundary conditions (8.103) and (8.104) takes the form (Officer, 1976):

$$u(z) = \frac{3}{2}\frac{Q}{h}\left[1 - \left(\frac{z}{h}\right)^2\right] + \frac{1}{48}\frac{g\lambda h^3}{\rho_w A_z}\left[1 - 9\left(\frac{z}{h}\right)^2 - 8\left(\frac{z}{h}\right)^3\right],$$ (8.105)

in which Q is the total discharge per unit width. In particular, the velocity at the surface is given by:

$$u(0) = \frac{3}{2}\frac{Q}{h} + \frac{1}{48}\frac{g\lambda h^3}{\rho_w A_z},$$ (8.106)

and the mean velocity becomes:

$$\bar{u} = \frac{Q}{h}.$$ (8.107)

To illustrate the velocity distribution (8.105) we adopted the flow parameters as measured by Hamilton and Wilson (1980) in the Lower Potomac Estuary. Thus, we have: $A_z = 3.6 \times 10^{-4}$ m^2/s, $i = 0.295 \times 10^{-6}$, $h = 17$ m, $Q = 5.1$ m^3/s per unit width, and $\lambda = 4.6 \times 10^{-5}$ kg/m^4. The resulting distribution is shown in Fig. 8.16.

If river discharge is very small, the first term in Eq. (8.105) is much smaller than the second term and is usually neglected, $i.e.$:

$$u(z) = \frac{1}{48}\frac{g\lambda h^3}{\rho_w A_z}\left[1 - 9\left(\frac{z}{h}\right)^2 - 8\left(\frac{z}{h}\right)^3\right].$$ (8.108)

Velocity distribution (8.108) is also shown in Fig. 8.16. In this case maximum positive and negative velocities occur at the surface $(z = 0)$ and at $z = -3/4h$, respectively. Therefore, even in the case of weak stratification, there is a net flow down estuary in the upper layer of the water column and a net flow upstream in the lower layer of the water column. In Fig. 8.16, the vertical distribution of velocity is also given for the case when the river flow dominates over the net circulation, $i.e.$:

$$u(z) = \frac{3}{2}\frac{Q}{h}\left[1 - \left(\frac{z}{h}\right)^2\right].$$ (8.109)

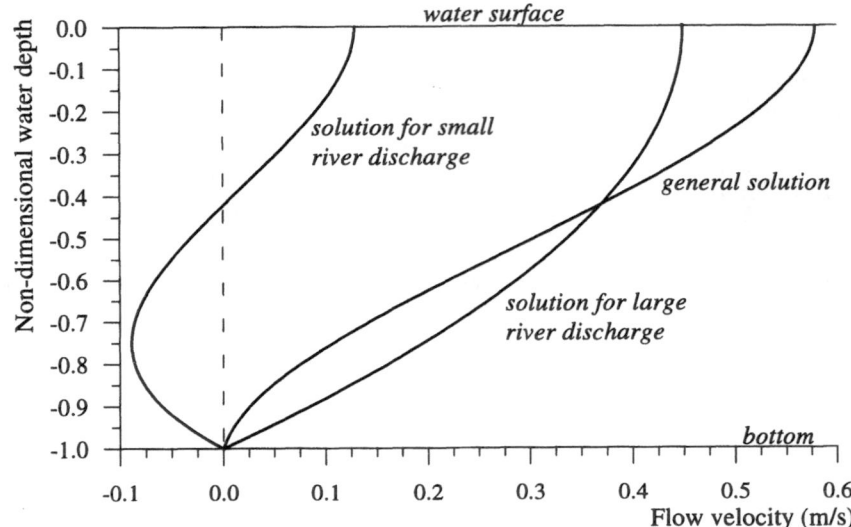

Fig. 8.16: Vertical distribution of the normalized flow velocity

Let us now determine the vertical distribution of salinity, S, for small river discharge. We assume both the diffusion, K_z, and gradient, $\partial S/\partial x$, to be constant as a function of depth. Thus, Eq. (8.30) simplifies as follows (Officer, 1976):

$$u\frac{\partial S}{\partial x} = K_z\frac{\partial^2 S}{\partial z^2}. \tag{8.110}$$

At the sea bottom, the vertical salt flux vanishes, *i.e.*:

$$K_z\frac{\partial S}{\partial z} = 0 \quad \text{at} \quad z = -h. \tag{8.111}$$

Substituting Eq. (8.108) into Eq. (8.110) and integrating we obtain:

$$S(z) = \frac{u_1 h^2}{K_z}\frac{\partial S}{\partial x}\left[\frac{1}{2}\left(\frac{z}{h}\right)^2 - \frac{3}{4}\left(\frac{z}{h}\right)^4 - \frac{2}{5}\left(\frac{z}{h}\right)^5 + C\right], \tag{8.112}$$

in which:

$$u_1 = \frac{g\lambda h^3}{48\rho_w A_z}. \tag{8.113}$$

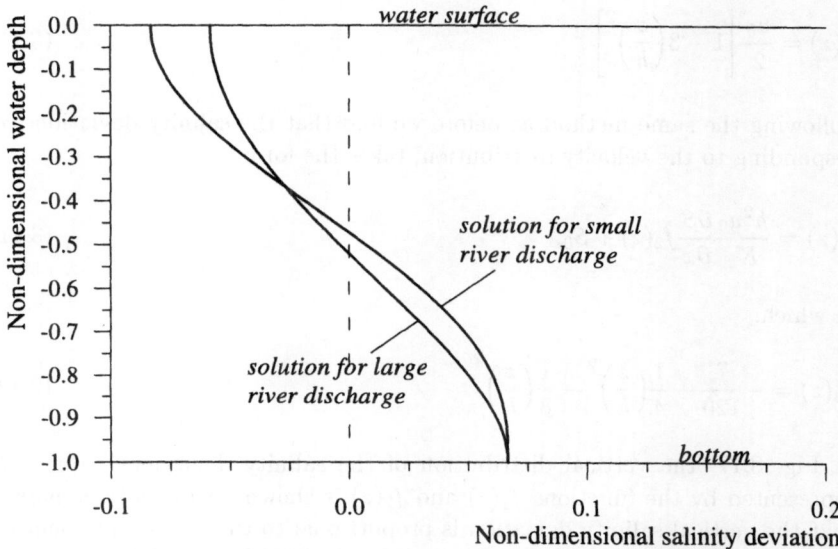

Fig. 8.17: Vertical distribution of the normalized salinity deviation in partly mixed estuary

Assuming that mean salinity is equal to S_0, we have for the integration constant C:

$$C = S_0 \left(\frac{u^2}{K_z} \frac{\partial S}{\partial x} \right)^{-1} - \frac{1}{12}. \tag{8.114}$$

Thus, the final salinity distribution becomes:

$$S(z) = \frac{h^2 u_1}{K_z} \frac{\partial S}{\partial x} f_1(z) + S_0, \tag{8.115}$$

in which:

$$f_1(z) = \left[-\frac{1}{12} + \frac{1}{2}\left(\frac{z}{h}\right)^2 - \frac{3}{4}\left(\frac{z}{h}\right)^4 - \frac{2}{5}\left(\frac{z}{h}\right)^5 \right]. \tag{8.116}$$

Let us examine the salinity when the river flow is dominant. From Eq. (8.109) for the corresponding mean velocity and velocity deviation we have:

$$\bar{u} = u_0, \tag{8.117}$$

and

$$u(z) = \frac{u_0}{2}\left[1 - 3\left(\frac{z}{h}\right)^2\right].$$
(8.118)

Following the same method as before we find that the salinity deviation, corresponding to the velocity distribution, takes the form:

$$S(z) = \frac{h^2 u_0}{K_z}\frac{\partial S}{\partial x}f_2(z) + S_0,$$
(8.119)

in which:

$$f_2(z) = -\frac{7}{120} + \frac{1}{4}\left(\frac{z}{h}\right)^2 - \frac{1}{8}\left(\frac{z}{h}\right)^4.$$
(8.120)

In Fig. 8.17, the vertical distribution of the salinity deviations, $S(z) - S_0$, represented by the functions $f_1(z)$ and $f_2(z)$ is shown. All equations indicate that the vertical salinity deviation is proportional to the salinity gradient and inversely proportional to the coefficient of vertical diffusion. Therefore, in an estuary with strong diffusion, the vertical salinity deviation is very small.

We will revisit the dynamics of a partly mixed estuary in Chap. 13, where we discuss the ecological implications of salinity intrusion and pollution dispersion in estuaries.

8.5.5   Dynamics of a Well-Mixed Estuary

When tidal energy dominates over river input, the total water column is completely mixed by the turbulent motion of eddies. The salinity field is mostly governed by tidal motion, river flow, channel geometry and bed roughness. The shear effects tend to increase vertically and transverse salinity gradients result in an increase of gravity induced currents. On the other hand, turbulent diffusion works towards reducing density gradients. Observed density gradients strongly depend on the relative magnitudes of both mechanisms.

Field observations are the best method to investigate the structure of turbulence in well-mixed estuaries. Mean velocity and salinity measurements in well-mixed flows by Anwar (1983) shows that the velocity profile is logarithmic (see Eq. 2.54), $i.e.$:

$$\frac{\bar{u}}{u_*} = \frac{1}{\kappa}\ln\left(\frac{z}{z_0}\right).$$
(8.121)

Field measurements were undertaken in the mouth of the River Carron, Scotland. The freshwater discharge of the river was small (about 30 m^3/s in winter and less than 1 m^3/s in summer) while the tidal discharge at peak spring ebb tide was about 400 m³/s. Roughness length, z_0, is in general, the function of intensity of flow. When mean flow velocity increases, the flow becomes stratified and well-mixed in the decelerating phase. This is due to the turbulent mixing process that increases during the decelerating phase of the flow.

Moreover, field experiments show that the unsteady phases of the flow have no appreciable effect on the distribution of salinity profiles which can be approximated by the relationship:

$$\frac{S_m - S}{S_m - S_s} \sim \left(\frac{z + h}{h}\right)^2, \tag{8.122}$$

in which S_m and S_s are the salinities near the bed and the surface, respectively.

In the case of well-mixed estuaries, the mixing and concentration of substances can be estimated using the random walk method described in Sect. 8.2.2. For applications see, for example, de Swart *et al.* (1997), and Chap. 13.

8.6 Sediment Transport in the Coastal Zone

At the end of this chapter we consider the transport of inorganic matter such as sediments due to waves and currents. Sediment transport is the main mechanism inducing beach erosion (or accretion) and changes in beach bottom topography. Sediment transport at any point in the nearshore zone may be viewed as a vector with both longshore and cross-shore components (see Fig. 8.18). However, this separation is not always valid in a strict sense because it is based on the assumption that bottom contours are parallel.

Cross-shore sediment transport encompasses both offshore transport, such as those occurring during storms and cyclones, and onshore transport which dominates out of storm season. Cross-shore coastal sediment transport is relevant to a number of coastal problems such as: beach and dune response to storms, profile nourishment (when the sand is placed in the nearshore with the expectation that it will move landward nourishing the beach), shoreline response to sea level rise, seasonal changes of shoreline positions, overwash, landward transport due to overtopping of the normal land mass (caused by high tides and waves), scour immediately seaward of shore parallel structures, and three-dimensional flow of sand around coastal structures (Kriebel and Dean, 1985; Kriebel, 1986; Nielsen, 1992; Dean, 1995).

Current knowledge of the physical processes involved in cross-shore sediment transport at spatial scales of centimetres and temporal scales of fraction of second is limited due to the turbulence of breaking waves and due to the complicated motion of sediment particles in spatially varying currents. However, changes in sediment transport and changes in beach profile, at scales of metres and hours, are much smoother and easier to predict (Larson and Kraus, 1989). Therefore, most numerical models of beach change are based on relationships (mainly empirical) between sediment transport rate and smooth

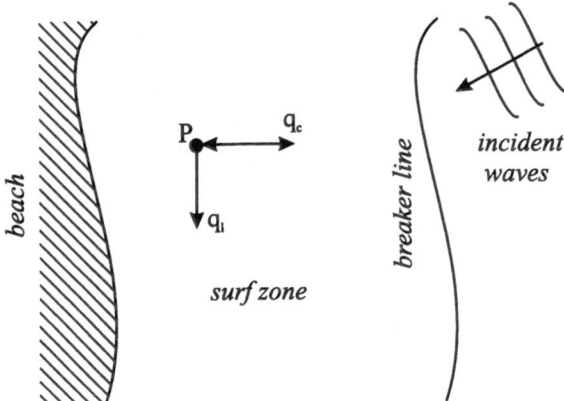

Fig. 8.18: Sediment transport as a vector with longshore and cross-shore components

change of wave energy at above scales. In such models, it is assumed that beach profile change is mainly governed by breaking waves. Larson and Kraus (1989) assumed that the sediment transport rate is proportional to the excess dissipation rate along the beach profile, from the offshore depth to the waterline. Wave dynamics vary along a beach profile due to wave shoaling, wave decay, and intensity of small- and large-scale motion. In particular, the region located immediately shoreward of the wave break point, up to plunging point, is characterized by a rapid transition in wave height and decaying of larger-scale flow which turns into small-scale turbulent fluctuations. The region shoreward of plunging point extends to the location where wave run-up begins. The identification of regions with different wave characteristics in the near-shore zone implies different characteristics of sediment transport. Following Larson and Kraus (1989, 1995), four different zones of transport have been distinguished (Fig. 8.19):

- Zone I: from the seaward depth of effective sand transport to the break point (pre-breaking zone),
- Zone II: from the break point to the plunge point (breaker transition zone),
- Zone III: from the plunge point to the point of wave reformation or to the swash zone (broken wave zone),
- Zone IV: from the seaward boundary of the surf zone to the shoreward limit of run-up (swash zone).

Relationships for the net transport rate, based on physical considerations and observations from the data, can be found elsewhere (Larson and Kraus, 1989). In Fig. 8.20 a schematic representation of sediment transport rate and sediment

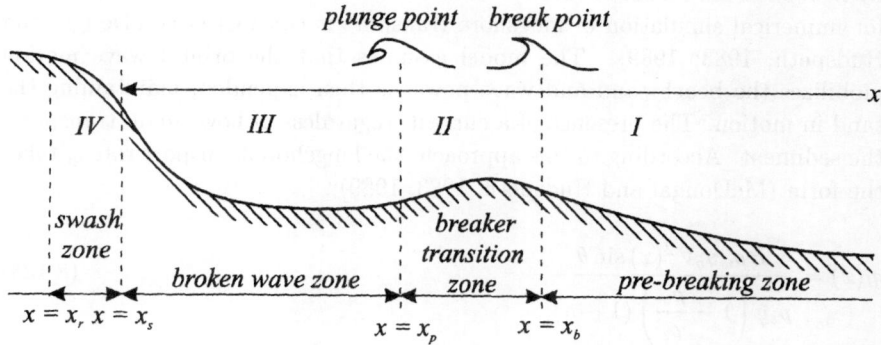

Fig. 8.19: Sediment transport in four different near-shore zones

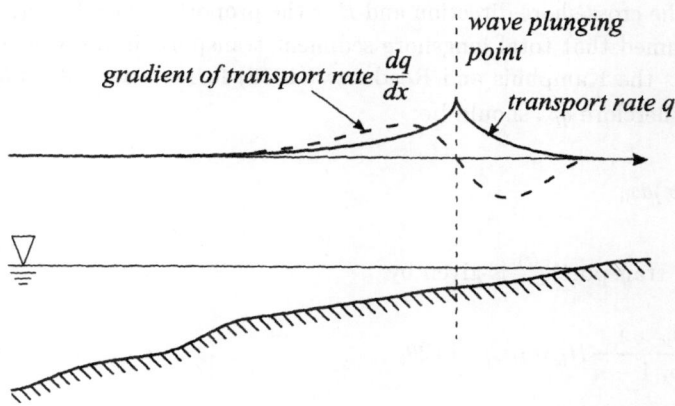

Fig. 8.20: Schematic representation of sediment transport rate and sediment transport gradient

transport gradient are shown. The peak of the cross-shore sediment transport rate is located at the plunging point, and transport decreases from that point in both directions. Therefore, the transport gradient becomes negative between plunging point and shore line, and is positive offshore from the plunging point.

8.6.1 Longshore Sediment Transport

The longshore movement of sand on beaches manifests itself whenever this natural movement is prevented through the construction of jetties, breakwaters or groynes. Such structures prevent or retard littoral drift causing a build-up of the beach on its updrift side and simultaneously starting erosion in the downdrift direction. There are a number of conceptual models predicting the longshore drift. A comprehensive review of these models has been published by Schoonees and Theron (1994). In this section, a so-called 'energetic' model for numerical simulation of longshore transport is considered (McDougal and Hudspeth, 1983; 1989). The model assumes that the orbital wave motion mobilizes the beach sand and wave power is then expended, maintaining the sand in motion. The presence of a current, regardless of how small, transports the sediment. According to this approach, the longshore transport rate q_l takes the form (McDougal and Hudspeth, 1983, 1989):

$$q_l(x) = \frac{B \rho_w u_b V^2(x) \sin \theta}{\rho_s g \left(1 - \frac{\rho_w}{\rho_s}\right)(1 - n)}, \tag{8.123}$$

in which ρ_s is the density of sediment, $q_l[\text{m}^3 \text{ m}^{-1} \text{ s}^{-1}]$ is the transport rate per unit width in the longshore direction and unit time; n is a porosity of sediment = volume of pores/total volume; θ is the angle of wave direction with respect to the cross-shore direction and B is the proportionality factor. To find it, it is assumed that total longshore sediment transport, in a given transect, $q_l^{(t)}$, satisfies the Kamphuis and Readshaw (1978) formula for total longshore transport; therefore $q_l^{(t)}$ should be:

$$q_l^{(t)} = \int q_l(x) dx, \tag{8.124}$$

where total transport $q_l^{(t)}$ is given by:

$$q_l^{(t)} = \frac{K_{tr}}{32} \frac{\rho_w}{\rho_s} \frac{1}{1 - n} H_{br}^2 m_{br} C_{br} \sin 2\theta_{br}, \tag{8.125}$$

in which $q_l^{(t)}$ is given in m^3/s; H_{br}, C_{br} and θ_{br} are wave height, phase speed and wave direction at a breaking line. The coefficient m_{br} is given by Eq. (4.22) for breaking depth h_{br}. For a proportionality coefficient, K_{tr}, a form suggested by Kamphuis and Readshaw (1978) is used:

$$K_{tr} = \begin{cases} 0.70\xi_b & \text{for } \xi_b < 1.4 \\ 1.24\xi_b & \text{for } \xi_b \geq 1.4, \end{cases} \tag{8.126}$$

where:

$$\xi_b = \frac{\tan \beta}{\sqrt{H_0/L_0}}, \tag{8.127}$$

in which β is beach slope seaward of the breaking point and H_0 and L_0 are height and length of waves in deep water, respectively. By solving the above

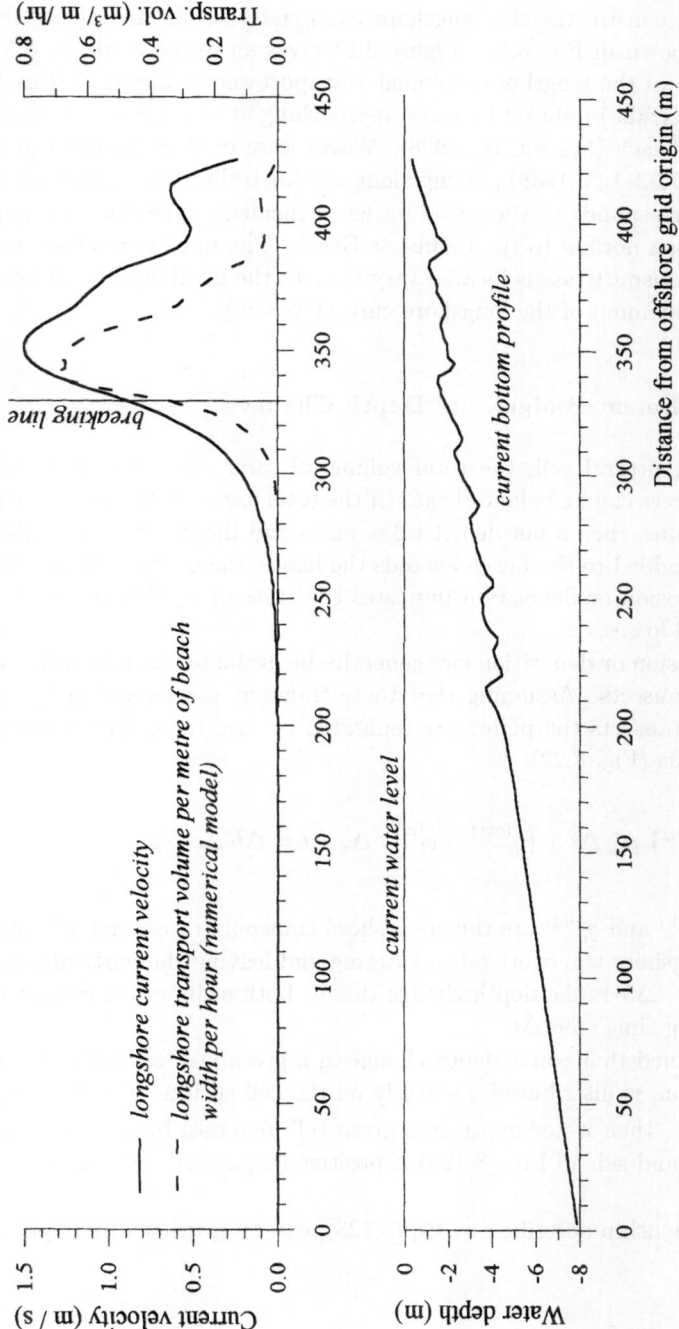

Fig. 8.21: Calculated distribution of the longshore transport across the beach induced by wave approaching at the angle 20^0 (adapted from Massel, B., 1998)

equations we obtain a sediment transport distribution $q_l(x)$ along a beach transect, which constitutes the longshore component of the sediment transport vector as shown in Fig. 8.18. Figure 8.21 gives an example of the calculated distribution of the longshore sediment transport across Lamberts Beach (east coast of Australia) induced by wave approaching at an angle of 20° to the normal to the beach (Massel, B., 1998). Waves were induced by tropical cyclone 'Charlie' (27.02–1.03.1988) moving along the Australian coast southwards. Figure 8.21 corresponds to the situation when incident waves form an angle 20° with direction normal to the Lamberts Beach. The maximum of the resulting sediment transport rate is located very close to the breaking line and coincides with the maximum of the longshore current velocity.

8.6.2 Sediment Budget and Depth Changes

For a given littoral cell, the total volume of sand added to the beach from various sources can be balanced against the total losses. If the losses are greater than the gains, then a net deficit takes place and the beach erodes. Similarly, if the sand added to the beach exceeds the losses, the beach accretes. The lack of either erosion or deposition indicates the state of equilibrium between the sources and losses.

Beach erosion or deposition can generally be evaluated by comparing a series of beach transects. Assuming that these transects are separated by Δy, and along the transects the points are separated by Δx, the sediment budget can be written as (Fig. 8.22):

$$\left(q_c^{(\mathrm{out})} - q_c^{(\mathrm{in})}\right) \Delta y\, \Delta t + \left(q_l^{(\mathrm{out})} - q_l^{(\mathrm{in})}\right) \Delta x\, \Delta t = \Delta h\, \Delta x\, \Delta y, \tag{8.128}$$

in which $q_c^{(\mathrm{in})}$ and $q_c^{(\mathrm{out})}$ are the cross-shore transport rates, and $q_l^{(\mathrm{in})}$ and $q_l^{(\mathrm{out})}$ are the longshore transport rates, entering and leaving the particular grid cell, respectively; Δh is the depth change due to both sediment transport components during time step Δt.

It is assumed that water depth change in a given grid cell, due to erosion or deposition, is distributed uniformly on the cell surface. If $q_c^{(\mathrm{out})} > q_c^{(\mathrm{in})}$ and $q_l^{(\mathrm{out})} > q_l^{(\mathrm{in})}$, then water depth in a given cell increases by Δh. More general, if the left-hand side of Eq. (8.128) is positive (negative), a given cell is eroded (deposited).

The relationship described by Eq.(8.128), can be rewritten more precisely as follows:

$$\frac{\partial h}{\partial t} + \frac{\partial q_c}{\partial x} + \frac{\partial q_l}{\partial y} = 0. \tag{8.129}$$

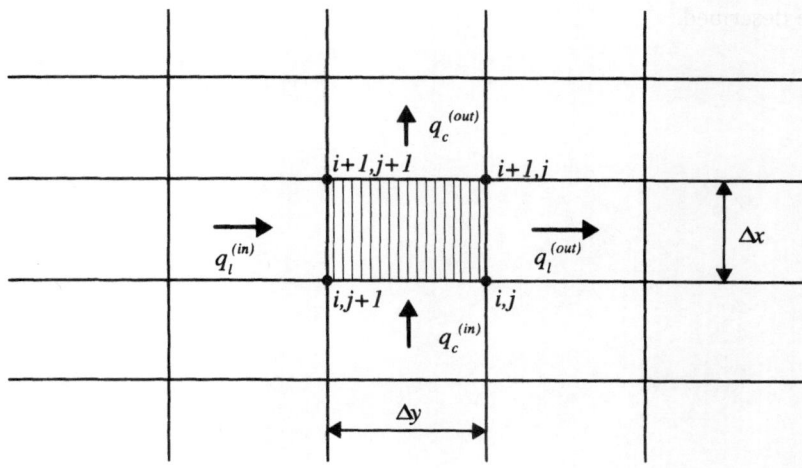

Fig. 8.22: Sediment budget

For parallel isobaths, the longshore sediment transport component is uniform in the y-direction, $i.e.$:

$$q_l^{(in)} = q_l^{(out)} \quad \text{or} \quad \frac{\partial q_l}{\partial y} = 0, \tag{8.130}$$

and the sediment budget equation takes the form:

$$\frac{\partial h}{\partial t} + \frac{\partial q_c}{\partial x} = 0. \tag{8.131}$$

The determination of longshore and cross-shore transport rates, described above, was based on a deterministic description of the incident wave field. Instead of treatment of all wave spectrum, as is observed in nature, the characteristic wave parameters (significant wave height, H_s, and spectral peak frequency, ω_p) have been chosen to represent the incoming wave energy. However, inclusion of full spectral representation can provide an effective tool to identify the influence of short and long waves on the beach morphology (Larson, 1995).

At present, the existing models of longshore current and longshore sediment transport are only valid for parallel contours. Although they are also applied to locally non-parallel contours, the accuracy of such approximation is not fully explored. It should be noted that the Larson and Krauss' approach to determine the cross-shore sediment transport can be extended to the accretion stage when the beach is restored under the non-storm action. Their model gives the opportunity to include also beach protection structures, when appropriate boundary conditions are formulated in the vicinity of structures. For example, the influence of reflected waves and the scour at the structure toe (or wall) has to be described.

9 Experimental Methods in Fluid Mechanics

9.1 Introduction

All previous chapters have clearly illustrated the importance of observations and experiments for understanding physical processes in the ocean. Although substantial progress has been made in theoretical and numerical modelling of many fluid dynamics phenomena, experimental data are still recognized as a basic source of information. As Mary Nicol Leakey said, *theories come and go, but fundamental data always remain the same.* This statement is especially true for oceanographic data.

Measurements and theory are complementary rather than competitive. Peregrine (1990) listed a few examples of interactions between theory and measurements which are fully applicable for ocean flows, namely: a) observations in nature or in the laboratory lead to a theory being developed to explain it, b) new phenomena predicted by theory lead to experiments being made to verify the prediction, and c) discrepancies between theory and measurements stimulate further development of both.

In this chapter, we discuss measurement techniques and data processing used in fluid mechanics. The classical, small scale laboratory experiments have been described in numerous papers and books (see, for example, Hughes, 1993). Here we concentrate mostly on the methods and instrumentation used in field experiments. In particular, in the next section we will provide a brief review of instruments used for measurement of waves, currents and tides *in situ*.

The launch of oceanographic satellites has provided a new opportunity to probe the ocean surface features. Satellite measurements are especially important in the study of the deep oceans, which comprise the vast majority of the globe, where reliable observations are almost nonexistent. At present, there are a number of satellite remote sensing systems designed especially for observing the oceans. Some satellites, carrying instruments specifically for detecting the ocean surface are described in Sect. 9.3.

One of the first steps in experimental data analysis is to decide which physical variables have an effect on the process being studied and to establish the functional relationship between all the important variables. The number of variables that need to be considered can be reduced by combining physical variables into dimensionless products. A rational procedure which forms the dimensionless products is known as dimensional analysis.

Experimental data on many ocean physical parameters reflect their random character. Therefore, special techniques are required to process these data. At the end of this chapter, a brief review of the most important statistical and spectral techniques in application to ocean data is given.

9.2   Field and Laboratory Measurement Techniques

There are many various instruments, types of observations and methods of analysis developed in fluid mechanics and oceanography which have undergone considerable change over the last few decades. A very comprehensive and practical compilation of the essential information on instrumentation and data analysis was recently published by Emery and Thomson (1997). The description of older types of instrumentation is provided in many books on physical oceanography (see, for example, von Arx, 1977). Therefore, in this book we only briefly describe the most important measurement methods and data analysis techniques.

9.2.1   Temperature

During the world's first oceanographic expedition (1872–1874), the *Challenger* investigators made observations at 362 locations taking tens of thousand of samples of the Atlantic and Pacific oceans. At each sampled depth, a temperature reading was made by breaking the capillary column in a bulb thermometer by remote control. For the next 90 years this time-consuming method of temperature measurement prevailed, with only modest improvements. Although this method is good for establishing the seasonal variations of sea-water temperature at various depths, its sampling accuracy is far too coarse to detect small-scale changes in temperature.

At present, there are a few methods for measurement of the ocean temperature. They include convenient mercury thermometers, simple and reversing, and more sophisticated bathythermographs and temperature profilers. Special methods have been developed to measure the sea surface temperature (SST). Here we will describe the principle of temperature measurement using bathythermographs and profilers. Measurement of sea surface temperature is described in the next section.

In the expendable bathythermograph (XBT), the dependence of a metal's electrical resistance on temperature is used. The XBT is a free-falling probe which provides an upper ocean temperature profile. The most commonly used metals are copper, platinum and nickel. For example, platinum thermometers have accuracies of $\pm 0.001°C$. Another class of resistive materials used for temperature measurements are semiconductors, known in oceanographic applications as thermistors, which have some advantages over metal due to their higher temperature resistance and very small size. In typical oceanographical applications, thermistors are fixed along a mooring cable forming a thermistor string used for stationary measurements or used in expendable bathythermographs (XBT). A thermistor is placed in the nose of the probe as the temperature sensing element. The accuracy of an XBT is ±0.1°C (Emery and Thomson, 1997). Resistance thermometers are also widely used in CTD (conductivity-temperature-depth) profilers.

9.2.2 Salinity

As we showed in Chap. 1, salinity is one of the major agents determining the density of ocean water. A first measure of salinity is related to another measured parameter 'chlorinity' by Knudsen's equation:

$$S(\text{ppm}) = 1.805 \, \text{Cl(ppm)} + 0.03. \tag{9.1}$$

At present, water conductivity is used as a measure of salinity rather than the titration of chlorinity. Conductivity of ocean water depends on the ion content of the water and therefore is directly proportional to the salt content. Modern conductivity-temperature-depth (CTD) profiling systems record conductivity directly as the water flows through the instrument. Then, an empirical relationship between observed conductivity and salinity is used to compute salinity. This new salinity definition, so called 'practical salinity scale', or PSS78 is now recommended as the scale in which salinity data should be reported (Lewis and Perkin, 1981).

9.2.3 Sea Level Measurement

The periodic rise and fall of the sea surface has been observed for thousands of years. Pytheas in the fourth century B.C. probably recorded the tides during his journey from Massilia in the Mediterranean northward into the English Channel and southern part of the North Sea (von Arx, 1977). Since then, tidal variations of sea level have been observed by a variety of methods. The simplest is the float stilling well gauge, in which the float rises and falls with the water level. Modern recording systems use sensitive and accurate pressure or acoustic gauges.

However, conventional sea level measurement systems require a fixed installation platforms and therefore are only possible at or near the coast. Over the vast ocean, away from any islands, sea level is measured using satellite-borne radar altimetry. We will describe this technique in Sect. 9.3.

9.2.4 Ocean Current Measurements

There are a variety of techniques and instruments used for measuring current velocity and direction. They can be measured at fixed locations (Eulerian currents) or by tracking tracers such as a surface float or dye patch (Lagrangian current). Since the launch of oceanographic satellites, tracking of surface buoys has provided reliable, long-term current trajectories for many different parts of the world's oceans. Various types of drifters and their use for particular applications are described by Emery and Thomson (1997) and will not be repeated here. In this section we briefly describe current measurements at fixed-locations. The methods available can be divided into two groups: single point current meters and current profilers.

Single Point Current Meter. A conventional single point current meter provides information on water flow velocity and direction at the point at which it is located. These devices must be moored at depths exceeding 10 m, in order to avoid 'contamination' induced by the oscillating flow due to surface waves. Data are either recorded aboard the instrument or telemetered to shore or ship. There are a few types of current meters used in oceanographic practice, namely rotor type current meters, acoustic current meters, and electromagnetic current meters. As was mentioned by Emery and Thomson (1997), the first modern current meter was the Aanderaa Recording Current Meter (RCM) developed by Ivar Aanderaa in Norway in the early 1960s under sponsorship of the North Atlantic Treaty Organization (NATO). Various versions of this current meter, such as the RCM4 or RCM5, are still in use in the oceanographic practice.

Aanderaa type current meters use a Savonious rotor to measure current speed. The Savonious rotor consists of six axisymmetric, flat blades enclosed in a vertical housing oriented normally to the direction of flow. Allowable sampling periods are 3.75, 7.5, 10, 30 and 60 min., while speed is obtained from the number of rotor revolutions for the sample interval.

The major shortcoming of rotor-type current meters is their inability to record currents in regions affected by surface motion. In the upper ocean wave regime, non-mechanical current meters (acoustic and electromagnetic) are better suited. Acoustic current meters measure the difference between an acoustic source and receiver separated by a fixed distance, L. A three-axis current meter determines the three-dimensional velocity by simultaneously measuring time differences along three orthogonal axes.

Examples of electromagnetic current meters are the InterOcean S4 current meter or Marsch-McBirney 512 current meter. Both instruments are commonly used for coastal and offshore oceanographic measurements. Current meters measure the voltage induced from the motion of a water through a magnetic field generated by the instrument. An electromagnetic force is induced which is directly proportional to the speed of the ocean current according to Faraday's law of electromagnetic induction. A two-axis electromagnetic current meter is equipped with an internal compass to produce horizontal components referred to the Earth coordinates.

The S4 current meter is spherical in shape with a diameter of 25 cm. Electrodes are located on the surface of the sphere; the compass and all electronics, including data storage and the power supply, are located within the sphere. At very low water velocity, flow around a sphere is laminar and separation from the rear of the sphere does not occur. At higher speeds, the flow becomes turbulent. In order to avoid separation of flow from the sphere, the surface was grooved vertically. S4 current meters can be equipped with a pressure sensor to estimate water depth and measure waves (see next section).

Ocean Current Profilers. As we found in Chap. 7, current velocity changes with depth, and measurement at a single point is not sufficient to characterize the flow field. It is possible to construct a string of current meters and to measure velocity at a few points simultaneously. Although this approach is popular, such instrument moorings are expensive and time consuming to deploy.

An alternative solution is to use an Acoustic Doppler Current Profiler (ADCP). As these instruments have only recently been introduced into oceanographic practice, we will describe them in some detail. The measurement principle is based on the Doppler effect. This term describes a change in the observed sound pitch that results from relative motion. An example of this effect is the sound made by a train as it approaches. As the train passes the observer, the pitch of the sound changes to a lower frequency. The change in pitch is directly proportional to the speed of the train. In other words, if the change of pitch is known, then the velocity of the train can be calculated.

The ADCP uses the Doppler effect by transmitting sound at a fixed frequency and listening to echoes returning from sound scatterers in the water. The sound scatterers in the water are any small particles or plankton which reflect the sound back to the instrument. These particles are usually neutrally buoyant and they move at the same velocity as the water. When scatterers move toward the ADCP, the sound heard by the scatterers is Doppler-shifted to a higher frequency and this shift is proportional to the relative velocity between the ADCP and scatterer. Some part of this sound is reflected back towards the ADCP. There is a double Doppler frequency shift, Δf_D, as the ADCP both transmits and receives sound, *i.e.*:

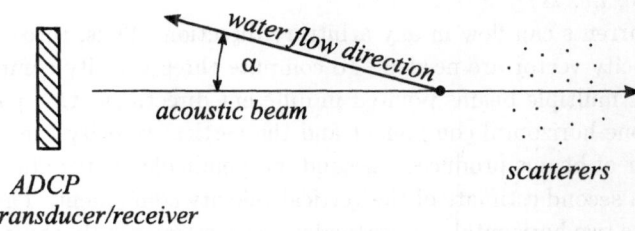

Fig. 9.1: Definition of angle α

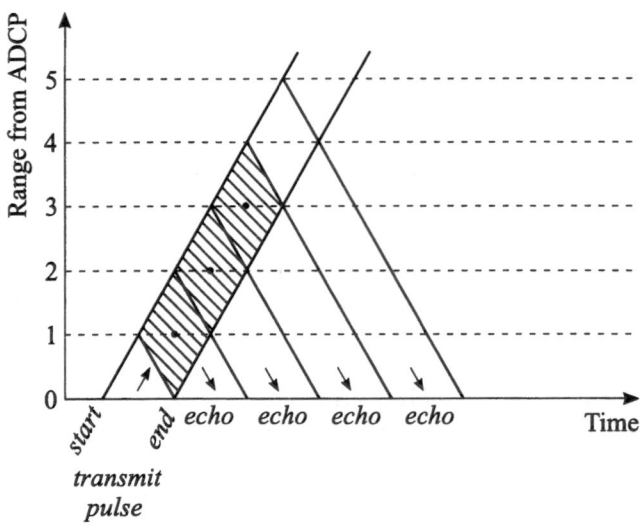

Fig. 9.2: Schematic representation of transmit pulses and echoes from successive ranges from ADCP (adapted from RD Instruments, 1989)

$$\Delta f_D = 2\frac{U}{C}f_s \cos\alpha, \tag{9.2}$$

in which f_s is the sound frequency in still water, U is the scatterer (current) velocity, C_s is the speed of sound in water, and α is the angle between the relative velocity vector and the line between the ADCP and scatterers (Fig. 9.1). The speed of sound, C_s, is about 1500 m/s, depending on water temperature, salinity and depth. For example, when the frequency of sound generated by an ADCP is $f_s = 300$ kHz, sound speed $C_s = 1500$ m/s, and scatterers move toward the ADCP with a speed of $U = 1$ m/s, then the Doppler frequency shift $\Delta f_D = 400$ Hz. When scatterers move away from the instrument this shift will be negative, *i.e.* $\Delta f_D = -400$ Hz.

Ocean currents can flow in any arbitrary direction. Thus, three components of the velocity vector are needed. To compute three velocity components, the ADCP uses multiple beams pointed in different directions. One pair of beams produces one horizontal component and the vertical velocity component. The second pair of beams produces a second, perpendicular horizontal component, as well as a second estimate of the vertical velocity component. Therefore, AD-CPs provide two horizontal and vertical velocity estimates. In theory only three beams are required to compute three dimensional velocity, but ADCPs have a fourth, redundant beam which allows the instrument to evaluate whether the velocity field within the measurement area is horizontally homogeneous. Such current homogeneity is one of basic assumptions used in ADCP instruments.

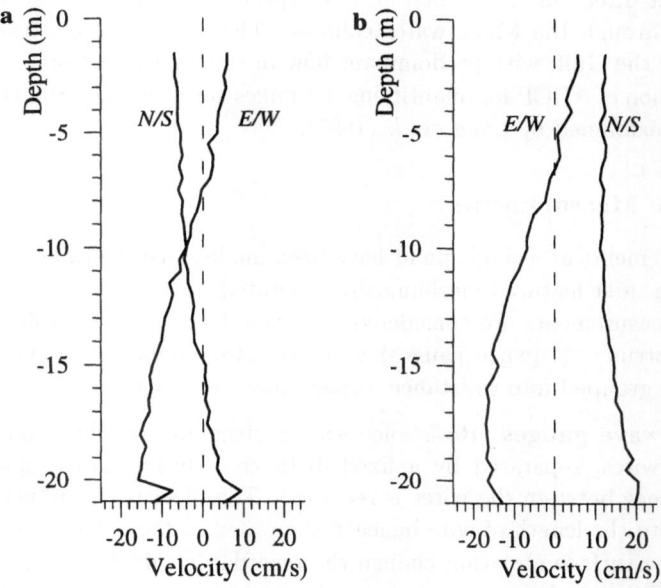

Fig. 9.3: Vertical velocity profiles measured by ADCP: **a** just before the beginning of ebb tide, and **b** velocities half an hour later

The most important feature of an ADCP is its ability to measure current profiles. The velocity profile is divided into uniform segments or cells, and the instrument measures average velocity over the depth range of each cell. Thus, the ADCP mimics a string of current meters uniformly distributed on a mooring. Profiles are produced by breaking the signal into successive segments and processing each segment independently of the others. Echoes from further away from the instrument take longer to return to the ADCP, compared to echoes from close ranges (Fig. 9.2). Thus, successive times of receiving pulses correspond to echoes from increasing distances away from the instrument. The water depth range and depth sampling of measurement cells depend on the frequency of the transmitted pulse. For example, an ADCP operating at a frequency of 1200 kHz can be used in water depths up to 20 m, and it provides velocity components every 0.25 m. For a lower frequency of 300 kHz, the water depth range is 110 m with sampling of 1 m. In Fig. 9.3, an example of the vertical profile of horizontal velocity components is shown. Velocities have been recorded in the entrance to Exmouth Gulf in North West Australia. The Gulf is about 80 km long and approximately 30 km wide, with its longer axis directed North-South. During measurement, there was no wind, so recorded velocities are totally due to tidal motion. Profiles **a** correspond to the time just prior to the beginning of ebb flow from the Gulf. Velocity at surface is small (~ 5 cm/s) and directed into the Gulf with the bottom velocity of the same order directed out of the Gulf. Some shearing of velocity is also observed in

the East-West direction. After half an hour (profiles **b**), water is flowing out
of the Gulf through the whole water column. There is still some transverse
circulation in the Gulf, with predominant flow in the west direction.

An application of ADCP for quantifying net fluxes across a macrotidal estuary
was recently examined by Lane *et al.* (1997).

9.2.5 Wave Measurements

Wave measurements at a single point have been made since the time when laboratory flumes first featured mechanically generated wave motion. At present,
such wave measurements are considered a routine laboratory capability. The
variety of instruments (wave gauges) used to obtain information on surface
waves can be grouped into resistance, capacitance and pressure type gauges.

Resistance wave gauges. Resistance wave gauges are usually formed from
two parallel wires, separated by a fixed distance. During gauge operation,
the conductance between the wires is recorded. The measured conductance is
proportional to the length of wire beneath the water surface. This means that
changes in sea surface elevation change the recorded conductance. This type
of gauge exhibits good linear response and possesses a resolution of about ±
0.1 mm.

Capacitance wave gauges. In capacitance wave gauges, the principle of
linear variation of capacitance with changes of sea surface elevation is used.
The capacitor is formed from an insulated wire held taut by a supporting rod,
with water serving as the 'ground'. The single-wire capacitance wave gauge
demonstrates good linearity and dynamic response over a reasonable length,
and is stable over a sufficiently long time.

Pressure transducers. Measurement of surface waves with pressure transducers has been practiced since 1947. Although measuring subsurface pressure
for evaluation of the surface wave height is practically feasible, an explicit transfer relationship between the wave pressure and the wave height is necessary.
The simplest relationship between wave pressure head, $H_p = p/\rho g$, and the
surface wave height, H, is:

$$H = \frac{H_p}{K_p},\qquad\qquad(9.3)$$

where K_p is the pressure response factor:

$$K_p = \frac{\cosh k\,(h+z)}{\cosh kh},\qquad\qquad(9.4)$$

where z is the submergence of the pressure sensor below the still water level,
and h is the water depth.

To account for the difference between theory and observation, an empirical correction factor, N, is usually introduced:

$$H = N \frac{H_p}{K_p}. \tag{9.5}$$

In engineering application, the typical value of N is equal to 1.25. Recent studies by Kuo and Chiu (1994), and Townsend and Fenton (1995) indicated that linear theory is adequate to compensate pressure records to give surface wave heights to within five per cent.

However, for wind-induced waves, the spectral analysis of the data can be used. Then, the relationship between surface wave elevation, ζ, and subsurface pressure, p, can be expressed in terms of the single-input/output system relationship:

$$H(\omega) = \frac{\mid S_{p\zeta}(\omega) \mid}{S_\zeta(\omega)}, \tag{9.6}$$

where $H(\omega)$ is the frequency response function, $S_\zeta(\omega)$ is a frequency spectrum of surface elevation, and $S_{p\zeta}(\omega)$ is the cross-spectrum of surface elevation and wave pressure (Massel, 1996a).

Wave staffs. For the wave staff, other principles apart from resistance and conductance, are used. One of them is the so called 'contact wave staff' where a set of contacts is distributed along a vertical pole. The contacts are closed when they are covered by water.

Zwarts (1974) proposed a wave staff for field application, based on a principle commonly used in the telephone industry to detect the location of faults in coaxial cables. The wave measuring staff consists of two pipes, one inside the other, forming a coaxial cable. Slots in the outer pipe allow the movement of water into the space between the outer and inner pipes. The configuration of the pipes in a coaxial cable form acts as the tuning element of an electronic oscillator. An electromagnetic wave, propagating down the pole, is generated by electronics located at the head of the pole. This wave reflects off the discontinuity (in the dielectric contact) at the air-water interface. The length of the unimmersed section of the staff is directly proportional to the period of the oscillation of the electronic signal. The output of the system is a number of reflections of the signal during a very small time interval.

Wave buoys. In deeper waters, where the supporting structures for wave staffs are not available, wave buoys are used. They are usually a small sphere, or small platform, which follows the movement of the water surface. In the simplest case, the internal sensor measures the vertical acceleration of the buoy. The acceleration is integrated twice with respect to time, to produce a measure of the instantaneous wave height profile about the still water level. The response of the buoy varies with wave frequency. For the most popular wave rider buoy manufactured by Datawell, Haarlem, the Netherlands, the transfer function shows that the buoy response is essentially flat for wave periods between 5 and 10 s. Some attenuation is observed for periods between 10 and 25 s (Massel, 1996a).

When a moored wave rider follows the waves, the force on the mooring line will change. This force is produced by a change in buoy's immersion. The wave rider buoy does not follow the wave surface if the wavelength is less than 5 m (the wave period is less than 1.8 s). To avoid measurement of unwanted accelerations due to roll and pitch of the buoy, the sensitive axis of the accelerometer is mounted on a stabilized platform. To keep a moored wave rider at the correct position, a rubber cord is used as a part of the mooring system. The stiffness of the rubber cord allows the buoy to follow waves up to 20 m. A buoyancy of approximately 900N keeps the wave rider from submerging under the combined action of an 18 m wave height and a 1m/s current.

Buoy displacement records are internally filtered at a high frequency cut-off of 0.6 Hz. Transmission of data is to Argos satellite or through standard 27–40 MHz radio-link to shore. The buoy measures heave in the range ±20 m with 1 cm resolution for wave period 1.6–20 s. For the directional period buoys, the direction is measured with a resolution of 1.5°.

For large wave buoys, the response of the buoy itself must be taken into account when calculating the wave parameters, especially those related to directional spreading. The buoy response depends on mooring constraints which are varied by a number of factors, including current and wind speed (Steele et al., 1992; Tucker, 1989; Gnanadesikan and Terray, 1994). A comprehensive comparison of various types of wave buoys is given by Allender et al. (1989).

In recent years, new methods of ocean wave measuring, based on satellite techniques have been developed. We will describe them in the next section.

9.3   Remote Sensing Techniques

9.3.1   Introduction

The launch of oceanographic satellites has provided a new and extensive set of data on the state of the ocean waters. These observations are of special importance in the study of the deep oceans, which comprise the vast majority of the globe, where reliable observations are almost nonexistent. Oceanographic remote sensing is usually subdivided into environmental satellites or more specialized satellites. The most important satellites designed for oceanographic applications include SEASAT, GEOSAT, TOPEX/POSEIDON, RADARSAT, ERS-1, ERS-2 and SeaWiFS. Although SEASAT suffered a disappointingly short life of 106 days, it attracted great interest within the oceanographic community. The first successful space mission specifically designed for studying the circulation of the ocean waters was TOPEX/POSEIDON, launched on August 1992. To meet the stringent measurement accuracy required for ocean circulation studies, a number of instruments have been installed on the satellite platform, such as a dual-frequency radar altimeter, a three-frequency microwave radiometer and a precise satellite tracking system (Fu *et al.*, 1994). The performance of the measurement system has been tested extensively and the results indicate that the root-mean-square accuracy of a single-pass sea level measurement is about 5 cm. In the next sections we will show some application of TOPEX/POSEIDON data for studying waves, tides and currents.

Other important oceanographic satellites are ERS-1 and ERS-2, launched by the European Space Agency (ESA). Satellite ERS-1 (launched in July 1991) occupies a sun-synchronous orbit of altitude 785 km, and has a 35-day repeat cycle. It is equipped with a high resolution imaging system, known as Synthetic Aperture Radar (SAR). It is based on the emission of short microwave pulses at an incidence angle of 20° to 25°. The typical swath scanned by SAR is of the order of 100 km, with a resolution of 20 m × 20 m. Thus, the SAR is able to detect and characterize surface and internal wave fields.

The ERS-1 satellite is also equipped with an Along Track Scanning Radiometer (ATSR), specifically designed to provide precise measurements of sea surface temperature (SST) for application in climate research, with an accuracy of better than 0.3 K. For measurements, the ATSR uses the same wave lengths as the Advanced Very High Resolution Radiometer (AVHRR) which is the operational instrument on the NOAA meteorological satellites.

In July 1997, the Orbview satellite was launched with the Sea-viewing Wide Field-of-view Sensor (SeaWiFS) on board. This system is designed primarily for the study of ocean biochemistry and is a joint venture between NASA and private industry. The data can be used for the study of such phenomena as ocean primary production and phytoplankton processes; cycles of carbon, sulfur and nitrogen; and ocean influences on global climate, including heat storage in the upper ocean and marine aerosol formation. Moreover, the data will be of great importance for commercial fishing, navigation and weather prediction.

In next section we briefly summarize the contributions of these dedicated satellites to observe and measure various ocean water features.

9.3.2   Surface Waves Observed by Satellites

When studying surface waves, two typical satellite sensors are usually used. The application of the satellite altimeter is obvious, as a radar altimeter is a nadir looking instrument. The measurement of the travel time of the reflected microwave pulse yields the position of the sea surface relative to the orbit of the satellite. The distortion of the mean shape of the return pulse provides an estimate of the sea variance, σ_ζ, and hence of significant wave height, $H = 4\sigma_\zeta$. To estimate the mean pulse shape, several hundred pulses need to be averaged, resulting in one significant wave height estimate about every 7 km along the satellite track.

The comparison of GEOSAT estimates of significant wave height, H_s, against values from buoys in NOAA's National Data Buoy Center network, showed that GEOSAT underestimated the significant wave height by 13% (Carter *et al.*, 1992). Cotton and Carter (1994) provided a satellite calibration through the use of monthly means of significant wave height derived for 2° latitude and 2° longitude bins between 60°S and 60°N. Monthly means from TOPEX, ERS-1 and GEOSAT altimeters have been compared against data acquired from the NOAA's buoy network.

Comparisons of means from ERS-1 and from TOPEX with those from GEO-SAT, for the same calendar month but 5 years apart, indicate inter-year variability in the observed bins. Recent studies for the ERS-1 altimeter showed a satisfactory agreement for wave heights up to 4 m. Higher waves are underestimated by the altimeter, relative to the buoy data. When compared with the WAM numerical prediction model (Massel, 1996a), the altimeter wave height measurements showed the best agreement in the Northern Hemisphere. In the Southern Hemisphere, the WAM model predicts wave heights lower than the altimeter wave heights (Komen *et al.*, 1994).

Hwang *et al.* (1998) compared the wind speed, wave height and wave period derived from satellite altimeter (TOPEX/POSEIDON) and ocean buoys in the Gulf of Mexico. The comparison showed that when separation between the buoy location and altimeter footprint is less than 10 km and the measurement time lag is less than 1 hour, the root-mean-square difference of the significant wave heights is approximately 0.1 m. For the wind speed, the root-mean-square difference is of the order of 0.8 m/s.

As was mentioned in Chap. 4, mean monthly significant wave height climatologies for the globe, based on 3-year GEOSAT radar altimeter data, were reported by Young (1994). Recently, Young (1998) compared the ocean mean monthly values of significant wave height H_s and mean wind velocity V_{10} obtained from altimeters on GEOSAT, TOPEX and ERS-1 with NDBC buoys in the Pacific and Atlantic Ocean. The following relationships result from the comparison:

$$\left.\begin{aligned} H_s &= 1.144 \cdot H_s(\text{GEOSAT}) - 0.148 \\ H_s &= 1.067 \cdot H_s(\text{TOPEX}) - 0.079 \\ H_s &= 1.243 \cdot H_s(\text{CERSI}) - 0.040 \end{aligned}\right\}, \tag{9.7}$$

and

$$\left.\begin{aligned} V_{10} &= V_{10}(\text{GEOSAT}) \\ V_{10} &= 0.99 \cdot V_{10}(\text{TOPEX}) + 1.61 \\ V_{10} &= V_{10}(\text{ERSI}) \end{aligned}\right\}. \tag{9.8}$$

Reflected SAR pulses from a moving, random sea surface contain information on the spatial characteristics of the wave surface and its spectrum. Extraction of the wave spectra from SAR image spectra is a complicated procedure, due to processes modulating the backscattering signal. The most important processes are:

1. Variation in the local angle of incidence associated with variation in the facet orientation and position.
2. Variation in the energy of Bragg scattering, caused by interactions between the short ripples and longer waves.
3. Oscillation of the Doppler shift of the return signal which results in variations of the apparent facet density in the SAR image plane.

A full discussion of the extraction processes and comparison with the WAM model can be found in the book by Komen *et al.* (1994).

9.3.3 Tides Observed by Satellites

Satellite altimeter data are extensively used to improve ocean tide models (see Chap. 5). The current methodologies for this improvement can be categorized in three groups: hydrodynamic modelling, empirical methods which use only data, and assimilation methods which use both data and hydrodynamic modelling. The first and second methods, and the resulting tidal pattern of the world oceans have been described in Chap. 5. Here we consider an application of satellite altimeter measurements to improve the empirical tide models. In particular, Cartwright and Ray (1990) used one year of altimetry data from GEOSAT to derive estimates of the diurnal and semidiurnal oceanic tides for latitudes between 58°N and 60°S. Global maps for M_2 and S_2 constituents compare well with ground truth at 66 open ocean sites.

The TOPEX/POSEIDON mission provided another opportunity to reduce the error of tidal prediction when comparing with ground-truth data. Ma *et al.* (1994) used one year of TOPEX/POSEIDON altimeter measurements to correct the Cartwright and Ray (1990) model. The corrections were determined on a $3° \times 3°$ grid. Comparison of tide gauge data and improved model predictions showed that the root-mean-square differences for M_2 constituents reduced from 3.9 cm to 2.7 cm. The corresponding reductions for S_2 and K_1 constituents are from 2.7 cm to 1.7 cm, and from 2.0 cm to 1.7 cm, respectively.

A combination of hydrodynamic modelling and TOPEX/POSEIDON altimeter measurements was used by Le Provost *et al.* (1995, 1998b) to calculate tides in the world oceans (for more details see Sect. 5.3.3). The increased accuracy of tidal prediction is required not only for better representation of tides but also for the study of general ocean circulation.

9.3.4 Ocean Circulation Observed by Satellites

The importance of accurate sea surface height measurements for large-scale ocean circulation results from the fact that the sea surface topography, relative to the geoid, represents the pressure field of the geostrophic part of the general circulation. However, sea surface height is influenced not only by surface geostrophic currents, but also by surface waves, tides and atmospheric pressure variations. Furthermore, the altimeter measurement of sea surface height is affected by the ionosphere, the components of the atmosphere and by the orbit of the satellite. Many of these effects can be eliminated by corrections introduced for each of the effects.

Geostrophic ocean currents flow with a velocity proportional to the local transverse slope. For example, the Gulf Stream which transports about 100 million m^3 of water per second has a transverse slope of a meter over several hundred kilometres. Maps of the global one-year mean topography also show the seasonal variability due to monsoon influence in the Indian Ocean (Tapley *et al.*, 1994). During the northern winter, due to cooler and denser air over Southeast Asia than air over the Indian Ocean, a northeasterly wind (the North-East Monsoon) blows from the continent towards the ocean. After some months, when the air over the land warms up, the pressure gradient reverses and warm and moist air (the South-West Monsoon) starts to blow from the ocean to the continent. This shift in wind is responsible for generation of the current system in the Indian Ocean, which is seasonally variable and clearly evident in the 10-day dynamic ocean topography anomaly maps from TOPEX/POSEIDON.

The sequence of topography maps (Tapley *et al.*, 1994) illustrates the development and propagation of equatorial Kelvin waves (with speed of 3 m/s) in the Pacific, beginning in December 1992. As we described in Sect. 7.6.4, the travel of Kelvin waves eastward is a key element for the 'delayed action oscillator' scenario of the El Niño event. These TOPEX/POSEIDON observations are also consistent with observed changes in the equatorial trade winds and other *in situ* observations of strengthening of the 1992 El Niño (Tapley *et al.*, 1994).

Similar analysis carried out for the East Auckland Current by using the TOPEX/POSEIDON and ERS-1 radar altimeter showed patterns of variability in dynamic sea-surface height consistent with *in situ* measurements (Laing, 1996). In particular, a strong east-southeastward flowing current between 33.5° and 34°S, with a return flow which meandered between 32° and 33°S has been detected. Also, an anticyclonic eddy to the north-east of New Zealand has been identified moving south-eastward during the spring of 1993.

9.3.5   Sea Surface Temperature Measured by Satellites

Sea surface temperature is an important indicator of the heat content of the upper ocean, and a tracer of surface velocities. It also controls air-sea exchange of gases and influences biological activity. Monitoring of Sea Surface Temperature (SST) by satellite observations is derived from a split window algorithm applied to the brightness temperatures issued from the infrared channels.

Another satellite sensor, known as Advanced Very High Resolution Radiometer (AVHRR) provides an opportunity to identify and describe an upwelling and frontal activity, which are the cause of high biological productivity (see Chap. 15). The AVHRR is a four- or five-channel radiometer with channels in the visible (0.6–0.7 μm), near infrared (0.7–1.1 μm) and thermal infrared (3.5–3.9 μm, 10.5–11.5 μm, and 11.5–12.5 μm). The channel centered at 3.7 μm, combined with the 11 and 12 μm channels is used to correct for various atmospheric factors (such as water vapour content) which can contaminate the temperature readings. The correction procedure is known as the 'split-window' method. According to Bernstein (1982), AVHRR data can be used to study climate variations with an accuracy of 0.5–1.0°C. The comparisons of the satellite SST with precision radiometric measurements made from a ship published by Schluessel *et al.* (1987) showed that satellite derived SST, when computed using a multichannel algorithm and corrected by HIRS (High Resolution Infrared Sounder) provide a reliable estimate of SST in the absence of visible cloud.

The monitoring of exchange or fluxes between the atmosphere and ocean, and in this regard, surface wind measurements and SST observations, together with estimates of rainfall, are important parameters as input to a global climate model. Other elements are the surface currents which provide a reference velocity *in situ*. Examples of these are westward-moving equatorial long wave patterns observed in both the Atlantic and Pacific oceans. They have a spatial scale of about 1000 km, and move westward with an average phase speed of 40 km/day (Barrett and Curtis, 1992). Sea surface temperature observations in the equatorial Pacific, using the Along Track Scanning Radiometer (ATSR) flying on ERS-1 satellite reveal the dynamics of these waves. Several eastward propagating signals have been discovered during January, March and April 1992 (onset of the 1992–1993 El Niño). The phase speed of these observed Kelvin waves was about 2.4 m/s (Llewellyn-Jones *et al.*, 1996).

9.4   Similitude and Dimensional Analysis

9.4.1   Introduction

In the previous chapters we demonstrated that numerous problems in marine fluid mechanics can be solved using theoretical and/or numerical models. However, there are still many problems for which such solutions are not possible. They are studied using experimental data, both from field and laboratory experiments. In the case of laboratory experiments, the basis of physical modelling is the idea that the model behaves in all aspects in a manner similar to the prototype it is designed to emulate. Requirements for similarity usually are expressed in terms of *criteria of similitude* or *conditions of similarity*.

Field and laboratory experiments provide a large quantity of various types of data. If no theoretical formulation of the problem is known, only various combinations of experimental data give us the possibility to establish the relationship between measured variables. Obviously, the number of such combinations can sometimes be very large. However, if several variables can be combined in the form of a single dimensionless variable, then the number of combinations can be significantly reduced. This approach is known as *dimensional analysis*. In next sections we will briefly describe an application of the similitude concept and dimensional analysis for marine fluid mechanics.

9.4.2   Dimensional Analysis

Dimensional analysis is a rational method for combining physical variables into a dimensionless product, thereby reducing the number of variables that need to be considered (Hughes, 1993). This analysis gives qualitative, rather than quantitative relationships and after performing this analysis, theoretical or experimental work is needed to establish the functional relationship between the dimensionless variables. Following Hughes, we identify three steps for applying dimensional analysis to a given problem:

1. identification of important independent variables of the process,
2. determination of how many independent dimensionless products can be formed from the variables,
3. reduction of the system's variables to the proper number of independent dimensionless variables.

The above three steps of methodology are based on the theorem known as the Buckingham Π Theorem, which states that *in a dimensionally homogeneous equation involving 'n' variables, the number of dimensionless products that can be formed from 'n' variables is 'n − r' where 'r' is the number of fundamental dimensions encompassed by the variables* (Hughes, 1993). Thus, if

$$x_1 = f\left(x_2,\, x_3,\, x_4, ..., x_n\right), \qquad\qquad (9.9)$$

is the equation involving n variables, then according to the Buckingham Π theorem, this equation can be rearranged into a new equation expressed in terms of dimensionless products:

$$\Pi_1 = \psi\left(\Pi_2,\, \Pi_3,\, \Pi_4, ..., \Pi_{n-r}\right). \qquad\qquad (9.10)$$

The number of terms in the equation is now less than the number of original variables by the number r. If the original n variables are independent, then the $n - r$ dimensionless products will also be independent when all original variables appear at least once in one of the dimensionless products.

The most important step in the dimensional analysis is an identification of important variables which should be included in the analysis and is guided by a sound understanding of the underlying physical processes. We have to make judgement as to which variables strongly influence the result and must be included, and which do not affect the result and can be omitted. We demonstrate the dimensional methodology by three examples involving the predominant influence of viscosity, gravity and buoyancy.

Sphere Moving in Viscous Fluid. The first example demonstrates an application of dimensional analysis for the determination of the drag force for a small particle moving in a fluid at low Reynolds number. We have considered this problem in Appendix C.3.2 using the Navier-Stokes equation. Now we will show that the same result can be obtained from dimensional analysis. Let us consider a small sphere with diameter D, moving with velocity v, in a fluid with dynamic viscosity μ. Our aim is to find a relationship between the drag force, F_d, and parameters D, v and μ. It is convenient to present all dimensions involved in the form of a dimension matrix (Hughes, 1993):

	F_d	D	v	μ
L	1	1	1	-1
T	-2	0	-1	-1
M	1	0	0	1

in which L, T and M denote the basic units of length, time and mass (see Appendix B). Thus, for example, the dimension of drag force F_d involves units of length, time and mass. In our case $n = 4$ and $r = 3$; thus from the Buckingham theorem the number of non-dimensional Π terms that can be formed is $n - r = 1$, i.e.:

$$\Pi_1 = F_d^{k_1} D^{k_2} v^{k_3} \mu^{k_4},$$ (9.11)

or:

$$\Pi_1 = \left(M \times L \times T^{-2} \right)^{k_1} (L)^{k_2} \left(L \times T^{-1} \right)^{k_3} \left(M \times L^{-1} \times T^{-1} \right)^{k_4}.$$ (9.12)

Equation (9.12) can be rewritten as:

$$\Pi_1 = L^{(k_1+k_2+k_3-k_4)} \times T^{(-2k_1-k_3-k_4)} \times M^{(k_1+k_4)}.$$ (9.13)

As the Π_1 term is non-dimensional, the powers of all basic units should equal zero. Thus, from Eq. (9.13) we obtain:

$$\left. \begin{array}{rcl} k_1 + k_2 + k_3 - k_4 & = & 0 \\ -2k_1 - k_3 - k_4 & = & 0 \\ k_1 + k_4 & = & 0 \end{array} \right\}.$$ (9.14)

The above set of equations is over-determined, and one coefficient is arbitrary. Let $k_1 = 1$. Solving the set for k_2, k_3 and k_4 yields: $k_2 = -1$, $k_3 = -1$ and $k_4 = -1$. Therefore:

$$\Pi_1 = F_d D^{-1} v^{-1} \mu^{-1} = \frac{F_d}{\mu D v}.$$ (9.15)

In general, the dimensionless variable $F_d/\mu D v$ is equal to the constant C, i.e.:

$$F_d = C \mu D v,$$ (9.16)

which is similar to the Stokes' solution (see Eq. C.56). From Stokes' solution we know that $C = 3\pi$, thus finaly:

$$F_d = 3\pi\mu Dv. \tag{9.17}$$

Prediction of Height of Wind-Induced Waves. The second example is related to surface wave generation by wind. From our discussion in Chap. 3, it follows that the significant wave height, H_s, depends on wind speed V_w, wind fetch X, water depth h and acceleration of gravity g. Therefore, these five variables can be arranged into a dimensions matrix as follows (Hughes, 1993):

	H_s	V_w	X	h	g
L	1	1	1	1	1
T	0	-1	0	0	-2
M	0	0	0	0	0

The unit of mass does not appear in the matrix, thus in this case $n = 5$ and $r = 2$, and the complete set of dimensionless products contains $n - r = 3$ Π terms of the form:

$$\Pi = H_s^{k_1} V_w^{k_2} X^{k_3} h^{k_4} g^{k_5}, \tag{9.18}$$

The independent exponent equations follows from the matrix as:

$$\left.\begin{array}{r} k_1 + k_2 + k_3 + k_4 + k_5 = 0 \\ -2k_5 - k_2 = 0 \end{array}\right\}. \tag{9.19}$$

We have two equations for five unknowns, and the procedure has to be different from the first example. We have to select r variables (and r dimensions among them) and use them as repeating variables in each of the Π terms, together with one of the remaining quantities. For example, we choose g and V_w as the repeating variables. Thus, H_s, h, and X are to be determined. Starting with H_s we get:

$$\Pi_1 = H_s^{k_1} V_w^{k_2} g^{k_5}, \tag{9.20}$$

and

$$\left.\begin{array}{r} 1 + k_2 + k_5 = 0 \\ -2k_5 - k_2 = 0 \end{array}\right\}. \tag{9.21}$$

Thus, $k_2 = -2$ and $k_5 = 1$, and the first Π term becomes:

$$\Pi_1 = H_s^1 V_w^{-2} g^1 = \frac{gH_s}{V_w^2}. \tag{9.22}$$

Now we use g and V_w together with the variable h:

$$\Pi_2 = V_w^{k_2} h^{k_4} g^{k_5}. \tag{9.23}$$

The equation set k_2 and k_5 is similar to the set (9.21). Therefore:

$$\Pi_2 = V_w^{-2} h^1 g^1 = \frac{gh}{V_w^2}. \tag{9.24}$$

Using the similar arguments we include the fetch X and third Π term becomes:

$$\Pi_3 = V_w^{-2} X^1 g^1 = \frac{gX}{V_w^2}. \tag{9.25}$$

Now we can write:

$$\frac{gH_s}{V_w^2} = f\left(\frac{gh}{V_w^2}, \frac{gX}{V_w^2}\right). \tag{9.26}$$

Fish Tail-Beat Frequency. As we showed in Chap. 2, the drag force which a fish has to overcome in order to move is a function of water density and velocity of fish movement. On the other hand, this velocity depends on the frequency at which the fish is beating its tail. If the fish beats its tail more rapidly, the stress exerted by the fish's muscles increases with increasing frequency. The stress needed to move the fish through the water depends also on the mass of the fish which we characterize by fish length. A more detailed discussion on the relationship between fish length and its mass is given in Chap. 11.

Summarizing, we have four variables relevant to this problem: frequency of tail beating, f; muscle stress, σ; water density, ρ_w; and fish length, l. The dimension matrix takes the form:

	f	σ	ρ_w	l
L	0	-1	-3	1
T	-1	2	0	0
M	0	1	1	0

As in the first example, there is only one non-dimensional Π term:

$$\Pi_1 = f^{k_1} \sigma^{k_2} \rho_w^{k_3} l^{k_4}. \tag{9.27}$$

The independent exponents have to satisfy the following equation set:

$$\left.\begin{array}{rcl} -k_2 - 3k_3 + k_4 &=& 0 \\ -k_1 - 2k_2 &=& 0 \\ k_2 + k_3 &=& 0 \end{array}\right\}. \tag{9.28}$$

The unknown frequency f has to be expressed with power $k_1 = 1$. Thus, solving Eq. (9.28) for k_2, k_3 and k_4 we obtain: $k_2 = -1/2$, $k_3 = 1/2$ and $k_4 = 1$. Hence:

$$\Pi_1 = f\sigma^{-1/2}\rho_w^{1/2}l = \frac{f\rho_w l}{\sigma^{1/2}} = \text{Const}, \tag{9.29}$$

and:

$$f = C\sqrt{\frac{\sigma}{\rho_w}}\frac{1}{l}. \tag{9.30}$$

Therefore, assuming that stress in the fish muscles, σ, and water density, ρ_w, are approximately constant, then the frequency of tail beating is inversely proportional to fish body length. We will return to this problem again in Chap. 11.

9.4.3 Similitude Principles

If the theory and field data do not exist for some physical phenomenon, or when we try to investigate the influence of a particular parameter on the process, laboratory experiments are the only solution. The great advantage of laboratory modelling is the fact that it is repeatable. However, a physical model of a given phenomenon can only be useful when it behaves in a similar manner to the prototype, in such aspects as the geometry, velocity, acceleration, and mass transport of the fluid. Correspondence between the model and prototype is expressed in terms of scales. For example, the length scale, N_L, is:

$$N_L = \frac{L_p}{L_m}, \tag{9.31}$$

where L_p is some distance in the prototype and L_m is the corresponding distance in the model.

In the same manner, we can define scales for velocity, acceleration, force and other parameters. In particular, the dynamic similitude requires that scales for all forces, such as inertial force, gravitational force, viscous force and pressure force should be constant and equal. However, such a requirement can not be fulfilled except at prototype scale. On the other hand, it is very rare that all forces are of equal importance for a given phenomenon. Usually one or two forces are dominant, and the effects of other forces can be neglected.

In problems of flow, inertial forces are always present and the ratio of the inertial force to any other force provides the relative influence of the two forces for a given flow problem. These ratio have the form of dimensionless numbers which are supposed to be the same on the model as on the prototype, in order to satisfy the principle of similitude. There is a long list of such numbers (also known as criteria) and the most important are given in Table 9.1.

Table 9.1: Similitude criteria

Name	Form	Application area
Froude Number (Fr)	$U/\sqrt{gL}$	free surface flow
Reynolds Number (Re)	UL/ν	viscous dominated flow
Euler Number (Eu)	$p/\rho U^2$	flow under pressure
Mach Number (Ma)	U/C_s	compressible flow
Weber Number (We)	$\rho U^2 L/\sigma$	flow when influence of surface tension is important
Richardson Number (Ri)	$(g/\rho)\dfrac{\partial \rho/\partial z}{\left(\partial \bar{u}/\partial z\right)^2}$	flow in stratified fluid
Péclet Number (Pe)	UL/D	flow with diffusion effect

Most flows with a free surface are under the influence of gravity forces, and consequently the Froude Number, Fr, is the most important criterion to be considered when modelling such flows. The Froude criterion requires that:

$$(Fr)_p = (Fr)_m,\tag{9.32}$$

or:

$$\frac{U_p}{U_m} = \sqrt{\left(\frac{g_p}{g_m}\right)\left(\frac{L_p}{L_m}\right)},\tag{9.33}$$

in which variables with subscript p are related to the prototype and variables with subscript m correspond to the model. The relationship (9.33) can be expressed in terms of scale ratios as follows:

$$N_u = \sqrt{N_g N_L}.\tag{9.34}$$

Assuming now that the model is built on the Earth, gravity accelerations, g_p and g_m, are the same; thus $N_g = 1$ and:

$$N_u = \sqrt{N_L}.\tag{9.35}$$

Therefore, if our model dimensions are one hundredth of the prototype dimensions (N_L), Froude criterion implies that velocities on the model are one tenth of velocities in the prototype ($N_u = 10$). In other words, velocities measured on the model have to be multiplied by 10 to obtain expected velocities in the prototype.

Equations (9.31) and (9.35) yield the following relationship for time scaling under the Froude criterion as:

$$N_t = \sqrt{N_L}. \tag{9.36}$$

Other similitude criteria and approaches to model similitude are discussed in more detail by Le Méhauté (1990) and Hughes (1993).

9.5 Spectral and Statistical Analysis of Time Series

Most experimental data on flows in the ocean are time series of surface elevation, velocities, pressure and other parameters for a given point or for a given profile (for example ADCP records). Except maybe tides, these data represent random oscillations and special statistical and spectral methods have to be used to analyze data. A very comprehensive overview of the analytical and numerical methods of data processing is given by Emery and Thomson (1997). Therefore, in this section we briefly summarize the practical methods for evaluation of experimental frequency spectra and statistics.

9.5.1 Data Sampling

Let us assume that a record is of duration t. The digital data consists of N data values with an equally-spaced sampling Δt. Thus:

$$\zeta_n = \zeta(t_0 + n\Delta t) \quad n = 1, 2, ..., N, \tag{9.37}$$

where t_0 is an arbitrary initial time. During the experimental planning stage, particular care should be taken in the adoption of the Δt value. Let us assume that the frequency band of interest ranges from 0 to f_c Hz ($\omega_c = 2\pi f_c$ rad/s).

The number, N, of discrete samples, required to describe $\zeta(t)$, should satisfy the following relationship:

$$N = \frac{t}{1/2f_c} = 2f_c t = \frac{\omega_c t}{\pi}, \tag{9.38}$$

or:

$$\Delta t = \frac{t}{N} = \frac{1}{2f_c} = \frac{\pi}{\omega_c}. \tag{9.39}$$

The requirement (9.38) indicates that N should be such that there are at least two samples in the shortest component. The fundamental increment $\Delta t = 1/2f_c = (\pi/\omega_c)$, is called the Nyquist sampling interval, and f_c or ω_c is called the Nyquist frequency. For example, a typical wave rider buoy records the

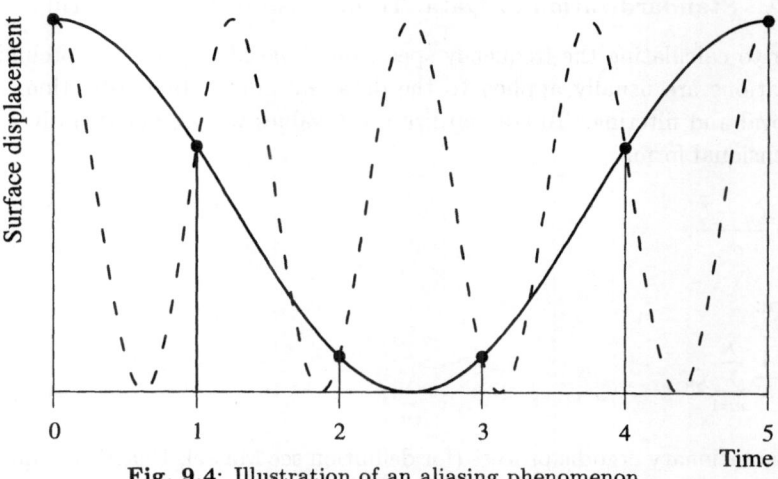

Fig. 9.4: Illustration of an aliasing phenomenon.

surface elevation with a sampling interval of $\Delta t = 0.3906$ s. Therefore, the corresponding Nyquist frequency is 1.2801 Hz or 8.043 rad/s. The sampling interval (9.39) is the maximum interval required to properly describe the data $\zeta(t)$. Frequencies in the original data above ω_c will lead to aliasing errors, which are inherent in all digital processing, but not present in direct analog data processing. The use of discrete data points introduces a cut-off frequency in the spectrum, at frequency ω_c. This means that all variance in the data which must be accounted for, must be distributed amongst the bands below ω_c. However, some of this variance may come from the higher unresolvable frequencies. Such a situation is illustrated in Fig. 9.4, in which two curves of different frequencies have been fitted to the same data. Only the continuous curve may be resolved, and the variance must be attributed to that, yet the data may be generated completely by the dashed curve. For any frequency, f, in the range $0 \leq \omega \leq \omega_c$, the higher frequencies which are aliased with ω_c are defined by:

$$(2\omega_c \pm \omega), (4\omega_c \pm \omega), \ldots, (2n\omega_c \pm \omega), \ldots \tag{9.40}$$

If $\Delta t = \pi/\omega_c$, the harmonic with frequency ω becomes:

$$\cos(\omega t) = \cos\left[(2n\omega_c \pm \omega)\frac{\pi}{\omega_c}\right] = \cos\left(\frac{\pi\omega}{\omega_c}\right). \tag{9.41}$$

Thus, all data at frequencies $(2n\omega_c \pm \omega)$ have the same cosine function as data at frequency ω, when sampled at points π/ω_c apart. For example, the wave rider buoy data ($\Delta t = 0.3906$ s, $\omega_c = 8.043$ rad/s) at frequency $\omega = 6$ rad/s would be aliased with data at the frequencies 10.086 rad/s, 22.086 rad/s, 26.172 rad/s, and so forth. Similarly, the power at these higher frequencies is aliased with the power in the lower frequencies. Thus, the true spectrum would be folded into the aliased spectrum.

9.5.2   Standardization of Data, Trend Removal and Filtering

Prior to calculating the frequency spectrum of variable ζ, various preliminary operations are usually applied to the data, *i.e.* data standardization, trend removal and filtering. To standardize the ζ values we present them in a non-dimensional form:

$$\xi_n = \frac{\zeta_n - \bar{\zeta}}{\sigma_\zeta}, \tag{9.42}$$

where:

$$\bar{\zeta} = \frac{1}{N} \sum_{n=1}^{N} \zeta_n. \tag{9.43}$$

For a stationary ergodic process (for definition see Massel, 1996a), the quantity $\bar{\zeta}$ is an unbiased estimate of the true mean value. The unbiased estimate of the standard deviation of data ζ_n is given by:

$$\sigma_\zeta = \left[\frac{1}{N-1} \sum_{n=1}^{N} \left(\zeta_n - \bar{\zeta} \right)^2 \right]^{1/2}. \tag{9.44}$$

Especially for the case of wave orbital velocity measurements in the coastal zone, in regions with large tidal motion, removal of the spurious trend or low frequency components with wavelengths longer than the record length, is usually required. The most common technique for trend removal is to fit a low-order polynomial to the data using the 'least squares method'. Thus, we assume that the original data $\{\zeta_n\}$ can be approximated by a polynomial of order K:

$$\tilde{\zeta}_n = \sum_{k=0}^{K} b_k (n\Delta t)^k \quad n = 1, 2, \ldots, N. \tag{9.45}$$

A 'least squares' fit provides a system of equations for unknown coefficients b_k as (Bendat and Piersol, 1986):

$$\sum_{k=0}^{K} b_k \sum_{n=1}^{N} (n\Delta t)^{k+m} = \sum_{n=1}^{N} \zeta_n (n\Delta t)^m, \quad m = 0, 1, 2, \ldots, = K. \tag{9.46}$$

Assuming that $K = 1$, we obtain:

$$b_0 = \frac{2(2N+1) \sum_{n=1}^{N} \zeta_n - 6 \sum_{n=1}^{N} n\zeta_n}{N(N-1)}, \tag{9.47}$$

$$b_1 = \frac{12 \sum_{n=1}^{N} n\zeta_n - 6(N+1) \sum_{n=1}^{N} \zeta_n}{\Delta t N(N-1)(N+1)}. \tag{9.48}$$

The de-trended time series $\hat{\zeta}_n$ now becomes:

$$\hat{\zeta}_n(n\Delta t) = \zeta_n(n\Delta t) - \sum_{k=0}^{K} b_k(n\Delta t)^k. \tag{9.49}$$

Filtering of data prior to detailed analysis is desirable for various reasons. If we are particularly interested in wind-induced waves, the swell component should be filtered out. On the other hand, if we concentrate only on the most energetic part of the wave spectrum, the high frequency components with negligible energy should be removed by filtering.

Digital filtering can be performed in either the time domain or the frequency domain. A detailed description of those filtering procedures can be found elsewhere, so they are not repeated here (Otnes and Enochson, 1972).

9.5.3 Calculation of Frequency Spectra

Spectral analysis is used to determine the partition of the variance of a time series as a function of frequency. For stochastic wind wave time series, contributions from the different frequency components are expressed in terms of the frequency spectral density. In practice, the term spectrum is applied to all spectral functions such as autospectrum for one time-series, or cross-spectrum for the two time-series. Frequency spectra are usually estimated by either of two methods. The first is based on the Wiener-Khinchine theorem and is called the Blackman-Tukey procedure. The Wiener-Khinchine relations link variance functions in the time domain to those in the frequency domain (Massel, 1996a; Emery and Thomson, 1997). In the second method, called the Cooley-Tukey method, the direct Fast Fourier Transformation is used (Massel, 1996a). In the following, we briefly discuss both methods.

Blackman-Tukey Method. This procedure requires calculations in the following steps:

1. Subtracting the mean value from digital data $\left\{\hat{\zeta}_n\right\}$, trend removal and filtering (if necessary).
2. Calculating the autocorrelation function:

$$K(r\Delta t) = \frac{1}{N-r} \sum_{n=1}^{N-r} \hat{\zeta}_n \hat{\zeta}_{n+r}, \quad r = 0, 1, 2, \ldots, m, \tag{9.50}$$

where r is called the lag number and m is the maximum lag number $(m < N)$. Selection of the m value, which provides the optimum estimate for the autocorrelation function, will be discussed later. A finite value of m implies that surface elevations $\zeta(t)$, at times $t > m\Delta t$ are uncorrelated. A typical normalized autocorrelation function, $K(\tau)/\sigma_\zeta^2$, for surface waves is shown in Fig. 9.5a.

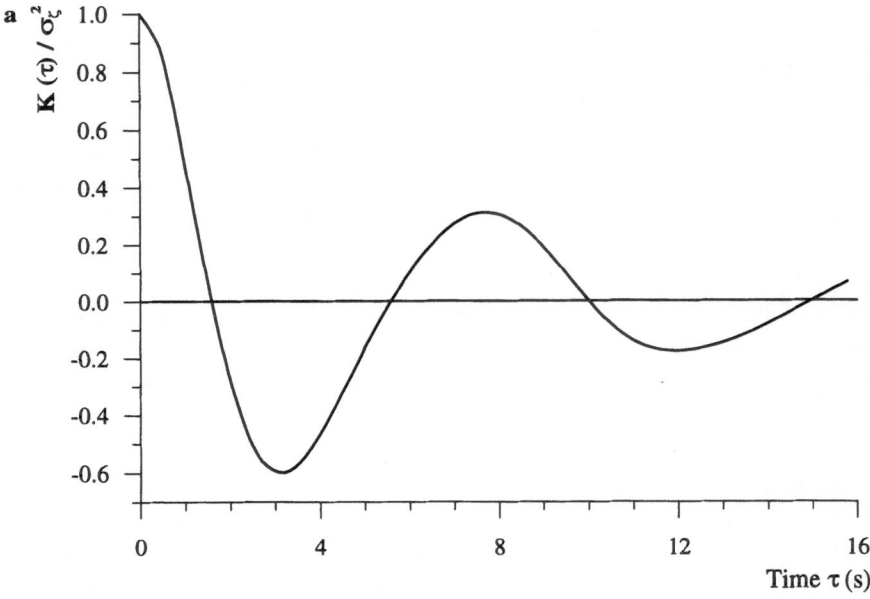

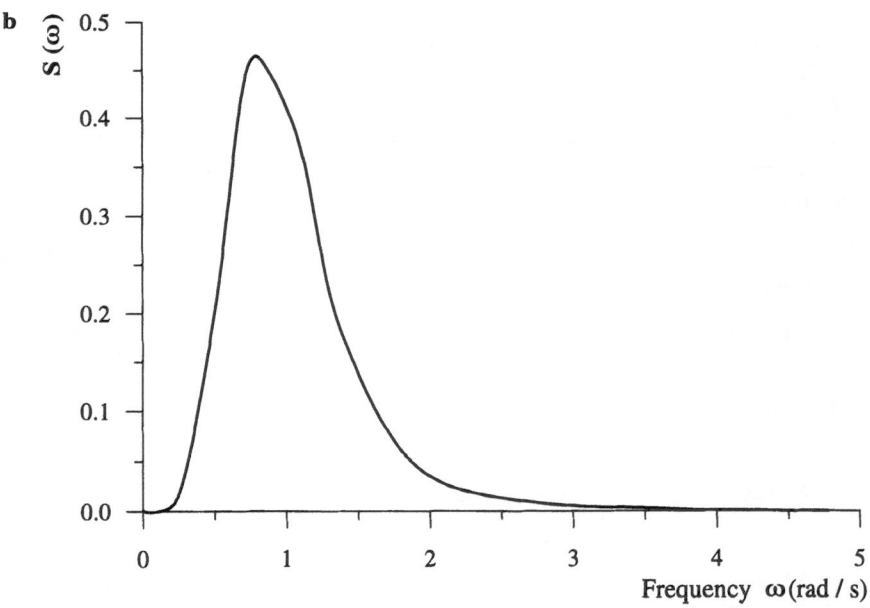

Fig. 9.5: Frequency spectra: **a** typical normalized autocorrelation function for surface waves, **b** corresponding frequency spectral density

3. Suppression of the spectrum leakage using a window for the autocorrelation function. Such window tapers the autocorrelation function to eliminate the discontinuity at the end of function K. There are numerous such windows in use. A typical window is the Hanning window:

$$u_{hr} = \frac{1}{2}\left(1 + \cos\frac{\pi r}{m}\right). \tag{9.51}$$

The modified autocorrelation function becomes:

$$\tilde{K}(r\Delta t) = K(r\Delta t)u_{hr}. \tag{9.52}$$

4. Calculating the frequency spectral density by numerical integration of the autocorrelation function $\tilde{K}(r\Delta t)$:

$$S(\omega_k) = \frac{\Delta t}{\pi}\left\{\tilde{K}(0) + 2\sum_{r=1}^{m-1}\tilde{K}(r\Delta t)\cos\frac{\pi r k}{m} + \tilde{K}(m\Delta t)\cos(\pi k)\right\}, \tag{9.53}$$

for frequencies $\omega_k = k\,\Delta\omega = \dfrac{K\pi}{m\Delta t}$, $k = 0, 1, 2, \ldots, m$.

The frequency spectral density, $S(\omega)$, corresponding to the autocorrelation function, $K(\tau)$, is given in Fig. 9.5b.

The estimate, $S(\omega)$, describes the time average of $\left[\hat{\zeta}(t)\right]^2$ in terms of its frequency components lying inside the frequency band: $\omega - (B_e/2)$, $\omega + (B_e/2)$, divided by the resolution bandwidth B_e (rad/s). Equation (9.53) gives $(m/2)$ independent estimates of the spectrum ordinates. The ordinates separated by frequency increments smaller than $2f_c/m$ are correlated.

For a given bandwidth B_e, the required maximum lag number m is:

$$m = \frac{2\pi}{B_e\Delta t}. \tag{9.54}$$

The spectrum of a random process is itself a random function and the standard error, ϵ, of the spectrum estimation is usually presented as a function of the number of degrees of freedom, $n = 2N/m$:

$$\epsilon = \sqrt{\frac{2}{n}} = \sqrt{\frac{m}{N}}. \tag{9.55}$$

It should be noted that, for a given N, when the maximum lag number, m, is small, the error is also small. The minimum total record length, t, required to achieve a desired error, ϵ, in terms of other parameters, is given by:

$$t = \frac{2\pi}{B_e\epsilon^2} = \frac{m\,\Delta t}{\epsilon^2}. \tag{9.56}$$

For example, the typical parameters for the measurement of surface waves using the wave rider buoy, are:

$$\Delta t = 0.3906s, \quad N = 3072, \quad m = 30,$$

$$n = \frac{2N}{m} = 205, \quad \epsilon = \sqrt{\frac{2}{n}} = 9.8\%, \quad B_e = \frac{2\pi}{m\Delta t} = 0.534 \text{ rad/s}.$$

Method Based on the Fast Fourier Transform of Original Data. The direct transform of the data is an alternative method for calculating the frequency spectra. We first assume that the wave record, $\zeta(t)$, is given for a finite time interval $(0, T)$ and is sampled at N equally spaced points, Δt apart. The Fourier representation of the $\zeta(t)$ record is:

$$X(\omega, t) = \Delta t \sum_{n=0}^{N-1} \zeta_n \exp\left[-i\omega n \Delta t\right]. \tag{9.57}$$

The usual selection of discrete frequency values for the computation of $X(\omega, t)$ is:

$$\omega_k = \frac{2\pi k}{T} = \frac{2\pi k}{N\Delta t}, \quad k = 0, 1, 2, \ldots, N-1. \tag{9.58}$$

Fast Fourier Transform techniques are designed to compute the quantities $X(\omega, t)$, and publications on FFT algorithms are very numerous. For example, the popular algorithm, appropriate for binary digital computers, was introduced by Cooley and Tukey (1965). This algorithm applies when the number of data samples, N, is a power of 2, *i.e.* $N = 2^p$. However, there are many other algoritms for more general N. If necessary, zero is added to the data sequence to satisfy this requirement. The iterative procedure to determine quantities $X(\omega, t)$ requires the sum of p terms, where every term involves $(N/2)$ Fourier transforms, requiring 4 operations each. This gives a total of $2Np$ complex multiply-add operations. A full discussion of these matters is given by Cooley and Tukey (1965), Otnes and Enochson (1972), Bendat and Piersol (1986), Emery and Thomson (1997), and many others.

For practical calculations, the record, $\zeta(t)$, is usually divided into K segments, each of length $L\Delta t$. The Fourier Transform of $\zeta(t)$ for each segment can then be viewed as the Fourier Transform of an unlimited time history record multiplied by a rectangular data window:

$$\tilde{\zeta}(t) = \zeta(t)v(t), \tag{9.59}$$

where:

$$v(t) = \begin{cases} 1 & \text{for } 0 \leq t \leq L\Delta t, \\ 0 & \text{otherwise.} \end{cases} \tag{9.60}$$

Fourier analysis of finite length records results in inherent side lobes in the spectral domain. The large side lobes allow leakage of energy at frequencies well separated from the main lobe. To reduce the leakage problem, it is common practice to introduce a time window that tapers the time-history data, to eliminate the discontinuities at the beginning and end of the records being analyzed. There are numerous such windows in current use. For example, the Hanning data window has the form:

$$
v(t) = \begin{cases} \dfrac{1}{2}\left[1 - \cos\left(\dfrac{2\pi t}{L\Delta t}\right)\right] = 1 - \cos^2\left(\dfrac{\pi t}{L\Delta t}\right) & \text{for} \quad 0 \le t \le L\Delta t, \\ 0 & \text{otherwise.} \end{cases} \tag{9.61}
$$

If the total number of points, N, cannot be made arbitrarily large to get a more accurate spectrum estimation, then a reasonable procedure is to overlap the segments by one half of their length. The number of overlapping segments now becomes:

$$
K_1 = \frac{2N}{L} - 1 = 2K - 1, \tag{9.62}
$$

in which K is the number of non-overlapped segments.

The final estimate of the one-sided spectral density $S(\omega)$ becomes:

$$
\tilde{S}(\omega_k) = \frac{2}{K_1 L\Delta t} \sum_{i=1}^{K_1} |X(\omega_k)|^2, \quad k = 0, 1, 2, \ldots, \frac{L}{2}, \tag{9.63}
$$

in which $\omega_k = \dfrac{2\pi k}{L\Delta t}$.

Spectra of Vector Series. Vector type time series such as currents and wind are recorded as speed and direction by rotor-type meters and as orthogonal components by acoustic and electromagnetic meters. Therefore, to calculate the spectra of current and wind, the data has to be resolved into orthogonal components related to the Earth-referenced Cartesian coordinate system with two orthogonal horizontal components and a vertical component. In the open sea, the horizontal velocity vector is usually resolved into eastward (u) and northward (v) component, whereas in the coastal region the v axis is oriented approximately parallel to the coastline and u axis is directed normally to the coastline.

When investigating currents over abrupt topography or other forms of narrowband oscillatory flows, it is useful to separate the velocity vector into clockwise and counterclockwise rotating circular components. Resulting spectra are known as rotary spectra (Gonella, 1972; Mooers, 1973).

9.5.4   Calculation of Statistical Characteristics of Waves

The basic statistical ocean flow parameters, namely surface displacement, wave-induced velocities and pressure as well as all kinds of turbulent oscillations, are random variables subjected to various statistics which can be evaluated using experimental data.

Let us consider N digital data values $\{x_n\}$, $n = 1, 2, \ldots, N$ with an equally spaced sampling interval of Δt seconds. We assume that record $x(t)$ is stationary with $\bar{x} = 0$. The estimate of the probability density function of $x(t)$ can be expressed as:

$$\tilde{f}(x) = \frac{N_x}{\Delta x N}, \tag{9.64}$$

where Δx is a narrow interval centred at x, and N_x is the number of data values that fall within the range $x \pm \Delta x/2$. To find the number N_x, the full range of x is divided into a number of classes with intervals of equal widths. The number of data in each class is then tabulated. This procedure implies that the estimate $\tilde{f}(x)$ is dependent on the number of class intervals and their width Δx. In an analysis of wave data, the number of class intervals is about 10–20.

In addition to the estimation of the probability density function, estimates of statistical moments are also required. Equations (9.43) and (9.44) provide estimates of the first two moments. The estimates of the third and fourth central moments can be written as:

$$\mu_3 = E\left[(x - \bar{x})^3\right] = \frac{N^2}{(N-1)(N-2)} m_3, \tag{9.65}$$

$$\mu_4 = E\left[(x - \bar{x})^4\right] = \frac{N\left(N^2 - 2N + 3\right) m_4 - 3N\left(2N - 3\right) m_2^2}{(N-1)(N-2)(N-3)}, \tag{9.66}$$

where:

$$m_n = \frac{1}{N} \sum_{k=1}^{} (x - \bar{x})^n. \tag{9.67}$$

Equations (9.65) and (9.66) are unbiased and consistent estimates of true central moments.

Part III

Marine Environment and Ecology

10 Mechanical Properties of Biological Materials

10.1 Introduction

The preceding two parts of this book were dedicated to the description of the fundamentals of fluid mechanics (Part I), and oceanic physical processes of various scales, which are not usually considered in fluid mechanics textbooks (Part II). This later field of fluid mechanics is known as geophysical fluid mechanics (Cushman-Roisin, 1994).

Part III of this book is totally focused on the description of various linkages between ocean water motions, given in the previous parts, and many aspects of biological life in the ocean. However, as this book is focused on fluid mechanics, the biological details will only be described to such an extent that is needed to understand the role of the physical environment for the functioning of marine animals. It is assumed that the reader has a basic knowledge of marine ecology.

At the beginning of Part III we will consider the mechanical properties of marine animals. The final size and shape of species are the result of many factors, such as animal physiology, availability of food and location in the food chain, as well as the characteristics of the physical environment in which the organism lives. Marine animals adapt in different ways to live in turbulent, or laminar environments. However, in each case an animal is subject to various environmental loadings (forces and moments). It should withstand these loadings by developing appropriate body strength. In the next section we will describe the main mechanical properties of marine animal bodies which determine the animal's strength.

For marine species, whose weight is supported by immersion in water, the relationship between the organism's density and the density of the surrounding water is fundamental in determining the animal's motion with minimum usage of energy. Aquatic organisms swim in a variety of ways and at a wide range of speed, operating under notably different Reynolds number regimes. As a result, they have developed remarkable methods of adaptation for different purposes. In particular, to avoid sinking, the marine species must either use buoyancy strategies or generate dynamic lift. However, prior to considering the locomotion of aquatic animals, we will firstly discuss the mechanical properties of biological material from which marine organisms are constructed.

10.2 Definition of Mechanical Properties of Biological Materials

Mechanics of materials is a branch of applied mechanics dealing with the behaviour of solid bodies subjected to various types of loading. Many methods used in the mechanics of materials are very useful for the determination of mechanical properties of biological material. Therefore, we will start with definitions of material properties and its behaviour commonly used in mechanics.

10.2.1 Stress and Strain

Let us consider a prismatic bar that is loaded by axial forces, P, at the ends (Fig. 10.1a). The axial forces, directed towards the bar, produce a uniform **compression** of the bar. When forces, P, are directed out of the bar, the bar is said to be in **tension**. In both cases, the intensity of force (the force per unit area) is known as the **stress** and is commonly denoted by the letter σ. Thus, if the cross-sectional area of the bar is A, the stress becomes:

$$\sigma = \frac{P}{A},\qquad(10.1)$$

in which stress σ is expressed in the same units as pressure, namely newtons (N) per area in square metres (m^2) or pascals (Pa).

Depending on force direction, the resulting stresses are **compressive** or **tensile stresses**. If stresses act in a direction perpendicular to the surface, they are referred to as **normal stresses**. Stresses acting parallel to the surface are known as **shear stresses**. By convention, tensile stresses are defined as positive and compressive stresses as negative.

When the bar is loaded axially, it becomes longer when in tension and shorter when in compression. The change in length or elongation per unit length is known as **strain** and is usually denoted by the letter ϵ. The relationship between stress, σ, and strain, ϵ, illustrates the behaviour of various materials as they are loaded statically in tension or compression. For some materials, when the load is slowly removed, they exactly follow the same curve back to the origin (see Fig. 10.1b). Such a property of a material, where it returns to its original dimensions after unloading, is known as **elasticity**, and the material itself is said to be **elastic**. The stress-strain curve from point 0 to A in Fig. 10.1b does not need to be linear in order for the material to be elastic. In Fig. 10.1b it is shown that in an initial region on the stress-strain diagram, many materials behave both elastically and linearly. This means that stress, σ, is linearly proportional to the strain ϵ, *i.e.*:

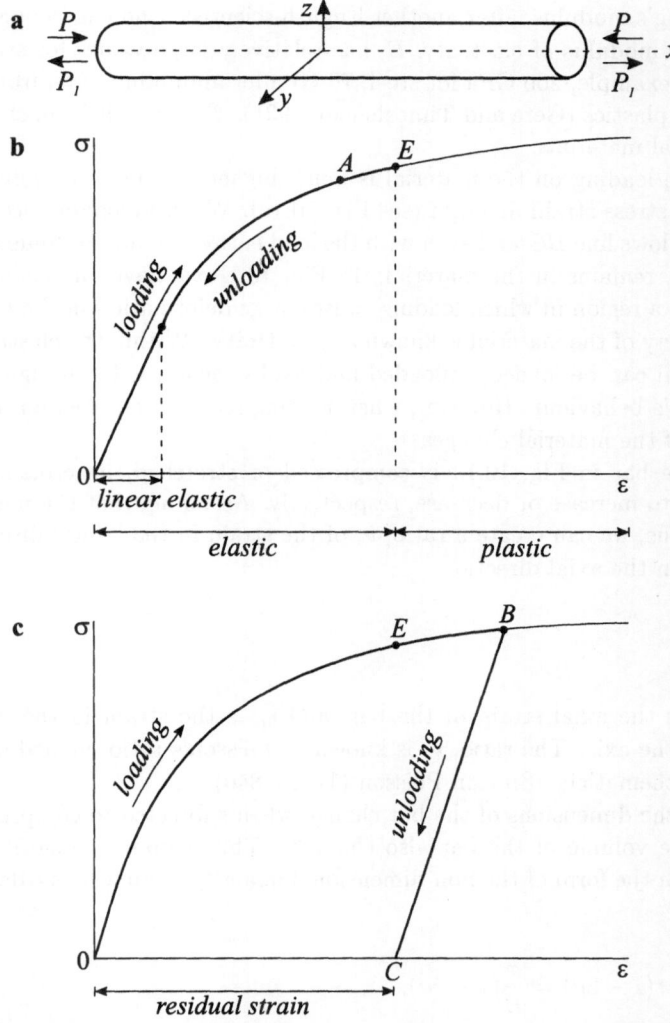

Fig. 10.1: Stresses and strains in a prismatic bar: **a** bar in compression and tension, **b** stress-strain diagram for elastic material, and **c** for partly elastic material

$$\sigma = E \, \epsilon, \tag{10.2}$$

where E is the constant of proportionality known as the **modulus of elasticity**. As strain, ϵ, is a non-dimensional quantity, modulus of elasticity has the same units as stress, σ, namely N/m^2 (or Pa).

Equation (10.2) is commonly known as Hooke's law, named after the English scientist Robert Hooke (1635–1703) who investigated the elastic properties of materials. A large number of materials that exhibit a linear relationship (see Eq. 10.2) are known as a Hookean materials. The modulus of elasticity, E, is called Young's modulus, after another English scientist, Thomas Young (1773–1829). The modulus of elasticity, E, has relatively large values for structural metals, for example, 200 GPa for steel, 70 GPa for aluminium, and from 0.7 to 14 GPa for plastics (Gere and Timoshenko, 1991). However, it is much smaller for biological materials.

When the loading on the material is much higher, the material follows line EB on the stress-strain diagram (see Fig. 10.1c). When unloading occurs, the material follows line BC and even with the load entirely removed, some residual strain, OC, remains in the material. In Fig. 10.1b we have an elastic region followed by a region in which loading causes large deformation in the material. This property of the material is known as **plasticity**. Within the elastic range, the material can be loaded, unloaded and loaded again, without significantly changing it's behaviour. However, when loading reaches the plastic range, the structure of the material changes.

When the bar in Fig. 10.1a is compressed or stretched, its cross-sectional area tends to increase or decrease, respectively. Assuming that the material is homogeneous, we can create a ratio, ν, of the strain in the lateral direction to the strain in the axial direction:

$$\nu = \frac{\epsilon_y}{\epsilon_x}, \tag{10.3}$$

where ϵ_x is the axial strain of the bar, and ϵ_y is the strain in the direction normal to the axis. The ratio, ν, is known as Poisson's ratio, named after the French mathematician Simeon Poisson (1781–1840).

Because the dimensions of the bar change when subjected to compression or tension, the volume of the bar also changes. This volume change is usually expressed in the form of the non-dimensional quantity e, known as **dilatation**, *i.e.*:

$$e = \frac{\Delta V}{V_0} = \epsilon(1 - 2\nu) = \frac{\sigma}{E}(1 - 2\nu). \tag{10.4}$$

From Eq. (10.4) it follows that the maximum possible value of ν for ordinary materials is $1/2$; any larger value of ν yields the volume decreasing when the material is stretched, which is an unlikely event. Most materials are characterized by Poisson's ratio, ν, of about $1/4$ or $1/3$. In the perfectly plastic region, no volume change (dilatation) occurs, when Poisson's ratio is taken as $\nu = 1/2$.

10.2.2 Shear-Stress and Strain

The shearing stresses, τ, acting parallel to the surface of the material are defined as force per unit area:

$$\tau = \frac{F}{A}, \tag{10.5}$$

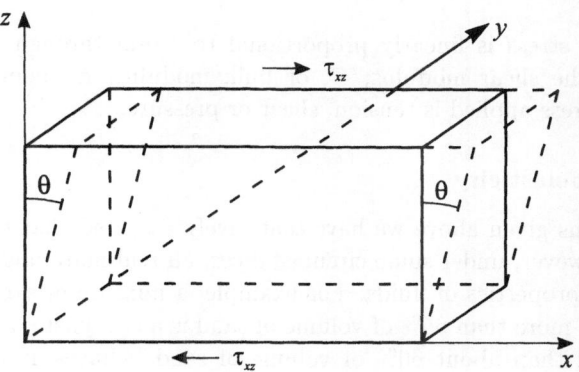

Fig. 10.2: Shear strain

in which F is the force and A is the area over which the force is applied and not the cross-sectional area. For example, as shear stress, τ, produces shear strain, the originally rectangular block is distorted into the parallelepiped by translation of the corners through the angle θ (Fig. 10.2). Assuming that the angle, θ, is measured in radians, the shear stress for the linear elastic material becomes:

$$\tau = G\,\theta, \tag{10.6}$$

where G is the **shear modulus**. It has the same units as the Young's modulus, E, namely, pascals (Pa). Both moduli are related by the following equation:

$$G = \frac{E}{2(1+\nu)}, \tag{10.7}$$

in which ν is Poisson's ratio. Because the value of ν for ordinary materials is between zero (for an ideally rigid material) and one-half (for a fluid), modulus G must be from one-third to one-half of E.

Let us assume that the material is subjected to equal stresses acting along two perpendicular axes, x and z, ($\sigma_x = \sigma_z = \sigma$). Now, dilatation takes the form (Gere and Timoshenko, 1991):

$$e = 3(1-2\nu)\frac{\sigma}{E}. \tag{10.8}$$

Usually Eq. (10.8) is expressed using the new quantity K, known as the bulk modulus of elasticity:

$$e = \frac{\sigma}{K}, \qquad K = \frac{E}{3(1-2\nu)}. \tag{10.9}$$

Therefore, the stress is linearly proportional to strain through the Young's modulus, E, the shear modulus, G, or bulk modulus, K, depending upon whether the stress applied is tension, shear or pressure.

10.2.3 Viscoelasticity

In all definitions given above we have tentatively assumed that the materials are 'solid'. However, under some circumstances, all real materials behave as if they have the properties of fluids. For example, a mixture of sand and water which contains more than 65% of volume of sand is a stiff paste, but a mixture containing less than about 60% of volume of sand behaves as a Newtonian liquid (Alexander, 1968). When we relax the stress on a Hookean material over time, t, we obtain:

$$\frac{d\sigma}{dt} = E\frac{d\epsilon}{dt}. \tag{10.10}$$

If some viscous flow occurs while the body is loaded, the stress tends to decrease at a rate depending on the initial stress value (Wainwright et al., 1976):

$$\frac{d\sigma}{dt} = E\frac{d\epsilon}{dt} - \frac{\sigma}{t_r}, \tag{10.11}$$

in which the constant, t_r, having the dimension of time is known as the relaxation time of the material.

To clarify the physical meaning of the relaxation time, let us extend the body by a fixed strain and then hold it constant ($d\epsilon/dt = 0$). Integration of Eq. (10.11) gives:

$$\sigma(t) = \sigma_0 \exp\left(-\frac{t}{t_r}\right). \tag{10.12}$$

Equation (10.12) indicates that the relaxation time is the time required for the stress to decrease to about 0.36 of its initial value. As the relaxation time involves both the elasticity and the viscosity, it is defined as (Wainwright et al., 1976):

$$t_r = \frac{\mu}{E}, \tag{10.13}$$

in which μ is the coefficient of dynamic viscosity. Behaviour of material depends on the relationship of the relaxation time, t_r, to the loading time, t_l. For example, silicon putty under slow loading, $t_l \gg t_r$, flows like a liquid, while under impact loading, $t_l \ll t_r$, no relaxation occurs and the material bounces elastically. As we will see later, all biomaterials exhibit some relaxation phenomena.

10.2.4   Bending

Pure compression or tension are relatively uncommon states of stress in biological materials. By far the most common situation is that of bending. This loading regime is experienced by many marine organisms subjected to wave or current loading, such as branches of hard corals, sea anemones, macroalgae, barnacles and many others.

To describe the behaviour of such biological materials under a bending regime, we apply methods developed in structural engineering, namely beam theory. A beam is an element designed to resist forces acting normal to its axis. In Fig. 10.3a, a simple cantilever beam loaded by force F at its end is shown. To keep the beam in equilibrium the **shear force**, F_1, equal to force F is applied at the fixing point O. Because of bending of the beam by the force F, the so called **bending moment**, M, occurs. In general, force tends to move a body along the line of force action, but the force also tends to rotate the body about an axis. The ability of a force to cause a body to rotate is measured by the moment of the force, which is a product of force and a perpendicular distance from the moment centre to the line of the force action. Thus, at the cross-section AB of the beam in Fig. 10.3a, the bending moment becomes:

$$M(x) = -Fx. \tag{10.14}$$

At point P, the moment $M = 0$, and it becomes maximum at point O, $i.e.$ $M = -Fl$. In this equation, the sign convention was used according to which a positive bending moment elongates the lower part of the beam and compresses the upper part (see Fig. 10.3b). For a negative bending moment, the situation is reversed. Also a positive shear force tends to deform the element by causing the right-hand face to move downward with respect to the left-hand face (Gere and Timoshenko, 1991).

Let us consider a cross-section AB of the beam (Fig. 10.3c). The neutral axis, x, denotes the point where there is no stress – the material here neither extends nor contracts. At points not on the neutral axis, tension or compression stresses in the beam are related to the bending moment, M, by a formula known as the flexure formula (Gere and Timoshenko, 1991):

$$\sigma = \frac{My}{I} = \frac{M}{S}, \tag{10.15}$$

in which y is the distance from the neutral axis, $S = I/y$ is the section moduli, and I is the moment of inertia (or second momentum of area) of the cross-section of the beam with respect to the neutral axis:

$$I = \int_A y^2 dA, \tag{10.16}$$

where A is the cross-section area. From Eq. (10.15) it follows that the maximum stresses appear at the edges of the beam (see Fig. 10.3c).

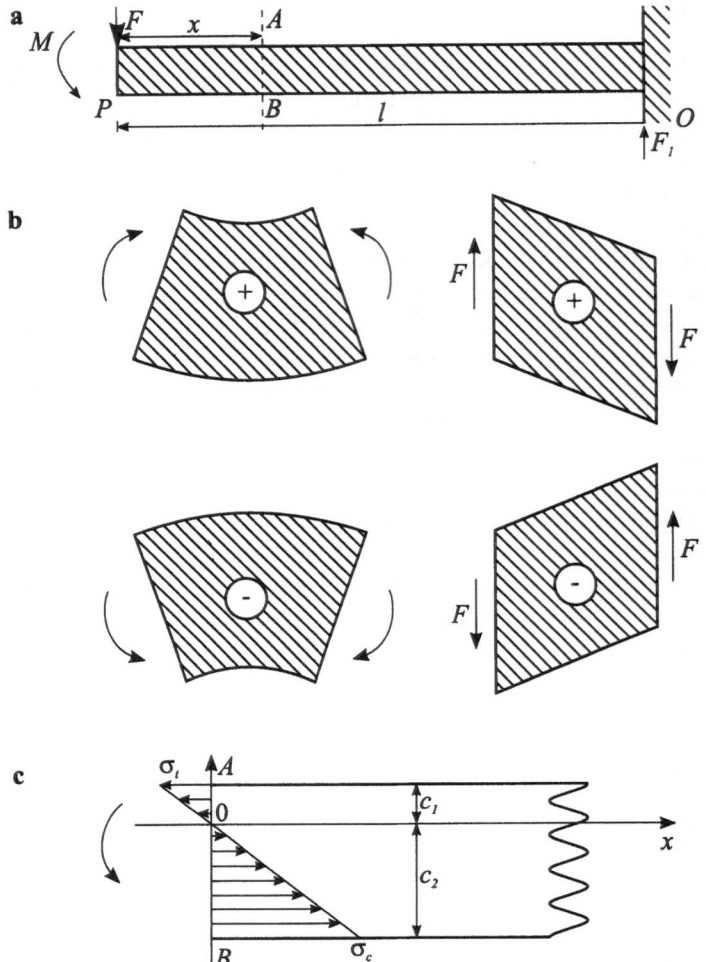

Fig. 10.3: Elastic beam: **a** force loading a beam, **b** sign convention for loading moment M and shear force F, **c** distribution of stresses σ in a beam: σ_t denotes tension, σ_c denotes compression

For some specific symmetric cross-sections, $c_1 = c_2 = c$, expressions for I, section moduli S, and maximum stress σ_{max} (see Eq. 10.15) can be simplified as follows:

- rectangular cross-section beam (Fig. 10.4):

$$I = \frac{bh^3}{12}, \quad S = \frac{bh^2}{6}, \quad \sigma_{max} = \pm\frac{6M}{bh^2}, \tag{10.17}$$

where b and h are the width and height of the beam, respectively;

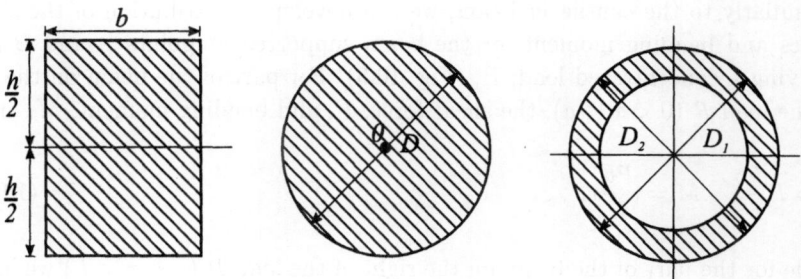

Fig. 10.4: Doubly symmetric cross-sectional shapes

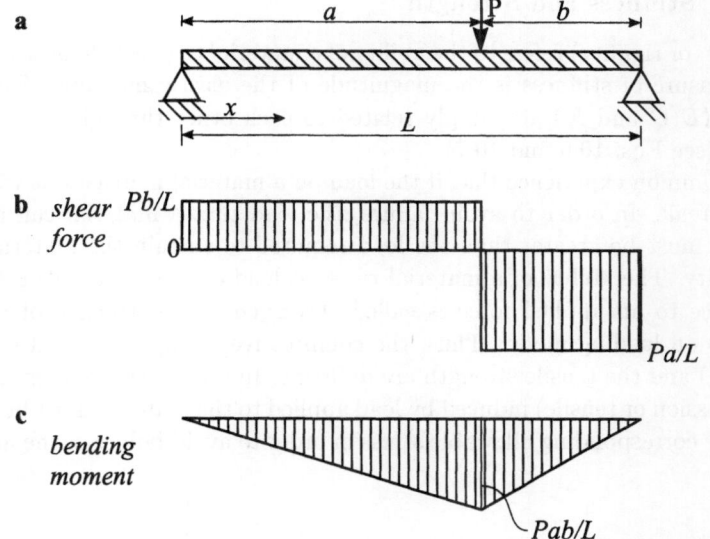

Fig. 10.5: Simple beam loaded by force F: **a** beam and force location, **b** shear force distribution, **c** bending moment distribution

- circular cross-section beam (Fig. 10.4):

$$I = \frac{\pi D^4}{64}, \quad S = \frac{\pi D^3}{32}, \quad \sigma_{max} = \pm\frac{32M}{\pi D^3}, \tag{10.18}$$

where D is the beam diameter.

- beam with hollow circular cross-section (Fig. 10.4):

$$I = \frac{\pi}{64}\left(D_2^4 - D_1^4\right); \quad S = \frac{\pi}{32}\left(D_2^3 - \frac{D_1^4}{D_2}\right); \quad \sigma_{max} = \pm\frac{32D_2M}{\pi\left(D_2^4 - D_1^4\right)}, \tag{10.19}$$

Similarly to the cantilever beam, we can develop a distribution of the shear forces and bending moment for the beam supported at points A and B and carrying a concentrated load, P, (Fig. 10.5). For part of the beam on the left of the load P $(0 < x < a)$, the shear force, F, and bending moment, M, are:

$$F = P\frac{b}{L}, \quad M = \frac{Pb}{L}x, \tag{10.20}$$

while for the part of the beam on the right of the load P $(a < x < L)$ we have:

$$F = -P\frac{a}{L}, \quad M = Pa\left(1 - \frac{x}{L}\right). \tag{10.21}$$

10.2.5 Stiffness and Strength

Stiffness (or rigidity) refers to the ability of a material to resist changes in shape. The measure of stiffness is the magnitude of the elastic modulus. The three moduli $(E, G$ and $K)$ are simply related to each other through the Poisson's ratio ν (see Eqs. 10.6 and 10.7).

It is common experience that if the load on a material increases, it will eventually break. In order to avoid failure, the load that a material can actually support must be greater than the load required to sustain the material's serviceability. The ability of a material to resist load or the material's maximum resistance to an applied force is called **strength**. The strength of material depends on loading mode. Thus, the compressive strength (strength in compression) and the tensile strength are different. In both cases, maximum stress (compression or tensile) induced by load applied to the material must be smaller than the corresponding allowable stress in order to avoid failure of the material. Thus:

$$\sigma_{max} < \sigma_{allow}, \tag{10.22}$$

in which:

$$\sigma_{allow} = \frac{\sigma_u}{n}, \tag{10.23}$$

where σ_u is the ultimate stress and n is the factor of safety. The allowable stresses used in the design of a structure are often specified by particular design codes (Cheng, 1985; Gere and Timoshenko, 1991).

For biological materials we are interested in determination of the critical loading under which the material breaks. Therefore, for compression or tensile modes of loading we have:

$$\left(\frac{F}{A}\right)_{max} = \sigma_u. \tag{10.24}$$

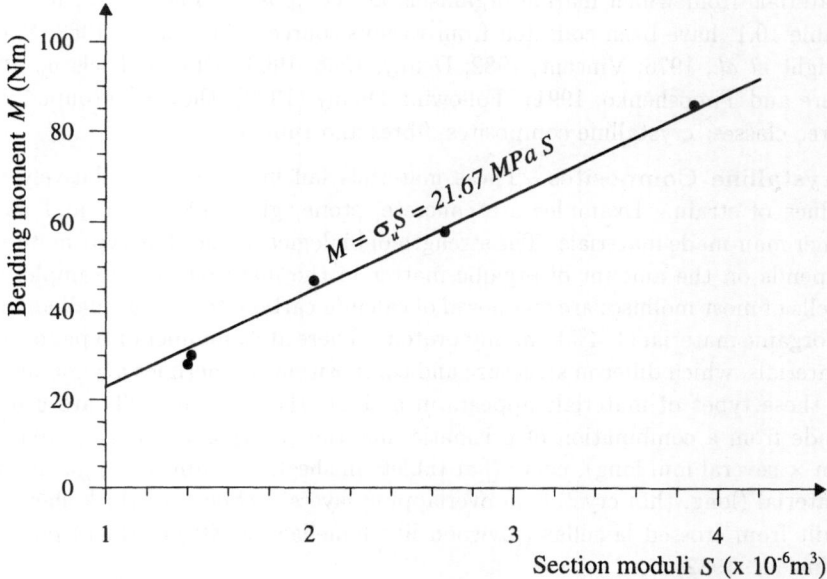

Fig. 10.6: Relationship between breaking moment M and section moduli S for samples of hard coral of genus *Acropora formosa*

The ultimate stress, σ_u, is the measure of the material's compressive strength, σ_c, or tensile strength, σ_t. In the case of bending, the material will break if the maximum tensile stress My_{max}/I (see Eq. 10.15) exceeds the tensile strength, σ_t, or if the maximum compressive stress My_{max}/I exceeds the compressive strength, σ_c.

An example of the results of a bending test for a hard coral sample of genus *Acropora formosa* is shown in Fig. 10.6. Cylindrical coral samples of length of about 200 mm and diameter between 20 and 35 mm were selected and breaking force, F, was measured during mechanical tests similar to that shown in Fig. 10.5. For the known breaking moment and section moduli $S = \pi D^3/32$, the breaking stress, σ_b, can be calculated from Eq. (10.15) as $\sigma_b = M/S$. The relationship between moment M and section moduli S for particular samples is given in Fig. 10.6. From this figure it follows that the average breaking stress, σ_b, for the hard coral of genus *Acropora formosa* is about 22 MPa. Also, it should be noted that the average density of the coral samples was about 2.15×10^3 kg/m^3. The values of ultimate stresses for other marine organisms are given in the next section.

10.2.6 Overview of Mechanical Properties of Biological Materials

Using the above definitions of various mechanical properties of materials, we will briefly list and describe values of modulus and ultimate stresses of some

materials from which marine organisms are composed. The values, listed in
Table 10.1, have been collected from various sources (Alexander, 1968; Wainwright *et al.*, 1976; Vincent, 1982; Denny, 1988, 1993; Schmidt-Nielsen, 1989;
Gere and Timoshenko, 1991). Following Denny (1988), they are grouped into
three classes: crystalline composites, fibres and rubbers.

Crystalline Composites. These materials fail in tension at relatively low
values of strain. Examples are concrete, stone, glass, cast iron, and many
other man-made materials. The strength of biological crystalline type materials
depends on the amount of organic matter in the material. For example, the
shells of most molluscs are composed of calcium carbonate with a small amount
of organic material (1–4%), mainly protein. There are a number of types of shell
materials, which differ in structure and composition. Sometimes a combination
of these types of materials appears in a single shell. Thin shells are usually
made from a combination of prismatic material (polygonal columns, 100–200
μm × several mm long), nacre (flat tablets in sheets, 0.3 μm thick), or foliated
material (long, thin crystals in overlapping layers), while very thick shells are
built from crossed lamellar (plywood like lamellae, 20–40 μm thick) material
(Vincent, 1982).

The strength of mollusc shell material depends on the relative contribution
of the various shell material components. For example, the ultimate bending
stress for the prismatic component is 140 MN/m^2, while for the nacre component it is 220 MN/m^2. Therefore, in Table 10.1, the range of σ_t, σ_c, σ_b
and E values are a result of various compositions of the particular material
components.

The breaking stresses of strong, relatively non-porous aragonite mollusc shells
are about 220 MN/m^2. However, for porous crystalline stony skeletons such as
hard corals, the strength is much smaller. In the previous section, for *Acropora
formosa* we found that $\sigma_b \approx 22$ MPa. A similar result was obtained for three
specimens of an unidentified seleractinian by Eliat, Israel, namely 27, 27, and
29 MN/m^2 (Wainwright, 1976).

Material strength drops considerably when some holes are present, according
to Ryskevitch's empirical formula:

$$\sigma = \sigma_0 \exp(-np), \tag{10.25}$$

where σ_0 is the strength of non-porous material, p is the porosity, and n is a
constant between 4 and 7. Assuming the mean porosity as 50% and taking
$\sigma_0 = 200$ MN/m^2, we obtain the strength of coral material equal to 16 MN/m^2
and 27 MN/m^2, for $n = 5$ and $n = 4$, respectively. These values are very
close to the observed values reported above. Bucher *et al.* (1998) measured
the porosity for branch tips from a range of *Acropora* species. They used
some modification of traditional methods including soaking the dried samples
in acetone to displace air from skeletal voids. Resulting porosity ranged from
33 to 70%.

Table 10.1: Mechanical properties of some biological and man-made materials; data collected from various sources (see text)

Material	σ_t (MN/m^2)	σ_c (MN/m^2)	σ_b (MN/m^2)	E (GN/m^2)
Brittle materials				
Steel	300–800			200
Glass	70			100
Cephalopoda:				
Nautilus (shell)	62	139	207	44
Bivalvia:				
Pecten (shell)	42	88	113	30
Mollusc (shell)	30–80	250–380	80–220	30–60
Hard corals:				
Acropora formosa			22	
Acropora reticulata	40			60
Acropora cervicornis			27–30	30
Jade			70	
Flint			180	
Slate			100	
Chalcedony			165	
Granite			150	
Fibres				
Fibreglass	300–1000			30–100
Collagen	50–100			2
Chitin	100			40
Keratin	100–200			4
Rubbers				
Resilin	4–6			0.002
Abductin	8–12			0.004
Elastin	4–6			0.002

An important factor in reducing the strength of coral structure is bioerosion of coral skeletons (Highsmith, 1981). Bioerosional damage to corals primarily depends on the amount of skeletal surface not covered by live tissue as most excavations in coral skeletons occur within 2 cm of a dead surface. Even small amounts of bioerosion in the basal region of corals result in a substantial decrease of skeletal attachment strength.

Tunnicliffe (1979) examined the Caribbean stony coral *Acropora cervicornis* and found that 75% of the corals had sponge holes at their bases. The most abundant boring sponge was *Cliona aprica*. The breaking strength of corals with basal erosion by boring sponges was one order of magnitude lower than the 'healthy' coral strength.

Density of massive coral skeleton varies during the course of a year. The resulting annual density bands have been recognized as a retrospective means of analyzing coral growth and as a possible source of proxy environmental information for shallow tropical waters (Barnes and Lough, 1993; Lough *et al.*, 1996). The comparison of density for forty colonies of *Porites* in the central Great Barrier Reef, Australia, showed that maximum density varies from 1.47×10^3 kg/m^3 to 2.23×10^3 kg/m^3, while minimum density changes from 0.82×10^3 kg/m^3 to 1.62×10^3 kg/m^3. Average overall density of skeleton is about 1.4×10^3 kg/m^3 (Lough and Barnes, 1992). Variations of skeleton density through the cross-section of coral cause changes in skeleton strength. However, at present there are no sufficient data to develop the relationship between these two quantities.

Fibres. This material consists of a base material (or matrix) in which are embedded high strength fibres or filaments. The resulting composite material is much stronger than the base material. For example, glass and resin are each substantially brittle. However, when combined into an engineering material known as 'fibreglass', they show very high strength (see Table 10.1). Fibreglass, synthetic textile fibres, as well as animal fibres such as collagen, chitin and keratin, contain both crystalline regions, where the molecules are arranged in an orderly pattern, and amorphous regions, where they are arranged randomly. The crystalline component provides material with a stiffness with Young's modulus of the order of 1 GPa (see Table 10.1). The material stiffness is only apparent when the fibre is loaded in tension. Under other loading modes, the fibre material buckles quickly. For example, when a tensile stress is applied at right angles to the fibres, they contribute little or nothing to the strength and stiffness. To provide some isotropy of material properties, the fibres are arranged in some sort of matrix. An example of the connective collagen fibres in a mainly polysaccharide matrix is the body wall mesogloea of sea anemones. The mesogloea of *Metridium senile* is composed of about 86% water, 5% salt, 6.7% collagen, and 2% matrix (Vincent, 1982). Young's modulus of collagen is of the order of 1–2 GN/m^2 and the elastic strain limit is about 0.04. However, the mesogloea of *Metridium* strains up to 3 times its initial length. This means that collagen in the matrix is discontinuous and the mechanical properties are mainly controlled by the matrix modified by collagen.

Fibres such as in mesogloea of *Metridium* belong to the class of materials known as composites. The theory of these materials is beyond the scope of this book and the interested reader should consult Wainwright *et al.* (1976) and Vincent (1982) for an in-depth discussion.

Rubbers. Rubbers or elastomers are materials which function in many biological systems when being deformed. They have a low modulus, but can be substantially deformed without breaking. They often show reversible elastic properties, similar to those of rubber. The rubber maintains a linear relationship between stress and strain up to very large strains.

Biological rubber type materials, such as resilin, abductin and elastin, are mostly proteins organized in cross-linked polymer chains. Generally, rubber type materials and protein rubbers have Young's moduli about 1 MPa. They are not stiff enough to be primary structural elements in organism skeletons and their main use is to be energy storage devices. For example, resilin provides an elastic mechanism that stores the kinetic energy of the moving wings of insects and helps to decelerate the wing at the end of its stroke and then accelerate it in the opposite direction (Wainwright *et al.*, 1976). However, not all energy put into deformation is recoverable. The ratio of energy recovered to energy put in is the resilience. Resilin has resilience as high as 96–97%, which is as good as, or better than very high quality synthetic rubbers (Vincent, 1982).

Another type of protein rubber, abductin, has been found in the inner hinge ligament of the bivalve mollusc shell. Swimming bivalves open and close their shells several times per second and require an efficient hinge ligament and high resilience level. The resilience level is lower (about 80%) for digging bivalves which do not move their shells so often. The abductin in the inner ligament of scallops is compressed when the shell is closed and its compressed material provides the force to reopen the shell. During shell closing, these animals also expel water, creating a thrust to escape from predators. We will consider locomotion by jet propulsion in the next chapter.

10.3   The Density of Marine Organisms

In Sect. 1.2, we have defined the density of fluid and provided a computer program (D11) to calculate density dependent on water temperature, T, and salinity, S. Now, we will examine the density of organisms living in the ocean. The previous discussion of forces imposed on marine animals by flowing water, or drag induced by moving, showed that both densities should be taken into account. More important than the absolute densities of both the organism and water is the difference between these two densities. For example, a fish can hang motionless above the sea bed without moving a fin, by bringing its specific weight close to that of the sea. At the same time, a hovering bird must frantically beat its wings to stay aloft.

As was shown in Sect. 1.2, the density of sea water, ρ_w, is very close to 1026 kg/m^3. For example, for salinity $S = 35$ ppm and temperature $T = 10°$, the

Table 10.2: Densities of some inorganic and biological materials (adapted from Denny, 1993)

Material	Density (kg/m^3)
Inorganic materials:	
Sea water	1026
Calcium carbonate	2700
Quartz	2700
Dense sand	1895
Iron	7870
Organic materials:	
Coral skeleton	2000
Nautilus shell	2700
Whale ear bone	2470
Proteins	1350
Muscle	1050–1080
Fats and oils	915–945
Body fluids (blood, hemolymph)	1010

density, ρ_w, is equal to 1026.9 kg/m^3. Generally, density changes with submergence in the water column, as water temperature and salinity change (see Chap. 6). However, for simplicity of calculations, we assume that for some depth range, the density, ρ_w, remains constant.

Marine animals, taken as a whole, have densities between 1060 and 1090 kg/m^3. The most dense part of an animal is its skeleton, primarily made of calcium carbonate of density 2700 kg/m^3 (for example, shells of snails and clams). Densities of various inorganic and organic materials are listed in Table 10.2.

As the density of marine animals, ρ_a, is very close to the density of sea water, the effective density, $(\rho_a - \rho_w)$, is very sensitive to small changes in the density of either the animal or the water. An animal with density of 1080 kg/m^3, immersed in sea water, has an effective density $\rho_{eff} = 54$ kg/m^3, while the effective density of the same animal in freshwater will be $\rho_{eff} = 80$ kg/m^3. Therefore, a decrease in water density on 2.6% results in a 32% increase in effective density, ρ_{eff}. The changing of effective density of marine animals has important biological consequences. These will be discussed in the next chapters.

11 Locomotion of Marine Animals

11.1 Introduction

In Part 1 we discussed all kinds of forces acting on a body submerged in flowing water and forces induced by the surrounding fluid on a body moving in calm water. A body can move only when sufficient energy is available to overcome resistance forces. The availability of energy denotes the work that has to be done by a body to overcome its drag. Organisms may either oppose drag by being rigid or they may avoid excessive drag by being flexible and bending with the flow, thus reducing the area perpendicular to flow. When the force, or thrust, by which the body tends to move itself is larger than the sum of the resistance forces, then the body can achieve an acceleration which is inversely proportional to the body's mass. This is essentially Newton's second law of motion. According to Newton's first law, when the body thrust is equal to its drag, the body moves at a constant speed for as long as the thrust is maintained.

There are many ways in which animals move through the ocean. We begin our discussion with the simplest mode of motion, when aquatic animals utilize their special properties to stay in mid-water or move up and down without any work. This situation is closely related to the physical law of buoyancy.

Probably the most interesting locomotion mode of marine animals is swimming. Lift and drag are the mechanisms by which the necessary thrust is produced. It is intuitively correct to expect some relationship between speed of swimming and animal shape and size. Another way to produce thrust is through propulsion, when some mass is forced away, similar to gases expelled by a jet engine. To see how thrust is created by moving animal, we will examine the physics of swimming and jet propulsion.

11.2 Buoyancy in Marine Animals

The physics of the buoyancy force has been explained in Sect. 2.2. Now we will examine how the buoyancy phenomenon is used by aquatic organisms to minimize their energy expenditure for transport in the water column. If the density of an animal is exactly equal to the average density of water, then its weight exactly balances the buoyancy – the animal is neutrally buoyant. When the density of animal body exceeds the density of water, then the animal is negatively buoyant and tends to sink. When it is lighter (positively buoyant), it tends to rise. The animal can remain at a particular depth in two possible ways: by controlling its own density to achieve neutral buoyancy, or using lift forces, generated by swimming, and controlling depth.

The animal's average density is the resulting density of the body components which tend to sink, and the components which are less dense than sea water and tend to float. From Table 10.1 it follows that sinking components are principally proteins, animal skeleton and the muscles. The principal floating components of marine animal body are fat, certain body fluids and chambers filled with gases. The relative proportions of the different sinking and floating components varies from animal to animal (Denton, 1974).

Some animals with bodies denser than sea water, such as octopi or lobsters, live on the sea bottom and being neutrally buoyant is not a great need for them. Another example is mackerel which has no special buoyancy mechanism. In order to overcome its negative buoyancy, mackerel must swim continuously to stay above the sea bed. To minimize the energy expenditure during swimming, it has a streamlined shape with a drag coefficient, C_d, as small as 0.0043 (see Fig. 2.25).

However, when animals swim quietly in the water column, they have to develop some mechanisms to bring them to neutral buoyancy. In general, the buoyancy mechanisms developed by aquatic animals can be divided in three groups: the use of gas chambers, the use of body fluids less dense than sea water, and the use of fat (Denton, 1974). Some marine fish, for example, cod, use the first mechanism maintaining a gas space within an organ called the swim bladder. The swim bladder volume usually amounts to about 5% of the total volume of the animal. By varying the volume of this organ, the fish can adjust its effective density to that of the surrounding water. It is now known that swim bladder is used for controlling buoyancy down to 2000 m water depth.

To explore the effect of changing the swim bladder volume, consider a simple example of a fish which, at a particular water depth, is neutrally buoyant. The force balance is then:

$$\rho_w g V - (1 - \alpha_{eq}) \rho_b g V = 0, \tag{11.1}$$

in which V is the overall volume of the animal (inflated swim bladder included), ρ_b is the density of the non-swim bladder body tissue, and α_{eq} is the fraction of the animal volume used by the swim bladder. The first term is the buoyancy force and the second term represents the weight of the animal in sea water.

The necessary volume of the swim bladder (as a fraction of total animal volume) to support the animal in the state of neutral buoyancy becomes:

$$\alpha_{eq} = \frac{\rho_b - \rho_w}{\rho_w}. \tag{11.2}$$

For example, when non-swim bladder components have an average density of $\rho_b = 1080$ kg/m^3 and sea water density is $\rho_w = 1026$ kg/m^3, then the fraction $\alpha_{eq} = 0.05$.

When an animal starts to swim upwards, the surrounding fluid pressure decreases. According to the ideal gas equation $pV = $ Const (p is the pressure and V is the volume in which the gas is confined), the volume of the swim bladder must increase to some fraction $\alpha > \alpha_{eq}$ of the total animal volume. Therefore, the animal weight becomes:

$$
\begin{aligned}
(1 - \alpha)\rho_b gV &= \left[1 - (\alpha_{eq} + \alpha - \alpha_{eq})\right]\rho_b gV = \\
&= (1 - \alpha_{eq})\,\rho_b gV - (\alpha - \alpha_{eq})\,\rho_b gV.
\end{aligned} \tag{11.3}
$$

Equating animal weight to buoyancy force gives an effective force, F_{eff}:

$$
\begin{aligned}
F_{eff} &= \rho_w gV - (1 - \alpha)\,\rho_b gV = \overbrace{\left[\rho_w gV - (1 - \alpha_{eq})\,\rho_b gV\right]}^{=0} + \\
&\quad + (\alpha - \alpha_{eq})\,\rho_b gV = (\alpha - \alpha_{eq})\,\rho_b gV > 0.
\end{aligned} \tag{11.4}
$$

As the fraction $\alpha > \alpha_{eq}$, the additional force $(\alpha - \alpha_{eq})\,\rho_b gV$ causes the animal to ascend further. The ascent is even faster if the animal moves into a region of decreasing ambient pressure, which again induces an increase in the volume of the swim bladder α. Further movement of the fish towards the surface may be catastrophic for the animal, unless quickly controlled by some means.

The situation is quite opposite if the animal initially starts to move down. As surrounding pressure increases, the swim bladder is compressed and its volume α decreases below the α_{eq} value. Therefore, the effective force, F_{eff} in Eq. (11.4), becomes negative, causing the animal to sink further. Because the volume of the swim bladder is inversely proportional to the surrounding pressure, the change in volume is less for higher pressures. Thus, for a fish living at great depth, a difference in swim bladder volume $(\alpha - \alpha_{eq})$ is easier to correct. In general, the time for proper adjustment of the swim bladder is of the order of hours (Denny, 1993). However, details of the physiology of the adjustment mechanisms are beyond the scope of this book.

A second method of adjustment of the buoyancy mechanism which involves a gas chamber is to have a chamber with rigid walls. The chamber walls have to sustain the difference in pressure between the external hydrostatic pressure and the pressure of gas within the chamber. The chamber contains both gas and liquid. For example, the cuttlefish uses its cuttlebone, and the pearly *Nautilus* uses its shell to control its buoyancy. When they try to sink, liquid is pumped into the cuttlebone in the case of the cuttlefish, or into the shell in the case of the *Nautilus*, and the volume of the gas decreases. The liquid is pumped out of the chamber when the animal ascends. In fact, this mechanism is very similar, in principle, to that used by submarines. The mechanism for pumping water into the ballast tank of a submarine is different from the 'osmotic' mechanism used in the case of cuttlefish and into the shell in the case of the *Nautilus*, but this problem will not be discussed here. The cautious readers should consult Alexander (1968) or Denton (1974) for an in-depth discussion.

Table 10.1 shows that fat and oil densities are less than that of sea water. Some aquatic animals, for example shark *Centrophorus squamosus*, and deep-diving whales or dolphins, make use of low-density fat to give buoyancy. Sharks of this genus have enormous fat reservoirs in their liver, accounting for about 20 percent of their total volume (Denton, 1974). Hence, the balance of vertical forces for shark can be written as:

$$\rho_w g V - [0.2\rho_f g V + (1 - 0.2)\rho_b g V] = 0, \quad \text{or}: \tag{11.5}$$

$$\rho_w - (0.2\rho_f + 0.8\rho_b) = 0, \tag{11.6}$$

in which ρ_f is the density of fat, while ρ_b is the density of non-fat components of the body. Let us assume that $\rho_f = 930$ kg/m^3, $\rho_w = 1026$ kg/m^3, and $\rho_b = 1060$ kg/m^3. The effective force acting on the shark body is only -78 N/m^3 of the animal's volume. This small negative buoyancy is easy rectified by the shark due to the high amount of the low density hydrocarbon, squalene, in the liver's fat.

In the balance of forces given by Eq. (11.6), we tactically assumed that both forces act along the same vertical line. However, in a resting shark the centre of buoyancy (B) and the centre of weight (G) are shifted with respect to each other. Thus, similarly to the case of stability of the cylinder examined in Sect. 2.2.3, a pair of weight and buoyancy forces tends to rotate the animal in a counterclockwise direction. This makes the shark unstable and it has to use its oily liver to rectify the buoyancy force and improve stability.

Instability of marine organisms can also be caused by ocean waves. As was shown in Chap. 4, dynamic pressure due to waves is transmitted from the surface into the ocean body and is changing in time. Pressure becomes highest under the wave crest and smallest under the wave trough. The increase in pressure under the wave crest compresses the swim bladder and the fish starts to sink. Under the wave trough, due to decreasing pressure, the swim bladder expands and increasing buoyancy forces the fish to ascend. As was indicated by Denny (1993), fish that rely on a swim bladder alone are as unstable under waves as they are in still water.

In order to improve stability while swimming slowly near the surface, blue marlin, swordfish, and perhaps other istiophorids, have a large swim bladder providing sufficient buoyancy. Blue marlin, for example, have an unusual swim bladder in the form of a thin-walled, multichambered sack extending from the level of the first pectoral fin to the first anal fin (Block *et al.*, 1992). However, a large swim bladder creates swimming problems at the beginning of a descent and at greater water depths.

Some aquatic animals control their buoyancy using a tissue fluid, less dense than sea water, instead of fat as described above for the case of the shark. Pelagic ocean squid replace some of the sodium ions by ammonium. Many gelatinous planktonic animals, such as jellyfish, comb jellies, sea elephants and sea butterflies keep body fluids of density lower than that of the surrounding water by replacement of sulfate by chloride ions. Fish eggs often have tissue fluids with a content of salt less than half of that in sea water. This allows them to float close to surface and be readily dispersed by waves and currents. The biochemical processes involved in ion transformation and preservation of less dense fluid in animal tissues are complicated, and their details are beyond the scope of this book.

11.3   Mechanics of Animal Swimming

11.3.1   Introduction

Animal swimming is a difficult subject for analysis due to the great morphological diversity in equipment used for swimming. Moreover, the dynamic processes which are involved in transmission of forces between fish and water are very complicated and not yet fully understood. We will start an examination of fish swimming with classification of the swimming modes.

In each swimming mode, different swimming apparatus is involved. Efficient swimming requires that most of the energy expended by the swimming muscles be turned into appropriate motions of the propulsive surfaces with minimum energy loss through elastic deformation and as little as possible kinetic energy left in the water. The majority of fish species swim by undulation of body and tail, powered by the lateral musculature, while other species use paired or unpaired fins moved by intrinsic muscles. The complex architecture of fish muscles is not treated here.

During steady fish swimming, tail movements are rhythmic and the distance forward covered for every stroke is usually constant. Therefore, the speed is determined by the stride frequency. Usually fish do not swim in a steady fashion, but instead rapid starts and swimming with bursts of activity are usually followed by coasting with a straight body.

A sound knowledge of the kinematics of swimming is fundamental for the formulation of hydrodynamical models from which such quantities as thrust, power and efficiency can be calculated. In the 1970s, hydrodynamicists developed reactive theories of swimming based on the inertial forces generated by the propulsive surfaces of fish in a perfect fluid (for example, Lighthill, 1969, 1971, 1975; Wu, 1971). Basic results of the theory are described here for straight forward swimming.

In the world of ciliary and flagellar propulsors, viscous shearing is the sole mechanism available for generating thrust. For example, protozoan flagella are small (0.2 μm in diameter) and swim slowly (10–10^4 μms^{-1}) with the corresponding Reynolds number, Re, being of the order of 10^{-6} (10^{-3}). Locomotion in this low Reynolds number environment is quite successfully predicted by the method known as 'resistive force theory' (Holberton, 1977; Daniel *et al.*, 1992).

Locomotion is an energetically costly activity that may comprise a significant component of an animal's overall energy budget. Thus, it is advantageous for animals to use a locomotion strategy which minimizes energy expenditure. Many fish species spend most of their time cruising only within a restricted area, such as a territory or a feeding range. They also try to adopt some optimum speed when the amount of energy per unit distance covered reaches a minimum.

11.3.2   Classification of Fish Swimming Modes

The propulsive movements of fish are usually classified as anguilliform, named after the eel, *Anguilla*, carangiform, named after the jack, *Caranx*, and ostraci-iform, named after the boxfish, *Ostracion* (Blake, 1983; Vogel, 1994).

The anguilliform mode of swimming (Fig. 11.1a) is characterized by the whole body bending into backward moving waves. The length of these body waves is less than the animal's body length. Anguilliform mode is typically used by long, thin fish with a cylindrical exterior section, such as aquatic snakes.

The carangiform mode is the most popular mode of fish swimming and it is used by perch, trout, cichlid and mullet. A wave of bending is passed backwards and less than one half of a wavelength is present on the fish's body at any given time (Fig. 11.1b). The amplitude of the waveforms increases rapidly over the posterior third to half of the body length and becomes maximum at the tail. Within the carangiform mode some submodes, known as 'carangiform with lunata tail' (Fig. 11.1c), or 'thunniform' (named after the tuna *Thunnus*), are usually distinguished. These species, such as tuna, mackerel, marlin and some

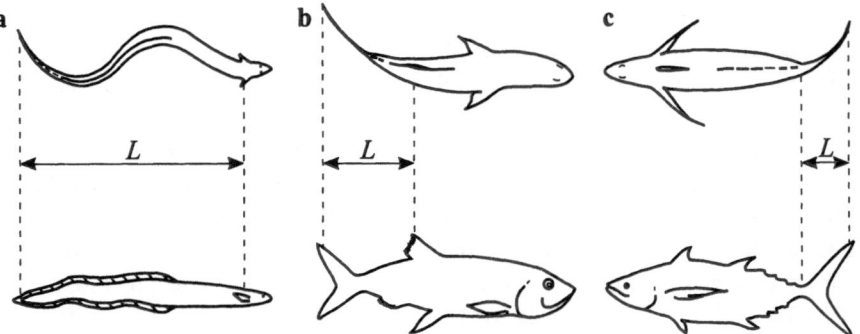

Fig. 11.1: Modes of swimming in fishes: **a** anguilliform mode, **b** carangiform mode, **c** Thunniform mode; the L denotes body wavelength (adapted from Vogel, 1994)

sharks, whales and dolphins, are characterized by a distinctive, high aspect ratio and lunate candal fin. Undulations of the body are confined to the candal peduncle and fin. They have attracted a great deal of attention from biologists and hydrodynamicists because of the high speeds they attain.

Ostraciiform mode is used by species which are unstreamlined. They are not propelled by undulations of the body and candal fin, but they employ the action of their median or paired fins for most of their swimming behaviour.

11.3.3 Kinematics, Speed and Size

The swimming speed of fish is related to the body size of the fish. Large fish swim faster than small fish. There is rather extensive literature on various aspects of fish swimming (see, for example, the books by Childress, 1981; Blake, 1983; Schmidt-Nielsen, 1989; Videler, 1993; Vogel, 1994, and many papers in various professional journals). Therefore, in this section we will consider only some specific problems, closely related to fluid mechanics. First, let us examine the relationship between animal speed, U, and its length, L.

For marine ecologists dealing with systems that span an enormous size range, the Reynolds number $Re = UL/\nu$ is one of the basic scaling parameters commonly used for categorizing the flow induced by these various systems. To illustrate this, in Table 11.1 the range of Reynolds numbers associated with swimming of various marine organisms is shown.

As the Reynolds number is linearly proportional to the product of size and speed, it becomes very high for large organisms moving fast and very low for small organisms moving very slowly. Therefore, the Reynolds number varies over thirteen orders of magnitude, while the length of organisms vary only over seven orders of magnitude. Although whale and bacteria swim in the same water, they live in different environments. The first lives in an environment dominated by inertial phenomena, while the second is totally dominated by viscous phenomena.

Using the data from Table 11.1, the approximate relationship between velocity, U, length, L, and Reynolds number, Re, can be established, *i.e.*:

$$U \approx 1.94 \times 10^{-3} \times Re^{0.49}, \tag{11.7}$$

$$U \approx 2.3 \times L^{0.93}, \tag{11.8}$$

in which length, L, is in metres and velocity, U, is in metres per second. Equation (11.7) can be rewritten as:

$$U = 1.94 \times 10^{-3} \times Re^{0.49} = 1.94 \times 10^{-3} \left(\frac{UL}{\nu}\right)^{0.49}, \quad \text{or:} \tag{11.9}$$

$$U = 4.83 \times 10^{-6} \times L^{0.96} \times \nu^{-0.96}, \tag{11.10}$$

Table 11.1: Typical swimming velocity, approximate size and corresponding Reynolds number of self-propelled marine organisms (adapted from Videler, 1993)

Marine organism	Velocity (m/s)	Length (m)	Re
Blue whale	10	30	3×10^8
Tuna	10	3	3×10^7
Mackerel	3.3	0.3	1×10^6
Herring – Adult	1	0.2	2×10^5
– Larvae	0.5	0.1	5×10^4
– Larvae	0.16	0.04	6×10^3
– Larvae	0.06	0.02	1×10^3
– Larvae	0.02	0.01	2×10^2
Copepods	0.002	0.001	2
Sea urchin sperm	0.0002	0.00015	3×10^{-2}
Bacteria	0.00001	0.000001	1×10^{-5}
Human	1.7	1.8	3×10^6

in which ν is the coefficient of kinematic viscosity. Comparison of Eqs. (11.8) and (11.10) indicates that swimming velocity, U, based on animal length, L, is consistent with velocity determined using Reynolds number Re.

A similar relationship between swimming velocity, U, and Reynolds number, Re, for a large variety of marine organisms, from bacteria to whales, was established by Okubo (1981) as:

$$Re = \frac{UL}{\nu} = 1.4 \times 10^6 L^{1.86}. \tag{11.11}$$

Thus:

$$U = 1.4 \times 10^6 \nu L^{0.86}, \tag{11.12}$$

which is comparable with the relationship (11.10).

The relationships from (11.7) to (11.12) are based on a variety of organisms which use different mechanisms to swim. For example, bacteria or copepoda are too small to swim independently against ocean currents. Buoyancy is most likely the only mechanism they use actively, otherwise they are passive swimmers. On the other extreme, there are large animals, powerful enough to swim independently by developing a thrust to overcome the drag forces.

Let us now be more specific and restrict our attention to one species, for example, rainbow trout of different lengths, swimming in a flume tank. Using the data published by Videler (1993), the following relationships between swimming velocity, U, and tail beat frequency, f, can be established:

$$\left. \begin{array}{llll} U & = & 0.039f & - & 0.070 & \text{for} & L = 0.55\,\text{m} \\ U & = & 0.090f & - & 0.120 & \text{for} & L = 0.116\,\text{m} \\ U & = & 0.234f & - & 0.312 & \text{for} & L = 0.249\,\text{m} \\ U & = & 0.400f & - & 0.44 & \text{for} & L = 0.433\,\text{m} \end{array} \right\} , \tag{11.13}$$

in which tail beat frequency, f, is expressed in Hz.

If swimming velocity, U, is expressed in body length of the fish rather than in metres per second, all four relationships in (11.13) can be combined in one form as:

$$\frac{U}{L} \approx 0.587f. \tag{11.14}$$

Physically, the U/L is the fraction of body length that a fish moves in 1 second, and it has the dimension of time to the power -1, the same dimension as the frequency. For simplicity, the linear function (11.14) has been developed as crossing through the origin of the coordinate system.

In a similar way, using the data published by Schmidt-Nielsen (1989), we obtain, for a small freshwater fish, the dace:

$$\frac{U}{L} = 0.616f. \tag{11.15}$$

Bluefin tuna were observed in sea cages on the east side of the Spanish town of Centa, close to the Strait of Gibraltar. The body lengths of the tuna were between 1.7 and 3.3 m, the U/L value varied from 0.6 and 1.2 s^{-1}, and the relationship between tail beat frequency, f, and swimming speed, U, took the form:

$$\frac{U}{L} = 0.65f. \tag{11.16}$$

Equations (11.14), (11.15) and (11.16) indicate that velocity of swimming is linearly proportional to animal length and frequency of tail beat.

According to Eqs. (11.8) and (11.10), the speed, U, is proportional to length, L, to a power of less than 1. The same relationship for salmon is (Brett, 1965):

$$U = 19.5L^{0.5}, \tag{11.17}$$

or (by Wardle *et al.*, 1989):

$$U \approx 0.07L^{0.43}. \tag{11.18}$$

For the maximum 60-min swimming speed, Brett (1965) obtained:

$$U_{max} \approx L^{0.63}. \tag{11.19}$$

In Sect. 9.4 we illustrated an application of dimensional analysis to develop a fish tail beat frequency dependence and we found that (see Eq. 9.31):

$$f = C\sqrt{\frac{\sigma}{\rho_w}}L^{-1}, \tag{11.20}$$

in which σ is the stress exerted by the fish's muscle, and C is a dimensionless constant. If it is true that the same species exert much the same stress, σ, in their muscles, and if the density of sea water, ρ_w, is regarded as constant, then fish of different size should beat their tails at frequencies universally proportional to their body lengths.

Using expression (11.20) in Eq. (11.14) or Eq. (11.15), we obtain:

$$U \approx C_1\sqrt{\frac{\sigma}{\rho_w}}, \quad \sigma \approx C_2\rho_w U^2, \tag{11.21}$$

where C_1 and C_2 are non-dimensional constants. Thus, fish muscle stress is proportional to the swimming velocity to power 2. As will be shown in the next section, the dynamic pressure caused by the tail pushing water at velocity U is equal to $1/2\rho_w U^2$, which is exactly equal to the muscle stress σ. According to Lighthill's (1971) large-amplitude elongated theory, which is a basic tool for understanding the mechanism of swimming, the total pressure force induced by a tail blade is equal to the product of the above dynamic pressure and the area of the circle around the end of the tail blade. The theory of swimming will be briefly described in the next section.

Swimming at uniform velocity along a straight path is rather exceptional among fish. More common are unsteady movements with fast starts and rapid turns. Acceleration or deceleration are measured by a calibrated accelerometer attached to the fish. Acoustic telemetry was used to monitor the swimming speed of three blue marlins, 60 kg, 75 kg and 125 kg (Block et al, 1992). The measurements showed that marlins spent most of their time swimming slowly, about 1 m/s. Short bursts of speed up to 2.25 m/s were associated with changes of depth.

There is a large scatter in the estimation of maximum swimming speed of pelagic fish given in literature. According to data collected by Block *et al.*

(1992), the highest speeds of blue marlin was 36 m/s, while yellowfin tuna (*Thunnus albacares*), much smaller than blue marlin, can accelerate to 21 m/s. Usually continuous swimming speed ranges from $U/L = 0.3$ s^{-1} in wahoo, to 1.6 s^{-1} in bluefin tuna, while burst swimming speed varies from $U/L = 8$ s^{-1} in the bonito, to 27 s^{-1} in yellowfin tuna.

Fish, by their muscles, generate the power needed for steady swimming. Most of this power comes from a muscle adjacent to the tail, and relatively little comes from the outerior musculature (Rome *et al.*, 1993). However, the structure of fish muscles is out of the scope of this book (for details see Videler, 1993).

11.3.4 Transmission of Forces Between Fish and Water

Drag Coefficient of Marine Organisms. Because fish are nearly buoyant, they use little or no energy to support themselves. However, energy is needed to overcome drag. When fish swim with a constant speed, U, the drag, F_d, can be calculated as follows (see Sect. 2.6.2):

$$F_d = \frac{1}{2}\rho_w C_d S U^2. \tag{11.22}$$

To minimize the drag force induced by fluid at a given swimming speed, most marine swimmers adopt a streamlined form with a very small drag coefficient C_d. In Table 11.2, measured values of C_d for some marine organisms have been collected.

Marine organisms, except maybe Cephalopod, have much smaller drag coefficients than some artificial shapes given in Fig. 2.25. However, when comparing the C_d values, one should take into account the definition of the reference surface, S. In Fig. 2.25, the values of C_d for artificial shapes have been calculated using 'frontal', or projecting area of an object. This is the maximum projection of the object onto a plane normal to the direction of flow. It is particularly useful for non-streamlined objects of relatively high drag.

For marine organisms listed in Table 11.2, a distinction between various definitions of area S is noted by the symbols v and w, where v denotes area based on the animal volume. Assuming that the volume of an animal is V, then surface $S \approx V^{2/3}$. This is probably the most appropriate measure of surface, S, for organisms and perhaps the easiest to measure.

The symbol w in Table 11.2 indicates that the surface S is defined as a 'wetted area', which is the total surface exposed to flow. It is especially useful for streamlined bodies where drag is largely due to viscosity. There is some difficulty in measurement of the wetted surface of real marine organism. In more complicated cases, the shape of an animal can be approximated with cylinder, spheroid and other forms for which surface is easily determined. The

Table 11.2: Drag coefficients for some marine organisms (adapted from Chamberlain, 1991; Vogel, 1994)

Animal	Re	C_d	Area
Cephalopod, *Nautilus*	100,000	0.48	v
Sturgeon, *Acipenser fulvescens*		0.0053	w
Trout, *Salmo gairdneri*	50,000–200,000	0.015	w
Mackerel, *Scomber*	100,000	0.0043	w
Mackerel, *Scomber*	175,000	0.0052	w
Saithe, *Pollachius virens*	500,000	0.005	w
Pike, *Esox sp.*		0.008	w
Sea lion, *Zalophus californ.*	2,000,000	0.0041	w
Seal, *Phoca vitulina*	1,600,000	0.004	w

so called 'plan form area', or 'profile area' are also used as a measure of surface, S. It is area that one would see from above.

Outline of Elongated Body Theory. When a fish swims at a constant speed in a straight line its momentum is constant and any momentum given to the water in one direction must be balanced by an equal momentum in the opposite direction. Many marine animals use this principle of propulsion by giving the fluid backward momentum. Let the animal velocity be U; the corresponding thrust is F, and a backward fluid velocity induced by the animal is $-U'$. Thus, the accelerated mass of fluid (in unit time) due to thrust F becomes F/U', and the kinetic energy given to the fluid (in unit time) is $1/2(F/U')U'^2 = FU'/2$. This is the wasted power used for driving the fluid backwards and it is not used for driving the animal forward. The useful power is simply $Fdx/dt = FU$. A measure of the efficiency of the propulsion mechanism is known as Froude efficiency (Alexander, 1977):

$$\text{Froude efficiency} = \frac{\text{useful power}}{\text{useful power} + \text{power lost to fluid}} = \frac{U}{U + \frac{1}{2}U'}. \tag{11.23}$$

Thus, the efficiency becomes higher when the animal accelerates a large mass of fluid (per unit time) to a low velocity (small U'), rather than accelerating a small mass to a high velocity (high U').

Aquatic organisms swim by transferring momentum from moving portions of their body to the fluid surrounding them. The amount of generated thrust depends on the rate at which the animal transfers momentum. The complicated mechanism underlying the exchange of momentum depends on size and speed of the moving portions of the animal body and the density of the fluid. Lighthill (1969) developed a theory applicable to small amplitude motion of fish swimming in the anguilliform mode (see Fig. 11.1). In a further study, Lighthill extended the theory to include large amplitude motions. This theory is known as large-amplitude elongated body theory. Following Alexander (1977), we will briefly describe the basic outcomes of Lighthill's theory in application to typical fish with fairly large candal fins, such as trout *Salmo*. We assume that a fish swims with velocity U, passing waves bending posteriorly along its body with velocity U_w. The transverse component of velocity of the posterior end of the fish body is W, and the water being left behind has transverse velocity W'. It is also assumed that this velocity is associated with all the water passing through the circle drawn (in a transverse plane) around the candal fin. Thus, water mass which passes through it in unit time is $\rho_w \pi D^2 U / 4$, and D is equal to the span of the candal fin.

On the other hand, the momentum given to the water in unit time becomes $\rho_w \pi D^2 U W' / 4$, while power output of the tail is:

$$P_{\text{out}} = \frac{\rho_w \pi D^2 U W\, W'}{4}.$$ (11.24)

Power output generated by movement of the tail is only partly used to overcome the drag on the body. Another part of the power, $\rho_w \pi D^2 U W'^2 / 8$, is transferred to the water, into the vortex wake. Therefore, the useful power becomes:

$$P_{\text{useful}} = \frac{\rho_w \pi D^2 U W'}{4} \left(W - \frac{W'}{2} \right),$$ (11.25)

and

$$\text{Froude efficiency} = \frac{1}{W} \left(W - \frac{W'}{2} \right).$$ (11.26)

The velocity of water left behind the fish body is (Lighthill, 1971):

$$W' = \frac{W(U_w - U)}{U_w}.$$ (11.27)

Substituting Eq. (11.27) into Eq. (11.25) gives the useful power at a given time t:

$$P_{\text{useful}} = \frac{\rho_w \pi D^2 U W^2 (U_w^2 - U^2)}{8U_w^2}.$$ (11.28)

As the power, P_{useful}, changes over a cycle of swimming moments, it is more appropriate to consider the mean values. Assuming harmonic motion of the tail with amplitude a, and frequency f, we obtain (Alexander, 1977):

$$\overline{P_{\text{out}}} = \frac{\rho \pi^3 D^2 a^2 f^2 U (U_w - U)}{2U_w}, \tag{11.29}$$

$$\overline{P_{\text{useful}}} = \frac{\rho \pi^3 D^2 a^2 f^2 U (U_w^2 - U^2)}{4U^2}, \tag{11.30}$$

in which $\overline{P_{\text{out}}}$ and $\overline{P_{\text{useful}}}$ are the mean values of total output power and useful power, respectively.

The power needed to overcome the drag on the body of the fish is the drag multiplied by the velocity. Thus, using Eq. (11.22) we have:

$$\text{Expected power requirement} = \frac{1}{2} \rho_w C_d S U^3, \tag{11.31}$$

in which S is the wetted area of the fish body, and C_d is the drag coefficient based on the wetted area.

To illustrate Lighthill's theory, in Table 11.3 the results of calculation are summarized for 28 cm rainbow trout (*Salmo gardineri*) swimming with various speeds. This fish has mass of 0.22 kg and wetted area 3.1×10^{-2} m^2.

It might be expected that the useful power should match the power requirement given by Eq. (11.31). However, from Table 11.3 it follows that the $\overline{P_{\text{useful}}}$ is about 7 times higher than the expected power requirement. As was shown by Webb (1984), the measured drag coefficient, C_d, was much higher (0.02) than 0.006 used in calculation of the expected power requirements in Table 11.3. A higher value of C_d implies a higher value of expected power requirement in the same proportion. Lighthill (1971) suggested that a drag augmentation could be caused by boundary layer thinning due to the oscillation of the body. When a flat plate moves with velocity U in its own plane, the laminar boundary layer thickness is about $5.5\sqrt{\nu x/U}$ (see Eq. 2.47) at a distance x from the leading edge. However, when the plate is moved perpendicular to itself, the transverse motion results in a much thinner boundary layer (of the order of $\sqrt{\nu D/U}$), where D is the span of the plate. Assuming that the values of $\sqrt{\nu x/U}$ and $\sqrt{\nu D/U}$ are of the same order, the two boundary layer thickness values will differ by a factor of about five. A reduction in boundary layer thickness by a factor of five results in an increase in the drag of about the same magnitude.

The elongated-body theory, outlined above, is not appropriate to tuna (*Thunnus*) and similar fish which have stiff candal fins in the form of hydrofoils of rather high aspect ratio. The candal fin of tuna, moving from side to side, adjusts its angle of attack so the resultant force on it has a forward component.

Table 11.3: Calculations of the mechanical parameters of swimming rainbow trout based on elongated-body theory (adapted from Alexander, 1977)

Parameters	Case 1	Case 2	Case 3
Swimming velocity, U (m/s)	0.10	0.21	0.52
Wave velocity, C (m/s)	0.45	0.49	0.86
Span of candal fin, D (m)	0.042	0.048	0.061
Frequency of tail beat, f (1/s)	2.1	2.3	4.0
Amplitude of tail beat, a (m/s)	0.011	0.02	0.021
Mean total power, $\overline{P_{out}}$ (W)	0.0011	0.0090	0.084
Mean useful power, $\overline{P_{useful}}$ (W)	0.0007	0.0064	0.063
Froude efficiency	0.61	0.71	0.81
Reynolds number, Re	2.8×10^4	5.9×10^4	1.46×10^5
Expected power requirement, (W)	0.0001	0.0008	0.0074

The propulsive force is generated by the lift on the fin; however, the work is done against the drag. Thus, it is advantageous to have as high aspect ratio (= span/chord) as possible. The high aspect ratio of the fin, for a given lift force, induces low drag (see Eq. 2.79).

Whales swim in the same manner but they have horizontal tail flukes which beat up and down. On the downstroke, flow over the fluke moves faster over the top than the bottom. As was shown in Sect. 2.6.3, the velocity gradient induces the distribution of pressure over the fluke which integrates over the fluke length, results in the lift force. The mechanics of swimming by tuna and similar fish and whales will not be discussed here. A detailed explanation is given by Lighthill (1969, 1971, 1975), Blake (1983), or Alexander (1982).

11.3.5 Jet Propulsion Mechanism

Drag and lift are not the only mechanisms by which thrust can be produced. Soft bodied aquatic animals, such as squid, octopod, cuttlefish, and *Nautilus* use jet propulsion mechanisms to swim in water. Jet propulsion allows rapid expulsion of water to produce thrust and elastic expansion for the next power stroke. The thrust produced is in the opposite direction from the ejection of water. The largest possible mass expelled with the greatest possible velocity drives the organism the furthest, which is important during an attack or an escape. However, in normal swimming, ejecting mass as rapidly as possible wastes energy. Rapid ejection is less effective than ejecting the same mass at a velocity only slightly higher than the animal's speed. Many of the basic swimming mechanisms of soft bodied animals have been described in books by Trueman (1975) and Alexander (1982). Here, we will concentrate only on some examples of the use of jet propulsion.

In jet propulsion, a thrust is generated by pushing away a mass of fluid. The rate at which momentum, M_{prop}, is pushed away by the animal is a product of the change in mass per time, $\rho_f dV/dt$, and velocity, $(dV/dt)/A$, in which A is the area of orifice, V is the volume of fluid pushed away, and ρ_f is the density of fluid, $i.e.$:

$$M_{\text{prop}} = \rho_f \left(\frac{dV}{dt}\right) \cdot \left(\frac{1}{A}\frac{dV}{dt}\right) = \frac{\rho_f}{A}\left(\frac{dV}{dt}\right)^2. \tag{11.32}$$

The rate of change of momentum is simply a force, as we have:

$$\frac{d}{dt}(mu) = m\frac{du}{dt} = ma = F, \tag{11.33}$$

where m is the mass, u is the velocity, a is the acceleration, and F is the force. Therefore, Eq. (11.32) represents the thrust created by an animal by ejecting fluid. The thrust:

$$F = \frac{\rho_f}{A}\left(\frac{dV}{dt}\right)^2, \tag{11.34}$$

is, by Newton's third law, in the direction opposite to the direction of the ejecting fluid. When the mass of the animal is known, the acceleration due to jet propulsion can be calculated as F/m.

Squids are the fastest swimmers of all the aquatic invertebrates. They use a combination of fin undulations and a jet which can direct thrust at any angle through a hemisphere below the body plane (O'Dor, 1988). Their performance is comparable to fish of roughly the same size. Among squids the *Loligo* are particularly fast. Squid of mass 100 g can accelerate from rest to 2 m/s by a single squirt and a maximum velocity of 8 m/s has been reported (Alexander, 1977). The Froude efficiency of a squid is low and squid swimming at 0.1 m/s have an efficiency of 0.12.

Medusae are another example of jet-propelled swimmers, but their maximum swimming velocities are low. For example, *Stomolophus* reaches a maximum velocity of only about 13 cm/s, and *Plyorchis* can swim at only of 7 cm/s. The primitive cephalopod, *Nautilus*, is a purely jet-propelled animal swimming as a result of pressurizing the water in its mantle cavity. *Nautilus* reaches a maximum velocity of about 25 cm/s (O'Dor *et al.*, 1990; De Mont, 1992), and the Froude efficiency for *Nautilus* is likely to be in the order of 0.29.

11.3.6 Swimming in Low Reynolds Number Environment

Up to this point we have discussed animal swimming mechanisms when Reynolds number was high and inertial effects dominated. However, for ciliary and flagellar propulsors, viscous shearing is the only mechanism generating thrust. It is presently believed that for flagellum, the progressive activation of a molecular motor, dynein, results in adjacent microtubular sliding. The resulting motion is a bending wave propagating along the flagellum. Flagella are about 0.2 μm in diameter and 10–100 μm in length, and there is between 1.3 to 5 waves on a flagellum (Daniel et al., 1992). They move slowly with a velocity of $10–10^4 \mu$m/s, and the corresponding Reynolds number is of the order $10^{-6}–10^{-3}$.

For a steady flow with very small Reynolds number, the Navier-Stokes equation (C.32) for a motion in the x direction becomes:

$$\frac{\partial p}{\partial x} = \mu \left(\frac{\partial^2 u}{\partial x^2} + \frac{\partial^2 u}{\partial y^2} + \frac{\partial^2 u}{\partial x^2} \right). \tag{11.35}$$

The most familiar analytical solution of Eq. (11.35) is the Stokes' result for the force, F_x, acting on a moving sphere of diameter D (see Eq. C.56):

$$F_x = 3\pi \rho_w \nu D U, \tag{11.36}$$

Regardless of how ciliar or flagellar motion is generated, the thrust is generated by a viscous shearing mechanism. Thus, similarly to Eq. (11.36), the drag force can be represented in a more general form as:

$$F_x = \rho_w \nu C_d U L, \tag{11.37}$$

where L is the characteristic length of the object.

Approximating the geometry of cilia or flagellum by a long thin cylinder (see Fig. 11.2), the sectional thrust can be determined as the resolution of the forces produced by fluid motion normal and tangential to the segment. In terms of drag coefficients, the tangential and normal forces on a cylinder of length L and radius D are, respectively:

$$F_t = \rho_w \nu C_d^{(t)} U_t L, \tag{11.38}$$

$$F_n = \rho_w \nu C_d^{(n)} U_n L. \tag{11.39}$$

Assuming that $U_t = U_n$, $C_d^{(n)} \to 2C_d^{(t)}$ as $L \to \infty$ (see Sect. 2.6.2). In general, it can be shown that (Holberton, 1977):

$$C_d^{(t)} = \frac{2\pi}{\ln \left(\frac{2L}{D} \right) - \frac{1}{2}}. \tag{11.40}$$

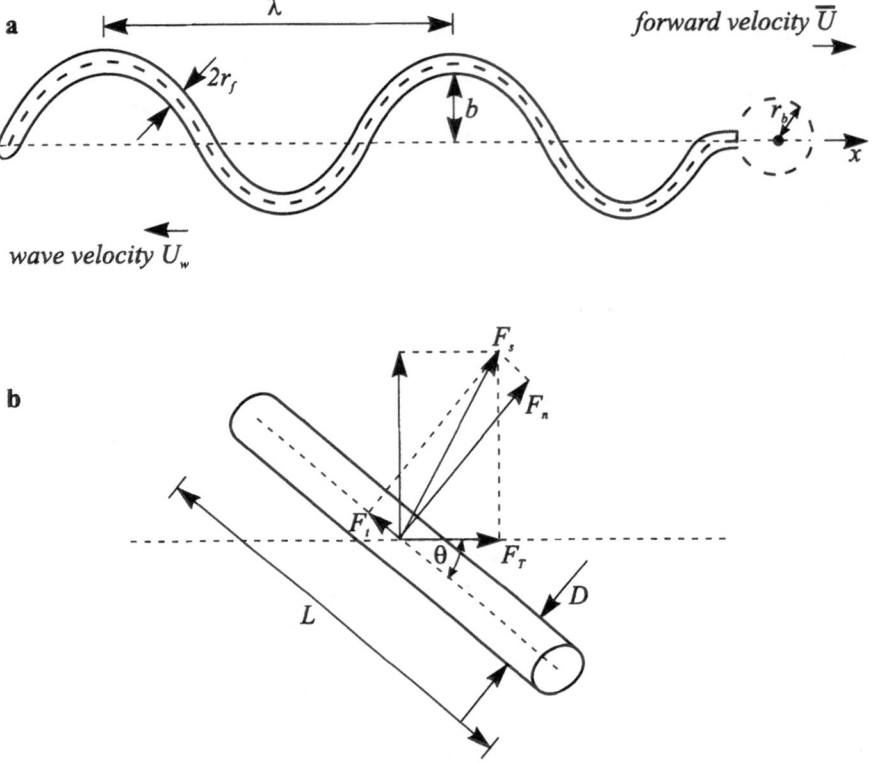

Fig. 11.2: Diagram of forces acting on swimming flagella: **a** flagella propels itself at speed U, **b** force components on tail segment

The total thrust generated by a wavy flagellum can be represented as the sum of the forces produced by many small cylindrical segments (Fig. 11.2). Assuming that a particular segment is oriented at an angle, θ, to the horizontal plane, the thrust, F_T, produced by that segment becomes:

$$F_T = \rho_w \nu \left(C_d^{(t)} U_t \cos\theta + C_d^{(n)} U_n \sin\theta \right) L. \tag{11.41}$$

As the drag coefficient $C_d^{(n)}$ is always greater than $C_d^{(t)}$, an inclined cylinder moving through a fluid generates a force normal to the direction of motion. This simple method of calculation of the thrust is known as the resistive theory, developed originally by Gray and Hancock (1955). There are a number of extensions of the resistive theory to some flagellar waveforms and other motile structures, some of which are listed by Holberton (1977). For example, for a plane sine wave on a smooth flagellum propelling a spherical body, Gray and

Hancock (1955) obtained:

$$\frac{\bar{U}}{U_w} = \frac{0.5(kb)^2}{1 + (kb) + 2 - \sqrt{1 + 0.5(kb)^2}\,[\ln(r_f/2\lambda) + 0.5]\,(3r_b/n\lambda)},\qquad(11.42)$$

where $\bar{U}$ is the mean forward velocity, U_w is the apparent wave velocity, b is the wave amplitude, λ is the wave length, $k = 2\pi/\lambda$, r_f is the radius of the flagellum, and r_b is the radius of the spherical body. The observation showed that the drag on the head is small compared to that of the tail. For headless flagellar propulsion with planar waves, the maximum Froude efficiency becomes (Lighthill, 1975):

$$\text{Froude efficiency} = \left[1 - \sqrt{\frac{C_d^{(t)}}{C_d^{(n)}}}\right]^2.\qquad(11.43)$$

Thus, when the ratio of drag coefficients is unity, the efficiency and thrust are zero. When $C_d^{(t)}/C_d^{(n)} \to 0.5$, efficiency becomes its maximum of 0.085.

Ciliates are larger than flagellated organisms (of the order of 25–1000 μm). They have hundreds to thousands of short, 15 μm long simple cilia (Daniel et al., 1992). Cilia typically beat in an asymmetrical rowing fashion, stiff and straight during their power stroke, and sliding tangentially during their recovery stroke. Two Reynolds numbers are appropriate to ciliary propulsion, one based on the cilia $Re^{(\omega)} = \omega L^2/\nu$, and other on the whole organism $Re^{(b)} = UL/\nu$. Both are usually much less than one, indicating that ciliated organisms swim also by using viscous mechanisms. Efficiency of swimming decreases rapidly as the body increases due to the vast number of cilia.

In the 1970s, three different modes – the envelope, sublayer and traction layers – of ciliary propulsion were developed (Daniel et al., 1992). They predict the velocity distributions around organisms, and the forces they generate. A detailed description of these models is given in the papers by Brennen (1975), Blake (1972) and Keller et al. (1975).

Other aquatic organisms which inhabit the low Reynolds number environment are larval fishes. They are small (of the order of 0.5–1.0 cm) and swim slowly. Therefore, Reynolds numbers are of the order of about 10–200, which corresponds to a transitional zone, when dependence of drag on flow velocity can not be determined in a straightforward manner. Vlymen (1974) developed a hydromechanical model of larval swimming which combines resistive and reactive terms. According to this model, the total work required for swimming is the sum of the work required to generate the resistive forces acting on the head and body $dW_{\text{resistive}}$, plus the work required to overcome the inertia of the head, dW_{head}, and to move the mass of the body, dW_m, and its associated added mass, dW_a, i.e.:

$$dW_{\text{tot}} = dW_{\text{resistive}} + dW_{\text{head}} + (dW_a + dW_m).\qquad(11.44)$$

Application of the model to a 1.4 cm larvae of *E. Mordax* showed that the resistive forces account for only 8% of the total energy required, and the inertial energy required to move the head is only 0.2% of the total energy.

11.4 Swimming Strategy

Locomotion is an energetically costly activity that comprises a significant component of an animal's overall energy budget. Animals use various locomotion strategies to minimize energetic expenditure. For example, fish have the option to cover a given distance by swimming steadily at a constant velocity, or by alternating periods of active swimming with periods of passive gliding. The selection of optimum cruising velocity is important when considering the long-range movements of fish such as feeding and spawning migrations. The optimal cruising velocity should be selected with respect to maximum range or to minimize the energy required to cross a given distance.

Let us determine the optimum velocity for long-range cruising fish. The total available energy of a fish can be represented as a sum of the component related to the swimming power P_s and the rate at which energy is expended on basal metabolism P_m, *i.e.* (Weihs, 1973):

$$E_{\text{tot}} = (P_s + P_m)t, \tag{11.45}$$

in which t is the swimming time. The basal metabolism is the minimum rate of energy expenditure needed to keep the fish alive. In fish, apart from the basal metabolism rate, the standard and active metabolic rates are distinguished (Videler, 1993). The standard metabolic rate includes the basal metabolism and the extra energy needed to bring the animal to an increased activity level. The active metabolic rate is the total energy used during swimming. All these levels depend on species, size, temperature and velocity of swimming.

The swimming power, P_s, is the product of the thrust required to overcome the drag and velocity of swimming, U:

$$P_s = \frac{1}{2\eta_p}\rho_w C_d S_w U^3, \tag{11.46}$$

in which η_p denotes the Froude efficiency of swimming, and S_w is the wetted surface of the fish. Weihs (1973) related a swimming efficiency to the swimming velocity:

$$\eta_p = \chi U, \tag{11.47}$$

in which χ is an empirical constant. Hence, using Eqs. (11.46) and (11.47) in Eq. (11.45) gives:

$$E_{\text{tot}} = \left(\frac{\rho_w C_d S_w U^2}{2\chi} + P_m\right)t. \tag{11.48}$$

During time t, the fish travels distance l, *i.e.*:

$$l = \frac{E_{tot}U}{\frac{\rho_w C_d S_w U^2}{2\chi} + P_m}. \tag{11.49}$$

This means that for some optimal swimming velocity, U_{opt}, the distance, l, becomes maximum when

$$\frac{dl}{dU} = 0 \quad \text{at} \quad U = U_{opt}. \tag{11.50}$$

After differentiating we obtain:

$$U_{opt}^2 = \frac{\chi P_m}{\frac{1}{2}\rho_w C_d S_w}, \quad P_m = \frac{\rho_w C_d S_w U_{opt}^2}{2\chi} = P_s. \tag{11.51}$$

Equation (11.47) shows that the maximum range (for a given energy store) of:

$$l_{max} = \frac{\chi E_{tot}}{\rho_w C_d S_w U_{opt}} \tag{11.52}$$

is reached when the swimming power and basal metabolic power are equal.

Using the ratio $\xi = U/U_{opt}$, a corresponding ratio of swimming range can be expressed as follows:

$$\frac{l}{l_{max}} = \frac{2\xi}{1 + \xi^2}. \tag{11.53}$$

The ratio l/l_{max} as a function of U/U_{opt} is shown in Fig. 11.3. There is a rather wide range of velocities at which the swimming range is reasonable large. For example, for swimming velocities varying from half of the optimum velocity to twice of that velocity, the swimming range is larger than 80% of the maximum range, l_{max}.

Experiments with migrating sockeye salmon, *Oncorhynchus nerka*, in the ocean between Vancouver Island and mainland Canada, showed that fish of an average length, L, of 66.3 cm had an average velocity, U, of 66.7 cm/s, which gives $U/L \approx 1$ s^{-1} (Quinn, 1988). Two bluefin tuna with estimated weights of 225 kg and 170 kg, tagged in the Gulf of Mexico were recaught 118 and 119 days later off Norway. They swam the distance of 7778 km at a velocity of 0.76 m/s, or $U/L = 0.3$ s^{-1} (Mather, 1962). It is probably not realistic to assume that the fish swam all 118 (or 119) days with the same velocity. It is more likely that the swimming velocity varied.

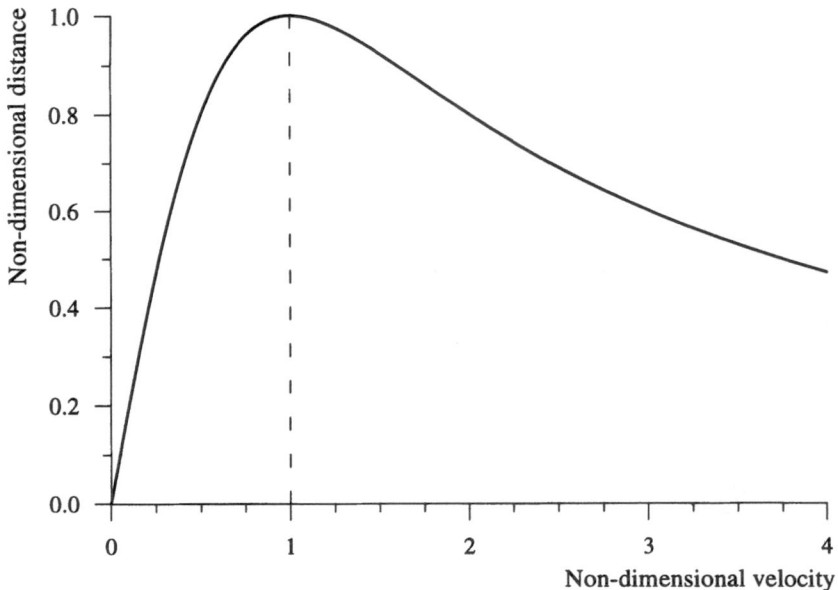

Fig. 11.3: Non-dimensional swimming distance as a function of non-dimensional swimming velocity

Some theoretical results (Weihs, 1974; Videler and Weihs, 1982) suggest that there is an advantage to using the so called burst-and-coast swimming behaviour instead of swimming at a constant velocity. For such behaviour, the fish starts off at some initial velocity, U_i, which is lower than the average velocity, $\bar{U}$. During a burst, the fish accelerates to a final velocity, U_f, higher than $\bar{U}$. Following the burst, the fish decelerates to its initial velocity, U_i, during the coast phase. The energy expenditure of burst-and-coast swimming and steady swimming can be compared by using the ratio:

$$R = \frac{E_i}{\bar{E}} = \frac{\int_0^{t_1} TU\,dt}{T\bar{U}(t_1 + t_2)}, \tag{11.54}$$

in which E_i is the energy expended during the burst phase of intermittent swimming, $\bar{E}$ is the energy required for crossing the same distance at a constant average velocity $\bar{U}$, t_1 and t_2 are the burst and coast times, respectively, T is the thrust produced by the fish during the burst phase, $\bar{T}$ and $\bar{U}$ are the mean thrust and velocity during steady swimming. Burst-and-coast swimming is more efficient than steady swimming when the ratio, R, is smaller than unity.

An extension of the Weihs (1974) model for intermittent swimming provides an expression for ratio R (Videler and Weihs, 1982). Fish can chose many combinations of initial velocity, U_i, and final velocity, U_f, which will result in R values lower than 1. Observations of cod and saithe in a large tank and observations in the field showed that intermittent swimming is used for feeding purposes. Cod of 26 cm and 30 cm length used the intermittent swimming mode to save energy, with the ratio, R, as low as 0.42 and 0.36, respectively. Saithe of 35 cm length swim at a mean velocity of $U/L = 5$ s^{-1} but can accelerate during the burst phase to about $a/L = 10$ s^{-2}, wherle a is the acceleration. As a result, intermittent swimming in this case is 2.5 times cheaper than constant swimming.

The proximity of the sea surface induces extra drag on swimming animals. This is especially true for air-breathing aquatic mammals, such as whales and dolphins, which experience higher drag when touching the sea surface (Hertel, 1969; Newman, 1977). The energy generated by a swimming animal is partly used to produce surface waves. If the height of surface waves is equal to H, then the energy wasted on wave formation (per unit area) is equal to $1/8\ \rho g H^2$. Webb *et al.* (1991) found that about 70% of the mechanical energy used for propulsion by trout in deep water was dispersed in the form of waves when the fish was just below the surface.

At the end of this section we will consider another mode of swimming, used by dolphins, which is very interesting from a fluid mechanics point of view. Dolphins exhibit at least three modes of swimming (Au and Weihs, 1980). In unhurried motion, they break the surface gently, often showing little more than the blowhole. At cruising speed of the order of 3–3.5 m/s, they swim just beneath the surface with a little splashing. However in the fastest 'running' mode, the dolphins clear the water in sequential, parabolic leaps. Leaps are accompanied by considerable splashing and interspersed with brief subsurface swimming. Let us compare, following Au and Weihs (1980), the efficiency of dolphin swimming with and without leaping.

The energy, E_s, required for swimming continuously under water with its blowhole just out of the water is:

$$E_s = F_d l, \tag{11.55}$$

in which l is the travelled distance, F_d is the drag force given by Eq. (11.22) as:

$$F_d = \frac{1}{2}\rho_w C_d S U^2, \tag{11.56}$$

in which the wetted surface S is:

$$S = \xi_1 V^{2/3}, \tag{11.57}$$

where coefficient ξ_1 is the correction due to proximity to the surface, and V is the volume of the animal's body.

The energy, E_l, needed for leaping, when air resistance has been neglected, can be estimated approximately as follows:

$$E_l = W h_{\max} (1 + \xi_2), \tag{11.58}$$

in which W is the dolphin's weight, h_{max} is the maximum height of the centre of gravity of the dolphin's body above still water level, and ξ_2 is the correction coefficient due to spray when the dolphin emerges from the water.

From ballistic theory we find that distance, l, and maximum height, h, are dependent on emergence angle, α (measured from the still water level):

$$l = \frac{U^2}{g} \sin 2\alpha, \tag{11.59}$$

and

$$h = \frac{U^2 \sin^2 \alpha}{2g}. \tag{11.60}$$

Equation (11.59) suggests that the longest leap, for a given velocity U, is obtained when $\alpha = 45°$. Let us now form the ratio, R, of energies E_l and E_s. After substituting all functions into Eqs. (11.55) and (11.58) we obtain:

$$R = \frac{E_l}{E_s} = \frac{\rho_d g \left(1 + \xi_2\right) V^{1/3}}{2\rho_w C_d \xi_1 U^2}. \tag{11.61}$$

When $R < 1$, swimming with leaping becomes energy saving. Setting $R = 1$ in Eq. (11.61), the corresponding crossover speed U_{cr} above which leaping will occur, can be obtained. Calculations by Au and Weihs (1980) indicate that for the usual size range (0.05–0.10 m^3) of most dolphin species, the speed U_{cr} is about 5 m/s, which is well within a dolphin's available range of speed. For example, when dolphin volume $V = 0.1$ m^3, the critical crossover velocity $U = 5.5$ m/s, and length, l, and maximum height, h_{max}, of leap are 3.09 m and 0.77 m, respectively. For larger dolphin ($V > 1$ m^3), leaping is probably not possible as the energy required increases rapidly with body size.

12 Internal Flows in Marine Organisms

12.1 Introduction

In contrast to the fluid mechanics of flows 'external' to the aquatic animal body, our knowledge of the 'internal' flows is rather limited. Aquatic organisms are filled with various pipes and channels through which fluids flow. Biological systems of internal flow are both diverse and complex, partly due to the non-Newtonian character of fluids themselves, partly due to the complicated network of the internal pipes and channels, and basically due to the unknown roughness of pipe walls.

In this chapter we will briefly concentrate on selected fluid mechanics problems. We restrict ourselves to laminar flow only, as within marine organisms very rarely do flows become turbulent where Reynolds numbers exceed the value of about 2000. Most of the circulatory systems of aquatic animals involve pulsative flow of non-Newtonian fluids in pipes with shapes and cross-sectional areas changing in time.

The metabolic rate of an animal is maintained through the steady consumption of fuel and oxygen. The amount of energy needed to propel an animal has been described in the last chapter and here we concentrate on gas-exchange organs, lungs and gills. These organs must have a size and diffusion capacity adequately scaled to the oxygen needs. We will demonstrate how the fractal approach may be useful to measure lung or gill area.

12.2 Flow in Pipes Revisited

From Eq. (2.99), it can be found that the velocity distribution in a circular pipe for laminar flow is (see Fig. 12.1):

$$u(r) = \frac{\Delta p}{4\mu l}\left[\left(\frac{D}{2}\right)^2 - r^2\right]. \tag{12.1}$$

At the pipe wall ($r = D/2$), velocity is zero, while at the pipe axis ($r = 0$), velocity reaches its maximum value of $(\Delta p/4\mu l)(D/2)^2$. The cumulative dis-

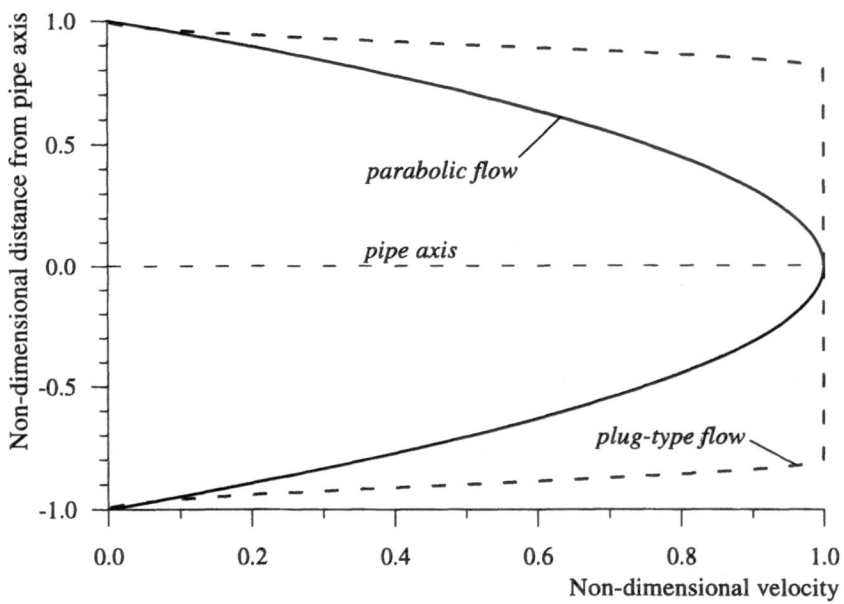

Fig. 12.1: Velocity distribution in a circular pipe

charge, $Q(r)$, of fluid transported through the pipe can be determined by multiplying the area of an annulus $2\pi dr$ by the local velocity $u(r)$ and integrating across the cross-section from the axis to a given distance, r, *i.e.*:

$$Q(r) = \int_0^r 2\pi r u(r)dr = \frac{\pi \Delta p D^4}{128\mu l}\left[2\left(\frac{2r}{D}\right)^2 - \left(\frac{2r}{D}\right)^4\right]. \tag{12.2}$$

The non-dimensional cumulative discharge $128\mu l Q(r)/\pi \Delta p D^4$ is shown in Fig. 12.2 as a function of non-dimensional distance from the pipe axis, $2r/D$. At $r = D/2$, the cumulative discharge is equal to the total discharge given in Eq. (2.103):

$$Q = \frac{\pi \Delta p D^4}{128\mu l}. \tag{12.3}$$

Let us rewrite Eq. (12.3) as follows:

$$\Delta p = \frac{128\mu l}{\pi D^4}Q. \tag{12.4}$$

The term $128\mu l/(\pi D^4)$ is the resistance of a steady laminar flow in a circular cylindrical pipe. Thus, the pressure drop, Δp, required to transport volume Q of the fluid is linearly dependent on the resistance of the flow.

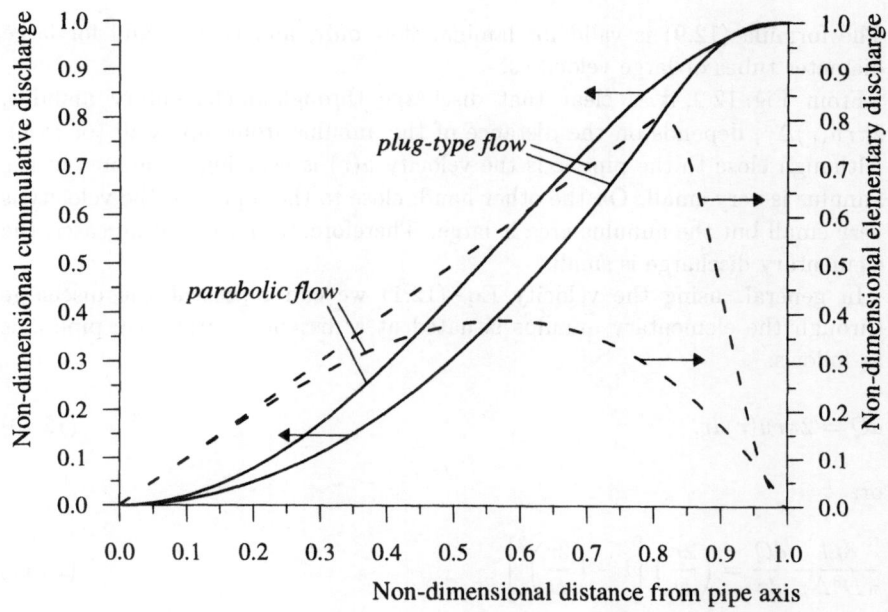

Fig. 12.2: Non-dimensional cumulative discharge for flow through pipe

For the parabolic velocity profile (Eq. 12.1) in the tube, the shear stress, τ, on the tube wall becomes:

$$\tau = \mu \frac{du}{dr} = -\frac{\Delta p D}{4l}, \tag{12.5}$$

or using Eq. (2.104) for average velocity $\bar{u}$ we get:

$$\tau = -\frac{8\mu}{D}\bar{u}. \tag{12.6}$$

After some manipulation we can represent the shear stress, τ, as:

$$\tau = \frac{16}{\frac{\bar{u}D}{\nu}}\frac{1}{2}\rho\bar{u}^2 = \frac{16}{Re}\frac{1}{2}\rho\bar{u}^2 = \frac{1}{2}C_f\rho\bar{u}^2, \tag{12.7}$$

in which C_f is the skin friction coefficient:

$$C_f = \frac{16}{Re}, \tag{12.8}$$

where Reynolds number, Re, is defined as:

$$Re = \frac{\bar{u}D}{\nu}. \tag{12.9}$$

The formula (12.9) is valid for laminar flow only, and is not valid for large diameter tubes or large velocities.

From Fig. 12.2, it is clear that discharge through an elementary annulus, $2\pi r u(r)\Delta r$, depends on the distance of the annulus from pipe wall (or axis). Although close to the pipe axis the velocity $u(r)$ is very high, the area of the annulus is very small. On the other hand, close to the pipe wall the velocity is very small but the annulus area is large. Therefore, in both limiting cases, the elementary discharge is small.

In general, using the velocity Eq. (12.1) we can represent the discharge through the elementary annulus situated at a distance r from the pipe axis as follows:

$$dQ = 2\pi r u(r)dr, \tag{12.10}$$

or:

$$\frac{8\mu l}{\pi D^3 \Delta p} \frac{dQ}{dr} = \left(\frac{2r}{D}\right)\left[1 - \left(\frac{2r}{D}\right)^2\right]. \tag{12.11}$$

Function (12.11) is shown in Fig. 12.2 as non-dimensional elementary discharge. The average distance of flow from the pipe axis, which can be considered as a 'centre of gravity of flow', becomes:

$$R_c = \frac{1}{Q_{tot}} \int_0^{D/2} r dQ. \tag{12.12}$$

Integration in Eq. (12.12) gives $R_c = 8/15(D/2) \approx 0.533(D/2)$. This means that the average distance of flow from the pipe wall is $0.467(D/2)$. This distance corresponds to the so called 'distance index', Di, suggested by Vogel (1994). Parabolic flow is not the most efficient form of internal flow in marine organisms in terms of enhancing the exchange of substances and heat across the walls of pipes, as intensity of exchange increases when the 'centre of gravity' of flow is closer to the pipe wall.

Let us consider a plug type velocity distribution in a pipe as follows (see Fig. 12.1):

$$u(r) = \begin{cases} U_{max}, & 0 < r < 0.4D \\ U_{max}\left[\dfrac{1}{2} + \dfrac{1}{2}\cos\left(\dfrac{10\pi r}{D} - 4\pi\right)\right], & 0.4D \leq r \leq 0.5D. \end{cases} \tag{12.13}$$

Using a similar approach as for a parabolic flow, we obtain non-dimensional cumulative discharge and non-dimensional discharge, through an elementary annulus, as shown in Fig. 12.2. The 'centre of gravity of flow' is at distance of $0.743(D/2)$ from the pipe axis, or $0.257(D/2)$ from the pipe wall. In this

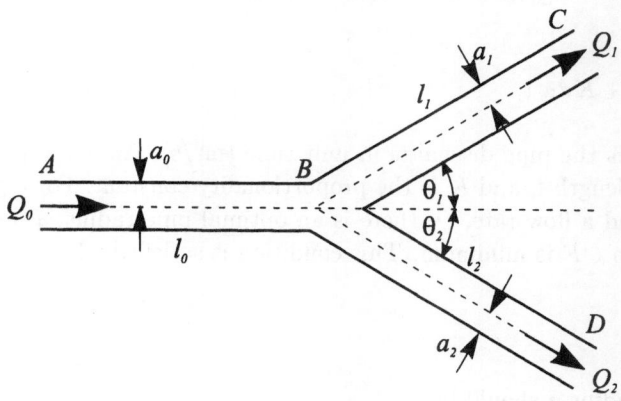

Fig. 12.3: Bifurcation of a blood vessel at point B

case, there is a better chance for more extensive exchange of material or heat through the pipe wall, compared to the case of parabolic flow.

Let us now briefly consider the flow between two parallel plates. In fact, a few biological situations involve closely spaced and parallel flat plates, for example, the gills of fish. The sieve units of the gill of a tuna can be idealized as a set of parallel plates. According to Stevens and Lightfoot (1986), the slots in tuna gills are 127 μm by 20 μm in cross-section and about 1.5 mm long in the flow direction. The Reynolds number, Re, remains low and less than 100. Using a velocity distribution between plates given by Eq. (2.107), we find that the 'centre of gravity of flow' is located at a distance of 0.625 $(D/2)$ from the walls. It should be noted that the flow between parallel plates is more uniform in cross-section than in the case of a circular tube. For example, the ratio of maximum and mean velocities is 1.5, while for a circular tube it is equal to 2.0. We will return to the area of fish gills in Sect. 12.5.

A typical pattern of internal flow in marine organisms shows that arteries and veins bifurcate many times before they become capillaries. Consider the bifurcation of a blood vessel AB into two branches BC and BD, supplying blood at a rate Q_0, from point A to points C and D, with outflow of Q_1 at point C, and Q_2 at point D (Fig. 12.3). The first condition of flow, which should be satisfied is the conservation of mass, *i.e.*:

$$Q_0 = Q_1 + Q_2. \tag{12.14}$$

The second condition results from minimizing the so called cost function. The cost function is similar to other minimum principles, such as minimum entropy in thermodynamics, the Fermat principle of least time of travel in optics, Hamilton's principle in mechanics, and others. For blood vessels, the cost function, CF, is the sum of the rate at which work is done on the blood and the rate at which energy is used up by the blood vessel due to metabolism (Fung, 1997):

$$CF = Q\Delta p + K\pi a^2 l, \tag{12.15}$$

in which Q is the pipe discharge in unit time (m^3/s), Δp is the pressure drop over a pipe length l, and K is the proportionality constant. For a given vessel length, l, and a flow rate, Q, there is an optimal pipe radius, a, at which the cost function CF is minimum. This condition is satisfied when:

$$\frac{\partial(CF)}{\partial a} = 0. \tag{12.16}$$

Thus, the radius a should be:

$$a = \left(\frac{16\mu}{\pi^2 K}\right)^{1/6} Q^{1/3}. \tag{12.17}$$

Now, from Eq. (12.14), the conservation of mass becomes:

$$a_0^3 = a_1^3 + a_2^3, \tag{12.18}$$

which is known as Murray's law. After substituting Eq. (12.17) into Eq. (12.15) we obtain a minimum value of the cost function:

$$CF = \frac{3\pi}{2} K l a^2. \tag{12.19}$$

The minimum value of CF for bifurcated blood vessels, as in Fig. 12.3, can be found by the variation of the lengths l_1, l_2 and l_3, and the radii a_1, a_2 and a_3, in such a way that the following condition is satisfied:

$$\frac{3\pi K}{2}\left(a_0^2 l_0 + a_1^2 l_1 + a_2^2 l_3\right) = \min. \tag{12.20}$$

The result is (Fung, 1997):

$$\cos\theta_1 = \frac{a_0^4 + a_1^4 - \left(a_0^3 - a_1^3\right)^{4/3}}{2a_0^2 a_1^2}, \tag{12.21}$$

$$\cos\theta_2 = \frac{a_0^4 - a_1^4 + \left(a_0^3 - a_1^3\right)^{4/3}}{2a_0^2(a_0^3 - a_1^3)^{2/3}}. \tag{12.22}$$

Equations (12.18), (12.21) and (12.22) form the necessary conditions for optimal bifurcation pattern of blood vessels.

12.3 Blood as Fluid and its Circulation in Marine Organisms

In this section we will briefly summarize a description of the mechanical properties of blood and basic information on blood circulation. More details can be found in Pedley (1980), and Fung (1993, 1997). Blood is a suspension of formed elements, namely the red cells, white cells and platelets. Usually, these elements occupy about 45% (by volume) of the blood. The remainder is made up of plasma. Plasma is a solution of large molecules, but on the scale of motion normally encountered in blood vessels, it can be regarded as a homogeneous Newtonian fluid of molecular dynamic viscosity, μ, of 0.0012–0.0016 kg/m/s.

However, in very small blood vessels, whole blood cannot be regarded as a homogeneous fluid, as the diameter and spacing of red cells is comparable with capillary diameters. In this case, blood is a non-Newtonian incompressible viscoplastic fluid. It means that for blood flow in small capillaries the Newton's relationship (1.2) does not apply and the coefficient, ν, of the dynamic viscosity is not a constant value. The anomalous behaviour of blood viscosity is mainly attributed to the shear-dependent deformation and agregation of red blood cells. Chien (1970) experimentally found that with an increase in shear rate (du/dz), the viscosity decreases suspension of red blood cells. The variation of blood viscosity with plasma protein concentration, hematocrit and shear rate can be explained as a function of the change in effective cell volume.

However, in most arteries blood behaves in a Newtonian fashion, and the viscosity can be assumed to be a constant (Ku, 1997). When the diameter of blood vessels exceeds 100 μm, and the scale of the microstructure is much smaller than that of the flow, blood is usually treated as a homogeneous fluid of density $\rho = 1.05 \times 10^3$ kg/m^3, dynamic viscosity $\mu = 0.004$ kg/m/s, and kinematic viscosity $\nu = 4 \times 10^{-6}$ m^2/s.

The blood pressure at any location is made up of two components: the hydrostatic component due to both atmospheric pressure and water hydrostatic pressure at a given depth (see Sect. 2.2), and the dynamic component due to the pressure generated by the heart and the frictional loss in the blood vessels. This last component is alone responsible for the motion of the blood and is commonly known as 'blood pressure'. The work required to circulate the blood is provided by the heart. In mammals (also aquatic), the heart makes up about 0.58% of the body mass. For fish hearts, experimental data (34 species of fish ranging from 0.005 to 32 kg) suggest the following relationship between heart and body size (Schmidt-Nielsen, 1989):

$$M_h = 0.0022 M_f^{1.026}. \tag{12.23}$$

Thus, there is an almost linear dependence of heart mass on fish body mass. A typical fish of 1 kg mass will have a heart of 2.2 g. The relative size of a fish's heart is about two and half times smaller than the relative size of a mammal's heart.

Generally, the flow in blood vessels may be laminar or turbulent, depending on the Reynolds number, Re. As we showed in Chapter 2, in circular tubes the critical Reynolds number is about 2000, and below this limit the influence of viscosity dominates and stabilizes the flow so that turbulence does not occur. However, the transition from laminar to turbulent flow in a tube may occur at Reynolds numbers between 2000 and 12000, depending on the roughness of the tube wall. For example, in humans the output through the cross-sectional area of the aorta of about 4 cm^2, is about 5 litres per minute, which gives the mean velocity of the blood of about 21 cm/s. Therefore, the Reynolds number will be $Re \sim 1500$. This indicates that the flow in aorta approaches the critical Reynolds number for turbulence.

To determine the Reynolds number of flow in the aorta of aquatic mammals, we use the relationships given by Schmidt-Nielsen (1989):

- cross-sectional area of the aorta, in cm^2:

$$A = 0.094 M^{0.82}, \qquad\qquad (12.24)$$

- discharge rate in aorta (cardiac output):

$$Q = 187\, M^{0.75}, \qquad\qquad (12.25)$$

in which M is the animal mass.

Thus, the flow velocity, $\bar{u}$, and the Reynolds number become:

$$\bar{u} \sim M^{-0.07}, \qquad\qquad (12.26)$$

and

$$Re \sim \bar{u}\sqrt{A} = M^{0.34}. \qquad\qquad (12.27)$$

As flow is not at the critical level for turbulence in the circulation in humans, it will be less so in any smaller aquatic mammals. Time is needed for some unstable modes of motion in the flow to grow into turbulence. Due to the pumping heart action, the blood flow in blood vessels is a pulsating flow with velocity changing in time. Therefore, the Reynolds number will also vary in time. We will describe briefly the propagation of pressure pulses in the blood vessels in the next section.

12.4   Propagation of the Pressure Pulse

At any location in the blood vessel, the motion of blood is driven by the local pressure gradient. This gradient is determined by the propagation of the pressure pulse. Description of the full complexity of pulse-wave propagation is beyond the scope of this book. We will consider here some idealized cases which, however, illustrate the basic principles of pulsative blood flow. Let us assume an infinitely long, straight cylindrical tube of circular cross-section. The excess pressure due to the heart action, p, and the tube cross-sectional area, A, are functions of distance, x, and time, t. Moreover, we will assume that the amplitude of pressure is small and the flow is essentially one-dimensional with a longitudinal component $u(x,t)$. Thus, the continuity principle (see Appendix C.2) requires:

$$\frac{\partial A}{\partial t} + \frac{\partial}{\partial x}(uA) = 0. \tag{12.28}$$

The momentum equation for one-dimensional motion in a tube becomes (see Eq. C.30):

$$\frac{\partial u}{\partial t} + u\frac{\partial u}{\partial x} + \frac{1}{\rho}\frac{\partial p}{\partial x} = 0. \tag{12.29}$$

For simplicity, we assume that in the elastic tube there is a single-valued relationship between the cross-sectional area, A, and pressure p:

$$p = P(A). \tag{12.30}$$

In reality, this relationship is much more complex due to viscoelasticity of the tube wall. However, in the case of an elastic tube, the tube radius, a, is linearly proportional to the blood pressure in the vessel:

$$a = a_0 + \alpha p, \tag{12.31}$$

in which a_0 is the radius, when $p = 0$. In the case when velocity is small, the term $u\partial u/\partial x$ in Eq. (12.29) can be ignored; thus:

$$\frac{\partial u}{\partial t} + \frac{1}{\rho}\frac{\partial p}{\partial x} = 0. \tag{12.32}$$

Because $A = \pi a^2$, Eq. (12.28) can be rewritten as follows:

$$\frac{\partial u}{\partial x} + \frac{2}{a}\frac{\partial a}{\partial t} = 0, \tag{12.33}$$

neglecting the gradient $\partial A/\partial x$. Differentiating Eq. (12.31) and substituting into Eq. (12.33) we obtain:

$$\frac{\partial u}{\partial x} + \frac{2\alpha}{a}\frac{\partial p}{\partial t} = 0. \tag{12.34}$$

Eliminating velocity u from Eqs. (12.32) and (12.34) yields equation for fluid pressure $p(x,t)$:

$$\frac{\partial^2 p}{\partial x^2} - \frac{1}{c^2}\frac{\partial^2 p}{\partial t^2} = 0, \tag{12.35}$$

in which:

$$c = \sqrt{\frac{a}{\rho\alpha}}. \tag{12.36}$$

Equation (12.35) is the well known wave equation, and the quantity c is the wave speed. If the tube is thin walled and the material obeys Hooke's law (see Sect. 10.2), wave speed is given by (Fung, 1997):

$$c = \sqrt{\frac{Eh}{2\rho a}}, \tag{12.37}$$

in which h is the wall thickness. The value of c is known as the Moens-Korteweg wave speed. For mammals the size of a dog, the velocity c is of the order of 5–8 m/s (Pedley, 1980).

To solve the wave equation (12.35), the boundary and initial conditions have to be known. However, even without complete knowledge of these conditions, it can be shown that the solution is a function that represents a progressive wave with the speed c:

$$p = f(x - ct), \tag{12.38}$$

where f is an arbitrary function. For illustration, let us assume that at $t = 0$, the pressure pulse has a form (Fig. 12.4):

$$p(x,0) = \begin{cases} \sin\dfrac{\pi x}{l}, & 0 \le x \le l \\[2mm] 0, & x < 0, \quad x > l. \end{cases} \tag{12.39}$$

Let the velocity c be 5 m/s. At the time $t = 5$ s later, the same pressure pulse is translated to the right. The value of the pressure pulses remains constant as long as value of $(x - ct)$ is the same. Thus, at $t = 5$ s, the point A will is at

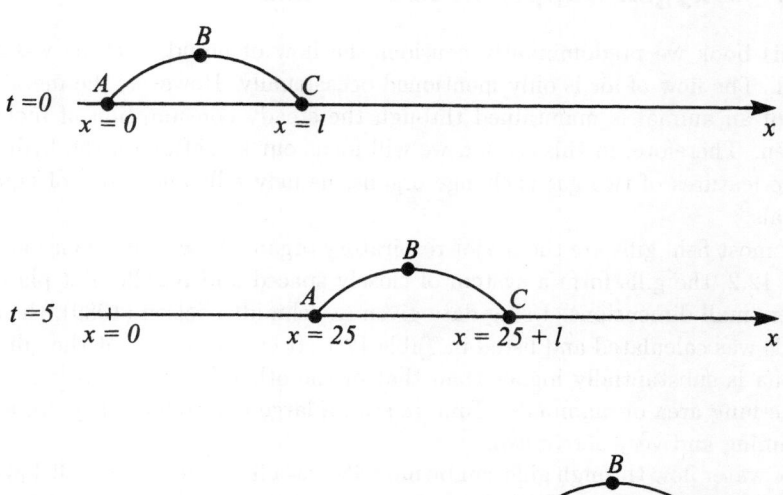

Fig. 12.4: Propagation of pressure pulse

distance $x = 25$ m, the point B is at distance $x = 25 + l/2$, and the point C is at distance $x = 25 + l$. In exactly the same manner we can find the position of the pressure pulse at an arbitrary time t_n.

Flow in a vein is similar to that in an artery. However, compared to arteries there are several important differences: the pressure in a vein is normally much lower than that in an artery at the same location, and veins have thin walls which may be collapsed in normal function. A description of flow in collapsible veins is given by Fung (1997). The pressure waves in veins attenuate with distance along the vein. There are a few reasons for this attenuation. The obvious is the viscosity of the blood. Another is the viscosity of the vessel wall.

The arteries and veins divide and divide again. The vessel diameter decreases with each division and the Reynolds number becomes very small, being determined by the balance of viscous stresses and the pressure gradient. Assuming the velocity flow of 1 mm/s, vessel diameter of 10 μm, and viscosity of blood of 4×10^{-6} m^2/s, then the Reynolds number becomes 0.005, which is typical for microcirculation in animal bodies. However, for blood flow, the Reynolds number is not the only characteristic of the microcirculation. In the capillary circulation, the exchange of fluid and other matters between blood and tissue surrounding the blood vessels occurs and the role of red blood cells must be recognized (Fung, 1997).

12.5   Oxygen Supply in Marine Animals

In this book we predominantly consider the flow of liquid, such as water or blood. The flow of air is only mentioned occasionally. However, the metabolic rate of an animal is maintained through the steady consumption of fuel and oxygen. Therefore, in this section we will focus our attention on the hydrodynamic features of two gas-exchange organs, namely gills and lungs of aquatic animals.

For most fish, gills are the major respiratory organ. As we have mentioned in Sect. 12.2, the gills form a system of closely spaced and parallel flat plates of rather small dimensions. Using data given in Schmidt-Nielsen (1989), the area of gills was calculated and listed in Table 12.1. It can be seen that the gill area of tuna is substantially higher than that of the other fish and it is very close to the lung area of mammals. Tuna require a large gill area as they are a fast swimming and very active fish.

The water flow through gills can be modelled as a flow between parallel plates. Assuming laminar flow, the volume flow per unit time is given by Eq. (2.108), *i.e.*:

$$Q = \frac{\Delta p w h^3}{12 \mu l}, \tag{12.40}$$

where h, w, and l are the spacing, width and length of the plates, respectively, while Δp is the pressure drop through gills, which provide a mechanisms for proper uptake of oxygen from the water. The gas exchange in the gills depends on the gill area, oxygen content in the water, and the diffusing capacity, which refers to the amount of oxygen that diffuses per unit time per unit pressure difference for the gill.

Lungs of aquatic mammals are the gas-exchange organs. The lung volume for small (dugong) and large (whale) aquatic mammals makes up almost the identical fraction of their body volume. Thus, we can write (Schmidt-Nielsen, 1989):

$$V_l = 0.0535 M_b^{1.06 \pm 0.02}, \tag{12.41}$$

where V_l is the lung volume (in litres), and M_b is body mass (in kg).

At the end of this section we will make some remarks on the accuracy of gill and lung surface area measurements. As we have discussed above, the efficiency of the respiratory systems depends on the surface of these organs. How accurately can we measure the surface area of lungs or gills? Mandelbrot (1983) noted that light microscopy yields 80 m^2 for human lung surface area, which is very close to value given in Table 12.1, while electron microscopy results in 140 m^2. When a more precise measurement device is available, the lung area will be even larger. The reason for these discrepancies is that the lung surface is fractal in character and cannot be accurately measured in square metres.

Table 12.1: Fish gill area (based on data from Schmidt-Nielsen, 1989)

Species	Size range (kg)	Gill area (m^2)
Toadfish	0.015-0.8	0.0047-0.11
Black bass	0.001-0.9	0.0008-0.18
19 species	0.071-6.4	0.045-1.83
Tuna:		
yellowfin & bluefin	4.0-40.0	4.40-32.9
skipjack	1.0-6.0	1.85-8.5
Mammals	0.025-25.0	0.09-77.6

To explain the fractal character of the surface, let us first consider the simple case of the area of a circle with radius a. The circle area can be approximated by the area of regular polygons incorporated into the circle. When the number of polygon sides is six (regular hexagon), its area is $2.598\ a^2$, and for polygon with 10 sides, the area becomes $2.938\ a^2$, while for 30 sides it gives $3.102\ a^2$. We can conclude that by increasing the number of sides, the polygon area quickly approaches the limit at πa^2, which is the circle area.

Let us consider a more complex case of measuring of area of a mountainside by 'tiling' it with flat tiles, all having the same area. The total unknown surface area is given by the sum of the area of all the tiles (Pennycuick, 1992). If we continue to repeat this operation with smaller and smaller tiles, we will be able to represent the details of all humps and gullies. However, we observe that the 'area' of a rugged surface is an undefined quantity, as it increases without limits as the area of the tiles used is reduced.

According to Mandelbrot (1983), such 'surface area' is a case of fractal extent with dimension 2.17. We note that the dimension 2 corresponds to the unit of surface, while the dimension 3 corresponds to the unit of volume. Using this fractal terminology we can say that a regular area, such as a circle, has a fractal dimension 2. Thus, for an irregular, rugged surface we cannot measure the surface area in square metres, but in fractal units. Pennycuick (1992) proposed a name 'metron' for this unit. It means that lung or gill area should be measured in metrons of dimension 2.17. Then the result of measurement becomes unique, and not dependent on the scale of measurement.

As the alveolar surface of aquatic mammals is rather smooth, it would be expected that surface area dimension is close to 2. However, the true dimension of lungs or gills of the aquatic animals should be checked by careful measurements.

13 Transport and Mixing in Coastal Ecosystems

13.1 Introduction

The basic mechanisms for transport in the deep ocean and in coastal waters, namely convection and diffusion, have been examined in Chap. 8. In this chapter we will describe some examples illustrating the influence of transport of organic and inorganic matter on the geomorphology and ecology of coastal ecosystems, being under the constant influence of the adjacent ocean waters. The rhythms of tides, severe storms and long-term climatic perturbations shape the coasts. Sediment is continuously moving, being deposited in one place and eroded in another, as the shoreline attempts to establish an equilibrium with prevailing hydrodynamic conditions.

A unique and important part of the coastal aquatic environment are the estuaries, where river flow meets the sea. These areas form the transition zone between inland areas of freshwater and the sea water lying offshore. Regardless of type, estuaries are characterized by the mixing of fresh and salt water. As we showed in Chap. 8, estuarine circulation is complex, and includes a variety of residual currents that depend on the relative mixing effects of river discharge and tidal energy. Due to changing gradients in temperature, salinity, and dissolved chemicals, and due to nutrients supplied by rivers, estuaries are able to support a thriving and abundant biota.

The energy which enters the biosphere principally in the form of sunlight flows through estuarine ecosystems, while the chemical constituents of organisms are continually cycled among organisms, water, sediments and the atmosphere. The exact nature of the chemical processes occurring in an estuary depends on the quantity and kind of materials transported by the fresh and salt water sources, the different chemical reactions that occur in fresh and salt water, and the residence time of river water in the estuary (Ketchum, 1983; Day et al., 1989).

The transport of organic and inorganic matter in any coastal ecosystem is primarily induced by the advection and diffusion mechanisms. In the next sections we will illustrate the relative importance of both these mechanisms by examples from various coastlines and estuaries.

13.2   Transport and Mixing in Estuaries

13.2.1   A Brief Orientation

As was discussed in Chap. 8, estuaries are the regions where fresh and salt waters mix. Because nutrient fluxes are spatially concentrated within the estuarine mixing zone, estuaries usually have a higher biological productivity than adjoining coastal and continental shelf water. Estuaries also provide convenient, sheltered locations for ever expanding human development. This trend yields increasing volumes of nutrients, human-derived materials and pollutants.

As the river flow rates, tides and winds are variable, the distribution of salinity, nutrients and plankton population within estuaries are rarely stable. In particular, during flood event, the discharge of freshwater may be so high that sea water cannot intrude into the estuary, and the mixing will take place on the boundaries of the flood plume (see Sect. 13.3). Many biological, chemical and geological effects in estuaries are dependent, in part, on the dynamics of circulation and mixing processes. Despite many efforts, a multidisciplinary perspective of how coastal estuaries function is only now emerging, with scientists realizing the crucial importance of the interaction between physical, chemical, geological and biological processes to sustain estuarine development. A complete summary of these interactions is beyond scope of this book. Rather, in this section we will describe selected examples which illustrate interesting phenomena in estuaries. Some of these are sufficiently well understood that analytical or numerical models have been developed. In other cases only a qualitative description is given. For better clarity of explanation, we will separately consider the transport and fate of sediments, and influence of hydrodynamics on nutrient transport and primary production.

13.2.2   Sediment Transport in Estuaries

An understanding of the distribution and movement of suspended particulate matter in estuaries is relevant to many other related studies, such as dredging/spoil dumping operations and transport and fate of particle-bound contaminants. In estuaries the transport may be inland or seawards due to tidal and freshwater flow variations. Sediment may be stored for long periods as bed material in intertidal zones. However, oscillatory motion produced by surface waves generates additional vorticity and turbulence in the bed boundary layer, and enhances the diffusivity and hence transport entrainment and suspension capacity of the flow. For example, Kana and Ward (1980) showed that in a shallow coastal zone, sediment transport may increase up to 60 times during storm wave conditions as opposed to calm conditions. For the estuarine environment this increase can be up to 40 times as was detected by Owen and Thorn (1978). A similar increase in sediment transport in the Tay Estuary, Scotland, was reported by Weir and McManus (1987). The turbulence and suspension capacity were particularly intensive during flood tides and opposing currents and wind.

Lindsay *et al.* (1996), using current meters and transmissometers, moored for 11 weeks, examined the influence of tidal range and river discharge on suspended particulate matter fluxes in the Forth Estuary, Scotland. On the semi-diurnal scale, suspended sediment concentration was closely related to current velocities. In the near bottom layer, the current velocity required for resuspension and deposition were 0.60 and 0.30 m/s, respectively. The dominant suspended sediment type was mud with a median particle diameter of less than 10 μm. However, there is also some evidence of particles with diameters of greater than 250 μm in the deeper channels. The data from the transmissmeters showed that during a flood tide, with a tidal range of 5.3 m, the concentration reaches 300 mg/l.

Clarke and Elliot (1998) developed a two-dimensional depth integrated transport model for the Forth Estuary in Scotland. The model is based on an advection-diffusion equation (see Eq. 8.30), which for the region of changing water depth takes the form):

$$\frac{\partial(h\bar{c})}{\partial t} + \frac{\partial(h\bar{u}\bar{c})}{\partial x} + \frac{\partial(h\bar{w}\bar{c})}{\partial z} = \frac{\partial}{\partial x}\left(hK_x\frac{\partial \bar{c}}{\partial x}\right) + \frac{\partial}{\partial z}\left(hK_z\frac{\partial \bar{c}}{\partial z}\right) +$$

$$+ \ E_r - D_r, \tag{13.1}$$

in which $\bar{c}$ is the depth-averaged suspended sediment concentration, h is the total water depth, $\bar{u}$ and $\bar{w}$ are the depth-averaged velocity components, K_x and K_z are the diffusion coefficients, E_r is the erosion rate, and D_r is the deposition rate.

To parameterize the erosion/deposition processes, a one-point approach was applied, which is a simplification of Eq. (13.1). Thus, when the advection terms are neglected, Eq. (13.1) becomes:

$$\frac{\partial(h\bar{c})}{\partial t} = E_r - D_r, \tag{13.2}$$

and the rates E_r and D_r are assumed in the form:

$$E_r = \alpha\left[\left(\frac{u_*}{u_{*e}}\right)^2 - 1\right], \tag{13.3}$$

and

$$D_r = w_s\bar{c}_b\left[1 - \left(\frac{u_*}{u_{*d}}\right)^2\right], \tag{13.4}$$

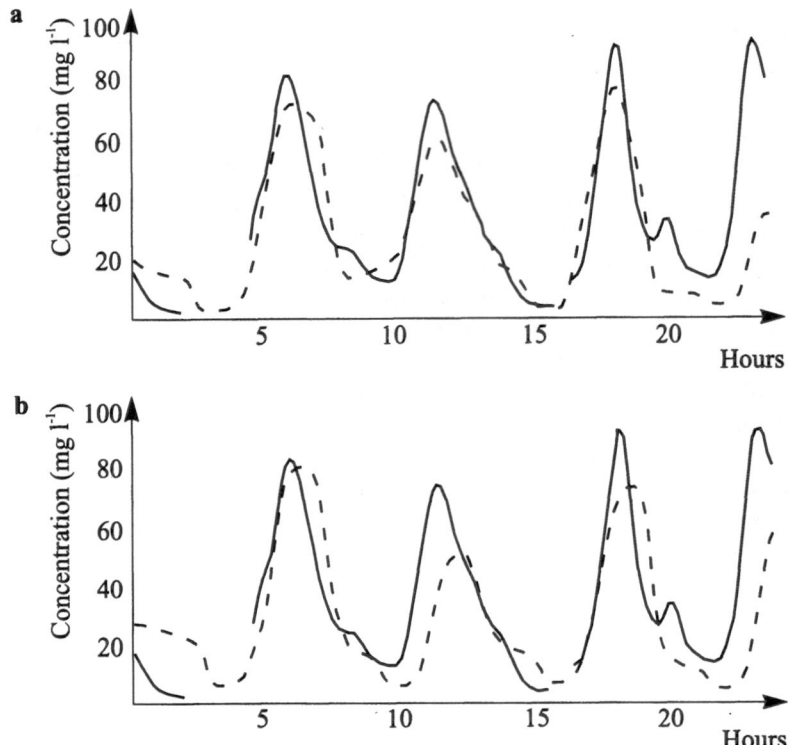

Fig. 13.1: Comparison of observed (solid line) suspended sediment concentration with model predictions (dashed line): **a** full two-dimensional sediment transport model, **b** simplified model with neglected horizontal advection and diffusion (adapted from Clarke and Elliott, 1998)

in which u_* is the friction velocity, u_{*e} and u_{*d} are the erosion and deposition threshold velocities, w_s is the falling velocity, $\bar{c}_b$ is the suspended sediment concentration near the bed, and α is the erodability coefficient.

To parameterize the friction velocities u_*, u_{*e}, and u_{*d}, the logarithmic profile of the tidal velocity was assumed and the vertical structure of the suspended sediment particulate matter concentration $c(z)$ was represented by a modified Rouse profile (Clarke and Elliott, 1998):

$$c(z_1) = c_a \left[\frac{ah}{(h-a)z_1} \right]^W , \tag{13.5}$$

in which z_1 is the height above the bed, a is the height above the bed of the reference level, and $c(z_1)$ and c_a are the sediment concentrations at height z_1

and a, respectively. The power W takes the form:

$$W = \frac{w_s}{\beta \kappa u_*}, \tag{13.6}$$

where w_s is the falling velocity of the suspended sediment, β is a constant, and κ is the Karman constant equal to 0.4.

When the time series of concentration, $\bar{c}$, at a given point is known and relationships for friction velocities are established, the erosion and deposition terms can be applied in Eq. (13.1). The comparison predictions with observations showed that both the hydrodynamics and the sediment load were satisfactory reproduced. The model confirmed that in the Forth Estuary, the horizontal processes are unimportant in comparison to the effects of erosion/resuspension and settling. In Fig. 13.1, a comparison is shown for observations and simulations using a full two-dimensional model (Eq. 13.1) and a simplified model in which horizontal advection and diffusion were neglected. The character of variation in time of the concentrations suggests that horizontal processes are relatively unimportant due to the strong quarter-diurnal signal that is characteristic of local resuspension and deposition.

Extensive data on fine sediment dynamics were obtained in the extremely turbid Jiaojiang River Estuary, China (Guan *et al.*, 1998). These data have been used to calibrate a two-dimensional width-integrated model. The Jiaojiang River Estuary is very shallow with depth 1–3 m at low tide. Semi-diurnal tides have a maximum tidal range of 6.3 m and a maximum vertically-averaged tidal current of 2.0 m/s. At the estuary bed, a cohesive sediment (clay and fine silt with a diameter less than 8 μm) prevails, and the suspended sediment concentration exceeds 40 kg/m^{-3} during spring tides in calm weather.

The governing equations of the model include the mass conservation and momentum equation for water, and conservation equations for salinity and sediments. Horizontal and vertical diffusion and the mass exchange of suspended sediment with the bottom have been parameterized using relationships known in literature. A comparison of measured currents and sediment concentration at three levels above the sea bottom (1.67 m, 0.55 m and 0.35 m) with model predictions is shown in Fig. 13.2. This figure demonstrates cycles of resuspension and settling in the bottom layer where concentration only rarely reaches 10 kg/m^3. A comparison of concentrations at 1.67 m and 0.55 m levels shows the occasional presence of a strong vertical gradient of concentration.

The field data also indicate that sediment concentration during a flood tide is 30% larger than during ebb tide. This asymmetry is responsible for an infilling rate of the estuary of about 0.1 m/year, while measured sediment load from the riverine inflow is too small to be a substantial contributor to the siltation of the estuary. Presumably the origin of such a large amount of sediment, which is infilling the estuary, is the Yangtze River, which has its mouth located 200 km further north (Guan *et al.*, 1998). The Yangtze River, the third largest river in the world, transports 9.3×10^{10} m^3/year of water and 4.7×10^7 tonnes/year

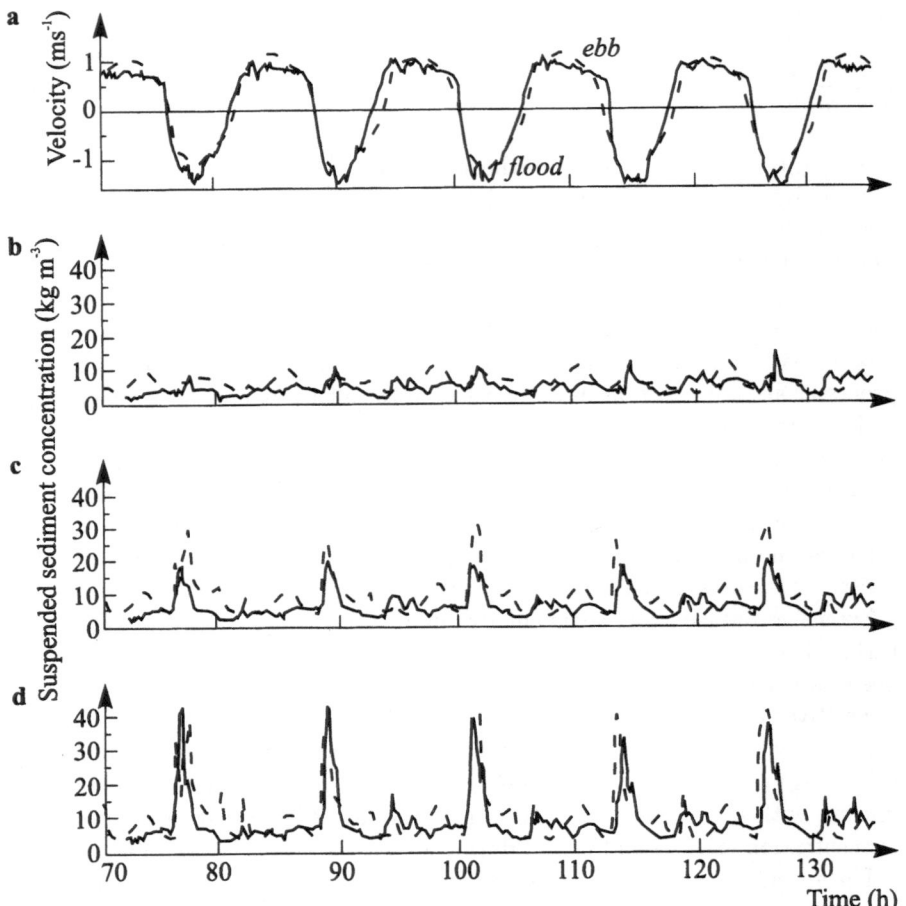

Fig. 13.2: Comparison of observed (solid lines) and predicted (dashed lines) water flow and sediment transport characteristics in Jiaojiang River Estuary at spring tides: a velocity at 1.75 m above sea bed, and suspended sediment concentration at b 1.67 m, c 0.55 m, and d 0.35 m above sea bed (adapted from Guan *et al.*, 1998)

of fine sediments into the sea (Yang, 1998). More than half of the sediment from the river is deposited in the estuary. As a result, wetland area is accreting at a rate of 15–20 km^2/year.

A level of turbidity and suspended sediment concentration, similar to that in Jiaojiang River Estuary, was observed in the Fly River Estuary, Papua New Guinea (Wolanski and Eagle, 1991; Wolanski *et al.*, 1995). High turbidity in the saline region of the estuary appears to be due to the simultaneous occurrence of tidal pumping and the baroclinic circulation due to freshwater discharge which varies little seasonally.

Sediments are important carriers of dissolved and particulate trace elements in the hydrological cycle. The concentration of trace metals depends on the sediment erosion and deposition processes, however, the geochemistry involved in the process of the dynamics of trace elements is beyond the scope of the book.

13.2.3   Dispersion in Non-Vegetated and Vegetated Estuaries

In Chap. 8, we have examined the basic features of diffusion and dispersion phenomena. In particular, as was shown in Sect. 8.3.4, diffusion experiments with the release of rhodamine dye are commonly used to determine the mixing characteristics in deep and shallow waters. Dye experiments have become more and more popular as part of environmental impact assessment studies for such things as industrial releases or the agriculture industry (Elliott *et al.*, 1997). However, the dependence of the mixing coefficients on the local environmental conditions, such as water depth, tidal current and wind speed, usually has to be parameterized, and the dye data are analyzed using the one-dimensional diffusion equation (8.22), which has a solution (for horizontal, x, coordinate):

$$c(x,t) = \frac{Q}{\sqrt{4\pi K_x t}} \exp\left(-\frac{x^2}{4K_x t}\right),$$
(13.7)

where Q is the amount of substance released at time $t = 0$ and K_x is the turbulent diffusion coefficient. Using the arguments from Sect. 8.2, we find that:

$$c(x,t) = \frac{Q}{\sqrt{2\pi(2K_x t)}} \exp\left(-\frac{x^2}{4K_x t}\right) = \frac{Q}{\sqrt{2\pi}\sigma_x} \exp\left(-\frac{x^2}{4K_x t}\right).$$
(13.8)

Thus, the concentration $c(x,t)$ takes the form of a Gaussian curve with variance:

$$\sigma_x^2 = 2K_x t.$$
(13.9)

The spreading due to diffusion can be measured in terms of variance, σ_x^2, with respect to the centroid of the dye patch:

$$\sigma_x^2 = \frac{1}{c_t} \int_{-\infty}^{\infty} c(x)(x - x_0)^2 dx,$$
(13.10)

in which c_t is the integrated concentration across the patch:

$$c_t = \int_{-\infty}^{\infty} c(x) dx,$$
(13.11)

and the centre of contaminant mass, x_0, becomes:

$$x_0 = \frac{1}{c_t} \int_{-\infty}^{\infty} c(x)dx. \tag{13.12}$$

When the diffusion coefficient, K_x, is not constant, it can be estimated from Eq. (8.34) as:

$$K_x = \frac{1}{2} \frac{d\sigma_x^2}{dt}. \tag{13.13}$$

More general, instead of Eq. (13.9) we can assume the variance σ_x^2 in the form:

$$\sigma_x^2 = at^m, \tag{13.14}$$

and respectively:

$$K_x = \frac{1}{2}mat^{m-1}, \tag{13.15}$$

in which a and m are constants that should be determined. A constant value of K_x (with $m = 1$) corresponds to Fick's diffusion process.

Dye patches usually elongate with time and two orthogonal axes can be defined in the along-patch and across-patch directions with corresponding turbulent diffusion coefficients K_x and K_y. As was discussed in Sect. 8.3.3, the elongation of a contaminant patch is the result of the interaction of a Fickian diffusion with the advection due to the tidal- or wind-driven currents. Okubo (1967) found that for steady shear ($\partial u/\partial y = $ Const) in the transverse direction, the along- and cross-patch concentration variances σ_x^2 and σ_y^2 are as follows:

$$\sigma_x^2 = 2K_x t + \frac{2}{3}K_y \left(\frac{\partial u}{\partial y}\right)^2 t^3, \tag{13.16}$$

and

$$\sigma_y^2 = 2K_y t. \tag{13.17}$$

By combining variances σ_x and σ_y we can form the so called radial variance σ_h^2:

$$\sigma_h^2 = \sigma_x \sigma_y, \tag{13.18}$$

and the corresponding turbulent diffusion coefficient, K_h, can be determined from Eq. (13.13).

In the case of vertical shear, the contaminant is usually vertically well-mixed and the variance σ_x^2 becomes (Okubo, 1967):

$$\sigma_x^2 = 2K_x t + \frac{\left(\frac{\partial u}{\partial z}\right)^2 h^4 t}{60 K_z}, \tag{13.19}$$

where K_z is the vertical turbulent diffusion coefficient and h is the water depth.

Elliott et al. (1997) used the above relationships to examine the dependence of the diffusion coefficients on the ambient tidal currents and winds around the coastline of Ireland. Most of the sites were characterized by strong currents and the dye was vertically well mixed. The experiments showed that:

$$\sigma_h^2 \approx 1.1 \times 10^{-6} t^{1.34}. \tag{13.20}$$

The growth of σ_h^2 like $t^{1.34}$ is faster than the $t^{1.0}$ growth that would be expected for a Fickian process. The ratio of the turbulent diffusion coefficients K_x and K_y, is $1/12$ and the patches are typically 3.5 times longer than they are wide.

In shallow waters, the flows are affected by shear due to lateral current differences near shore or due to vertical shear generated by tidal currents and winds. The dependence of coefficient K_h (in m^2/s) takes the form:

$$K_h = 0.03 + 1.03U + 0.04 V_w, \tag{13.21}$$

where U is the depth-averaged current (in m/s) and V_w is the wind speed (in m/s). At least for the Elliott et al. (1997) experimental conditions (generally light winds), the wind-induced mixing is of secondary importance.

In a case when vertical current dominates over the lateral shearing and the water is vertically well mixed, the diffusion coefficient K_x becomes (Elder, 1959):

$$K_x = aUh, \tag{13.22}$$

where a is a constant.

In estuaries, where the cross-sections and bathymetry vary in a complicated manner, the concepts of turbulent diffusion and shear dispersion usually yield dispersion coefficients which are smaller than observed values. The main reason for this discrepancy is the fact that the shear in the velocity field is not uniform over distances comparable with the tidal excursion and the tide-topography interaction generates significant horizontal residual circulation superimposed on the main water flow. To cope with these circulations, de Swart et al. (1997) proposed the tidal random walk model which relates the mixing properties of the flow to the velocity and length scales of the tidal current and the residual eddies. Using this model, the longitudinal and lateral dispersion coefficients have been computed for the 75 km long Ems Estuary which is a part of the border between the Netherland and Germany. The tidal range varies from

2.3 m near the tidal inlet, to 3.2 m upstream, near the town of Emden. The influx of freshwater is of the order of 115 m^3/s. As a result, the average salinity at the inlet is about 30 ppm and decreases to zero 75 km upstream.

The particle displacement in a random walk type model is a superposition of the net displacements of particles during the preceding ebb and flood phases. Due to interaction of the particles with residual eddies, particle trajectories move in an irregular way and are generally random functions of time. Using the random particle displacements, the longitudinal and lateral dispersion coefficients can be defined as follows (de Swart et al., 1997):

$$K_x = \frac{1}{2}\frac{d}{dt}\left(\overline{l_x^2}\right), \quad K_y = \frac{1}{2}\frac{d}{dt}\left(\overline{l_y^2}\right), \tag{13.23}$$

where l_x and l_y are random displacements in longitudinal and lateral directions, respectively. An upper bar denotes an ensemble average. The random walk model predicts longitudinal dispersion coefficients in the estuary as high as 200–12,000 m^2/s, while the lateral dispersion coefficients are much smaller, of the order of 5–30 m^2/s. If the tidal and residual flow characteristics in the estuary are known, the random walk model provides a simple method to estimate the dispersion characteristics in an estuary with complex bathymetry.

Many estuaries, especially in the tropics, are vegetated to various extents. When vegetation is present, hydrodynamics and sedimentation become affected. Due to highly irregular streamlines, the flow through a dense population of surface-piercing plants becomes slow. For example, the bottom velocity and suspended sediment concentration in the intertidal seagrass community of the Corner Inlet, Australia, decreases to 40% and 60%, respectively (Zhuang and Shebel, 1991). The stems and leaves of *Scripus* marshes in the Yangtze River Estuary are able to trap about 300 g/m^2 of sediments, and sediments in marsh area are much finer than those in the adjacent flat (Yang, 1998).

In Chap. 8, we showed that the most familiar form of dispersion is the shear-flow dispersion, described by Taylor (1953). The interaction of non-uniform advection and cross-stream diffusion enhances longitudinal spreading. Non-uniform advection changes the local distribution of contaminant and intensifies the vertical and lateral concentration gradients. On the other hand, molecular or turbulent diffusion tries to reduce these gradients. The resulting effect of these competing mechanisms is to increase the longitudinal length of the contaminant path. After some initial time, the effects of shear-flow dispersion become analogous to diffusion and can be modelled as a Fickian process.

It should be expected that the presence of vegetation changes the dispersion of the contaminant. Let us for simplicity assume that the estuary is laterally uniform and the one-dimensional, advection-diffusion equation (8.54) applies, *i.e.*:

$$\frac{\partial \bar{c}}{\partial t} + \bar{u}\frac{\partial \bar{c}}{\partial x} = K_x\frac{\partial^2 \bar{c}}{\partial x^2}, \tag{13.24}$$

in which $\bar{c}$ and $\bar{u}$ are the cross-sectional averaged concentration and stream velocities, respectively, and x is the longitudinal spatial dimension. In a laterally uniform flow, only vertical shear contributes to the shear dispersion. Thus, using Taylor's (1953) approach (see Sect. 8.3.3), we can express the dispersion coefficient, K_x, as follows: (Elder, 1959):

$$K_x = -\frac{1}{hK_z} \int_{-h}^{0} u'(z) \int_{-h}^{z} \int_{-h}^{z_2} u'(z_1)\, dz_1\, dz_2\, dz, \tag{13.25}$$

in which h is the water depth, u' is the deviation of the velocity, $u(z)$, from the mean velocity, and K_z is the turbulent diffusion coefficient.

In vegetated estuaries additional dispersion mechanisms arise due to the physical obstruction associated with plant stems and wakes behind the stems. These mechanisms are known as mechanical dispersion and they are common in porous media or in flow through random media, for example, flow in mangrove forests (Massel et al., 1998).

In general, the resulting dispersion coefficient, K_x, depends on the biological morphology and density of the vegetation in the estuary. Nepf et al. (1997a), studied the influence of vegetation on longitudinal diffusion in a laboratory flume, for various flow velocities and population densities. Spartina attermiflora, the dominant marsh grass in much of Eastern North America, was modelled with 0.6 cm diameter hard-wood dowels. Rhodamine was injected continuously upstream of the dowel array. For the no-dowel cases with velocities 2.9, 5.5 and 7.4 cm/s, respectively, the observed dispersion coefficient, K_x, was found to be 7.5, 7.3 and 8.4 cm^2/s. However, for 5.5% dowel density, the coefficient, K_x, drops to about 1.2 cm^2/s. The influence of the mechanical dispersion appears to be small in comparison with the shear dispersion.

Detailed laboratory velocity measurements of Nepf et al. (1997b) showed that the production of turbulence within a stand of emergent vegetation is dominanated by the stem wakes rather than by the bottom boundary shear. Using this observation, they formulated a random walk model which can be used to determine the contribution of stem wakes to the turbulent diffusivity within a plant canopy.

The influence of mangrove swamps on the longitudinal dispersion in mangrove-fringed tidal creeks was examined by Ridd et al. (1990). The effect of turbulent diffusion was found to be negligible compared with the dispersion due to the trapping effect of the mangroves. The longitudinal dispersion coefficient was proportional to the square of the water velocity. Therefore, at the creek head, the mixing rates are very small. The resulting residence time of contaminants becomes longer for water close to the head of the creek.

13.2.4 Influence of Mixing on Primary Production

Definition of Primary Production. Productivity of biomass is the basic biological quantity measured in the aquatic environment. In this chapter, as well as in Chaps. 14 and 15, we will examine the relationships between primary production and physical factors for different regions. Prior to this, we will define the basic quantities and briefly summarize the relationships between primary production and physical processes that affect the production.

For the purpose of this book we adopt a simplified definition of **primary production** as a process of building plant tissue by photosynthesis. As a result, the inorganic material (e.g. nitrate, phosphate) is converted into new organic compounds (e.g. lipids, proteins). The most popular method for measuring productivity in the ocean is the ^{14}C method. The details of this method can be found in many biological textbooks (for example, Valiela, 1995). Here we only note that the productivity is expressed as the amount (in mg) of carbon fixed in new organic material per volume of water and per time. The rate of plant material production varies from zero to as much as about 80 mg $C/m^3/hr$ (Lalli and Parsons, 1997).

As chlorophyll a is universally present in all species of phytoplankton, one obtains a measure of the growth rate in units of time (mg C per mg chlorophyll a per hour). Because phytoplankton, which is the dominant primary producer, vary greatly in size, primary production is sometimes defined in terms of biomass, $i.e.$ as the total weight of all organisms in a given area or volume.

A distinction must be made between gross and net primary production. The total amount of photosynthesis achieved by an organism during a certain period of time is the gross production. However, during that time the organism is also carrying out respiration. The difference between gross production and respiration in that time is the net production.

Hydrodynamic Factors and Aquatic Productivity. The intensity of light and concentration of nutrients are two principal mechanisms controlling primary production. Light intensity decreases with depth as the short wave radiation is absorbed in the water column, $i.e.$ (Fig. 13.3):

$$I = I_0 e^{\alpha z}, \tag{13.26}$$

in which I_0 is the light intensity at the sea surface, z ia the vertical coordinate directed upward from the sea surface, and α is the extinction coefficient (Kirk, 1994; Dera, 1995).

Nutrient concentration is generally higher in the lower layers and lower in the upper layers of the ocean. When the surface layer is mixed by wind, a nutricline exists at the boundary between the upper and lower layers. The nutricline is a zone where nutrient concentrations increase rapidly with depth. It can be located below the euphotic zone which is the region where there is sufficient light to support the growth and reproduction of plants. A well-mixed upper zone is typical for oceans in temperate latitudes during summer time, and almost permanent in tropical seas.

The light needed for photosynthesis is generally only available in the upper layer, but the nutrients needed for growth are concentrated in the lower

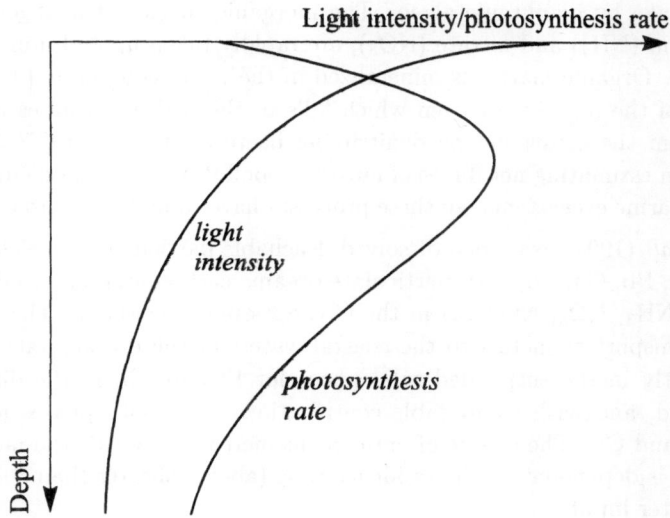

Fig. 13.3: Schematic vertical distribution of light attenuation and photosynthesis rate

layer. There are several processes which break down the barrier and transport nutrient-rich water from the lower layer to the surface. These include mixing, upwelling fronts and eddies formation. The influence of mixing on the primary production in coastal waters and estuaries is described in this chapter, while frontal and upwelling processes in the shelf and deep waters will be discussed in Chap. 15.

The resulting vertical distribution of photosynthesis is sketched in Fig. 13.3. Photosynthesis may occur to a depth of 100 m in very clear oceanic water, however, in very nutrient-rich waters photosynthesis may be restricted to the upper few centimetres only. The resulting phytoplankton production varies over the year, depending on light and nutrient availability.

Mixing Basis of Nutrient Transport and Primary Production. Hydrodynamic processes are the basic agents which alter the concentration and spatial distribution of biologically important nutrient materials, metals and pollutants. A transition from freshwater to salt water affects the concentration of many elements and ions carried by rivers (Furnas, 1996). Iron, which is a biologically essential element in terrestrial soils and rocks, is relatively soluble in freshwater, however, it is highly insoluble in sea water with a high content of oxygen. On the other hand, for example copper, a highly toxic element to marine life at some threshold concentration, is much less affected in the transition from fresh- to sea water.

Two other elements, phosphorus and nitrogen, which are of fundamental importance for biological life in estuaries, undergo a number of transformations in estuaries. Some portion of available phosphorus is dissolved and is available for uptake by planktonic algae. The inorganic species of nitrogen, such as ammonium (NH_4) and nitrate (NO_3), are rapidly taken up and mineralized by plankton. Organic matter is mineralized in the water column and the benthos. Half of the organic nitrogen which falls to the bed in estuaries may be removed from the ecosystem by denitrifying bacteria (Furnas, 1996; Alongi, 1998). When estimating net fluxes of nutrients, pollutants or other compounds in coastal marine ecosystems, all these processes have to be taken into account.

Muller *et al.* (1994) examined dissolved, leachable particulate trace elements (Fe, Mn, Zn, Pb, Cd, Co, Cu), particulate organic carbon and dissolved nutrients (PO_4, NH_3, NO_3, NO_2, Si) in the Clyde Estuary, Scotland. They found that the transport of metals to the coastal waters of the estuary takes place predominantly in the suspended solid phase for Fe and Pb, in the dissolved phase for Cd, and with comparable contributions from both phases for Mn, Zn, Co, Ni and Cu. The degree of enrichment increases towards the mouth of estuary and is dependent on the residence time (always shorter than 3–7 days) and freshwater input.

In Sect. 8.5.1 we have examined various types of estuaries. In particular, in shallow and narrow estuaries dominated by high river discharge, a salt-wedge occurs. The velocity shear between the freshwater and salt water flows induces an entrainment process. For the upward entrainment in estuaries, salinity increases in the upper layer. The dynamics and mixing of nutrients is a function of the entrainment mechanisms and the concentration of source nutrients. When a river has a lower nutrient concentration than the ocean water, entrainment results in an increase of nutrient concentration in the surface layer. However, when there is higher nutrient concentration in the river, the entrainment of the sea water results in dilution. The entrainment mechanism, which is supposed to play an important role in nutrient supply, was examined by Yin *et al.* (1995a, b, c) using data collected in the Fraser River estuary (British Columbia, Canada). The Fraser River is one of the largest rivers along the west coast of North America. River discharge reaches its maximum of 10 000 m^3/s in June, and gradually decreases to about 700 m^3/s in September, remaining at that level until March, when it starts to increase again. The river discharge is modulated by tides, when a salt wedge invades the river during tidal flood and retreats during tidal ebbs.

According to Yin *et al.* (1995a), in the Fraser River estuary there are three water masses: the riverine plume, the estuarine plume, and deep water (Fig. 13.4). The mixing of nitrate, NO_3, depends on tides and river discharge. During a flood tide, the salt wedge invades the river at depth and the discharge of freshwater decreases. The estuarine plume is pushed into the river at the surface (Fig. 13.4a). Little upward mixing takes place during the advance of the salt wedge. During a tidal ebb, when the salt wedge starts to retreat, intensive entrainment takes place between both layers, resulting in broadening of the halocline. After the salt wedge is broken down and swept out of the river,

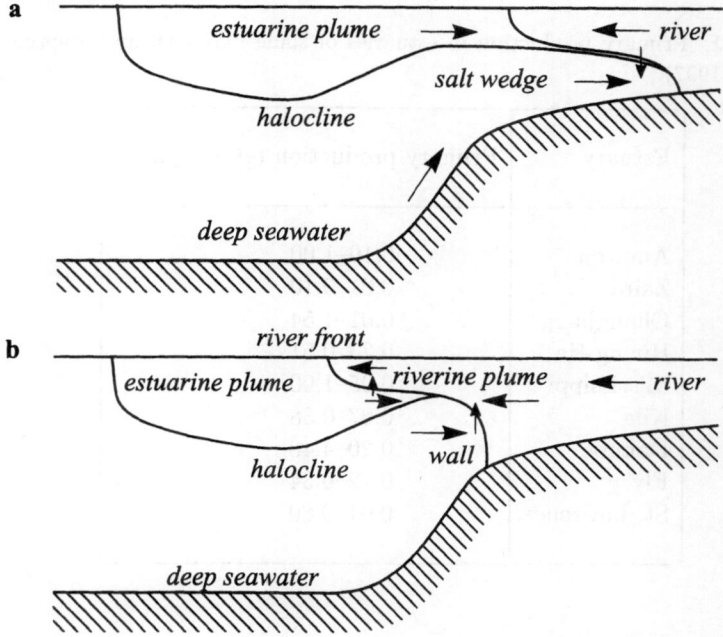

Fig. 13.4: Schematic representation of various water masses in estuary during: **a** flood, and **b** ebb tides (adapted from Yin *et al.*, 1995a)

a steep wall-like structure of sea water of high salinity is formed at the river mouth (see Fig. 13.4b).

Yin *et al.* (1995b) have found that more NO_3 was entrained during the spring tide (24 mmol/m^2) than during the neap tide (17 mmol/m^2). The reason of such a situation is the river discharge, pushing the estuarine plume seaward, exposing a larger area of deep sea water to the riverine plume for entrainment. The spring tides also cause stronger bottom stirring which results in higher NO_3 concentration in the deep sea water which is the source of NO_3 nitrate entrained upwards.

The surface wind may be another agent responsible for entrainment of NO_3 in the estuaries. It is expected that shear between the surface layer (the riverine plume or the estuarine plume) and the water beneath would be increased under strong winds. When this shear is strong enough, the stratification may be destroyed and mixing across the pycnocline would appear. Usually the gradient Richardson number Ri type criterion (see Chap. 8) is used to indicate the strength of the wind induced shear over the stability of a density gradient. In particular, the gradient Richardson number takes the form:

$$Ri = \frac{g}{\rho} \frac{d\rho}{dz} \left(\frac{du}{dz} \right)^{-2},$$

$$(13.27)$$

Table 13.1: Primary production in estuaries of some major rivers (adapted from Humborg, 1997)

Estuary	Primary production ($gC/m^2/day$)
Amazon	0.10–1.00
Zaire	0.01–0.10
Changjiang	0.01–1.54
Huang He	0.23–0.51
Mississippi	0.05–1.00
Nile	0.07–0.38
Danube	0.20–4.40
Fly	0.22–0.34
St. Lawrence	0.01–0.80

in which velocity, u, is assumed to be parallel to the surface flow. The Ri value of 0.25 is critical, below which the velocity shear is strong enough to overcome the density gradient, which may result in entrainment. Yin *et al.* (1995c) found that the amount of entrained NO_3 under windy conditions is about 44 $mmol/m^2$, while under weak winds it is 16 $mmol/m^2$.

Estuaries are considered as highly productive ecosystems because they are often nutrient rich and have multiple sources of organic carbon to sustain population of heterotrophs. However, primary production in estuaries varies widely throughout the world, as shown in Table 13.1. There are several factors which contribute to the observed high estuarine productivity, such as (Alongi, 1998):

- consortia of phytoplankton, macrophytes, and benthic microalgae that maximize available light and space,
- tidal energy and circulation,
- abundant nutrients, and conservation and
- retention and efficient recycling of nutrients among benthic, wetland and pelagic habitats.

These factors are dependent on the residence time of water within the estuary. Residence times for water and plankton are determined by the rates of freshwater input, the pattern of the estuary, the presence of adjoining wetlands, for example, mangrove swamps and the volume of the tidal prism.

Water mixing and associated turbidity reduce light availability via sediment resuspension, and phytoplankton photosynthesis is confined to a shallow photic zone. The phytoplankton biomass depends on the suspended particulate matter concentration and the light limitation prevents phytoplankton blooms when sediment concentration exceeds 50 mg/l (Cloern, 1957). The influence of resuspension of estuarine sediments on the microphytobenthos in the Ems Estuary was examined in detail by de Jonge (1992).

Ragueneau *et al.* (1996) have compared the dynamics of phytoplankton blooms in two coastal ecosystems of Western Europe, which differ in the influence of river discharge on the vertical stratification of the water column. In the Bay of Brest (a semi-enclosed ecosystem connected to the adjacent ocean and entered by two nutrient-rich rivers), where light is the triggering factor of production, the plankton blooms occur during neap tides. A relaxation of vertical mixing during those tides enables the utilization of nutrients originating from freshwater inputs. On the other hand, in the English Channel, the light is not a limiting factor and increased mixing during spring tides enables nutrient replenishment from the sea bottom. This difference in functioning of both ecosystems is reflected in the macrozoobenthos ecostructure, dominated by filter-feeders in the English Channel and suspension-feeders in the Bay of Brest (Ragueneau *et al.*, 1996).

In salt-wedge type estuaries, where river discharge floats on the denser sea water, light conditions in the upper layer improve. Nutrient concentrations remain one order of magnitude higher than in mixed sea water, providing the basis for substantial primary production. Humborg (1997) has found that the Danube estuary, due to its shallowness of the upper layer and strong stratification and high nutrient concentration, provides favourable conditions for high primary production. In fact, he obtained a long-term average value of about 0.5–1.0 gC/m^2/day, and up to 4 gC/m^2/day for the estuary plume (Table 13.1). In the salt-wedge estuary Danube tides are negligible and circulation is mainly governed by the river flow and prevailing winds. High nutrient concentration and sufficient light conditions result in very high primary production in the estuary, comparable to, for example, the Amazon estuary.

In contrast, the Zaire estuary is well mixed and the freshwater plume is instantaneously mixed by tidal currents. Hence, the maximum primary production is moved to the outer edge of the estuary at high salinities. However, there production is limited by nutrient concentrations from previous dilution (Humborg, 1997).

Primary production in large estuaries varies between the seasons of high and low nitrate concentrations, high and low river discharge, and shallow or deep location of the pycnocline. During a particular season, the biomass and production of phytoplankton depend on the nutrient availability, temperature and maintenance in the upper water column. Tremblay *et al.* (1997) have found in the Lower St. Lawrence Estuary, Canada, a positive correlation between phytoplankton biomass and the Brunt-Väisälä frequency $N(z)$:

$$N(z) = \left\{ -\frac{g}{\rho}\frac{d\rho}{dz} \right\}^{1/2}. \tag{13.28}$$

In particular, the correlation coefficient between frequency, N, and chlorophyll a at 2 m depth significantly increases with phytoplankton particle size. The mean Brunt-Väisälä frequency was calculated as an index of buoyancy restoring forces over the upper 50 m, $i.e.$:

$$\frac{d\rho}{dz} \approx \frac{\Delta\rho}{\Delta z}, \tag{13.29}$$

in which $\Delta z = 50$ m. A positive size-dependent relationship between phytoplankton biomass and the frequency, N, suggests that the observed variability in the large size fraction can be explained by their residence time in the euphotic zone.

For the case of more complicated ecosystems which comprise of sea, lagoon and rivers, two or three-dimensional coupled hydrodynamic and biochemical models are required. A classic example of such an ecosystem is the Venice Lagoon, Adriatic Sea system (Bergamasco et al., 1998). Venice Lagoon, located on the northern coast of the Adriatic Sea, has an area of about 450 km², 75% of which is very shallow (less than 2 m), and only three channels, deeper than 6 m, join the lagoon to the sea. Small rivers which flow into lagoon supply nitrogen and phosphorus of agricultural, industrial and municipal origin.

Hydrodynamics of the lagoon-sea ecosystem were modelled using the Princeton Ocean Model (Mellor, 1991). The resulting circulation model allows the simulation of the three-dimensional fields of velocity, temperature, salinity, and density under the forcing induced by astronomical and meteorological tides, winds, and thermal and evaporative fluxes. This model is coupled with the biochemical model which describes an evolution of the concentration of various substances (Bergamasco et al., 1995):

$$\frac{\partial c_i}{\partial t} + u\nabla c_i = \nabla\left(K_i\nabla c_i\right) + S_i(c, t), \tag{13.30}$$

in which c_i is the concentration of a particular substance i, K_i is the corresponding diffusion coefficient, and S_i are the parametric and functional expressions for modelling of primary production. Simulation results include the spatial and temporal variations of the concentration of nutrients (mg N/l), phytoplankton (mg C/l) and oxygen (mg O_2/l).

13.3   Dispersion of River Plumes

Rivers are the main conduits of water, particulate and dissolved organic and inorganic matter from the continents to the oceans. When freshwater and nutrient materials pass through the estuary, they do not disperse instantaneously into the adjoining ocean water, but rather form a buoyant plumes of river water which, under the influence of the longshore current, flow along the coastline.

The annual discharge of the world's rivers is about 38×10^3 km^3 of freshwater and about 26×10^9 metric tonnes of sediments (Alongi, 1998). More than half of the world's river water enters the tropical coastal regions in the Indo-Pacific margins or in northeastern South America and west-central Africa. A great variety of physical processes, driven by the tropical climate, make the tropical coastal margins unique compared to coastal zones in the temperate latitudes. The climate of the tropical regions is characterized by sustainable high rates of rainfall, sunlight and temperature. The enormous discharge of tropical rivers, such as the Amazon (5700 km^3/year), or Zaire (1292 km^3/year), and the transported sediment produce extensive river plumes, in some areas extending beyond the shelf edge.

Observations in the Caribbean Sea show regular, seasonal salinity fluctuations, with the fall of salinity coincident with increased concentrations of dissolved silicate (Froelich *et al.*, 1978). These salinity fluctuations are related to the Amazon River discharge. For example, Gibbs (1976) has observed the Amazon plume to extend northwestward from the mouth of the river along the continental shelf before turning offshore. These large lenses of Amazon River water may be subsequently transported by the North Equatorial and Guiana Current systems into the Caribbean Sea. Calculations by Froelich *et al.* (1978) indicate that as much as 60% of the freshwater seasonally entering the Caribbean Sea is of Amazon and Orinoco origin.

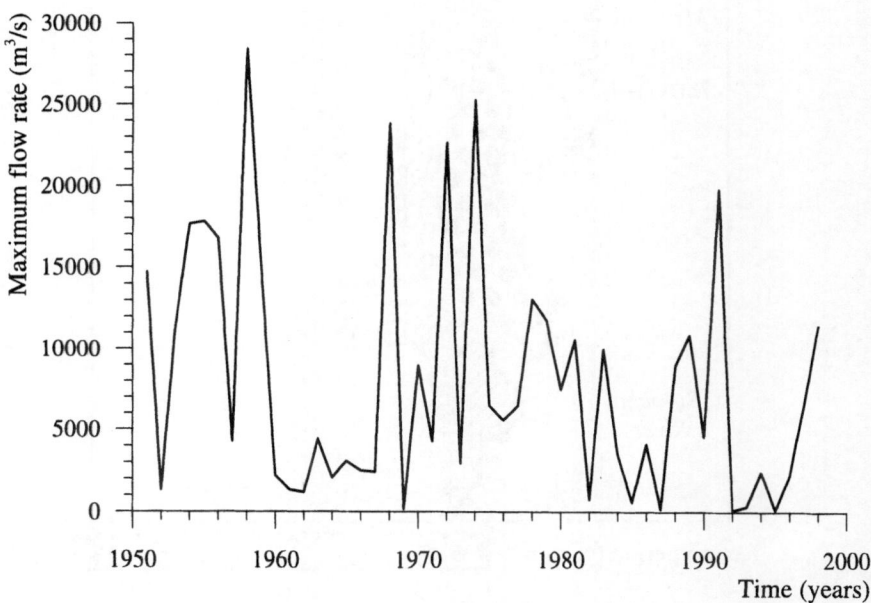

Fig. 13.5: Histogram of Burdekin River maximum flow rates (courtesy of King, 1998)

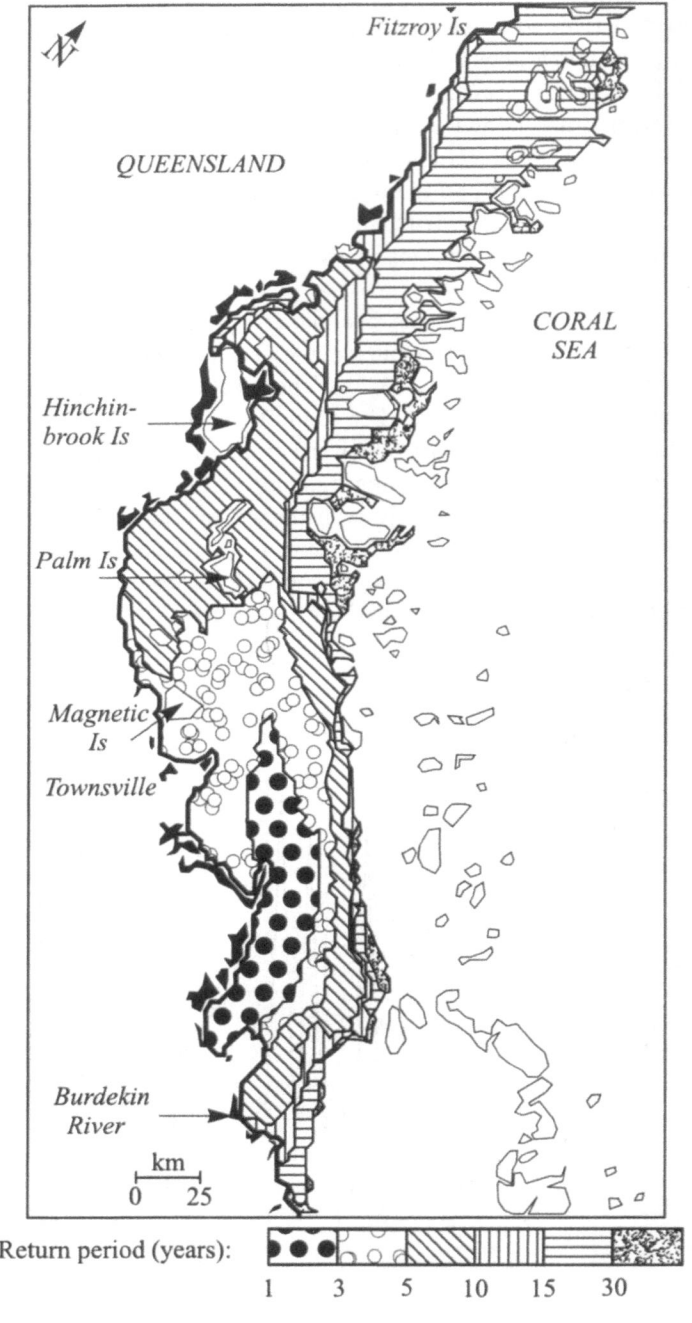

Fig. 13.6: Return period distribution for the event of 10% freshwater exceedance over 120 hour duration in the Great Barrier Reef lagoon (courtesy of King, 1998)

Transport of various matter and biochemical interactions within the river plumes are the key factors for primary production and global geochemical cycling. This transport controls the input of land-derived natural and pollutant chemicals into shelf seas (Morris *et al.*, 1995). In some locations in the tropics, such as the northern east coast of Australia, river catchments are small and major flood events are associated with the activity of monsoonal depressions or tropical cyclones. As tropical cyclones move onto the continental shelf and coastal regions, the whole water column is extensively mixed, and bottom sediments are resuspended together with dissolved and particulate nutrients. Nutrients mineralized by bacteria trigger regional phytoplankton blooms. The resulting plankton biomass and primary productivity can increase 5–10 times within a few days after a cyclone (Furnas, 1996).

The Burdekin River, Australia, is the largest river in the Great Barrier Reef region with mean annual discharge of 9.7×10^9 m^3 into the Great Barrier Reef lagoon (Wolanski, 1994). The river runoff is highly variable and limited to occasional flood events, usually occurring during the Australian summer months of December to March. The Burdekin River peak discharges are of the order of 10,000 m^3/s to 30,000 m^3/s (see Fig. 13.5).

Data on the fate of flood waters during and after large discharge events of the Burdekin River are limited. King *et al.* (1997) used the data set on salinity distribution after the 1981 flood event, reported by Wolanski and van Senden (1983), to calibrate and verify a three-dimensional hydrodynamic model of the Burdekin River in flood. They used the NOAA MECCA model which incorporates river plume dynamics into the governing equations. The model predicts flows due to tides, winds and density difference forcing, as well as the salinity and temperature distributions. The entire 1981 flood event period (90 days) has been simulated and compared with experimental data. Agreement between the observed and predicted salinity distribution was very good.

Detailed information on the fate of the Burdekin River plume is of particular importance for the determination of the impact of low salinity waters on the adjacent coral reefs. Low salinity water affecting corals for prolonged periods of time may cause irreversible coral damage. Therefore, a risk assessment analysis is needed to quantify the impact of river floods on the Great Barrier Reef. In Fig. 13.6, an example of preliminary results of the risk analysis is given. Isolines in the figure indicate the particular return periods (in years) of events when the exceedance of 10% freshwater is of the duration of 120 hours. Thus, close to the Burdekin River mouth, a return period of such event is about 3 years, while for reefs close to the Fitzroy Island it is about 30 years.

13.4 Larval Settlement and External Fertilization

13.4.1 Settlement of Larvae

In Chap. 11, we have examined various mechanisms used by aquatic organisms to move through the water column. Basically, organisms generate the thrust necessary to overcome the resistance forces. However, many aquatic organisms are not able to generate any thrust and they are passively transported by water movement. In this section we will examine mechanisms through which flow in the coastal zone results in the transport of the planktonic larvae of benthic species, such as a barnacle cyprid, the pluteus of a sea urchin, or gametes and spores. Especially in the surf zone where water motion is very turbulent, the sinking or swimming velocities of larvae and spores are small (from about 0.0001 to 0.01 m/s), compared to the ambient water fluctuations. The organisms have little or no control over the rate at which they are transported to the substratum. However, the biology of larvae requires that the organisms must return to the substratum before they can metamorphose and grow into adults (Denny and Shibata, 1989).

Let us consider a larva or spore travelling at a height, z_1, above the substratum. The water is turbulent and well mixed throughout the depth. The diffusivity is assumed to be uniform in the horizontal direction. Because the size of eddies is constrained by the presence of the nearby substratum, the vertical turbulent diffusion coefficient is much smaller than the horizontal diffusivity.

Using the logic by Berg (1983) in his examination of the mechanism of 'diffusion to capture', Denny and Shibata (1989) have found that the ensemble average time that it takes the larva to reach the substratum from height z_1 becomes:

$$T(z_1) = \frac{z_1 - z_1 \ln z_1 + z_1 \ln h}{\kappa u_*}, \tag{13.31}$$

in which u_* is the friction velocity (see Eq. 2.52) and κ is the von Kármán constant.

The solution (13.31) is valid under an assumption that when a neutrally buoyant larva reaches the water surface, it is immediately 'reflected' from the water's surface and mixed back into the water column. However, when larvae contact the substratum, they adhere to it. Also in development of the solution (13.31) it was assumed that the turbulent diffusion coefficient, K_z, increases with height above the substratum as:

$$K_z = \kappa u_* z. \tag{13.32}$$

For the surf zone, instead of Eq. (13.32), Svendsen (1987) suggested use of:

$$K_z = \kappa u_* z \exp\left(\frac{2z}{h}\right). \tag{13.33}$$

A typical settlement time, $T(z_1)$, for water depth $h = 2$ m, and for different values of velocity u_* is shown in Fig. 13.7. For a very small distance, z_1, it is only a matter of a few seconds, on average, to reach the substratum. The settlement time is much longer for a larva or spore originating at the water's surface.

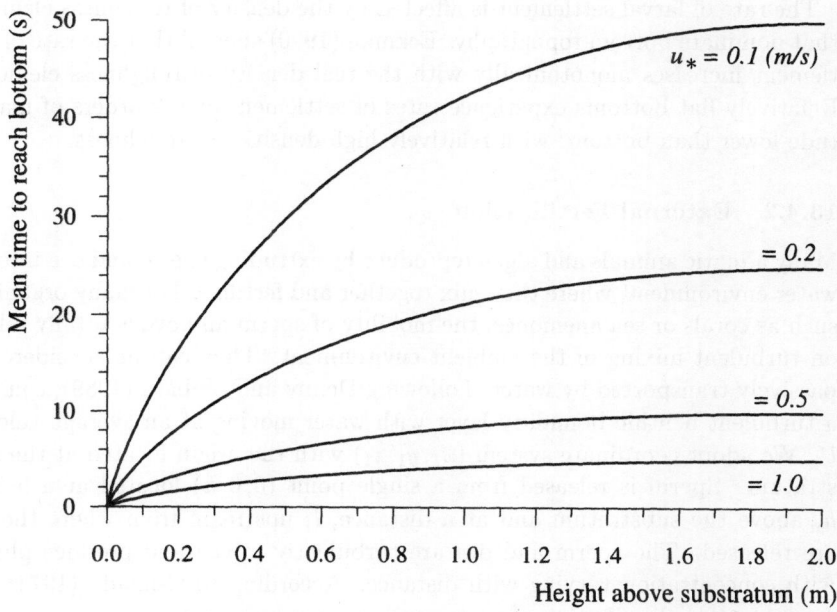

Fig. 13.7: Mean larvae settlement time as a function of height above substratum for various friction velocities

A mean time, $\bar{T}$, of settlement can be found by averaging over the heights from $z_1 = 0$ to $z_1 = h$:

$$\bar{T} = \frac{1}{h}\int_0^h T(z_1)dz_1 = 0.75\frac{h}{\kappa u_*}. \tag{13.34}$$

As should be expected, the average time, $\bar{T}$, increases directly with water depth and inversely with the friction velocity. For a surf zone, when the turbulent diffusion coefficient, K_z, is given by Eq. (13.33), the mean time $\bar{T}$ becomes:

$$\bar{T} = 2.15\frac{h}{\kappa u_*}. \tag{13.35}$$

In fact, the average velocity with which the larva or spore is transported to the substratum by turbulent mixing alone is:

$$\bar{V} = \frac{h}{\bar{T}} = 0.46\kappa u_*. \tag{13.36}$$

In the surf zone, the friction velocity κu_* varies from 0.7 m/s to 1 m/s, which is much higher velocity than the typical sinking velocity of planktonic larvae (typically 0.001–0.01 m/s).

The rate of larval settlement is affected by the density of roughness elements that dominate bottom topography. Eckman (1990) showed that the rate of settlement increases monotonically with the real density of roughness elements. Relatively flat bottoms experience rates of settlement of 1–2 orders of magnitude lower than bottoms with relatively high densities of roughness.

13.4.2   External Fertilization

Many aquatic animals and algae reproduce by extruding sperm and ova into the water environment, where they mix together and fertilize. For many organisms such as corals or sea anemones, the mobility of sperm and ova is totally reliant on turbulent mixing of the ambient environment. They can be considered as passively transported by water. Following Denny and Shibata (1989), consider a turbulent benthic boundary layer with water moving at an average velocity $\bar{U}$. We adopt coordinate system (x_1, y_1, z_1) with the origin located at the substratum. Sperm is released from a single point $(0, 0, h)$ located at a height, h, above the substratum and at a distance, l, upstream from where the ova are released. The sperm and ova are turbulently mixed and produce plumes with concentration varying with distance. According to Csanady (1973), the concentration distribution of sperm and ova can be represented as:

$$
c_S(x_1, y_1, z_1) = \frac{Q_s}{2\pi \bar{U} \sigma_y(x) \sigma_z(x)} \left\{ \exp\left[-\frac{y_1^2}{2\sigma_y^2(x)} - \frac{(z_1 - h)^2}{2\sigma_z^2(x)} \right] + \\
+ \exp\left[-\frac{y_1^2}{2\sigma_y^2(x)} - \frac{(z_1 + h)^2}{2\sigma_z^2(x)} \right] \right\} \tag{13.37}
$$

and

$$
c_E(x_1, y_1, z_1) = \frac{Q_e}{2\pi \bar{U} \sigma_y(x) \sigma_z(x)} \left\{ \exp\left[-\frac{y_1^2}{2\sigma_y^2(x)} - \frac{(z_1 - h)^2}{2\sigma_z^2(x)} \right] + \\
+ \exp\left[-\frac{y_1^2}{2\sigma_y^2(x)} - \frac{(z_1 + h)^2}{2\sigma_z^2(x)} \right] \right\}, \tag{13.38}
$$

in which c_S and c_E are concentrations of sperm and eggs, respectively, and Q_s and Q_e are the corresponding rates (in number per second) at which sperm and eggs are released, respectively. The standard deviations, $\sigma_y(x)$ and $\sigma_z(x)$, can be modelled in terms of the friction velocity, u_*, and the mean velocity, $\bar{U}$ (Csanady, 1973):

$$
\left. \begin{aligned}
\sigma_y(x) &= \frac{\alpha_y u_*}{\bar{U}} x^\beta \\
\sigma_z(x) &= \frac{\alpha_z u_*}{\bar{U}} x^\beta
\end{aligned} \right\}, \tag{13.39}
$$

in which α_y, α_z and β are empirical coefficients chosen to best match these relationships. Using the fact that $x = \bar{U}t$, the relationships (13.39) can be rewritten in terms of time. For short distances, the coefficient $\beta \approx 1$.

Denny and Shibata (1989) argued that the rate at which ova are fertilized is given by:

$$\frac{dc_E}{dt} = -\varphi c_S(t)c_E(t),$$ (13.40)

where $c_S(t)$ and $c_E(t)$ are instaneous concentrations of sperm and eggs, respectively, and φ is the proportionality coefficient. They used sea urchins as an example to provide some values for the biological parameters of the model. In particular, they assumed that the rate of sperm and egg release was $Q_s = 1 - 3 \times 10^6/s$ and $Q_e = 10^4/s$, respectively. Both have been released at 5 cm above substratum and the distance, l, separating male and female was equal to 0.1 m. Mean and friction velocities, $\bar{U}$ and u_*, were assumed as 1.1 m/s and 0.11 m/s, respectively. The results of the model showed that the fraction of eggs fertilized is low, less than 1% for the parameters used. Even when the distance, l, is as small as 1 cm, the fraction of eggs fertilized is only 3.9% and decreases with increasing values of mean velocity, $\bar{U}$. The model results compare favourably with experimental data of Pennington (1985).

As was shown in the previous section, turbulent mixing provides an effective mechanism for rapid transport of larvae to the substratum, even in the absence of sinking velocities. However, in the case of external fertilization, the rapid dilution of gametes by turbulent mixing may drastically reduce the efficiency of external fertilization with some negative consequences for animal reproduction.

13.4.3 Dispersion of Coral Eggs Following Mass Coral Spawning

The distribution and dispersion of planktonic larvae released by benthic organisms are of fundamental importance to the ecology of aquatic benthic communities. On the Great Barrier Reef, corals engage in mass spawning during a brief period each year. Clear patterns in the timing of mass spawning have been identified and this provides a basis for the annual larval influx prediction (Oliver and Willis, 1987; Willis and Oliver, 1990).

During the spawning periods, enormous quantities of eggs and larvae are injected into the reef system. The eggs and larvae of most coral species are highly buoyant and they accumulate at the surface. Under the assumption that currents are steady in time and spatial current gradients are negligible, the concentration of coral eggs and larvae can be predicted from the advection-diffusion equation (8.30) as:

$$\frac{\partial \bar{c}}{\partial t} + \bar{u}\frac{\partial \bar{c}}{\partial x} + \bar{v}\frac{\partial \bar{c}}{\partial y} = \frac{\partial}{\partial x}\left(K_x\frac{\partial \bar{c}}{\partial x}\right) + \frac{\partial}{\partial y}\left(K_y\frac{\partial \bar{c}}{\partial y}\right).$$ (13.41)

Let us assume additionally that coral eggs and larvae are released instantaneously at the origin of coordinate system ($x = y = 0$), settling and predation
are omitted, and the turbulent diffusion coefficients, K_x and K_y, are constant.
Thus, the solution of Eq. (13.41) takes the form:

$$\bar{c}(x, y, t) = \frac{N_o}{4\pi \sqrt{K_x K_y t}} \exp \left[-\frac{(x - \bar{u}t)^2}{4K_x t} - \frac{(y - \bar{v}t)^2}{4K_y t} \right], \qquad (13.42)$$

where N_o is the number of larvae released. Equation (13.42) is an extension
of Eq. (8.55) for two dimensional advection and diffusion and for $\sigma = 0$. The
program illustrating solution (13.42) is given in Appendix D (program D.82).

   The theoretical solution of Eq. (13.42) is only approximate as many simplified hydrodynamic assumptions have been used. Observations during spawning
events show that concentration of coral eggs and larvae near reefs is very high
but extremely variable for two or three days after mass coral spawning (Oliver
and Willis, 1987; Wolanski and Hamner, 1988; Willis and Oliver, 1990; Oliver
et al., 1992). The larvae of spawning corals form visible surface aggregations
of very patchy distributions related to the local hydrodynamic features. In
some cases, the aggregations of eggs and larvae occurred as coral slicks up
to a few kilometres long but only a few metres wide, drifted away from the
reef. Many of the slicks are closely associated with fronts between water mass,
and wakes and eddies behind reefs (Oliver and Willis, 1987; Wolanski et al.,
1989). A comparison between observed larval concentrations around a coral reef
and predicted concentration derived from hydrodynamic and dispersion models
suggests that the standard depth-averaged models should be used in such topographically complex environments with great care (Oliver et al., 1992). Fully
three-dimensional models with a small grid size are required to predict the
complex flow patterns and patchy coral eggs and larvae distributions around
reefs.

   Shapiro et al. (1988) argued that immediate fate of spawned eggs and the
location of spawning sites on a reef are likely to be much more complex and
empirical studies examining the relationship between water movement, egg
transport and time and place of spawning are needed.

14 Tides and Waves on Vegetated Coasts

14.1 Introduction

Quantifying coastal and ocean processes is a principal objective of modern practice in coastal and ocean engineering and oceanography. In particular we seek a predictive capability for (Sobey, 1993): a) wave and current patterns on the continental shelf, in the nearshore zone, in coastal bays and estuaries, and in tidal wetlands, b) wave and current interactions with marine structures, in applications ranging from offshore platforms, ocean outfalls and breakwaters to coral reefs and wetland restoration, c) contaminant transport in the coastal environment, including the transport and fate of surface oil slicks, transport of suspended solids (salinity, sediment, nitrogen, pesticides, etc.) and others.

However, the purposes for which we predict are often conflicting. For example, those between heavy engineering structures, recreational activities and environmental protection. We are living in the age of environmental awareness, with increasing demands on minimizing negative impact of human activity on the environment. The environmental factors become more and more important in traditional marine developments.

The necessity for inclusion of environmental requirements is clearly seen on vegetated coasts, especially those in tropical climates. In contrast to the non-vegetated coasts, the understanding of physical processes at vegetated coasts is not adequate enough to develop effective management plans or engineering designs of coastal developments which are today subjected to very stringent requirements to minimize their impact on the environment.

In tropical climates, most coastlines are vegetated. Many inter-tidal mudflats are colonized by mangrove trees, with branching root systems and twisted trunks and branches that create resistance to water motion and stabilize the coastline. Mangrove forests provide a variety of goods, such as timber for fuel and construction, drugs and beverages, but most of all they protect coasts against erosion and support fisheries. It has been demonstrated that there is a positive correlation between mangrove areas and nearshore fish and shrimp catches. The mangrove-fisheries connection lies in the nursery function through the provision of the food and shelter from predators (Primavera, 1998).

Tidal flats are environments of net accumulation within coastal lagoons. They play an important role in lagoon dynamics, heat exchange, and nutrient inputs. In tidal flats, the inhabitants, mostly different species of worms and shells, must be able to withstand strong environmental stresses, such as large variations of salinity and temperature, storms, sediment load, shifting sands and anaerobic conditions.

Coral reefs are a common and important feature in tropics, especially in the Pacific region. Although ultimately morphologic and biological variability of reefs involves complicated interactions between chemical, biological and physical parameters, wave energy has long been recognized as one of the most important controls of coral growth and subsequent reef development. Coral reefs are also vital for the protection of coastlines against the attack of cyclone waves.

Accumulations of gel-like fluid mud front many coastlines, from the humid tropics of the equatorial latitudes to temperate latitudes. Most coastal fluid muds occur as intertidal and subtidal mudshoals near major rivers. In addition to serving as temporary and transitory storehouses for littoral sediments, intertidal mudflats can be colonized by vegetation, such as seagrasses.

Hydrodynamics of vegetated coastal zones, and especially the processes associated with wave motion, are still very poorly understood. Various types of ecological models are used to simulate the impact of specific developments on vegetated environments to provide the physical and biological consequences of alternative solutions for coastal regions. Most existing ecological models concentrate on the flow pattern, transport and dispersion of the pollutants or other substances. However, in these models the surface wave impact is usually neglected. Any predictive models for sediment transport, coastline stability, coral growth and its physical degradation need a substantial knowledge of tidal and surface wave mechanics. In this chapter, we discuss some theoretical and experimental results of various aspects of tidal and wave motion and their impact on the ecology of vegetated coasts.

14.2   Tides and Waves in Mangrove Forests

14.2.1   Tide-Induced Water Motion in Mangrove Forests

Mangroves are densely vegetated mudflats that exist at the boundary of marine and terrestrial environments (Fig. 14.1, see colour plate p.566). Inherent in these habitats is their ability to survive in a highly saline environment (Robertson and Alongi, 1992; Alongi, 1998). In recent years it has been realized that mangroves may have a special role in supporting fisheries, stabilizing the coastal zone and protecting the lives and properties of the people living near the sea and offshore islands (Jackson and Winant, 1983; Jenkins and Skelly, 1987; Qureshi, 1990; Siddiqi and Khan, 1990; Mazda *et al.*, 1997a).

Hydrodynamic factors play a major role in the structure and function of mangrove ecosystems. Biogeochemical and trophodynamic processes, and forest structure and growth, are intimately linked to water movement. However, studies of physical processes in tropical mangrove swamps and mangrove-fringed estuaries are few, and far behind compared to those of temperate estuaries. Water circulation in riverine mangrove forests, which comprises tidal creeks and shallow mangrove swamps, has been studied somewhat more than other types of mangrove forests (Wolanski et al., 1992; Furukawa and Wolanski, 1996; Furukawa et al., 1997). Long waves, namely tidal waves, are the dominant cause of water movement and sedimentation in mangrove systems. Tidal currents in creeks often exceed 1 m/s, however, velocities within swamps rarely reach 0.1 m/s. Modelling of tidally induced water motion in mangrove creeks is based on vertically averaged, barotropic equations of motion for unsteady flow in an open channel with lateral storage in mangrove swamps (Wolanski et al., 1992).

Circulation within the creeks is characterized by a pronounced asymmetry between the ebb and flood tides. This asymmetry is caused by the phase changes that occur between the mouth and the head of the creek. When the water level reaches high tide at the head of the creek, it is already falling at the mouth. This provides the necessary water slope to accelerate the water back towards the mouth as the ebb phase commences. The ebb-flood tide asymmetry is crucial in the maintenance of the geometry of the system. The larger ebb currents tend to export sediments from the creek and maintain a deep, navigable tidal channel. Reduction in the size of the swamp reduces the tidal asymmetry and the peak ebb tidal currents. As a result the creek silts rapidly. This increased siltation and reduced navigability is a common occurrence in the South-East Asia region where prawn farms have been constructed in the mangrove forests.

The complex circulation through mangrove vegetation is friction dominated, with jets, eddies, stagnation regions and vegetation-scale turbulence. Observations in a mangrove area close to Cairns (the eastern coast of Australia), and computer modelling indicate that at small velocities (0.005 m/s), non-separated laminar flows prevail, while at high flow speed (0.2 m/s), eddies and wakes are generated around individual roots (Furukawa and Wolanski, 1996).

14.2.2 Storm or Cyclone-Induced Waves in Mangrove Forests

Numerical Modelling. During storms or tropical cyclones, the energy of surface waves substantially exceeds tidal energy. Due to the complexity of mangrove structures, the transmission of cyclone induced waves through mangrove areas is poorly understood. Assuming that the diameter of mangrove trunks is very small in comparison with wavelength, wave energy is dissipated mostly due to drag forces induced on trunks by waves.

Only recently, a series of laboratory and field experiments, as well as theoretical analysis, have been carried out to investigate surface wave energy dissipation in mangrove forests (Brinkman et al., 1997; Mazda et al., 1997a,b; Massel et al., 1998). There are two main energy dissipation mechanisms in mangrove forests: multiple interactions of wave motion with mangrove trunks and roots, and bottom friction. Bottom friction can be accommodated through the concept of a bottom friction coefficient. However, at this stage, the bottom friction will be omitted from consideration as the bottom friction coefficient for mangrove forests is not known.

In the simplest representation, mangrove trunks and roots can be treated as cylindrical elements located in the water (Fig. 14.2). In areas occupied by *Rhizophora* species, the density of mangrove trunks and roots is greater in the bottom layer than in the upper layer, where only the vertical trunks are observed (Wolanski et al, 1992). Wave induced forces on trunks and roots are inertial and drag type forces, and for typical mangrove trunks and roots the drag force dominates.

Because of the proximity of other trunks, some interactions between trunks can be expected. To include these interactions in the resulting drag force, an appropriate modification of the drag coefficient, C_d, depending on the density of the mangrove trunks, has been proposed using the discrete vortex method (Massel *et al.*, 1998).

Wave motion within the mangrove forest is subjected to strong dissipation due to the multiple interactions with mangrove trunks. Hence, the momentum equation for motion with dissipation can be written as follows (see Appendix C.3):

$$\frac{\partial \boldsymbol{u}}{\partial t} = \frac{1}{\rho} \nabla \left(p + \rho g z \right) - \frac{1}{\rho} \boldsymbol{F}, \tag{14.1}$$

in which $\boldsymbol{u} = (u, w)$ is the wave-induced velocity vector, p is the corresponding dynamic pressure, and $\boldsymbol{F}$ is the force vector (per unit volume).

Let us now consider a unit control area of mangrove and assume that in this area there are N_u trunks piercing the sea surface (usually N_u is of the order of 1-10 per m^2), each of mean diameter $\bar{D}_u$. In the bottom layer of thickness, h_l, (usually thickness h_l is of the order of 0.3 m–1.0 m) mangroves are very dense and smaller trunks and roots are randomly oriented. It is assumed that the number of trunks, N_l, each of mean diameter, $\bar{D}_l$, is of the order of 10–30 per m^2. The control area has to be selected sufficiently large to accommodate N_u and N_l trunks, where $N_u > 1$ and $N_l \gg 1$. On the other hand, this area has to be sufficiently small in order to neglect the variation of wave velocity within the control area and to neglect the exact location of each trunk within the control area. The spacing $\Delta x = \Delta y = 1$ m is probably a reasonable compromise for the above requirements.

As the mangrove roots are randomly oriented against the water flow direction, it is impossible to exactly reproduce the mangrove geometry. However, in order to get some insight into the problem, we consider the simpler problem of a mangrove forest where all trunks in the upper layer are vertical (see Fig. 14.2). The inclination of mangrove roots and trunks in lower layer is parameterized through the mean inclination angle, $\bar{\theta}$, measured with respect to the vertical axis, z. Observations suggest that the angle $\bar{\theta}$ is of the order of 30°.

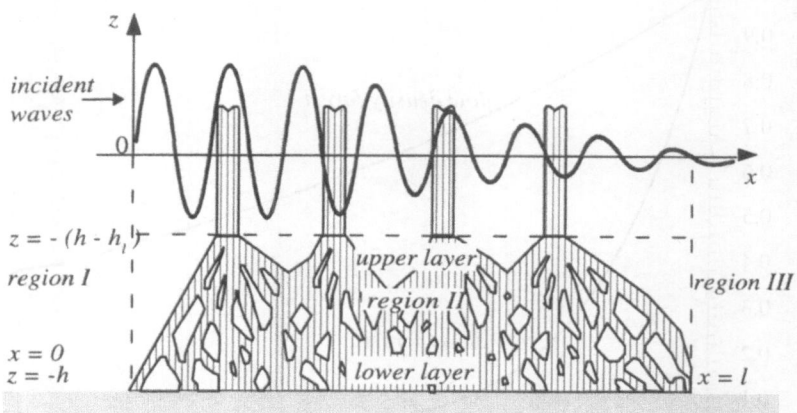

Fig. 14.2: Coordinate system in mangrove forest

Assuming that the drag force dominates over inertial force, the total force, F, per unit volume, can be represented as follows (Massel *et al.*, 1998):

1. *upper layer:* $-(h - h_l) < z < 0$,

$$F_u(x, z) = \frac{1}{2}\rho_w \bar{D}_u \sum_{j=1}^{j=N_u} C_d^{(m)}(Re)u_{n,j}(x, z)\,|u_{n,j}(x, z)|\,. \tag{14.2}$$

2. *lower layer:* $-h < z < -(h - h_l)$,

$$F_l(x, z) = \frac{\rho_w \bar{D}_l}{2\cos(\bar{\theta})} \sum_{j=1}^{j=N_l} C_d^{(m)}(Re)u_{n,j}(x, z)\,|u_{n,j}(x, z)|\,. \tag{14.3}$$

The vector $u_{n,j}(x, z)$ is water velocity normal to the longitudinal axis of the particular trunk j induced by wave orbital velocity $u(x, z) = [u(x, z), w(x, z)]$, $\bar{D}_u$ and $\bar{D}_l$ are mean diameters of trunks in upper and lower layers, respectively, $\bar{\theta}$ is mean angle of inclination of trunks and roots in the lower layer, and $C_d^{(m)}$ is the modified drag coefficient C_d due to possible interactions between trunks.

The momentum equation (14.1) can be solved when incident wave characteristics as well as mangrove forest parameters are known. Details of the proposed numerical model are given by Massel *et al.* (1998). To illustrate the model, let consider two mangrove forests with different trunk densities: for a dense forest, the number of trunks in the upper layer $N_u = 16/\text{m}^2$ and the number of trunks in the lower layer, $N_l = 49/\text{m}^2$, while for a sparsely populated forest $N_u = 1/\text{m}^2$ and $N_l = 9/\text{m}^2$.

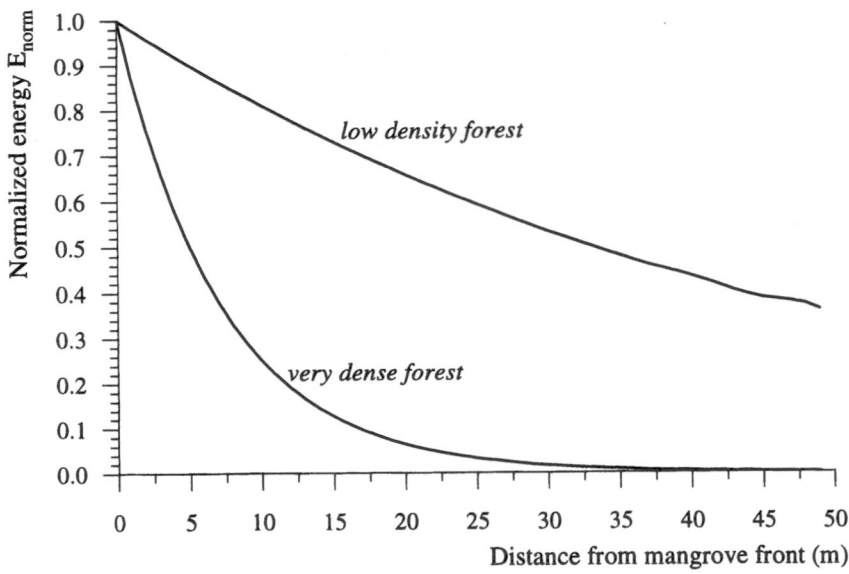

Fig. 14.3: Calculated normalized energy E_{norm} in densely and sparsely populated forests (adapted from Massel *et al.*, 1998)

In both cases, other forest parameters are the same, *i.e.* forest width $l = 50$ m, water depth $h = 1$ m, mean diameter of upper layer trunks $\bar{D}_u = 0.08$ m, and mean diameter of lower layer trunks $\bar{D}_l = 0.02$ m. The mangrove forest is subjected to wind induced waves of significant wave height, $H_s = 0.6$ m and wave period of $T = 5$ s. Numerical calculations indicate that wave energy attenuates very quickly with distance from the mangrove/ocean boundary, and behind the mangrove forest wave energy is negligible. Figure 14.3 illustrates the calculated normalized energy as a function of distance from the mangrove/ocean boundary:

$$E_{norm}(x) = \frac{E(x)}{E_0}, \tag{14.4}$$

in which $E(x)$ represents wave energy at a distance x from the front of the mangrove forest and E_0 is the incident wave energy. For a densely populated forest, almost total wave energy is dissipated within the mangrove forest; in a sparsely populated forest, a remaining 35% of the incident wave energy is observed behind the forest area.

Wave induced velocities in mangrove forests are of special interest, as water kinematics control the exchange of water, fluxes of nutrients and sediment transport. Both water velocity components change their magnitude and direction during one wave period, however for practical applications, the most useful characteristic of wave velocity is the mean amplitude. The vertical profiles of

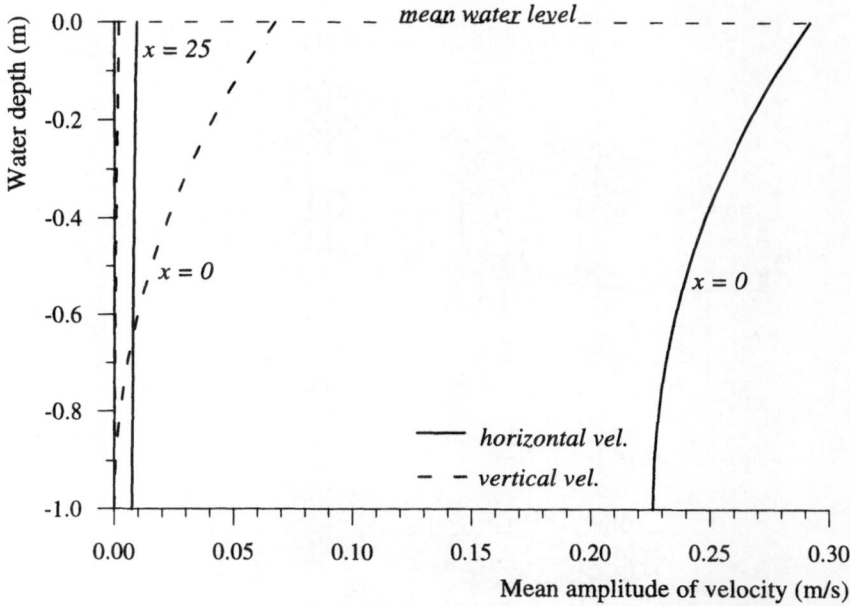

Fig. 14.4: Calculated vertical profiles of the mean amplitudes of horizontal and vertical components of orbital velocity at cross-sections ($x = 0$ m, 25 m, 50 m) in a very dense forest; at $x = 50$ m the vertical velocity is negligibly small and is not shown (adapted from Massel *et al.*, 1998)

mean amplitudes of horizontal, $\bar{u}$, and vertical, $\bar{w}$, components of velocity at three cross-sections in a mangrove forest are shown in Fig. 14.4. In this case, the ratio of wavelength to water depth is relatively large, $L/h = 16$ and the profiles of both velocity components are almost vertically uniform. They attenuate very quickly with distance from the mangroves front, and behind the mangroves they are negligible.

Field Experiments. Observations of physical processes in tropical mangrove forests are very sparse. Two field experiments were conducted at Cocoa Creek in Australia and at Nadara River on Iriomote Island, Japan, during January and February of 1997. Species distributions at these two locations are shown in Fig. 14.5, and a detailed description of these sites and the observational techniques used at each location are given by Brinkman *et al.* (1997).

Time series of water surface elevation recorded at various locations along the transects at both study sites were analyzed for basic statistical and spectral quantities (for details see Brinkman *et al.*, 1997). Here we examine the wave propagation through a mangrove forest in terms of the normalized energy, $E_{norm}(x)$, as defined by Eq. (14.4). Values of $E_{norm}(x)$ calculated from field data at various distances into the forest at Cocoa Creek are shown in Fig. 14.6. These data show an obvious decline in wave energy transmission with distance into the forest. Normalized energy value, $E_{norm}(x)$, was calculated using incident

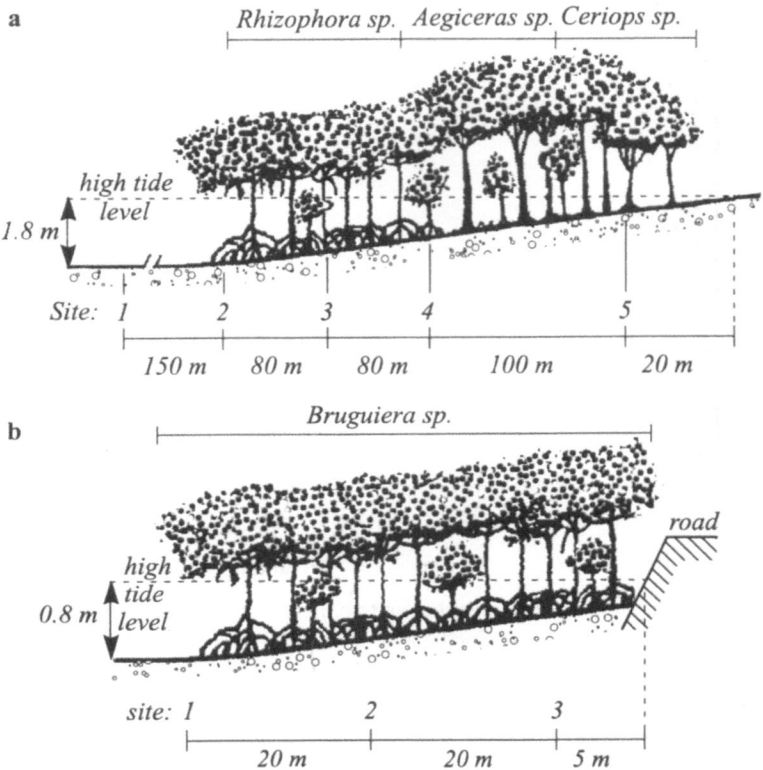

Fig. 14.5: Mangrove species distribution at: **a** Cocoa Creek, Australia **b** Iriomote Island, Japan (adapted from Brinkman *et al.*, 1997)

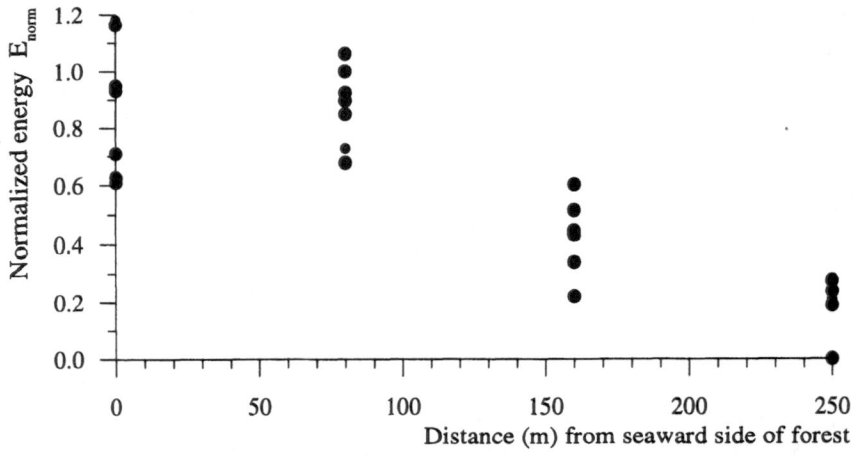

Fig. 14.6: Experimental normalized energy E_{norm} in the mangrove forest at Cocoa Creek (adapted from Brinkman *et al.*, 1997)

waves recorded approximately 150 m in front of the forest. Between this site and the seaward boundary of the forest, there is a gradually sloping mudflat resulting in some wave transformation in front of the forest. This could explain normalized energy slightly greater than unity at, and near, the sea/mangrove boundary.

At any location within the forest we would expect $E_{norm}(x)$ to be a function of the water depth, given the changing numbers of mangrove trunks and roots, and their diameters, with vertical distance from the sea bottom. In general, the amount of obstruction caused by trunks, prop roots, and pneumatophores decreases with distance upwards from the forest floor until the canopy is reached. This characteristic makes the drag coefficient, C_d, a function of both distance from substrate and the Reynolds number, Re.

14.2.3   Sedimentation in Mangrove Forests

Most coastal mangrove forests are connected to the sea via tidal creeks. Sediment transport in these creeks has been investigated in detail (for example, Wolanski, 1995). Asymmetry of the tidal current (mentioned earlier), baroclinic circulation and shear-induced destruction of flocs control the amount of sediment transported in tidal creeks.

Water flow in mangrove forests is impeded by the friction caused by the high vegetation density. Interaction of the water flow with mangrove roots creates complex systems of eddies, jets and stagnation zones. This micro-turbulence environment maintains sediment in suspension at flood tidal currents which transport fine sediment particles into the forest. Turbulent intensities are largest in the complex matrix of roots, such as *Rhizophora* species, and are much smaller for single trees such as *Ceriops* species (Furukawa and Wolanski, 1996).

Some data indicate that the sediment movement threshold velocities for sediments with modal diameter 0.2 mm are about 0.4 m/s. It has also been inferred that accumulation of fine sediment takes place on the mangrove forest flats by settling at high water spring tides. In some creeks, a net flux is mostly seaward because of higher ebb flows.

The sedimentation of suspended sediment occurs when turbulence vanishes at slack high tide. During ebb tide, there is a substantial attenuation of current speed which is not sufficient to resuspend the sediment. Furukawa *et al.* (1997) showed that about 80% of the suspended sediment brought into the mangroves from the coastal waters at spring flood tide became trapped in the mangroves. The maximum suspended sediment concentration in the mangroves at the creek edge was about 150 mg/l, which is a typical value for Queensland mangrove creeks at spring tides. Therefore mangroves are an important sink for fine sediment from rivers and coastal waters.

Although tidal currents through mangrove forests are weak, these currents influence water quality in terms of dissolved organic carbon (DOC). A study by Furukawa *et al.* (1997) shows that the import of DOC at flood tide, and

export at ebb tide, from creek to the mangrove are 360 and 260 gC/m of creek
length, respectively. The bulk of the organic carbon in the mangrove area is
in dissolved form with maximum concentration of about 2.5 mg/l. Dissolved
organic carbon accounts for 85% of the total organic carbon. These values are
much higher than those in nearby coastal waters. A study suggests that organic
carbon may be partitioned between creek and mangrove forest, with the major
change in the molecular weight of the DOC in the mangroves occurring between
flood and ebb tides.

Removal of mangroves may increase water turbidity and hence decrease primary productivity in tidal creeks and adjacent coastal waters. Unfortunately,
mangroves have declined worldwide, particularly in the South-East Asia, where
losses in the last three decades have reached 70–80% (Sasekumar *et al.*, 1994).
On the other hand, in some areas efforts to plant mangroves on an extensive
scale are observed. Mazda *et al.* (1997a) demonstrated the usefulness of mangrove reforestation for coastal protection in the Tong King delta, Vietnam.

14.3   Tidal Flats

14.3.1   Tidal and Wave Motion on Tidal Flats

Tidal flats are usually defined as sandy to muddy or marshy flats emerging
during low tide and submerging during high tide. The frequency of tidal flat
inundation, examined in a perspective of months or years, indicates that approximately the lower 25% of the extreme tidal range is intermittently exposed,
while the higher 25% is intermittently inundated, leaving only the central 50%,
corresponding to the above definition (Amos, 1995).

Tidal flats can be found in all climate regions, in arid and wet tropical or
subtropical regions as low-latitude tidal flats, in temperate regions as mid-
latitude tidal flats, and in areas influenced by ice as high-latitude tidal flats.
All of them exhibit a common relationship in grain size, benthic flora and
fauna diversity and abundance, surface morphology and slope. However, two
major groups of tidal flats are apparent: sandy tidal flats and muddy tidal
flats. The first group corresponds to the mean inorganic suspended sediment
concentration (SSC) of the inundating water of less than 1 g/l, while for muddy
tidal flats, the SSC is generally greater than 1 g/l (Amos, 1995).

The tides on tidal flats are characterized by a fast start of the flood current,
prolonged period of high water, then slack water followed by a short, fast ebb
current. Such a current pattern develops primarily in the small tidal creeks
adjacent to the flats. The current velocities on tidal flats are much smaller,
mostly below 0.5 m/s and they change with water depth, according to the
formula $C = \sqrt{gh}$, where h is the water depth. Tidal currents on the flats stop
almost immediately at high water, but in the creeks they often continue to
flow for up to half an hour after high water. Creeks are not filled (or emptied)
simultaneously by the tidal water, and filling time depends on the speed of
propagation of tidal wave in the creeks. As a result, a residual current pattern

exists over tidal flats, apart from the main oscillatory tidal water movement (Posma, 1988).

Since water depth on the tidal flats is very small, waves propagating towards shore are considered 'long' when the Ursell number is greater than 75 (see Sect. 4.2). Waves play an important role in the resuspension of sediments on tidal flats. They can either amplify or reverse the flux of suspended sediment. Thorne (1979) found that a near-bed oscillatory flow of only 5 mm/s could double the transport of fine sand by tidal currents. Also, the SSC is of an order of magnitude greater during storms than at other times. Amos (1995) observed, in the upper Bay of Fundy, that wave erosion prevails with a consequent export of suspended sediment when the significant wave heights reach 4–6 m.

Storm waves, due to their periodic character, may induce pore-pressure amplification and subsequent liquefaction of tidal flat sediment. However, little is still known about these processes, in contrast to liquefaction in offshore soft marine sediments (Mei and Foda, 1981).

14.3.2 Salinity and Sediment Transport on Tidal Flats

During the flooding stage of a tidal cycle, currents in the creek transport salt water further inward. This creates a salinity gradient between the creek and the tidal flat. When no freshwater is available and temperature is high, sea water evaporates and salinity increases above the normal level. In subtropical areas, when there is little or no rainfall, evaporation proceeds until the stage of salt saturation and salt flats are formed (Amos, 1995). Similar variations are also observed in temperature; high extremes can be found in tropical tidal pools, but flood tide causes a sudden drop of temperature.

The movement of tidal water affects complex processes such as the transport, erosion, deposition and consolidation of sediments. When intertidal sediments are exposed, the surface layer is drained of water. The depth of the drainage strongly depends on grain size and elevations. Beaches and sandflats drain more quickly than mudflats.

Tidal flat sands are extremely well sorted with the median grain size varying between 100 and 300 μm. On some tidal flats these limits are even narrower. However, the grain size distribution over a tidal flat is not totally homogeneous. The median grain size decreases over the flat with distance from the offshore boundary of the tidal flat, as should be expected. This is mainly caused by the loss of wave energy when travelling over the flat.

According to Groen (1967) and Amos (1995), the shallow-water asymmetry of the flood and ebb current controls vertical exchanges of sediment within the benthic boundary layer. The maximum ebb current is preceded by a period of low current velocities much longer than the flood current maximum. This means that during the period preceding maximum ebb flow, particles have a much longer time to settle out of suspension, and the suspended load decreases. Therefore, the suspended load at peak ebb is lower. This results in a shoreward sediment residual motion which may be up to 38% greater than the seaward motion. The intensity of the transport depends also on the gradient of suspended sediment concentration (SSC). Measurements of SSC indicate that the concentration decays exponentially with distance from the shoreline. For example, in the Bay of Fundy, the decrease of SSC takes the form (Amos, 1995):

$$SSC(x) = SSC(0)e^{-kx}, \tag{14.5}$$

in which the decay coefficient varies with the season between 0.023 and 0.091 m^{-1} (mean value is 0.049 m^{-1}).

A model for sediment accretion/erosion on the tidal flats of The Wash, England, and Minas Basin, Canada, was developed by Amos (1995). Computer simulation included the time of inundation, velocity of current, instantaneous bed shear stress, cumulative deposited and eroded mass, and suspended sediment concentration. The tide and current speed were determined for eighty M_2 tidal cycles. The starting SSC was set in turn to 10, 100, 1000 and 10,000 mg/l, and was assumed to be constant across the flats. The simulation for The Wash indicated that peak erosion appeared to exceed peak deposition during spring tides and the converse was evident during neap tides. The net predicted result was one of long-term erosion of the flats, and an overall increased in SSC, modulated by the spring-neap cycle.

The presence of plant cover along tidal flats may lead to modification in water circulation and sedimentation rates. For example, *Spartina alterniflora* marshes act at the surface through their stems and leaves, and at the subsurface through their roots and rhozomas. These marshes are typical features of the southern Brazilian estuaries and lagoons. Netto and Lana (1997), show that *Spartina alterniflora* is a major source of sediment heterogeneity between vegetated and unvegetated areas.

Most organic material accumulating on tidal flats is processed in the sediments rather than in the water column. The traditional concept of a closed carbon cycle consists of the microphytobenthos living on the flat as the producer, and the benthic fauna and micro-organisms living in the sediments as the consumer of organic matter and the suppliers of nutrients to the phytobenthos (Posma, 1988). However, recently Alongi (1998) pointed out the importance of the role of bacteria and protozoa in utilizing and mineralizing sediment organic matter. The bacterial biomass, growth, and productivity in sediments are greater than was previously believed, and the oxidation reactions and pathways of microbiological decomposition in sediment are very complex. The inquisitive reader should consult Alongi (1998) for an in-depth discussion.

14.4   Waves at Coral Reefs

14.4.1   A Brief Overview

Coral reefs are truly magical places. Although remote from the natural environment of humans, coral reefs are increasingly becoming an area of tourist

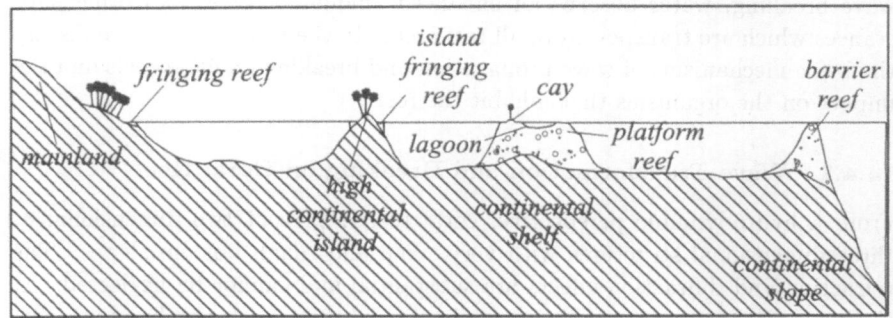

Fig. 14.7: Types of coral reefs (from Veron, 1996, with author permission)

expansion, enhanced fishery (coral reef productivity is extremely high) and exploration and exploitation of mineral resources (Veron, 1986). The most important factors controlling coral community composition are: light availability, wave action, sediment transport, salinity, tidal range, food and inorganic nutrients, temperature and bathymetry. Coral reefs may be: barrier (separated from the shoreline by a deep channel), fringing (adjacent to the shoreline or separated from the shoreline by a shallow moat or boat channel), of atolls type (reef rim with a central lagoon), or of reef island type (table or platform reef without a central lagoon) – see Fig. 14.7.

Many aspects of coral biology require the movement of water. Planktonic larvae depend on water flow to disperse over any substantial distance. Aquatic organisms need water movement to remove their waste and to bring in new supplies of oxygen (Denny, 1988). Unfortunately, water movements intensify the contamination of coral by pollutants. Wave energy and water turbulence associated with wave motion are responsible for reef zonation and segregation of organisms. Wave shoaling over reef slopes and transformation on reef platforms impose forces on the organisms that inhabit the wave-swept zone of reefs. Water flow is seldom laminar. Waves are one of the dominant mechanisms generating turbulence and mixing. It is trivial to say that turbulent mixing is important for every aspect of biology that relies on the transport of fluid (see Chap. 8). Wave motion is periodic, with horizontal and vertical components of orbital velocity, hence it produces extra advection and mixing in the water column.

The wave-dominated bottom boundary layer on reef slopes is usually turbulent. The presence of organisms on the substratum contributes to the production of turbulence due to increased bottom roughness. On steep slopes, the turbulence associated with wave breaking reaches the bottom. Thus, organisms living on such slopes are likely to experience turbulent flow and the consequent large accelerations and forces with every breaking wave (Massel, 1994).

Waves breaking on the reef slope and their overtopping produce an extra water mass flux over the reef top, towards the reef flat and reef lagoon. During

wave breaking, water absorbs an enhanced amount of gases and other substances which are transported to all reef areas. In the next sections, we discuss
the basic mechanisms of wave propagation and breaking on coral reefs and the
impact on the organisms that inhabit the reefs.

14.4.2   Wave Transformation and Breaking on Coral Reefs

From a hydrodynamic perspective, coral reefs represent two dimensional or
three dimensional structures with very steep and rough bottom slopes. The
refraction and diffraction effects are substantial and cannot be neglected. At
the reef edge, waves loose their stability and break, with the breaking processes dominating the wave transformation. As was shown in Chap. 4, when
the bottom slope is very gentle, wave height at a given water depth h can
be determined from a simple formula (4.50) resulting from the conservation
of energy. On the other hand, when the bottom slope is steeper, the transformation of waves becomes very rapid, with rapidly decreasing wave length
and increasing wave height. Equation (4.50) is not applicable for calculation of
wave energy transformation over steep slopes. Massel (1993a, 1996a) showed
that wave height can be calculated from the refraction-diffraction equation,
which for elongated reefs with waves approaching normally takes the form:

$$\frac{d^2\varphi}{dx^2} + (CC_g)^{-1} \frac{d(CC_g)}{dx} \frac{d\varphi}{dx} + \left[k^2(1+\psi) + i\gamma k\right] \varphi = 0, \tag{14.6}$$

in which:

$$\varphi = \frac{H(x)}{H_i}, \tag{14.7}$$

$$\psi = E_1(kh) \left(\frac{dh}{dx}\right)^2 + E_2(kh) \frac{g}{\omega^2} \frac{d^2h}{dx^2}, \tag{14.8}$$

and:

$$\gamma = \gamma_b + \gamma_f, \tag{14.9}$$

where H_i is the incident wave height, C and C_g are phase and group velocities,
respectively, the ψ term describes an influence of bottom slope dh/dx and bottom shape d^2h/dx^2, and the γ term represents energy dissipation due to wave
breaking and bottom friction. Possible methods of determining the amount of
energy dissipated during wave breaking have been described in Chap. 4. Equation (14.6) can only be solved numerically, and details of solution can be found
elsewhere (Massel, 1992, 1996a).

In Fig. 14.8, an example of solution of Eq. (14.6) for the fringing reef bottom
profile at Hayman Island (Gourlay, 1994) is shown. Incident wave height and

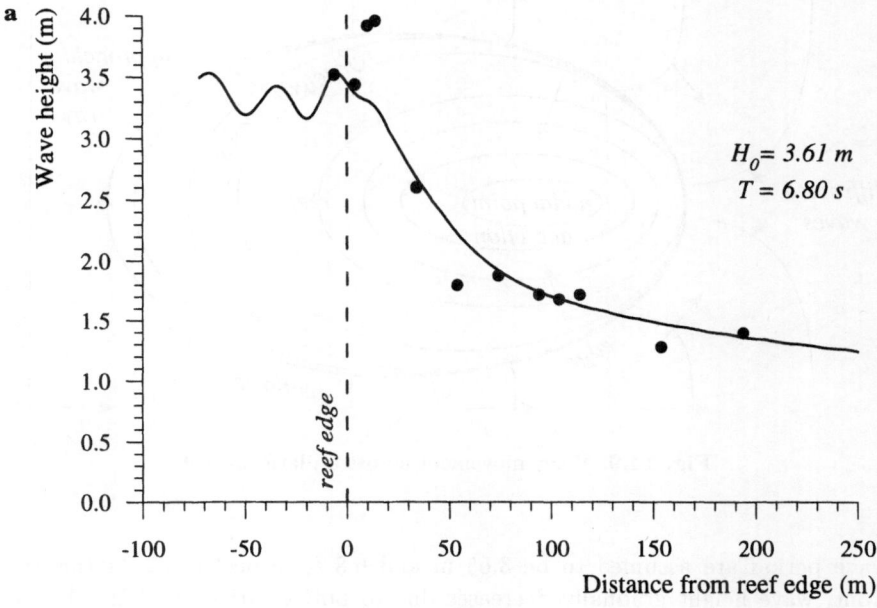

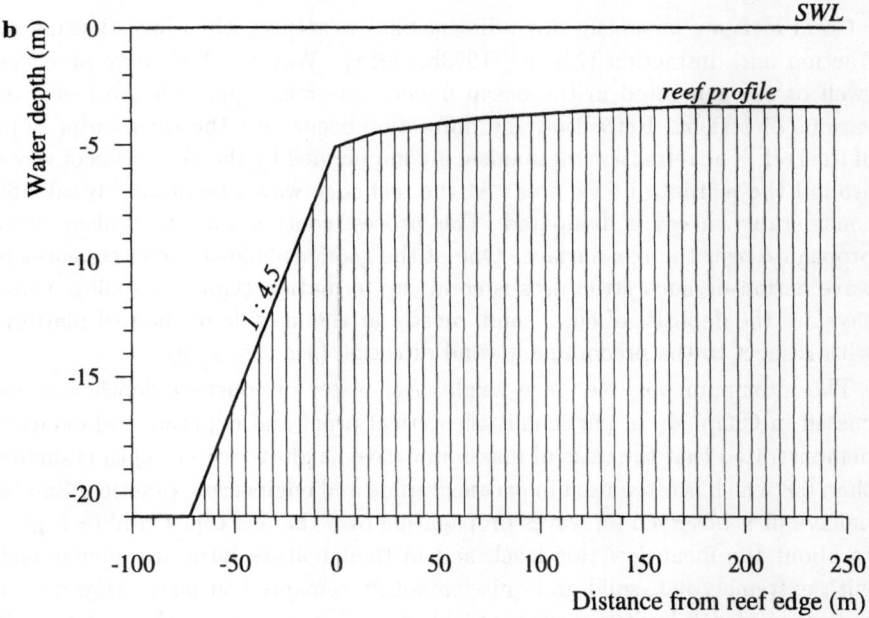

Fig. 14.8: Wave attenuation over a reef slope: **a** wave height attenuation; solid line denotes numerical model, dots denote experimental data, **b** reef bottom profile

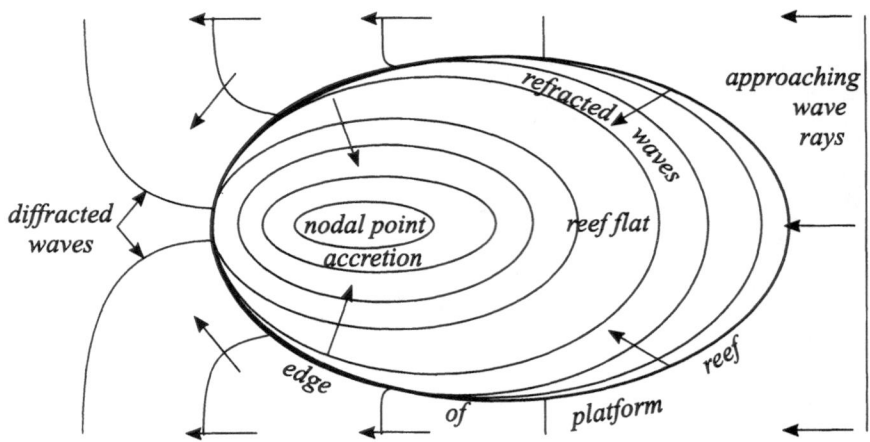

Fig. 14.9: Wave movement across a platform reef

wave period are assumed to be 3.65 m and 6.8 s, respectively. At the reef front, wave height gradually decreases due to bottom friction. Wave height oscillations observed at the reef front are due to reflection from the reef. At the reef edge, waves start to break and their height constantly decreases.

Coral reefs are obviously three-dimensional structures when investigating refraction and diffraction (Massel, 1993b,1996a). Waves, which may be either swell or sea generated in the ocean outside the reef, approach the reef from various directions. Refraction and diffraction occur over the steep outer slope of the reef. The refraction mechanism is compounded by the diffraction of waves around the reef into its lee area. At the reef edge waves eventually break and considerable energy is dissipated. This process continues as the broken waves propagate over the reef surface. One of the geomorphological consequences of wave action on such structures is coral cay formation (Gourlay, 1988). Coral cays are the deposits of coral sand, mostly at the lee side of the reef platform with respect to the predominant wind direction (see Fig. 14.9).

The maximum possible wave height over water of constant depth was discussed in Chap. 4. In particular, theoretical proof and experimental evidence demonstrated that the ratio of maximum wave height to water depth is smaller than 0.8 which is often used in oceanographic and engineering practice. Similar behaviour is observed for waves propagating over the reef top. Coral reefs grow to about the mean low tide level, and in their mature form, are planar reefs with extremely flat, solid and quite smoothly cemented surfaces. However, at high tide, depth-limited waves are able to propagate across the reef top. In the GBR region, tide ranges may be as great as 3 to 4 m, and water depths of 2 m to 3 m over the reefs at high tide are common. During the passage of a cyclone storm surge, these depths could increase by 0.3 m to 0.5 m, or much more, depending upon the intensity and path of the cyclone (Massel, 1996a).

Since wave heights are governed by prevailing water depths, depth limited wave height criteria are essential for determining wave heights for structures, and for sedimentological and geomorphological processes on coral reefs. Waves propagate from the deep ocean onto the reef slope, break on the reef as plunging breakers and then propagate over the reef top as a succession of bores, before reforming into stable oscillatory waves at some distance from the reef edge. Data collected during the REEF88 experiment (Hardy *et al.*, 1990) indicate that the largest stable wave had $H_{max}/h \approx 0.6$, achieved by an individual wave in a random wave train propagating over a horizontal bed. Again this is considerably lower than the usual oceanographic and engineering criterion.

Gourlay (1994) has published the results of laboratory experiments on wave transformation over a natural reef profile and provided further evidence that maximum depth-limited wave heights on a horizontal reef were consistent with the REEF88 experiment results. Sulaiman *et al.* (1994) has reported the results of experiments on maximum wave heights on the horizontal reef flat at Sanur Beach in Bali. For a non-dimensional water depth range $1.7 \times 10^{-4} < h/(gT^2) < 0.17$, the ratio of maximum wave height to water depth is always smaller than 0.6, except for a very few cases when it is about 0.65.

It is important to note that the lower limit of $(H/h)_{max}$ ratio is applicable to those locations on coral reef platforms where the distance from the reef edge is sufficient for waves to reform into oscillatory waves, after dissipating energy due to breaking. Close to the reef edge, waves with larger $(H/h)_{max}$ ratio can exist, but these would be turbulent breaking waves attenuating rapidly with distance from the reef edge.

14.4.3   Water Circulation on Coral Reefs

Energy of waves propagating over the reef slope, tides and wind contribute to the water movement through coral reefs and within coral reef lagoons. Wave breaking and a rapid transformation of energy at the reef edge result in a sea level set-up and drive strong currents. The features of the observed circulation are usually reproduced using a numerical model in which a pattern depth integrated circulation in the lagoon is predicted (Prager, 1991; Wolanski *et al.*, 1993; David and Kjerfve, 1998; Kraines *et al.*, 1998). However, the details of circulation are strongly controlled by relative importance of major forcing mechanisms, *i.e.* waves, tides, ocean currents, local lagoon pattern and lagoon bathymetry. For example, large Laguna de Términos, in Mexico, is mostly controlled by tides which are responsible for 70% of the water level variability, and approximately 95% of the current variability in the lagoon's inlets.

In smaller lagoons, the significance of wave motion can be more substantial. The measurements by Massel and Brinkman (1998) in the lagoon behind Ningaloo Reef, the largest fringing coral reef system in Australia extending 300 km along the Western Australian coast, showed no evidence of inflow into the lagoon other than via flow over the reef top. The continual outflow of water from the lagoon, independent of tidal phase, indicates that wave pumping is a more important process than tidally driven circulation in reef flushing.

The exposure time of resident biota to water-borne materials and accumulation of sediments or adsorbed substances in lagoon depend on the so-called lagoon residence time, τ. There are a few definitions of residence time, depending on the method used for the calculation of τ. Residence time, τ, can be calculated in the simplest way using the tidal prism approach (Prager, 1991; see also Chap. 8):

$$\tau = \frac{\text{lagoon volume}}{\text{daily volume input}}, \tag{14.10}$$

or:

$$\tau = \frac{\text{lagoon depth} \times \text{tidal period}}{\text{tidal range}}. \tag{14.11}$$

In particular situations, the residence time varies in the large extent. For shallow, narrow lagoons time, τ, may be relatively short, less than one day, while in more isolated and larger systems it may be of the order of one to four months. For example, in Enewetak Atoll lagoon, the average residence time of water in the lagoon is 28 days, and contribution from the wind-ward reef comprises 85% of the total inflow, when time, τ, is calculated according to Eq. (14.10) by dividing the lagoon volume by the rate of water input (Atkinson et al., 1981). On the other hand, Kraines et al. (1998) after measurements and numerical modelling of the lagoon at Bora Bay, Miyako Island, showed that in this case the residence time is in the range of 1.5 hour to 3.7 hours. This means that lagoon is flushed several times each tidal cycle. The residence time, τ, calculated using the tidal prism method for this lagoon was almost 10 times longer.

14.4.4 Impact of Waves on Physical Degradation of Reefs

Cyclone waves affect the density, structure and local distribution of coral assemblages by acting as agents of mortality and colony transport. In general, the probability that an organism will be destroyed (by breakage or dislodgement) is a function of organism size, its structural strength, wave-induced velocity and acceleration, and the probability of encountering waves with given parameters. We will illustrate two different mechanisms of coral damage, using as the example hard corals *Acropora formosa* and *Porites*. In the first case, we will examine a relationship between coral branch proportions and water flow. The second case is dealing with a probability of coral survival in a given wave environment.

Coral Branch Proportion. Consider a single vertical coral beam of the cylindrical shape with dimensions as in Fig. 14.10a. The axis z is vertical and oriented upwards from substratum, while x is the horizontal axis. We

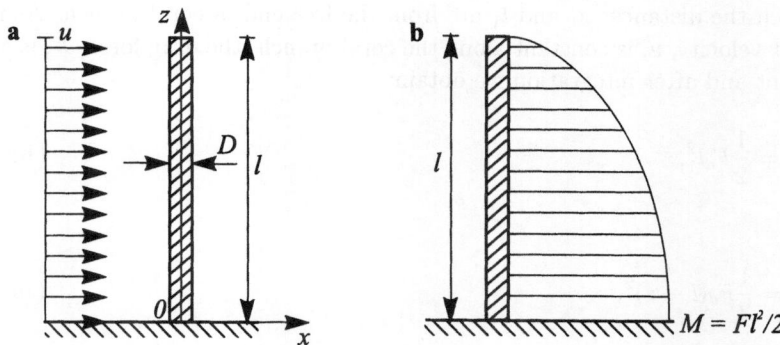

Fig. 14.10: Vertical coral branch: **a** coordinate diagram, **b** resulting bending moment

assume that coral branch is subject to uniform horizontal current of velocity u, directed along the x axis. For a given coral dimensions: diameter D, length l, and ultimate bending strength σ_u, there is some current velocity which induces stresses larger than the coral strength. As the coral can not withstand such stresses, it breaks. Thus, at breaking, the following condition has to be satisfied at some point of coral branch:

$$\sigma_{\max} \geq \sigma_u, \tag{14.12}$$

where $\sigma_{\max}$ is the maximum bending stress, and σ_u is the ultimate bending stress (see Chap. 10).

Horizontal current induces a drag force on a coral branch. For cylindrical shape of the branch, the drag force, per unit length of branch, is defined by Eq. (2.63), *i.e.*:

$$F_d = \frac{1}{2}\rho_w C_d D u^2. \tag{14.13}$$

In general, when in addition to current, waves are present, there are two components of force induced on coral branch: drag component, F_d, and inertial component, F_i. According to discussion in Sect. 4.2.8, a significance of the inertial component is examined using the Keulegan-Carpenter number $K = uT/D$. Usually, when $K > 100$, wave-induced force on a body is totally dominated by drag force, and oscillatory component of force can be neglected. For typical wave conditions, such as: wave period $T \approx 6$ s, wave-induced velocity $u \approx 1$ m/s, and coral branch diameter $D \approx 0.02$ m, the coefficient K is of the order of 300. Thus, the inertial components are not important and they can be omitted. Now, the bending moment induced by the current corresponding to the force F becomes:

$$M = \int_0^{l_1} F_d z_1 dz_1, \tag{14.14}$$

in which the distances z_1 and l_1 are from the free end of coral branch. As the current velocity, u, is constant along the coral branch, the drag force, F, is also constant and after integration we obtain:

$$M(l_1) = \frac{1}{2}F_d l_1^2, \tag{14.15}$$

or:

$$M(z) = \frac{1}{2}F_d(l - z)^2, \tag{14.16}$$

where z is the distance from substratum (see Fig. 14.10b). At the free end of coral branch, the moment M is equal to zero, as expected, and the maximum moment occurs at the substratum $(z = 0)$, i.e.:

$$M_{\max} = \frac{1}{2}F_d l^2. \tag{14.17}$$

Vertical variation of moment M with a distance along coral branch is shown in Fig. 14.10b. Using Eq. (14.13) in Eq. (14.17) we have:

$$M_{\max} = \frac{1}{4}\rho_w C_d D u^2 l^2. \tag{14.18}$$

The maximum bending stress $\sigma_{\max}$ is given by the flexure formula (10.15) which for circular cross-section of coral branch gives:

$$\sigma_{\max} = \frac{M_{\max}}{S} = \frac{8\rho_w C_d l^2}{\pi D^2}u^2. \tag{14.19}$$

The condition for coral branch breaking is given by Eq. (14.12). Thus, a critical current velocity u, needed for breaking becomes:

$$u \geq \sqrt{\frac{\pi}{8}\frac{\sigma_u}{\rho_w C_d}}\left(\frac{D}{l}\right). \tag{14.20}$$

Usually, the non-dimensional coral branch diameter D/l is of the order of 0.05–0.15. Thus, assuming for example, that $D/l = 0.1$, $\sigma_u = 22 \times 10^6$ Pa and $C_d \approx 1.2$, for $Re \approx 10^4$, we obtain the critical velocity $u \approx 8.4$ m/s. For a given current and wave environment, Eq. (14.20) provides the critical branch proportion (D/l) which guarantees the survival of the structure, i.e.:

$$\frac{D}{l} \geq \frac{u_{\max}}{\sqrt{\dfrac{\pi}{8}\dfrac{\sigma_u}{\rho_w C_d}}}. \tag{14.21}$$

At the frontal reef slope, wave-induced velocity is higher than current velocity and velocity $u_{\max}$ can be considered as maximum orbital velocity. In Fig. 14.11a, a variation of wave height and corresponding orbital velocity at sea bottom, over the reef slope, is shown. When the wave height is known (see previous section), wave-induced orbital velocity can be calculated from Eq. (4.25) as:

$$u_{\max}(0) = \frac{gkH}{2\omega}\frac{1}{\cosh kh}.$$

(14.22)

Using the values of stress, σ_u, and water density, ρ_w, as 22×10^6 Pa and 1023 kg/m³, respectively, and substituting Eq. (14.22) into Eq. (14.21), the minimal value of non-dimensional branch diameter, at each point of reef slope, is obtained. As was shown in Fig. 14.11a, the maximum value of $D/l \sim 0.025$ corresponds to the highest flow velocity at the vicinity of reef edge, as expected. This figure illustrates the variations of the D/l values, necessary for coral survival. The results are valid only for a given reef geometry and incident wave parameters. For example, during tropical cyclone wave heights and wave-induced velocity can be higher and higher value of D/l is needed for coral survival. Moreover, as the velocity u has been represented as bottom velocity, an estimated value of D/l is restricted to rather short coral branch, located close to sea bottom.

From Eq. (14.16) it follows that bending moment induced by uniform flow changes along coral branch, being zero at the free end of branch and reaching its maximum of $1/2\ F_d l^2$. Therefore, to keep the stress σ_u constant along coral branch, it is sufficient that dimension D of branch satisfies the following condition:

$$\frac{8\rho_w C_d u^2 (l-z)^2}{\pi D^2} = \sigma_u = \text{Const},$$

(14.23)

or:

$$D = \sqrt{\frac{8\rho_w C_d u^2}{\pi \sigma_u}}(l-z).$$

(14.24)

The stress remains constant along coral branch when its dimension increases in proportion to the distance from the free end. The greater is the velocity, u, the faster branch must grow. Conversely, the stronger is branch material (the larger value of σ_u), the slower branch growth can be.

In general, coral branches are not perfect circular cylinders and elliptical cross-sections are very common. Using a similar reasoning, it can be found that flow induced moment, M, and bending stress, σ_b, are:

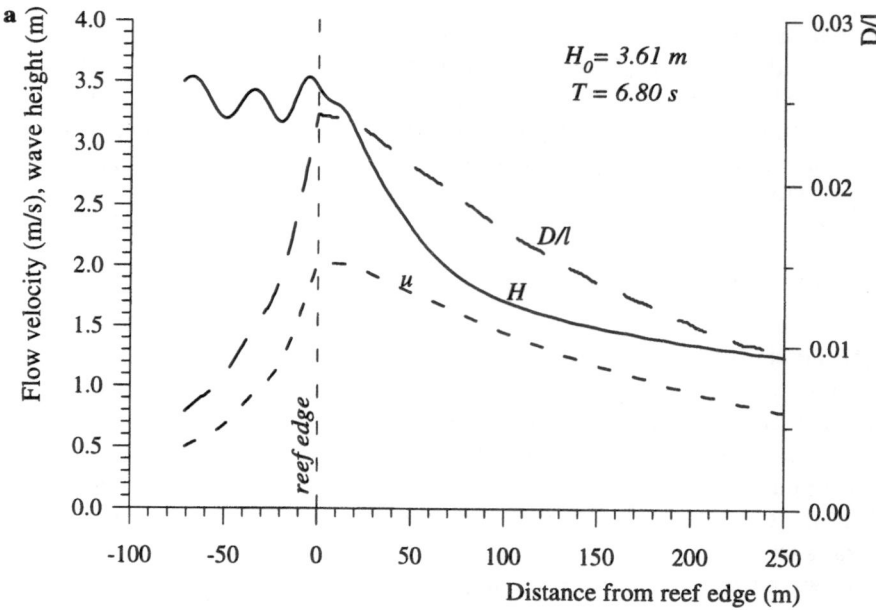

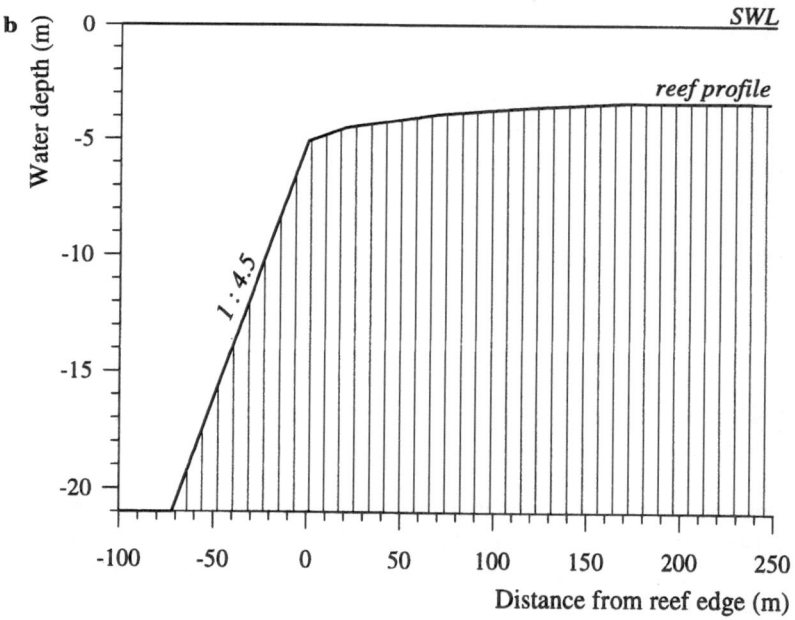

Fig. 14.11: Coral branch proportion over the reef slope: **a** variation of wave height, water velocity and dimensionless coral branch diameter, **b** reef bathymetry

$$M = \frac{1}{2}\rho_w C_d b u^2 (l - z)^2, \tag{14.25}$$

$$\sigma_b = \frac{2}{\pi}\rho_w C_d \left(\frac{l - z}{a}\right)^2 u^2, \tag{14.26}$$

where we assumed that the longer axis of elliptical cross-section of the branch was parallel to water flow, and a was the length of the half of this axis. From Eq. (14.26) we can find that flow velocity causing a branch breaking at the substratum cross-section is:

$$u \geq \sqrt{\frac{\pi}{2}\frac{\sigma_b}{\rho_w C_d}}\left(\frac{a}{l}\right). \tag{14.27}$$

The patterns of coral branches are much more complicated than a simple vertical cylinder. Many of hard coral species form a complex, tree-type structure with many joints and branches oriented in various directions. Elementary component of the coral structure is Y-shape branch (Fig. 14.12). For simplicity of derivation of the critical velocity required to break a branch, we assume that the branch is fixed firmly to the substratum and both higher branches are oriented in the plane (x, z) parallel to water flow, and the lower branch is vertical. The dimensions of particular branches and their lengths are shown in Fig. 14.12.

For calculation convenience, we start with determining of forces and moments at point B. In particular, the moment in point B is given by:

$$M_B^{(A-B),\text{drag}} = \int_0^{l_1} F_d^{(A-B)} s\, ds, \tag{14.28}$$

where s is the distance along branch $A - B$, measured from the point A; and the force $F_d^{(A-B)}$ becomes:

$$F_d^{(A-B)} = \frac{1}{2}\rho_w C_d D_1 (u \cos\theta_1)^2. \tag{14.29}$$

After substituting Eq. (14.29) into Eq. (14.28) and integrating, we obtain:

$$M_B^{(A-B),\text{drag}} = \frac{1}{4}\rho_w C_d D_1 (l_1 \cos\theta_1)^2 u^2. \tag{14.30}$$

In a similar way we can obtain the moment due to drag forces on the branch $(B - C)$ as:

$$M_B^{(B-C),\text{drag}} = \frac{1}{4}\rho_w C_d D_2 (l_2 \cos\theta_2)^2 u^2. \tag{14.31}$$

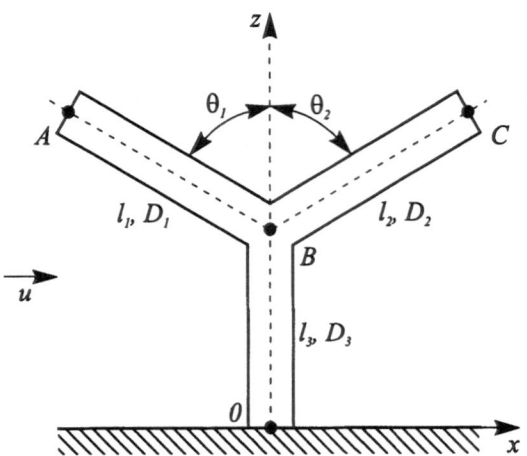

Fig. 14.12: Scheme of the Y-shape coral branches

Moments given by Eqs. (14.28) and (14.31) try to turn branches $(A - B)$ and $(B - C)$ around the point B in clockwise direction, considered as a positive.

Other forces imposed on coral branches are forces due to gravity. Assuming that coral density is uniform along the whole branch, the gravity force at branch $(A - B)$ produces the following moment at the point B:

$$M_B^{(A-B),\text{grav}} = -\frac{\pi(\rho_c - \rho_w)g}{8}(l_1 \sin \theta_1)^2 D_1^2, \tag{14.32}$$

in which ρ_c is the coral density.

The sign $(-)$ in Eq. (14.32) indicates that the moment $M_B^{(A-B),\text{grav}}$ is negative, *i.e.* gravity force moment tries to turn branch $(A - B)$ around the point B in the counterclockwise direction. In a similar way we can determine the moment $M_B^{(B-C),\text{grav}}$ as:

$$M_B^{(B-C),\text{grav}} = \frac{\pi(\rho_c - \rho_w)g}{8}(l_2 \sin \theta_2)^2 D_2^2. \tag{14.33}$$

The total moment at the point B is the summation of all moments acting on branches $(A - B)$ and $(B - C)$, *i.e.*:

$$M_B^{(\text{total})} = \left(M_B^{(A-B),\text{drag}} + M_B^{(B-C),\text{drag}}\right) + \left(M_B^{(A-B),\text{grav}} + M_B^{(B-C),\text{grav}}\right). \tag{14.34}$$

Another critical cross-section of Y-shape coral branch is the cross-sectional point O, at the substratum. Moment at the point O includes the contribution from the moments at the point B and from the drag force acting on the branch $(B - O)$. Thus finally, at the point O we have:

$$
M_O^{(\text{total})} = M_B^{(\text{total})} + \frac{1}{2}\rho_w C_d \left(D_1 l_1 \cos^3\theta_1 + D_2 l_2 \cos^3\theta_2\right) l_3 u^2 +
$$
$$
+ \frac{1}{4}\rho_w C_d D_3 l_3^2 u^2. \tag{14.35}
$$

Assuming for a moment that only vertical branches exist, such as: $\theta_1 = \theta_2 = 0$, $l_2 = 0$, and $D_3 = D_1$, we obtain:

$$
M_O^{(\text{total})} = \frac{1}{4}\rho_w C_d D_1 (l_1 + l_3)^2 u^2, \tag{14.36}
$$

as was expected.

Using the moment $M_O^{(\text{total})}$ in Eq. (14.12), we can calculate bending stress σ_b for a given flow velocity u or to determine critical flow velocity u_{cr} which causes a branch breaking for a given ultimate stress σ_u. Assuming, for example, that $l_1 = l_2 = 0.042$ m, $l_3 = 0.055$ m, $D_1 = D_2 = 0.01$ m, $D_3 = 0.012$ m, and $\theta_1 = \theta_2 = 30°$, we obtain critical velocity value $u_{cr} \approx 8$ m/s. Similar calculations can be carried out for bending sea anemones (*Metridium senile* or *Anthopleura xanthogrommica*). The details of derivation of critical stresses can be found, for example, in Denny (1988).

Probability of Massive Corals Dislodgement. There are two modes of response of massive corals which have been observed in cyclone-damaged areas on the GBR (Van Woesik *et al.*, 1991; Done, 1992) and Vanuatu in the south-west Pacific Ocean (Done and Navin, 1990):

1. *Dislodgement*, in which many massive *Porites* coral heads, up to several meters in diameter and height were overturned (Done and Potts, 1992). These corals, which had apparently been either unattached, or attached to a weak reef matrix, either remained at or near their position of growth and came to rest on a terrace some metres down-slope from their position of growth, or were moved up the slope onto an adjacent reef. Dislodgement is more likely to cause partial mortality and colony relocation than colony death. Tissues brought into contact with the substratum were killed, but remaining areas of the colony survived.

2. *Resistance* is the other response, in which massive corals remained standing, either sustaining or escaping obvious physical damage. These corals included some *Porites* colonies, which displayed evidence of polyp mortality from the impact of projectiles (Done and Navin, 1990) and others which were undermined during the storm, a distinct trench having been eroded in the surrounding rubble or coral assemblages. Other *Porites* and the majority of colonies of *Disploastrea heliopora* - a common massive coral in the family *Faviidae* - escaped apparently unharmed.

Using the meteorological record, hydrodynamic formulations and risk analysis, some demographic consequences of tropical cyclones for massive corals
growing in different regions of the (GBR), Australia, have been predicted (Massel and Done, 1993). Analysis of shear, compression and tension forces generated by waves indicate that corals firmly attached to a solid substratum, even
if only over a small proportion of their base, can resist all waves, regardless
of colony size shape, cyclone intensity or region. Waves are thus directly important as controls on colony size and frequency distributions only for weakly
attached or unattached colonies. For example, the relationships suggest that
massive corals of any size with only a small attachment to a consolidated substratum are invulnerable to even the most powerful wave, either unbroken
or broken, generated by the most severe tropical cyclone in GBR waters. It
appears that adhesion can resist the most powerful drag and the overturning moments which impose compression and tension forces on the basal plate.
Corals on the GBR which are dislodged in storms must, therefore, be either
unattached, or strongly attached to substrata which are themselves, weak, or
of small mass.

For unattached corals, coral heads will overturn about point O (Fig. 14.13)
and dislodge from substratum when:

$$F_h^{(t)} \frac{D}{2} \geq F_v(t) r_b, \qquad\qquad (14.37)$$

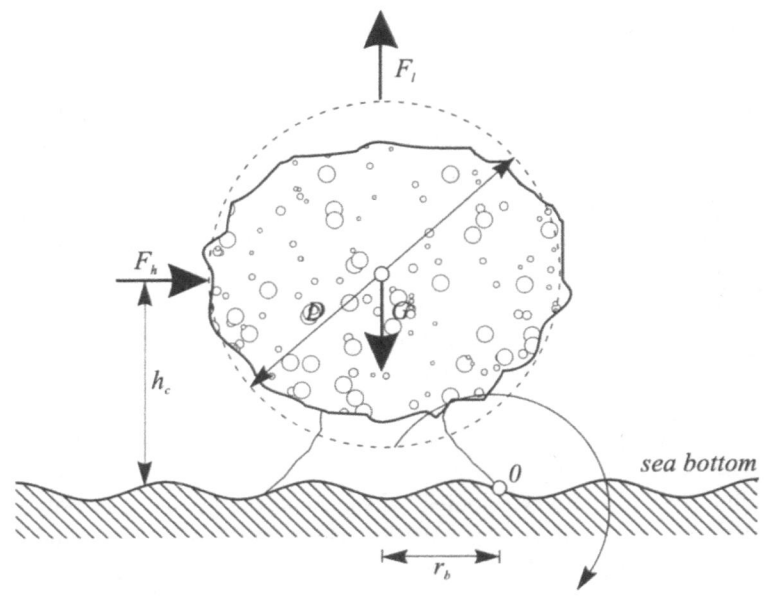

Fig. 14.13: Wave forces on the massive *Porites*

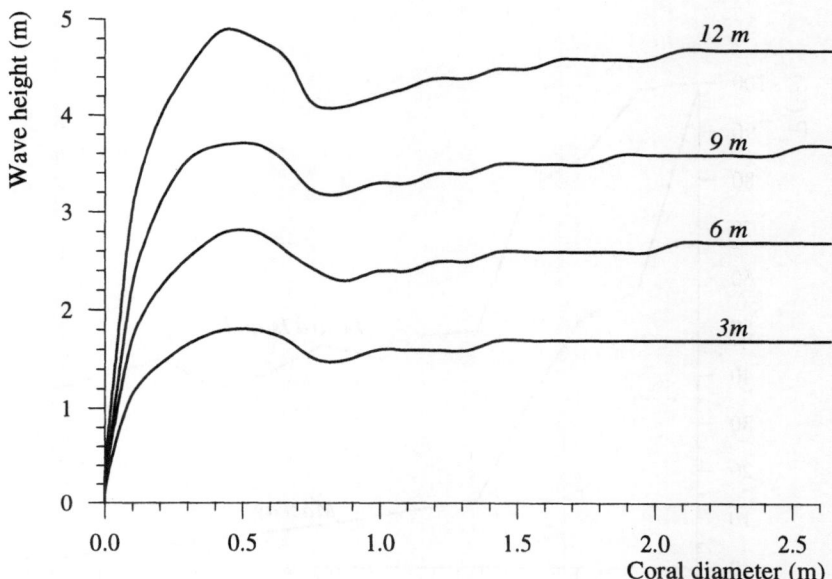

Fig. 14.14: Height of unbroken waves required to dislodge unattached coral of indicated diameter at indicated depth (adapted from Massel and Done, 1993)

where $F_h^{(t)}$ is the total horizontal force (drag force and inertia force), $F_v^{(t)}$ is the total vertical force (lift force and gravity force), and r_b is the arm of the vertical force with respect to the point O. The forces $F_h^{(t)}$ and $F_v^{(t)}$ change during wave period and time instant, at which the condition (14.37) is satisfied, has to be found (Massel and Done, 1993). Assuming, for example, that the reef front slope is 1:4, and the mean period of incident waves is $T = 7$ s, the dependence of threshold wave height on coral diameter and water depth is shown in Fig. 14.14.

The threshold wave height needed to dislodge a bigger coral is almost independent of its diameter. This rather unexpected result can be explained as follows. The inertia force due to wave oscillatory motion depends on the product of coral mass, which is proportional to the cube of its diameter and water acceleration. Also a coral weight is proportional to the cube of diameter. Thus, the overturning and stabilizing forces do not involve coral diameter and threshold wave height remains almost constant.

The threshold wave height, H, needed to overturn a coral of a given diameter is associated with the specific return period, $\bar{T}(H)$, (or average time between storms) of storm or cyclone conditions, which induce such waves. This period varies among regions. If a coral occupies a location for L years, the *encounter probability* $E(L,T)$ for a wave exceeding height H, during the L years of coral life (Borgman, 1963), is:

$$E(L,\bar{T}) = 1 - \left(1 - \frac{1}{\bar{T}}\right)^L. \tag{14.38}$$

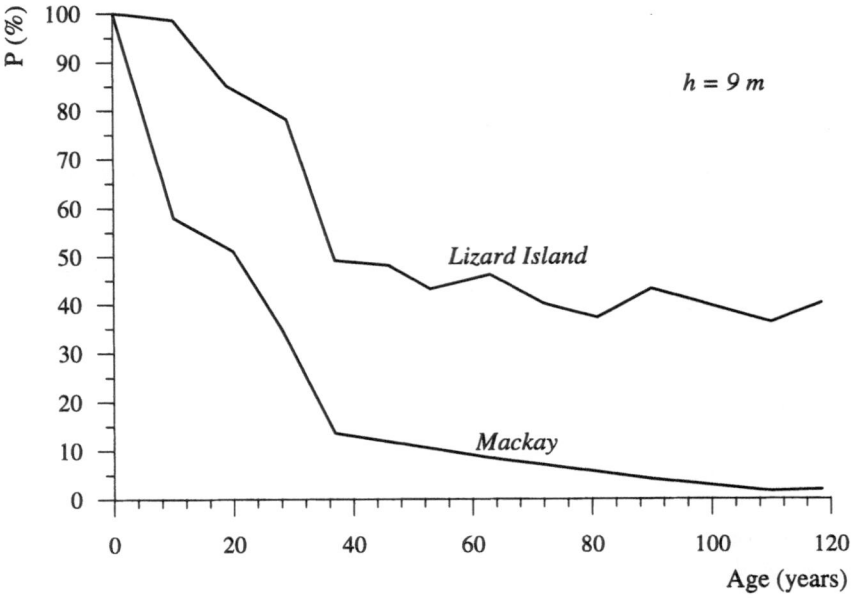

Fig. 14.15: Probability of unattached spherical colonies exposed to cyclone-induced waves, of attaining indicated ages for water depth $h = 9$ m, before encountering a wave capable of dislodging it from that depth (adapted from Massel and Done, 1993)

Let us define the probability of survival of a massive coral to a given age at a given water depth as P. Thus we have:

$$P = 1 - E = \left(1 - \frac{1}{\bar{T}}\right)^{L}. \tag{14.39}$$

These numerical relationships have been used to develop the survival probabilities of massive corals in various regions as follows (Massel and Done, 1993):

- Obtain the limiting wave height, H, for a given water depth and coral age, L. The wave height, H, creates sufficient force to overturn the coral head. The vector of force is a sum of three components: inertia, drag and lift force (see Sect. 2.6).

- Find the dependence of the limiting wave height on the return period $\bar{T}$ of cyclone, for the particular location. For a given cyclone severity, incident wave height should be determined and then this wave height should be transformed into shallower water on the reef slope (see Massel, 1996a).

- Calculate the survival probability from Eq. (14.39).

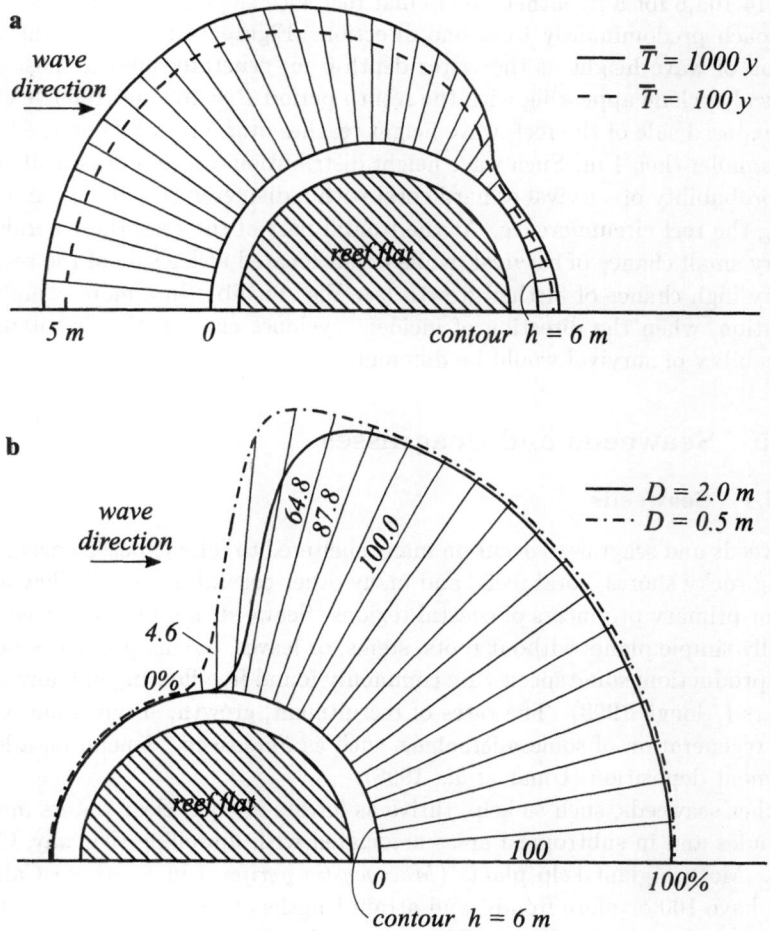

Fig. 14.16: Influence of sheltering effect on the probability of coral survival; **a** distribution of wave height along reef circumference, **b** distribution of probability of coral survival

The numerical simulations for various regions of the (GBR) showed that for habitats exposed to cyclone induced waves, survival probability generally decreased with coral age and the rate of this decrease strongly depends on both location and water depth (Massel and Done, 1993). For example, a coral in 9.0 m of water at Lizard Island (northern part of GBR) has about 70% chance of reaching 30 years age, while the same coral in the Mackay (central part of GBR) has only 20% chance (see Fig. 14.15).

Another important factor controlling the chance of survival for hard coral is the location of coral on the reef slope with respect to the direction of incoming waves. Coral on the exposed slope of the reef has a smaller chance of growing than coral located on the lee side of the reef. This effect is illustrated in

Fig. 14.16a,b for a hypothetical circular reef with slope about 1:6, when waves approach predominantly from one direction. Figure 14.16a shows the distribution of wave height at the water depth 6 m, generated by the most severe tropical cyclone appearing with the return period $\bar{T} = 100$ and 1000 years. On the exposed side of the reef, wave height reaches over 5 m, while on the lee side it is smaller then 1 m. Such wave height distribution causes a drastically different probability of survival of hard coral with a diameter $D = 0.5$ m, and 2.0 m along the reef circumference. At the frontal part of the reef, these corals have a very small chance of survival, while corals located at lee side of the reef have a very high chance of further growth (see Fig. 14.16b). In a more complicated situation, when the direction of incident cyclones change, the distribution of probability of survival would be different.

14.5   Seaweeds and Seagrasses

14.5.1   Seaweeds

Seaweeds and seagrasses occur on many sheltered to fully exposed coasts, occupying rocky shores, coral reefs, and many other coastal habitats. They are the major primary producers of coastal regions. Seaweeds are marine algae: structurally simple plants without roots, stems, or leaves, having primitive methods of reproduction; some species are commonly found free-floating in many coastal waters (Alongi, 1998). The rates of recruitment, growth, survival and vegetative regeneration of some microalgae, such as *Sargassum*, depend on a level of sediment deposition (Umar *et al.*, 1998).

Other seaweeds, such as kelp, thrive as forests in cool, clear waters in higher latitudes and in subtropical areas associated with upwelling (Norway, California). Mature giant kelp plants (*Macrocystis pyrifera*) in Southern California may have 100 or more fronds, and attain lengths of 15 to 50 m while attached to the seafloor by holdfasts. The average density of typical kelp beds varies from 6 to 12 plants per 100 m² (Elwany *et al.*, 1996). Under favourable conditions, maximum rates of primary production of seaweeds can exceed rates of production of all macrophytes, including mangroves and marsh grasses. Kelp in wave exposed locations have long stipes and thick, narrow blades with heavily corrugated surfaces; in more sheltered areas, the stipes may be as long, but laminar, wider and thinner (Hurd *et al.*, 1996). Physical factors that affect kelp growth and recruitment include water temperature, underwater light, nutrient levels, and the concentration of suspended particles near the bottom. Therefore, the deep-water limit of kelp growth is approximately 20 m or less. The inshore limit is mostly controlled by the largest breaking waves which cause the holdfasts of kelps to break loose from the bottom (Seymour *et al.*, 1989).

Other kelp species, *Laminaria hyperborea*, grow at several locations along the Norwegian coast at water depths between 2 and 20 m. Usually the kelp plant has 1–2 m height stipe, and average density of kelp forest is about 10–15 plants per square metre. Fronds have the same length as the stipe. The total kelp biomass along the Norwegian coast has been estimated at some 10 million tones and about 160,000 tones of kelp is harvested annually for chemical production. The harvesting of seaweeds has been a controversial issue, as there exists a public perception that kelp can somehow detrimentally or beneficially affect beach widths. It is reasonable to assume that a kelp bed must alter the incoming wave to have any effect on beaches, and seaweeds are known to reduce wave energy when waves propagate through them.

Laboratory and theoretical analysis by Dubi and Tørum (1994) proved that average damping of wave height by kelp *Laminaria hyperborea* can be expressed as:

$$\frac{H(x)}{H_0} = e^{-k_d x}, \tag{14.40}$$

in which H_0 is the incident wave height, and $H(x)$ is the wave height at the distance x from the offshore boundary of the seaweed area. The attenuation coefficient, k_d, depends on water depth, *i.e.*:

$$k_d = 0.0586 e^{0.5h}, \tag{14.41}$$

in which h is the water depth in metres. Therefore, at water depth $h = 6$ m ($k_d = 0.0029$ m^{-1}), the zone of kelp of 100 m width will reduce incident wave height by 25%.

No final conclusions have been made as to whether kelp harvesting influences beach erosion. Elwany *et al.* (1996) used two statistical approaches to examine the relationship between the width of kelp beds and the width of the beaches inshore in the San Diego region of Southern California. They did not find a clear correlation or consistent pattern indicating that offshore kelp beds had any direct influence on adjacent beach width. A more significant governing factor for beach erosion is the impact of high storm waves associated with high storm surges. This conclusion has been confirmed by experimental and numerical studies of the Norwegian coast (Tørum, 1996).

Efficiency of kelps in damping of wave energy strongly depends on their ability to withstand drag forces produced by strong current or waves. Flexible kelp stipes and fronds can reorient and, due to their high extensibility, they can take more streamlined position or move (to some extent) with a flow. In such way, the relative velocity between organism and the surrounding water is substantially reduced, and the drag and lift forces are also reduced. For example, a small, isolated population of the kelp *Laminaria ochroleuca* occurs in the Strait of Messina, separating the Italian mainland from Sicily, central Mediterranean (Drew *et al.*, 1982). Unusual hydrographic conditions produce almost continuous strong currents through the strait.

Photographs of kelps at 60 m depth showed that many narrow lamina digitations became arranged in a cylindrical bundle about 25 cm diameter and up to 3 m long. The presence of considerable number of 'decapitated' stipes suggests that they were damaged by strong currents in the strait. Mechanical test showed that large kelp (6–7 year old) break at or very near to the stipe-lamina junction under the force of 255 N. The junction is usually the narrowest part of the *Laminaria* plant and is composed of young meristematic tissue (Drew et al., 1982). A drag formula (2.63) suggests that the force of about 255 N is exceeded at current velocity of about 2.9 m/s. Seasonal variation of currents in the strait show that the highest value is about 3.07 m/s.

The situation is more complicated when the flexible organisms, such as kelps, have to withstand forces in the wave-swept environment when the velocities and accelerations may be large. Waves passing over the plants induce drag forces on the lamina, causing lateral swaying of both the stipe and lamina. There are a few mathematical models recently proposed to describe the dynamics of the kelp stipe/lamina system (Denny, 1988; Dubi and Torum, 1994; Denny et al., 1998). All models are based on the theory of vibrations of mechanical systems. For simplicity, the stipe is considered as a vertical cantilever with the entire effective mass, m, of the system concentrated at the free end of the stripe (Fig. 14.17). The effective mass includes the plant's lamina mass, the equivalent mass of the stipe and the hydrodynamic added mass of the plant, $\rho_w C_a V$, in which V is the volume of water displacement by the plant, and C_a is the added mass coefficient, which is the function of plant shape (for definition of C_a see Sect. 2.6).

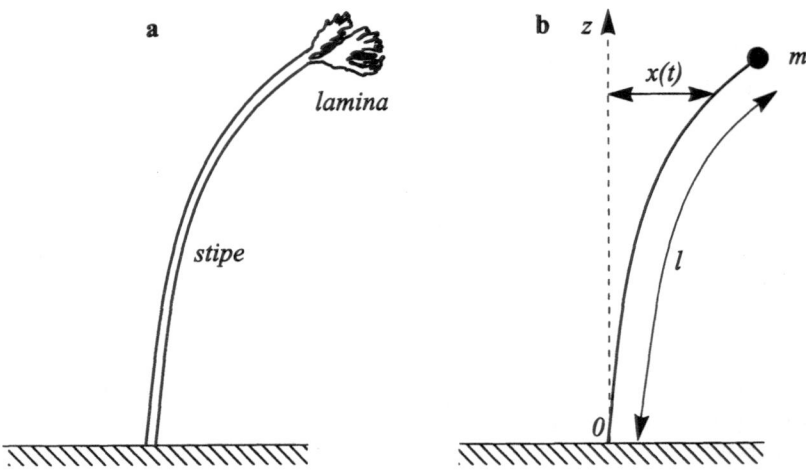

Fig. 14.17: Biological models: **a** actual structure of the kelp, **b** model of the structure

The basic equation for cantilever vibration has the form:

$$m\frac{d^2x}{dt^2} + kx = \text{forcing terms},\qquad(14.42)$$

where k is the stiffness of the stipe. When deflections of the stipe are small (less than a tenth of the stipe's length), stiffness k is approximately given by (Gere and Timoshenko, 1991):

$$k \approx \frac{3(EI)}{l^3},\qquad(14.43)$$

where EI is known as flexural stiffness of the stipe, E is the Young modulus of the stipe material, and I is the moment of inertia. In real marine conditions, kelp stipes experience large deflections and formula for stiffness k becomes more complicated (see, for example, Gere and Timoshenko, 1991, or Denny et al., 1998).

Let us assume for a moment that all forcing terms are zero. Equation (14.42) can be integrated to give the deflection $x(t)$ in the form of harmonic oscillations:

$$x(t) = \left(\frac{u(0)}{\omega_n}\right)\sin(\omega_n t) + x(0)\cos(\omega_n t),\qquad(14.44)$$

in which $x(0)$ and $u(0)$ are the deflection and velocity at time zero, and frequency ω_n is:

$$\omega_n = \sqrt{\frac{k}{m}} = \sqrt{\frac{3EI}{ml^3}}.\qquad(14.45)$$

Frequency ω_n is known as the natural circular frequency, and corresponding natural period T_n becomes:

$$T_n = \frac{2\pi}{\omega_n} = 2\pi\sqrt{\frac{ml^3}{3EI}}.\qquad(14.46)$$

As no damping forces oppose the motion, the cantilever continues to oscillate with the frequency ω_n forever.

The external forces at the right-hand side of Eq. (14.42) for oscillating kelp stipe can be summarized as follows (Denny et al., 1998):

$$\text{forcing terms} = \frac{1}{2}\rho_w S_d A u_r |u_r|^{\gamma-1} + \rho_w C_m V a_x,\qquad(14.47)$$

in which the first term on the right-hand side is the contribution from drag force and the second term is the contribution due to the inertia force. The coefficient S_d is the shape coefficient of drag for oscillating stipe, A is the maximum projected area of the object, and the coefficient γ expresses the dependence of drag force on the relative flow velocity u_r:

$$u_r = u - \frac{dx}{dt},\tag{14.48}$$

where u is the velocity of the water, and the dx/dt is the velocity of the organism. From Gaylord et al. (1994) measurement follows that $S_d \approx 0.041$ and $\gamma = 1.55$ for kelp Eisenia and 0.042 and 1.23, respectively, for kelp Pterygophora.

Dubi and Tørum (1994) measured drag forces on stipes of Laminaria hyperporea in the Norwegian coastal waters by sharp towing, and in laboratory using a shear plate on which 95 model kelp plants were fixed. The resulted drag force showed a linear dependence on the current velocity instead of quadratic dependence, which was expected, i.e.:

$$\text{drag force} \approx 9.33 u_c,\tag{14.49}$$

where u_c is the current velocity.

Under wave conditions with the periods $11 < T < 14$, the relationship between drag and maximum wave horizontal velocity u becomes:

$$\text{drag force} \approx 3u.\tag{14.50}$$

The second term on the right-hand side of Eq. (14.47) represents the contribution of inertia factor to the external force. Denny et al (1998) solved Eq. (14.42) with right-hand side given by Eq. (14.47). Resulted stipe deflection becomes non-linearly related to the applied force. Moreover, the maximum stress on the stipe appears for the external forcing frequency, ω, very close to the natural frequency, ω_n, i.e.: when $\omega/\omega_n \approx 1$. The fact that inertia forces reach the maximum values at certain frequencies provides some foundation for the possibility that the structure of organisms can be tuned to avoid potentially damaging loads.

More experimental data and theoretical results on dynamics of flexible marine organisms can be found in papers of Carrington (1990), Gaylord et al. (1994), Dubi and Tørum (1994), Friedland and Denny (1995), Bell (1996), Denny and Gaylord (1996) and Utter and Denny (1996). In particular, Dubi and Tørum (1994), unlike other studies, derived the model of the dynamics of kelp in wave dominated environment when boundary conditions at sea surface are present. This means that real vertical profile of wave-induced velocity is used in calculations of stresses on kelp.

14.5.2 Seagrasses

Seagrasses, being rooted angiosperms, flourish as meadows in soft sediments at and below the intertidal zone. Seagrasses are usually highly productive autotrophs with productivity rates somewhat less, on average, than those for marsh grasses, mangroves, and terrestrial plants. The growth and productivity of plants is greatly influenced by water movement which distributes nutrients and gases and removes wastes. In particular, currents and waves can increase primary productivity by enhancing water mixing. However, recent studies have demonstrated that this effect of water flow is clearly seen for only a narrow window of current speed. Higher current speeds and the associated higher levels of turbulence result in higher concentrations of suspended material which reduces light availability and rates of production.

Seagrasses have recently become of interest due to the role they play in increasing sediment stability in anthropogenically disturbed areas. With increasing numbers of vessels using waterways, the reduction of boat wake wave energy by these plants is suspected to be an important factor in reducing sediment resuspension. Reduction of anthropogenically derived wave energy by seagrasses should also lower shoreline erosion rates. Also seagrasses are able to modify current flow and sediment composition. Laboratory tests by Fonseca and Cahalan (1992) for the seagrass species *Halodule wrightii*, *Syringodium filiforme*, *Thalassia testudium* and *Zostera marina*, populating in American waters (Florida coast, Gulf of Mexico) showed that these seagrasses, when occupying 50%–100% of the water column, are able to reduce wave energy by approximately 40%. The behaviour of a seagrass under wave action is complex because of the turbulence generated along the interfacial boundary and among the vegetation stands. However, it has been observed that each stand does not move at random, but groups of stands sway in an organized manner. For short seagrass, it may be treated as a horizontal swaying motion. As a first attempt, the vegetation motion due to wave motion is simply modelled as a forced vibration with one degree of freedom. The restoring forces are the buoyancy and stiffness of the vegetation material. Such a type of solution was recently proposed by Asano *et al.* (1992).

14.6   Coastal Fluid Muds

Accumulations of gel-like fluid mud front many coastlines, from the humid tropics of the equatorial latitudes to temperate latitudes. Most coastal fluid muds occur as intertidal and subtidal mudshoals near major rivers. The Mississippi River supplies muds to the coast of Louisiana and the Amazon River is responsible for muds on the shorelines of French Guiana, Surinam, and Guyana. Mudflats of the Yellow Sea and Gulf Po Hai are derived from both the Yangtze and Hwangho rivers in China. As nearly 30% of the world's fluvial sediment is derived from rivers in Asia, the greatest concentrations of fine sediment are found in Asian countries such as India, Malaysia, Thailand, China and Korea.

Fluid muds are often described as thixotropic substances that possess the rheological properties of pseudoplastics and Bingham plastics. Bulk density ranges from 1.03 g/cm^3, slightly greater than that of sea water, to a yogurt like consistency with density greater than 1.30 g/cm^3. Mechanisms that lead to the accumulation of fluid muds are poorly understood. It is believed that the most important mechanisms are: trapping by waves and coastal currents and estuarine-type inner-shelf circulation.

Many observations indicate that the attenuation rates of surface waves propagated over muddy bottoms are much greater than normally attributed to bottom friction dissipation. An experiment at the front of the Surinam River showed that the wave energy loss between two stations at 22 km (water depth 7.1 m) and 4 km offshore (water depth 1.3 m) was about 96% (Wells, 1983). Although water depth at the inner station was only 1.3 m, waves did not break, nor did they undergo the typical shoaling and transformation that leads to a height increase just prior to the time of breaking. The spectra at the outer station show a combination of swell ($T = 9$–10 s) and wind-induced waves ($T = 3$–5 s). The ratio wave height/water depth remains nearly constant at 0.19, which is a much lower value than the 0.55 associated with the limiting wave height (see Chap. 4). Various theories have been proposed to describe the interaction between water waves and a soft bed. The interested reader should consult Massel (1996a) for a more detailed discussion.

15 Vertical Structure of Ocean Waters and Biological Productivity

15.1 Introduction

It is a widely accepted opinion that phytoplankton need light and nutrients for growth and reproduction. The source of light comes from above, while the source of nutrients is at depth. Therefore, to bring the nutrients to the surface, vertical water fluxes are needed. As we showed in the previous chapters, there is a variety of hydrodynamic processes, such as upwelling, turbulent and tidal mixing, and internal waves, all resulting in a vertical transport of water mass. A complete description of the link between vertical structure of ocean water and biological ocean productivity is out of the scope of this book. Rather, this broad subject deserves its own separate book. In this chapter we will only provide some examples of the influence of vertical ocean structure on primary production.

The obvious physical mechanism which provides vertical transport of water mass is upwelling. The upwelling results in the replacement of warm surface water with colder nutrient-rich water. There are three primary types of upwelling: coastal, polar and equatorial. The best known upwelling is coastal. The relationship between water movement and primary production associated with coastal upwelling is discussed in Sect. 15.2.

The distribution of ocean water temperature, salinity, density and other water properties is not uniform throughout the water column or in the horizontal plane. Instead, there are many regions where horizontal gradients are very large. They are known as fronts, which appear in many sizes, from large oceanic fronts down to small scale fronts within island wakes. As the larger fronts are relatively well known, in Sect. 15.3 the emphasis is on small-scale, less known fronts in shallow waters.

Finally, in Sect. 15.4, the problem of a lack of stratification due to tidal mixing and biological production is described. The lack of stratification would be expected to decrease phytoplankton productivity. However, vigorous mixing by tides in shallow and shelf waters brings the nutrients back into the water column and stimulates phytoplankton production (Sects. 15.5 and 15.6).

15.2   Upwelling and Biological Production

15.2.1   General Overview

Upwelling regions around the globe make up about 0.1% of the ocean's surface area, however, these areas provide more than 40% of all the commercial fish captured globally. In general, there are a few types of upwelling: coastal upwelling, upwelling driven by river runoff, upwelling due to tidal mixing and upwelling induced by large-scale circulations. In this section, we will concentrate on coastal upwelling. Other types of upwelling and their corresponding biological consequences will be summarized briefly.

As was shown in Sect. 13.2.4, the movement of freshwater towards the sea creates a shear which generates the progressive mixing of fresh and salt water in the estuary. Due to the entrainment mechanism, the freshwater flow carries out to the sea several times its own volume of salt water from below the pycnocline. Bottom water inflowing to the estuary is nutrient-rich through its contact with the decomposing organisms on the sea bottom. These nutrients are vertically transported into the surface waters. The phytoplankton take up nutrients, zooplankton eat phytoplankton which are in turn eaten by fish. This regular food web induces further sinking of organic matter which enriches the water moving up the estuary near the sea floor (Mann, 1991).

When tidal currents are very strong, they can break down the stratification in the water column and cause the complete mixing of the freshwater with the underlying salt water. Nutrient-rich water from the bottom is distributed throughout the water column and provides the basis for the phytoplankton growth. However, due to high turbidity, the penetration of light is limited and phytoplankton production may be much less than in a stratified estuary (see Sect. 13.2.4 and Bowman et al., 1986).

In Sect. 7.7.4 we have described two other examples of upwelling driven by large-scale circulation, namely polar upwelling and equatorial upwelling. Much of the high aquatic productivity around the Antarctica results from upwelling at the Antarctic Divergence. As much as 75% of the primary production close to the Antarctic continent is accomplished by microplankton ($20~\mu m$–2 mm), and other primary food sources include detritus and bacteria. A bio-optical algorithm based upon monthly climatological phytoplankton pigment concentration from the coastal zone colour scanner show that there is a large variation of annual primary production over the Southern Ocean (all area south of 50°S) from 630 TgC/year in the South Indian Ocean up to 1222 TgC/year in the Ross Sea, which accounts for nearly 28% of the annual production of the Southern Ocean (Arrigo et al., 1998).

The zone of maximum primary production advances poleward with the progression of the summer and the retreat of sea ice. Although the phosphate (PO_4) and nitrate (NO_3) concentrations persist in these waters due to the widespread upwelling, there are several factors limiting the primary production. The most important are the low angle of light incidence, strong winds and large waves (see, for example, Massel, 1996a). Turbulence induced by winds and waves drags phytoplankton down to depths where light intensity is too low to permit photosynthesis (Pinet, 1992). A full description of the food web in the Southern Ocean is beyond the scope of this book and the reader should consult the numerous biological books and papers on this subject (for example, Burton, 1977; Gray and Christiansen, 1985; Laws, 1985).

Equatorial upwelling is a result of the complex interaction of the westward-flowing North and South Equatorial currents and the eastward-flowing Equatorial Countercurrent and Equatorial Undercurrent. Coriolis deflection induces divergence of the equatorial water and an upwelling of cold water from depth. Despite seasonal variations in the nutrient content of the surface water, primary production remains remarkably stable throughout the year and food chains in the photic zone are long, from phytoplankton to large predators such as marlin and shark.

An oceanographic survey of the eastern tropical Pacific Ocean in August–November 1990, confirmed a productive high-chlorophyll surface layer in two upwelling regions, namely the equatorial divergence (east of the Galapagos) and the countercurrent divergence, more than 1000 km west of the Costa Rica Dome (Fiedler et al., 1991). The observed rates of ^{14}C phytoplankton productivity were of the order of 400–800 mg C/m^2/day in equatorial water between the Galapagos and longitude 140°W. However, the productivity is limited by NO$_3$ availability. Excess NO$_3$ persists in the euphotic zone and is advected to weak upwelling regions, where it can persist for longer than 200 days.

15.2.2 Physical Factors and Primary Production in Coastal Upwelling Ecosystems

Coastal upwelling regions have been well studied, partly due to important fish stocks which exist in these areas. The factors controlling high biological production have been extensively reviewed by Boje and Tomczak (1978), Richards (1981) and Mann and Lazier (1996) and the basic physical mechanisms involved in coastal upwelling were described in Chap. 7. The typical nutrient cycling in a coastal upwelling system is schematically shown in Fig. 15.1. However, there is a great variability of the physical processes occurring in the major upwelling areas, namely the Oregon and California coasts, west coasts of Southern Africa and Peru, Spanish Sahara, and Somalia coasts.

To illustrate the linkage between physical and biological processes in an upwelling ecosystem, we will concentrate on the Peruvian upwelling system at 15°S, and Chilean upwelling (the Gulf of Aranco). The first system was studied intensively in 1976–1977 during the JOINT II experiments (Mann and Lazier, 1996). Due to a relatively narrow shelf (20 km) with a steep slope at the shelf break, and constant and relatively weak winds, offshore transport occurs mainly in the top 20 m and the wind-induced mixing does not penetrate deeply. The poleward countercurrent, flowing beneath the equatorward coastal current is situated at intermediate depth over the continental shelf. MacIsaac et al.

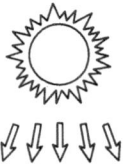

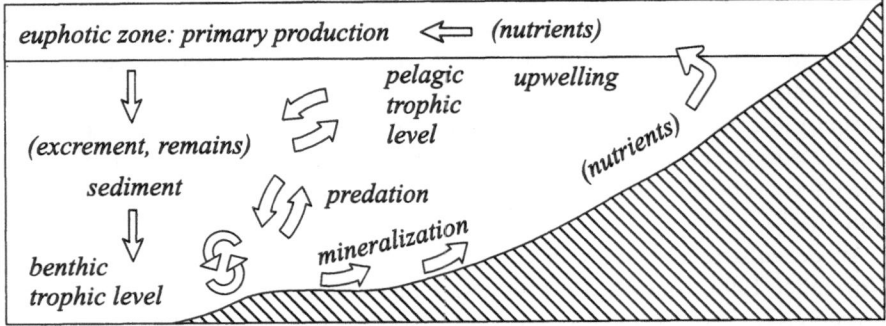

Fig. 15.1: Schematic representation of nutrient cycling in coastal upwelling system (adapted from Glantz, 1996)

(1985) reported results of comprehensive studies on primary production in an upwelling area at 15°S on the coast of Peru. They found that the rate of nitrate uptake in the shallow, as well as the deep, euphotic zone is controlled by light in the nutrient-rich, freshly upwelled waters. Once phytoplankton have been upwelled, it is rarely mixed below the euphotic zone because of the relatively strong stratification. The surface mixed layer in the area is rather shallow and only occasionally deeper than 20 m. The spatial distribution of the primary production in the surface waters is shown in Fig. 15.2. The pattern clearly demonstrates the areas affected by coastal upwelling as well as areas affected by the coastal band of equatorial upwelling.

Measurement by MacIsaac *et al.* (1985) showed that there is a consistent pattern of physiological transformation starting at the upwelling centre. In this sense, the coastal upwelling ecosystem has quasi time-space stability. However, the scale of this stability is rather small, when compare with oceanic scales, being of the order of 10 to 100 km and 1 to 10 days. Apart from this mesoscale variability, the magnitude of biological production varies interannualy.

There is experimental evidence that biological productivity decreases during El Niño events. The coastal upwelling continues but the water entrained is warmer and poorer in nutrients. The thermocline is progressively depressed toward and below the depth of entrainment (40 to 80 m) and smaller quantities of nutrients are transported to the surface. This in turn causes primary production of organic material to decrease proportionally. Also, the amount of light available to a phytoplankton population for the synthesis of organic

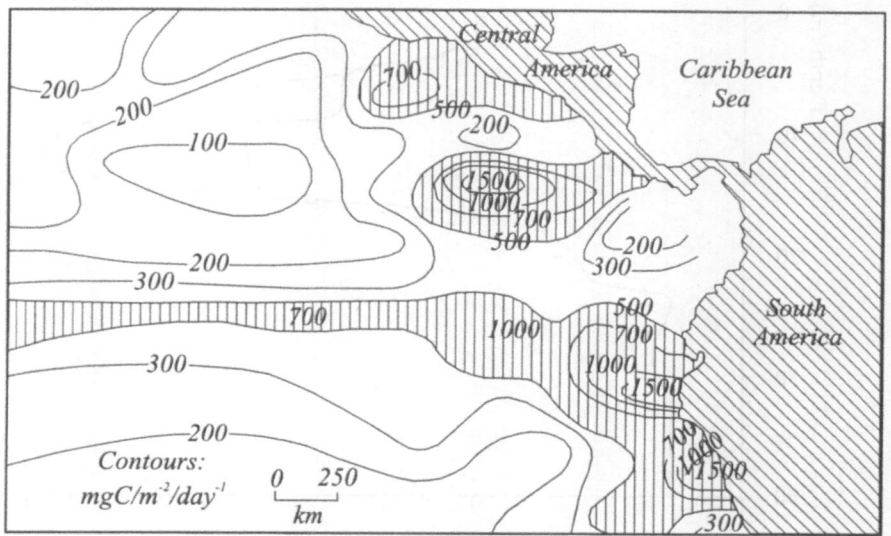

Fig. 15.2: Primary productivity in the surface waters of the eastern Pacific Ocean (adapted from Pinet, 1992)

material is decreased by a deepened mixed layer. Barber and Chavez (1983) found that mean surface nitrate and chlorophyll concentrations during the El Niño event in 1982–1983 reduced by 78 and 3.4 times, respectively. In the same time period, the primary production in the transect at 92°W and between the latitudes 2°N and 2°S decreased 20 times, compared with normal La Niña conditions.

Using fine-scale profiling of current, conductivity, temperature and chlorophyll a in an upwelling area also reveals the existence of high chlorophyll concentration well below the depth of optimum light intensity for photosynthesis. Djurfeldt (1994) examined the formation and maintenance of such subsurface chlorophyll maximum in the Gulf of Arauci, a highly productive coastal upwelling area on the Chilean coast at latitude 37°S. Several vertical profiles were taken during a 24 hour period from a ship following a freely floating drogue at the depth of a pronounced subsurface chlorophyll maximum.

To explain the observed changes of chlorophyll concentration, the advection-diffusion equation (8.30) has been used under the assumption that horizontal mixing is negligible, $i.e.$:

$$\frac{\partial \bar{c}}{\partial t} + u\frac{\partial \bar{c}}{\partial x} + v\frac{\partial \bar{c}}{\partial y} + w\frac{\partial \bar{c}}{\partial z} = \frac{\partial}{\partial z}\left(K_z\frac{\partial \bar{c}}{\partial z}\right) + S, \tag{15.1}$$

where $\bar{c}$ is the mean phytoplankton concentration, S is the source or sink term

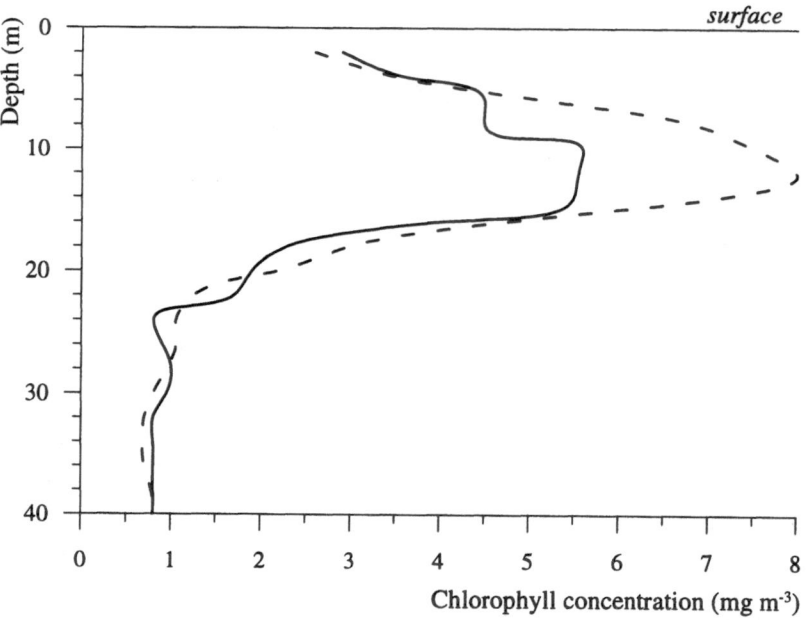

Fig. 15.3: Comparison of measured (dashed line) and calculated (solid line) vertical profiles of chlorophyll concentration in the Gulf of Arauco (adapted from Djurfeldt, 1994)

for phytoplankton, and other symbols have their standard meaning. If we rewrite Eq. (15.1) in the form:

$$\frac{\partial \bar{c}}{\partial t} = -\left[u\frac{\partial \bar{c}}{\partial x} + v\frac{\partial \bar{c}}{\partial y} + w\frac{\partial \bar{c}}{\partial z}\right] + \frac{\partial}{\partial z}\left(K_z\frac{\partial \bar{c}}{\partial z}\right) + S, \qquad (15.2)$$

we find that the first two terms on the right-hand side explain the influence of physical processes on the observed phytoplankton change, while the third term, S, is responsible for changes due to biological factors.

The unknown diffusion coefficient, K_z, can be parameterized using the gradient Richardson number, Ri, as follows (Djurfeldt, 1994):

$$K_z = K_z^{(0)}(1 + nRi)^p, \qquad (15.3)$$

where $K_z^{(0)} = 5 \cdot 10^{-3}$ m^2/s, $n = 5$, and $p = -2$. These values are based on measurements of the upwelling system in Peru (Shaffer, 1982).

Assuming that measured chlorophyll is a good representation of the phytoplankton biomass, the vertical profiles of chlorophyll concentration at particular time steps were obtain. In particular, effects of mixing and divergence, as well as the biological factors are distinguished. In Fig. 15.3, the comparison of a calculated and measured profile, taken before midday, is shown. Neglecting the biological factor, S, in the modelling causes the largest discrepancy for water depths between 6 and 16 m. The effect of the differential sinking of individual particles was found to be of negligible importance for observed profile changes.

A more comprehensive model of the chlorophyll a distribution, including production and loss of phytoplankton mass will be discussed in Sect. 15.5.

15.3   Oceanic Fronts and Biological Productivity

When the temperature, salinity or density do not vary gradually, regions of high gradients of these properties are observed. These regions are known as fronts, following the name of similar structures in the atmosphere. There is a large variety of sizes of oceanic fronts. For example, the front associated with the Gulf Stream is long, exhibiting a temperature change of 10°C in 50 km near the surface. On the other hand, there are small scale fronts of the order of kilometres, such as those associated with island wakes or tidal jets between reefs (Wolanski and Hamner, 1988; Mann and Lazier, 1996).

Fronts are important for the marine environment because of their higher biological productivity. Dense aggregations of phytoplankton and zooplankton are usually associated with fronts (Le Fevre, 1986). The physics of large-scale fronts and their biological productivity have been documented in many papers and a comprehensive review is given in Mann and Lazier's book (1996). The influence of tidally mixed fronts on phytoplankton production is also discussed in Sect. 15.6.

Small-scale frontal systems have not been studied in such detail as the large-scale fronts. However, fronts associated with island wakes are common in shallow waters. Eddies resulting downstream of the island generate a secondary circulation which brings the fine sediments and benthic organisms to the surface near the eddy centre (Wolanski and Hamner, 1988). On coral reefs, the fronts associated with reef wakes play an important role in the redistribution of coral eggs and larvae. Initially after spawning, eggs are distributed over the whole area of the reef. After a few hours, the eggs and larvae usually aggregate in slicks trapped along the outer edges of reef wakes.

Tide-induced currents in the passages between reefs are usually strong and generate localized upwelling on the upper continental slope elevating nitrate and phosphate to the surface (Wolanski et al., 1988). Simultaneously, a tidal jet through the passage advects the nutrients to the shallow water area. Wolanski et al. (1988) reported results of measurements and modelling of the tidal jets transporting nutrients towards the calcareous alga Halimeda banks situated several kilometres inshore of the reefs. Halimeda can gradually accumulate nutrients and over a certain time the quantity of upwelled nitrogen finally becomes more than sufficient to supply the total nitrogen requirements of the Halimeda vegetation.

Localized fronts can also be generated by buoyancy effects, boundary mixing, river plumes and rip currents. All of these topographically generated fronts affect the distribution of plankton, larvae and sediments. The resulting aggregation controls the distribution of benthic assemblages and pelagic predators.

15.4 Plankton Patchiness in Upper Ocean

Due to the variety of hydrodynamic structures such as currents, gyres, fronts and eddies, the distribution of plankton in the oceanic surface layer is very patchy. There is a long cascade of plankton concentration structures ranging from 1 to 100 km (Mackas and Boyd, 1979). For example, at mid latitudes, the transfer of variance (considered as a measure of plankton distribution in a plume) from the size of about 100 km to the scale 1 km occurs during approximately 10 days. One of the mechanisms which causes a transfer of structure variability from small to large length scales is diffusion. However, diffusion is insignificant at scales larger than centimetres, where only ensemble-averaged processes should be represented as diffusive (Abraham, 1998). An alternative representation of plankton patchiness can be obtained using the theory of turbulent transport (Bennett and Deuman, 1985; Holloway, 1986). Although the turbulent approach is very powerful, it is difficult to apply for a real plankton population.

To simplify the calculation, Abraham (1998) proposed a simplified two-dimensional model for turbulent flow. In the model, turbulence is represented as the sum of N randomly distributed circular eddies. Thus, the final stream function $\psi(\vec{x}, t)$ becomes:

$$\psi(\vec{x}, t) = A \sum_{n=1}^{N} (\pm) R_n^2 \exp\left[\frac{-|\vec{x} - \vec{x}_n(t)|^2}{2R_n^2}\right], \tag{15.4}$$

in which R_n and $\vec{x}_n = [x_n(t), y_n(t)]$ are the radius and centre position of particular eddies, respectively, while A is a calibration constant. The orientation of each eddy is a random variable and eddies move with the local velocity:

$$\frac{d\vec{x}_n}{dt} = \vec{u}(\vec{x}_n). \tag{15.5}$$

Using the definition of the stream function (see App. C.4), the velocity $\vec{u}(\vec{x}_n)$ becomes $\vec{u}(\vec{x}_n) = (\partial\psi/\partial y, -\partial\psi/\partial x)$.

Each water parcel carries a phytoplankton population density, P, and adult zooplankton population density, Z. The zooplankton are truly planktonic, drifting with their respective water parcels, and not able to swim independently. Additionally, water parcel carries some capacity, C, representing the effect of a limiting nutrient. The proposed model is not a fluid-dynamics model, but with a proper choice of parameters it provides a simple and easy to implement tool for quasi-geostrophic turbulence. The domain of the model is a periodic square with side of 256 km long. Although the model dynamics are relatively simple, the simulation showed very complex resulting phytoplankton and zooplankton patchy-like patterns. Phytoplankton have distributions similar to those of physical quantities, such as, for example sea surface temperature. On the other hand, zooplankton populations have marked a more fine scale structure. In general, observed spatial patterns of phytoplankton and zooplankton are the consequence of the time required to respond to changes of the environment due to turbulent advection.

Detailed simulations, incorporating fluid-dynamics models of turbulence and multi-component ecosystem dynamics (for example Smith *et al.*, 1996; Gallagher *et al.*, 1996) yield a similar conclusion, where zooplankton have a larger variety of spatial pattern scales than phytoplankton.

15.5 Phytoplankton Concentration in Stratified Ocean

In the previous section we considered the patchiness of plankton on the sea surface. However, the key to high biological productivity is the upwelling of nutrients from deep waters into the euphotic zone and the retention of phytoplankton by stratification in well lit waters. Phytoplankton cells are suspended motionless in the water and use the nutrients from the surrounding water. The rate of nutrient uptake depends on the rate at which the nutrients can diffuse towards them. Most phytoplankton cells are more dense than water and they sink passively or generate their own locomotion. In particular, the density of cytoplasm ranges from 1030 kg/m^3 to 1100 kg/m^3 and diatoms have a cell wall of hydrated silicon dioxide with density of about 2600 kg/m^3. Other elements of phytoplankton, such as coccolithophorids, have plates of calcite or aragonite with density as high as 2700–2950 kg/m^3. Thus, except where an upward movement of the sea water prevents it, the phytoplankton cells sink with velocity, w, which can be approximately estimated from the Stokes solution. Thus, Eq. (2.94) gives:

$$w_p = \left(\frac{\rho_p}{\rho_w} - 1\right)\frac{gD^2}{18\nu},\tag{15.6}$$

in which ρ_p and ρ_w are the phytoplankton and water densites, respectively, D is the diameter of a plankton particle, ν is the coefficient of kinematic viscosity.

Let us now assume that the horizontal gradient of the flow velocity, as well as the gradients of phytoplankton and nutrient concentrations, are negligible. Therefore, the basic equations for vertical dynamics take the form (Druet, 1994; Druet and Zieliński, 1994):

$$\left.\begin{array}{l} \dfrac{\partial \bar{c}_p}{\partial t} + \dfrac{\partial\left(w_p \bar{c}_p\right)}{\partial z} - \dfrac{\partial}{\partial z}\left(K_z \dfrac{\partial \bar{c}_p}{\partial z}\right) = \Pi_p \bar{c}_p \\[4mm] \dfrac{\partial \bar{c}_n}{\partial t} - \dfrac{\partial}{\partial z}\left(K_z \dfrac{\partial \bar{c}_n}{\partial z}\right) = \Pi_n \bar{c}_n \end{array}\right\},\tag{15.7}$$

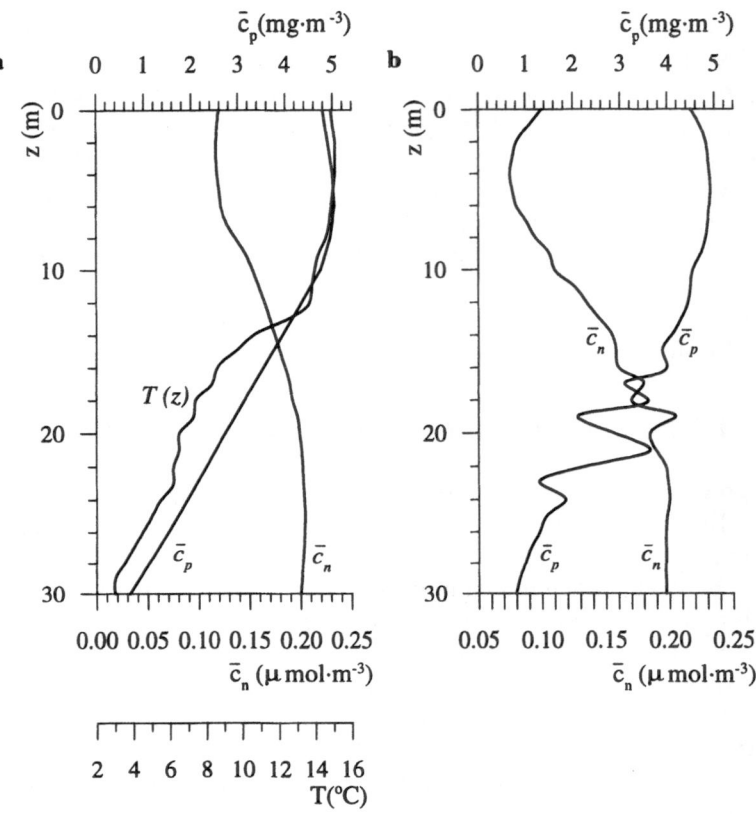

Fig. 15.4: Vertical profile of phytoplankton and nutrients concentration: **a** initial profiles, **b** profiles after 60 min (adapted from Druet and Zieliński, 1994)

in which $\bar{c}_p$ and $\bar{c}_n$ are the concentrations of phytoplankton and nutrients, respectively, and K_z is the turbulent vertical diffusion coefficient of the phytoplankton and nutrient particles. In the model, the coefficient K_z was assumed to be equal to the coefficient of turbulent viscosity A_z. Function Π_p describes the processes of biomass production during photosynthesis, natural phytoplankton mortality and grazing effects. Function Π_n expresses the difference between nutrients available in the system and nutrient uptaken by phytoplankton. Both Π_p and Π_n are parameterized using various experimental relationships.

The unknown phytoplankton and nutrient concentrations, $\bar{c}_p$ and $\bar{c}_n$, have to satisfy the following initial and boundary conditions:

$$\bar{c}_p(z,0) = \bar{c}_{p0}; \quad \bar{c}_n(z,0) = \bar{c}_{n0} \text{ for } t = 0, \tag{15.8}$$

$$w_p\bar{c}_p = K_z\frac{\partial\bar{c}_p}{\partial z}; \quad \frac{\partial\bar{c}_n}{\partial z} = 0 \text{ for } z = 0, \tag{15.9}$$

$$w_p \bar{c}_p = K_z \frac{\partial \bar{c}_p}{\partial z}; \quad \bar{c}_n = \bar{c}_{nh} \text{ for } z = h, \tag{15.10}$$

in which depth, h , equals approximately twice of the thickness of the euphotic zone, and $\bar{c}_{nh}$ is the nutrient concentration at depth $z = h$.

Druet and Zieliński (1994) applied the relationships (15.7)–(15.10) to simulate the changes of chlorophyll and nutrient concentration for the Baltic Sea conditions. The initial vertical distribution of concentrations $\bar{c}_p(z)$ and $\bar{c}_n(z)$ are shown in Fig. 15.4a, while Fig. 15.4b illustrates the evolution of these concentrations after 60 min. In the simulation, the level of turbulent mixing intensity was selected by adopting the velocity shear $\partial u / \partial z = \text{Const} = 0.01 \text{ s}^{-1}$. Thus, the Richardson number becomes $Ri = 10^4 N^2$, where N is the Brunt-Väisälä frequency. Changes in the phytoplankton concentration, c_p, are at all times at the expense of available nutrients.

15.6   Tidal Mixing and Phytoplankton Production

Many highly productive shallow regions of the world's oceans are characterized by waters which are well mixed all year. The Georges Bank in the Gulf of Maine, the southern North Sea, the English Channel, and the southern Irish Sea are examples of such regions. A marked discontinuity in the sea surface temperature (with a horizontal gradient up to 1°C/km) is observed in these regions, being a boundary between stratified and vertically mixed regimes. To support continual mixing of the water column for the full year, energy sufficient to overcome the barrier of stratification should be supplied to the system. Following Simpson and Hunter (1974), we estimate the amount of energy required as the difference between the potential energy of the water column before and after mixing. When density stratification before mixing has

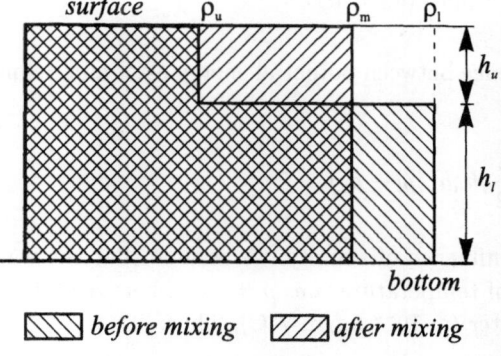

Fig. 15.5: Schematic vertical profiles of density before and after mixing

a simple two-layer shape, as in Fig. 15.5, the corresponding potential energy becomes:

$$PE_i = PE_u + PE_l, \tag{15.11}$$

where:

$$PE_u = mg\left(h_l + \frac{1}{2}h_u\right) = \rho_u g h_u\left(h_l + \frac{1}{2}h_u\right), \tag{15.12}$$

$$PE_l = \frac{1}{2}mgh_l = \frac{1}{2}\rho_l g h_l^2, \tag{15.13}$$

in which PE_u and PE_l are the potential energy of the upper and lower layers, respectively. Other symbols are noted in Fig. 15.5. Thus, the total potential energy before mixing is:

$$PE_i = \frac{1}{2}\rho_l g h_l^2 + \rho_u g h_u\left(h_l + \frac{1}{2}h_u\right). \tag{15.14}$$

After mixing of the water column, the potential energy can be written as:

$$PE_m = \frac{1}{2}\rho_m g(h_u + h_l)^2. \tag{15.15}$$

Using the fact that mixing does not cause a change of the water column's height, the conservation of mass requires that:

$$\rho_m\left(h_u + h_l\right) = \rho_u h_u + \rho_l h_l. \tag{15.16}$$

After substituting Eq. (15.16) into Eq. (15.15) we obtain:

$$PE_m = \frac{g}{2}\left(\rho_u h_u + \rho_l h_l\right)\left(h_u + h_l\right). \tag{15.17}$$

Thus, the difference between potential energy after and before mixing takes the form:

$$PE_m - PE_i = \frac{1}{2}gh_u h_l\left(\rho_l - \rho_u\right). \tag{15.18}$$

Neglecting the influence of salinity, we assume that the water density, ρ, is a linear function of temperature, i.e. $\rho = \alpha T$, where α is the thermal expansion coefficient of water ($\approx 10^{-4}$ kg/m^3/°C). Thus:

$$\Delta PE = PE_m - PE_i = \frac{1}{2}gh_u h_l \alpha \Delta T. \tag{15.19}$$

A change in water temperature depends on a change of heat, ΔQ, added to the water. In particular, the absorption of short-wave radiation by water in the upper layer results in an increase of temperature as follows:

$$\Delta T = \frac{\Delta Q}{m_u c},\tag{15.20}$$

in which ΔQ is the heat added, m_u is the water mass of the upper layer and c is the specific heat (≈ 4.2 kJ/kg/°C). Substituting Eq. (15.20) into Eq. (15.19) we obtain:

$$\Delta PE = \frac{gh_u h_l \alpha \Delta Q}{2m_u c} = \frac{gh_l \alpha \Delta Q}{2\rho_u c}.\tag{15.21}$$

Equation (15.21) represents the amount of mixing energy that must be supplied to overcome the stratification created by the addition of heat, ΔQ, into the upper layer. Simpson and Hunter (1974) postulated that the required energy is generated by the friction between tidal currents and the bottom, and Loder and Greenberg (1986) found that the rate at which mixing energy is extracted from the tidal currents can be written as follows:

$$\frac{dE}{dt} = \epsilon_t \rho C_d \overline{|u|^3},\tag{15.22}$$

where $|u|$ is the magnitude of the bottom velocity, C_d is the friction coefficient, and ϵ_t is the fraction of energy extracted from the tidal currents and used to change the stratification. The bar indicates that the velocity values are averaged over a tidal cycle. Comparing Eq. (15.22) and Eq. (15.21), and using the fact that the amplitude of tidal stream is sinusoidally varying, we obtain the criterion for the maintenance of mixed conditions in the form:

$$\frac{4\epsilon_t C_d \rho_u u^3}{3\pi} > \frac{g\alpha h \Delta Q}{2c},\tag{15.23}$$

or at the front, the inequality becomes an equality (Bowes and Simpson, 1987):

$$\frac{h}{u^3} = \frac{8\epsilon_t C_d \rho_u c}{3\pi g\alpha},\tag{15.24}$$

in which total depth $h = h_u + h_l$ can replace h_l, when water depth h_u is small relative to h_l. Simpson and Hunter (1974) found that for the western Irish Sea front, the critical value of h/u^3 was about 70 m²/s³. Calculation also indicates that only a very small fraction of the dissipated tidal energy is used for vertical mixing.

The Simpson and Hunter (1974) approach has been refined to include the effect of wind mixing. Bowers and Simpson (1987) compared various theoretical approaches with observed positions of tidal fronts on the European shelf seas and found that the models incorporating wind mixing are most accurate. The condition (15.24) with wind-induced mixing takes the form (Simpson et al., 1978):

$$\frac{h}{|u|^3} = \frac{4\epsilon_t C_d \rho_u}{3\pi} \left(\frac{\alpha g \Delta Q}{2c} - \frac{\delta C_d^{(s)} \rho_a V_w^3}{h} \right)^{-1}, \tag{15.25}$$

in which $C_d^{(s)}$ and ρ_a are the surface drag coefficient and air density, respectively, V_w is the wind velocity, and $\delta = 0.023$. For depths larger than 50 m, tidal stirring dominates over the wind influence. However, with decreasing water depth, wind mixing becomes increasingly important.

In highly mixed waters, dead matter is continually decomposing on and in the sediments, providing a source of nutrients which are transported into the water column. This can cause a homogeneous distribution of chlorophyll a in the water column, with primary production continuing throughout the year with no clear seasonal peak. The total annual primary production in the shallow part of the Georges Bank (< 60 m) is 450 g C/m^2, and 320 gC/m^2 in the deeper part. This level of production is higher than many other tidally mixed waters due to the specific water circulation pattern on the bank. There are two broad and permanent circulation gyres in the region. One rotates clockwise around the Georges Bank, the other counterclockwise around the Gulf of Maine. The clockwise gyre retains water on the bank, with mean residence times of about 2–3 months, which is sufficient for plankton to grow.

In contrast to the almost constant primary productivity on the Georges Bank, the productivity in the North Sea exhibits a variation between seasons. The tidally mixed areas have well defined spring and autumn peaks, and the coastal waters have a summer productivity almost twice as high as the offshore areas, but this productivity falls quickly in autumn (Mann and Lazier, 1996). This difference is related to the poor light penetration in the autumn. The total annual primary production for the North Sea is about 250 g C/m^2 for offshore waters and 200 g C/m^2 in coastal waters. Thus, although tidal mixing in relatively shallow coastal waters prevents stratification of the water column, which is postulated as a necessary condition for the spring bloom, the nutrient flux into the water column from the sediments is increased and annual primary productivity in tidally mixed areas appears to be higher than the average for coastal waters.

A Symbols and Notations

a	–	circle radius [m]
a_c	–	centrifugal acceleration [m·s^{-2}]
a_C	–	Coriolis acceleration [m·s^{-2}]
a_n	–	acceleration acting normal to the Earth's surface [m·s^{-2}]
a_t	–	acceleration acting tangentially to the Earth's surface [m·s^{-2}]
a_x, a_y, a_z	–	components of acceleration [m·s^{-2}]
A_x, A_y, A_z	–	turbulent viscosity coefficients [m^2·s^{-1}]
A	–	buoyancy index
A	–	wave amplitude [m]
b	–	span [m]
c	–	chord [m]
c	–	concentration of matter [kg·m^{-3}]
$\bar{c}$	–	mean concentration [kg·m^{-3}]
c'	–	fluctuation component of concentration [kg·m^{-3}]
C	–	phase velocity [m·s^{-1}]
C_a	–	added mass coefficient
C_d	–	drag coefficient
$C_{d,fric}$	–	frictional drag coefficient
C_g	–	group velocity [m·s^{-1}]
C_{g0}	–	deepwater group velocity [m·s^{-1}]
C_i	–	phase velocity of the internal wave [m·s^{-1}]
C_l	–	lift coefficient
C_m	–	inertia coefficient
C_0	–	deepwater phase velocity [m·s^{-1}]
C_s	–	sound speed in the ocean [m·s^{-1}]
Cx	–	Cox number
C_{10}	–	drag coefficient at level 10 m
d	–	displacement thickness [m]
D	–	diameter of cylinder or pipe [m]
D	–	molecular diffusion coefficient [m^2·s^{-1}]

E	–	energy [N·m] or [J]
E_k	–	kinetic energy [N·m] or [J]
E_p	–	potential energy [N·m] or [J]
Eu	–	Euler number
erf	–	error function
f	–	frequency [Hz]
f	–	inertial frequency [Hz]
f	–	friction factor
f	–	Coriolis parameter [rad·s^{-1}]
f_b	–	bottom friction [m^2·s^{-2}]
f_c	–	Nyquist frequency [Hz]
f_s	–	sound frequency in still water [Hz]
f_ζ	–	probability density function for wave surface displacement [m^{-1}]
$\boldsymbol{F}$	–	force [N]
F	–	tide form number
F	–	energy flux [N·m^2·s^{-1}]
F_d	–	drag force [N]
F_E	–	gravitational force acting on the body on Earth [N]
F_g	–	gravitational force [N]
F_i	–	inertia force [N]
F_M	–	gravitational force acting on the body on Moon [N]
F_0	–	densimetric Froude number
Fr	–	Froude number
F_ζ	–	cumulative probability distribution
$F_{1\zeta}$	–	exceedence probability distribution
g	–	gravitational acceleration [m·s^{-2}]
g'	–	reduced gravity [m·s^{-2}]
G	–	universal gravitational constant [N·m^2·kg^{-2}]
h	–	water depth [m]
h_E	–	Ekman layer thickness [m]
H	–	piezometric (or total) head [m]
H	–	wave height [m]
$\bar{H}$	–	mean wave height [m]
H_b	–	wave height at breaking [m]
H_i	–	incident wave height [m]
H_{max}	–	maximum wave height [m]
H_{rms}	–	root–mean–square wave height [m]
H_s	–	significant wave height [m]
i	–	imaginary unit
i	–	longitudinal surface gradient
k	–	wave number [m $^{-1}$]
k_s	–	Nikuradse roughness size [m]

K	–	Keulegan-Carpenter number
K_d	–	diffraction coefficient
K_p	–	dynamic pressure coefficient
K_p	–	pressure response factor
K_{tr}	–	proportionality coefficient in sediment transport calculation
K_τ	–	autocorrelation function
K_x, K_y, K_z	–	turbulent diffusion coefficients $[\mathrm{m^2 \cdot s^{-1}}]$
K_x, K_y, K_z	–	dispersion coefficients $[\mathrm{m^2 \cdot s^{-1}}]$
l	–	length of water basin [m]
L	–	wavelength [m]
L_0	–	wavelength in deep water [m]
L_b	–	characteristic length [m]
m	–	mass [kg]
m_b	–	mass of the body [kg]
m_E	–	mass of the Earth [kg]
m_M	–	mass of the Moon [kg]
Ma	–	Mach number
n	–	porosity
n	–	degree of freedom
$N(z)$	–	Brunt–Väisälä frequency [Hz]
N_L	–	length scale
p	–	pressure $[\mathrm{N \cdot m^{-2}}]$
$\bar{p}$	–	mean pressure $[\mathrm{N \cdot m^{-2}}]$
p_a	–	atmospheric pressure $[\mathrm{N \cdot m^{-2}}]$
p_c	–	central pressure $[\mathrm{N \cdot m^{-2}}]$
p_{norm}	–	normal atmospheric pressure [Pa]
p_0	–	ambient pressure $[\mathrm{N \cdot m^{-2}}]$
p'	–	fluctuation pressure $[\mathrm{N \cdot m^{-2}}]$
P	–	power [J/s] or [W]
Pe	–	Pécklet number
q_c	–	cross-shore sediment transport $[\mathrm{m^3 \cdot m^{-1} \cdot s^{-1}}]$
q_l	–	longshore sediment transport $[\mathrm{m^3 \cdot m^{-1} \cdot s^{-1}}]$
Q	–	discharge $[\mathrm{m^3 \cdot s^{-1}}]$
r	–	radius [m]
R	–	Rossby radius of deformation
R	–	modified deformation radius [m]
Re	–	Reynolds number
R_E	–	Earth's radius [m]
R_M	–	Moon's radius [m]
R_{EM}	–	distance between Earth and Moon centres [m]

Re_x	–	local Reynolds number
Ri	–	Richardson number
s	–	wave steepness
S	–	surface [m^2]
SOI	–	Southern Oscillation Index [N·s^{-2}]
S_p	–	platform area [m^2]
S_{xx}	–	cross-shore component of radiation stress tensor [N·m]
S_{xy}	–	longshore component of radiation stress tensor [N·m]
S_{xy}	–	cross-spectrum of variables x and y
S	–	water salinity [ppm]
$\bar{S}$	–	mean salinity [ppm]
$S(\omega)$	–	frequency spectrum of surface (internal) waves [m^2·s]
t	–	time [s]
t_f	–	flushing time [s]
T	–	wave period [s]
T	–	temperature [°C]
$\bar{T}$	–	mean wave period [s]
T_f	–	inertial period [s]
T_i	–	period of the internal wave [s]
T_n	–	period of n-seiche mode [s]
$\boldsymbol{u}$	–	velocity vector, $\boldsymbol{u} = (u, v, w)$
u	–	horizontal component of velocity [m·s^{-1}]
$\bar{u}$	–	mean horizontal component of velocity [m·s^{-1}]
u_0	–	free-stream velocity [m·s^{-1}]
u_*	–	friction velocity [m·s^{-1}]
u'	–	horizontal component of fluctuation velocity [m·s^{-1}]
u_t	–	tangential velocity [m·s^{-1}]
U	–	characteristic flow velocity [m·s^{-1}]
U	–	Ursell number
v	–	horizontal component of velocity [m·s^{-1}]
$\bar{v}$	–	mean horizontal component of velocity [m·s^{-1}]
v'	–	horizontal component of fluctuation velocity [m·s^{-1}]
v_r, v_θ, v_z		components of velocity in a cylindrical coordinate system
V	–	mean flow velocity [m·s^{-1}]
V	–	volume of water or animal body [m^{-3}]
V_e	–	entire volume of the estuary [m^3]
V_f	–	freshwater volume of the estuary [m^3]
V_g	–	geostrophic wind velocity [m·s-1]
V_w	–	wind speed [m·s^{-1}]
w	–	vertical component of velocity [m·s^{-1}]
w	–	terminal velocity [m s^{-1}]

$\bar{w}$	–	mean vertical component of velocity $[\text{m·s}^{-1}]$
w'	–	vertical component of fluctuation velocity $[\text{m·s}^{-1}]$
W	–	body weight $[\text{N}]$
We	–	Weber number
x	–	horizontal axis of the rectangular coordinate system
X	–	wind fetch $[\text{m}]$
y	–	horizontal axis of the rectangular coordinate system
z	–	vertical axis (directed upwards when positive) of the rectangular coordinate system
$z = x + i \cdot y$	–	complex number
z_0	–	roughness length $[\text{m}]$
α	–	angle $[°]$
α	–	Phillips' constant
β	–	bottom slope
γ	–	specific weight $[\text{N·m}^{-3}]$
γ	–	peak enhancement factor
Γ	–	circulation $[\text{m}^2\text{s}-1]$
δ	–	delta function
$\delta, \delta'_1, \delta''_1$	–	thickness of boundary layer $[\text{m}]$
Δp	–	increase of pressure $[\text{N·m}^{-2}]$
ζ	–	surface displacement $[\text{m}]$
$\bar{\zeta}$	–	mean surface displacement $[\text{m}]$
η	–	vertical component of water particle path $[\text{m}]$
θ	–	angle $[°]$
κ	–	von Kármán constant
λ	–	longitudinal density gradient $[\text{kg·m}^{-4}]$
λ	–	spatial step
μ	–	coefficient of dynamic molecular viscosity $[\text{N·s·m}^{-2}]$
μ_n	–	central statistical moments
ν	–	coefficient of kinematic viscosity $[\text{m}^2\text{·s}^{-1}]$
ν	–	Poisson's ratio
ξ	–	horizontal component of water particle path $[\text{m}]$
ξ	–	surf similarity parameter
ρ	–	liquid density $[\text{kg·m}^{-3}]$
ρ_a	–	air density $[\text{kg·m}^{-3}]$

ρ_c –   body density [kg·m^{-3}]

ρ_s –   solid particle density [kg·m^{-3}]

ρ_w –   water density [kg·m^{-3}]

σ –   surface tension [N·m^{-1}]

σ_ζ –   standard deviation of wave surface displacement [m]

σ_0 –   shape parameter

σ_t –   water density shorthand units [kg·m−3]

τ –   shear stress [N·m^{-2}]

τ –   time step [s]

τ_0 –   shear stress at body surface [N·m^{-2}]

Π –   non-dimensional products in the Buckingham Theorem

ϕ –   velocity potential [m^2·s^{-1}]

ϕ –   latitude [°]

ψ –   stream function [m^2·s^{-1}]

ω –   angular frequency [rad·s^{-1}]

ω_E –   frequency of the Earth's rotation [rad·s^{-1}]

ω_p –   peak frequency [rad·s^{-1}]

$\approx$ approximately equals

$curl$ rotation operator

rot rotation operator

$\sim$ proportional to

∂ partial differential operator

∇ gradient operator

∇^2 Laplace operator

$\Im$ imaginary part of complex quantity

$\Re$ real part of complex quantity

B International System of Units

B.1 Introduction

The mechanics of fluid is described by many functions and quantities which are expressed in units. The most widely used system is the *Système International d'Unitès* (International System of Units), commonly referred to as the **SI** or **Metric System**. This Appendix describes the *fundamental units* and other *related units*, as well as the conversion factors and other information relevant to water studies.

Table B.1: SI Prefixes

Prefix	SI symbol	Multiplication factors
tera	T	10^{12}
giga	G	10^{9}
mega	M	10^{6}
kilo	k	10^{3}
hecto	h	10^{2}
deca	da	10^{1}
deci	d	10^{-1}
centi	c	10^{-2}
milli	m	10^{-3}
micro	μ	10^{-6}
nano	n	10^{-9}
pico	p	10^{-12}

In SI system, the fundamental units are:

- **length** – metre (m),
- **time** – second (s),
- **mass** – kilogram (kg),
- **temperature** – kelvin (K).

Prefixes are attached to names or symbols of SI units to form powers-of-ten multiples. Table B.1 lists the prefixes and associated SI symbols likely to be used in marine science.

B.2   Base Units

B.2.1   Length

The metre was originally proposed by the French Academy of Science in 1791 as one ten-millionth of the distance from the North Pole to the equator on the meridian passing through Paris. However, the surveyors got their sums wrong and for 162 years (to 1960) the metre was defined as an arbitrary distance marked on a metal bar made of 90 percent of platinum and 10 percent of iridium and preserved in the International Bureau of Weights and Measures at Sèvres, near Paris.

At present, one metre is defined as the distance travelled by light in a vacuum in $1/299{,}792{,}458$ second. This value results from the value of the speed of light accepted in 1983 as $299{,}792{,}458$ metres per second.

In oceanography, sailing and shipping, distance is frequently expressed in nautical miles (Nm), where 1 nautical mile $= 1852\,\mathrm{m}$. On older nautical maps, water depth is not expressed in metres but in feet and fathoms where the following relationships exist:

$$1\,\mathrm{foot} = 12\,\mathrm{inches} = 30.48\,\mathrm{cm} = 0.3048\,\mathrm{m}, \tag{B.1}$$

$$1\,\mathrm{fathom} = 6\,\mathrm{feet} = 1.83\,\mathrm{m}. \tag{B.2}$$

B.2.2   Time

In the past, the basis of time units were astronomical events. For example, 1 second was defined as $1/86400$ of the time between two consecutive highest positions of the Sun. According to the present definition, the second is equal to $9{,}192{,}631{,}770$ periods of the light emitted by atom of Caesium ^{133}Cs. The number of periods of radiation was chosen to make the length of the defined second correspond as closely as possible to that astronomically determined. At present, the most accurate timekeeping device is a commercially available atomic clock which makes an error of one second in 1.6 million years.

B.2.3   Mass

The kilogram is the basic metric unit of mass, equal to the mass of the international prototype kept in the form of a platinum-iridium cylinder at the International Bureau of Weights and Measures at Sèvres. A kilogram is very nearly equal to the mass of 1,000 cubic centimetres of water. The mass per unit volume is called **density** and is discussed in detail in Sect. 1.2.1.

B.2.4   Temperature

For studies of marine systems, absolute temperatures are usually not of interest and for the purpose of this book, temperature is expressed in °C (Celsius), rather than in °K (Kelvin), where $273.15°K = 0°C$. The change of 1°C is the same as a change of 1°K. Apart from degrees Celsius, degrees Fahrenheit (°F) are sometimes used. The conversion between these units is as follows:

$$°C = \frac{5}{9} \left(°F - 32° \right),\qquad\qquad\qquad (B.3)$$

$$°F = \frac{9}{5} \left(°C \right) + 32°.\qquad\qquad\qquad (B.4)$$

B.3   Derived Units

Many other quantities useful in fluid mechanics have *derived units*, which are defined in terms of the base units. We list now some of them which are used in this book.

B.3.1   Frequency

Frequency is the number of oscillations of a system during one second and it is expressed in hertz (Hz), where 1 hertz = 1 cycle/1 second. In this text, angular frequency is also used. The units of angular frequency are 1 radian/1 second. As 2π radians = 1 full cycle = $360°$, therefore:

$$1 \text{ radian} = \frac{360°}{2\pi} = \frac{180°}{\pi} = \frac{180°}{3.1416} \approx 57.3°.\qquad\qquad (B.5)$$

B.3.2   Velocity

Velocity defines the rate at which the position changes with time and it is expressed in metres per second (m/s). Traditionally, in sailing and shipping, and sometimes in oceanography, the velocity of wind, ocean currents and the speed of the vessel is expressed in knots, where:

$$1 \text{ knot} = \frac{1 \text{ nautical mile}}{1 \text{ hour}} = \frac{1852 \text{ m}}{3600 \text{ s}} = 0.51 \text{ m/s} \approx 0.5 \text{ m/s}.\qquad (B.6)$$

Therefore,

$$1 \, \text{m/s} = 1.94 \, \text{knots} \approx 2.0 \, \text{knots} \quad \text{and} \quad 1 \, \text{km/hour} = 0.54 \, \text{knot}. \tag{B.7}$$

Velocity is a vector quantity; hence to define velocity completely, the magnitude of velocity as well as the velocity direction are needed.

B.3.3 Acceleration

Acceleration, a, is the rate at which velocity changes with time; it is expressed in m/s^2. In particular, the specific notation of g is reserved for gravitational acceleration, related to the attraction of any body by the Earth. In particular, mean gravitational acceleration $g = 9.806 \, \text{m/s}^2 \approx 9.81 \, \text{m/s}^2$, and varies within about 1% with position on the Earth's surface as:

$$g = 9.806 - 0.026 \cos 2\phi, \tag{B.8}$$

in which ϕ is the latitude.

B.3.4 Discharge

Discharge is the rate at which a volume of water passes a section over some unit of time. In the SI system it is expressed in m^3/s. Discharge is normally symbolized by Q.

B.3.5 Force

Force is described in terms of its effects. By Newton's second law of motion, force is proportional to the product of mass and acceleration. In the SI system, the unit of force is the newton (N), which is the force necessary to accelerate one kilogram of mass at an acceleration of one metre per second squared, $i.e.$:

$$\text{force} \, (\text{N}) = \text{mass} \, (\text{kg}) \times \text{acceleration} \, (\text{m/s}^2). \tag{B.9}$$

One newton is a very small force. For example, a small apple with a mass of 0.11 kg experiences a gravitational force on Earth of about one newton. Here we must stress the difference between the mass of a body and its weight. Mass is an expression of the amount of matter in something, a steel bar or a bucket of water. Therefore, the mass does not change with the changing of gravitational acceleration. On the other hand, weight is a gravitational force depending on gravitational acceleration. Hence, the mass of an astronaut's body walking on the Moon is the same as on the Earth, however, his weight is approximately 6 times smaller, as the gravitational acceleration on the Moon is about six times smaller than that of Earth.

It should be noted that kilograms are frequently used incorrectly in terms of weight. This is a result of the metric systems which expressed weight as kilograms. Therefore, when the weight is expressed using units of kilograms (or pounds), the value must first be converted from kilograms (pounds) to newton before beginning any calculations, $i.e.$:

$$1 \text{ kilogram(weight)} = 9.806 \text{ N}, \qquad (B.10)$$

and

$$1 \text{ pound(weight)} = 4.448 \text{ N}. \qquad (B.11)$$

However, in everyday life, the distinction between mass and weight does not matter. If an Englishman buys 2.2 pounds of apples at the supermarket and a Polish woman buys 1 kilogram of apples at the farmer's orchard, they both get the same amount of produce.

B.3.6 Pressure

The pressure at any point is the force per unit area at a point. In the SI system, pressure is expressed in N/m^2, and a pressure of $1 \ N/m^2$ is called a pascal (Pa), $i.e.$ $1 \text{ Pa} = 1 \ N/m^2$. This is a very low pressure. Therefore, for practical applications, the multiplications of pascal are used, for example, $1 \text{ hPa} = 10^2 \text{ Pa}$, $1 \text{ kPa} = 10^3 \text{ Pa}$, $1 \text{ MPa} = 10^6 \text{ Pa}$. In particular, atmospheric pressure is usually expressed in hPa, and normal atmospheric pressure at sea level, called the *normal atmosphere*, is equal to about $1000 \text{ hPa} = 10^5 \text{ Pa} = 1$ bar (the exact value is 1.01325×10^5 Pa at 15°C). This pressure corresponds to the pressure of 1 atmosphere (1 atm) in the old metric system, where:

$$1 \text{ atm} = 1 \frac{\text{kg (weight) of force}}{\text{cm}^2}, \qquad (B.12)$$

in which 1 kg (weight) is the force needed to induce an acceleration of $g = 9.806 \ m/s^2$ for a mass of 1 kg. Thus:

$$1 \frac{\text{kg (weight)}}{\text{cm}^2} = 1 \text{ atm} = \frac{9.80665 \text{ N}}{10^{-4} \text{m}^2} \approx 10^5 \frac{\text{N}}{\text{m}^2} = 10^5 \text{Pa} = 1 \text{ bar}. \qquad (B.13)$$

The magnitude of pressure determines it completely.

B.3.7 Shear Stress and Shear Force

Similarly to pressure, shear stress is force per unit area. However, now this force acts parallel to a surface. For example, the force due to water flow over the sea bottom induces shear stress at the bottom. As the shear stress force is divided by the area over which it acts, it has the same units as pressure, $i.e.$ N/m^2.

B.3.8 Energy and Power

Energy expresses the capacity for doing work. Therefore, energy and work have the same units. Work is expressed as follows: Work (N × m or J) = Force (N) × distance (m). The J is an abbreviation for joule, *i.e.*: 1 J = 1 N × 1 m = 1 kg × 1 m^2/s^2. In this book, energy is usually symbolized by E.

Power is the amount of work done per unit time, *i.e.*: Power (J/s or Watt) = Work (J)/time (s) and 1 Watt = J/s = 1 kg × m^2/s^3. In this book power is symbolized by P. For example, in Sect. 4.2 it was shown that wave energy per unit area is given by:

$$E = \frac{1}{8} \rho_w g H^2, \tag{B.14}$$

in which ρ_w is the sea water density and H is the wave height. Thus, if wave height is 1 m, wave period is 4 s and water density is 1023 kg/m^3, wave energy per m^2 is 1254 J/m^2 and wave power per m^2 is 314 W/m^2.

B.4 Conversion Factors Between BG and SI Units

In many textbooks, the old British system of units (BG - British Gravitational) are still used. The conversion factors for the most important dimensions are listed in Table B.2 for convenience.

Table B.2: Conversion factors

Dimension	SI units	BG units	Conversion factor
Mass	kilogram (kg)	slug	1 slug = 14.5939 kg
Length	metre (m)	foot (ft)	1 foot = 0.3048 m
Time	second (s)	second (s)	1 s = 1 s
Area	m^2	ft^2	1 m^2 = 10.764 ft^2
Volume	m^3	ft^3	1 m^3 = 35.315 ft^3
Velocity	m/s	ft/s	1 ft/s = 0.3048 m/s
Acceleration	m/s^2	ft/s^2	1 ft/s^2 = 0.3048 m/s^2
Force	N	lbf	1 lbf = 4.4482 N
Pressure	N/m^2	lbf/m^2	1 lbf/ft^2 = 47.88 Pa
Energy	J = N × m	lbf × ft	1 lbf × ft = 1.3558 J
Power	W = J/s	lbf × ft/s	1 lbf × ft/s = 1.3558 W
		hp	1 hp = 745.7 W

C Useful Theoretical Approaches and Formulas

C.1 Vector Notations Used in the Book

In fluid mechanics, two quantities are distinguished; **scalars** and **vectors**. A **scalar** is a real number or a quantity that has *magnitude*. For example, length, temperature, and blood pressure are represented by numbers. On the other hand, a **vector** usually is described as a quantity that has both, *magnitude* and *direction*. For example, wind velocity is a vector quantity, and has both speed and an associated direction. In this book, a vector is represented by an arrow, $\vec{u}$, or is written as a boldface symbol, $\boldsymbol{u}$. The bold type $\boldsymbol{u}$ denotes velocity at a specified time and position in space. To describe a vector analytically, its projections (components) on the x, y and z axes of the selected coordinate system are used. For example, in Fig. 2.1a, the components of vector $\boldsymbol{u}$ are u, v and w. In general, for vector representation, the following notation is used:

$$\boldsymbol{u} = (u, v, w). \tag{C.1}$$

In fact, the projections of vector $\boldsymbol{u}$ on the coordinate axes should also be treated as vectors. Therefore, instead of notation (C.1), another representation is frequently used, *i.e.*:

$$\boldsymbol{u} = u\vec{i} + v\vec{j} + w\vec{k}, \tag{C.2}$$

or:

$$\boldsymbol{u} = u\boldsymbol{i} + v\boldsymbol{j} + w\boldsymbol{k}, \tag{C.3}$$

where $\boldsymbol{i} = (1,0,0)$, $\boldsymbol{j} = (0,1,0)$, and $\boldsymbol{k} = (0,0,1)$.

The vectors $\boldsymbol{i}$, $\boldsymbol{j}$ and $\boldsymbol{k}$ are unit vectors that form a basis for the system of three-dimensional vectors. For example, the two-dimensional vector $\boldsymbol{a}$ in Fig. C.1 can be represented as $\boldsymbol{a} = 9\boldsymbol{i} + 4\boldsymbol{k}$ and vector $\boldsymbol{b}$ takes the form

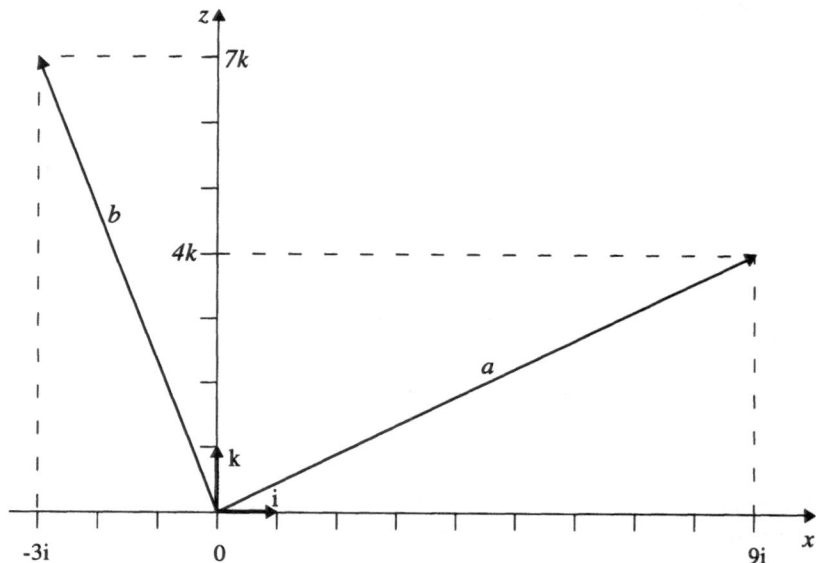

Fig. C.1: Two-dimensional vectors

$b = -3i + 7k$. Basic operations on vectors are described in many textbooks and they are not repeated here. However, for later convenience, some of the most frequently used in this book are defined below. In particular, the magnitude or length of a vector u is denoted as $|u|$. Motivated by the Pythagorean Theorem, it can be defined as:

$$|u| = \sqrt{u^2 + v^2 + w^2}. \tag{C.4}$$

The dot product of two vectors a and b is a scalar quantity:

$$a \cdot b = |a|\,|b|\cos\theta, \tag{C.5}$$

where θ is the angle between the vectors. Thus, for two parallel vectors, a and b, we have $a \cdot b = |a|\,|b|$, and for orthogonal vectors, when the angle between them is $\theta = \pi/2$, $a \cdot b = 0$.

If $a = (a_1, a_2, a_3)$ and $b = (b_1, b_2, b_3)$, then the dot product $a \cdot b$ can be written in terms of vector components as:

$$a \cdot b = a_1 b_1 + a_2 b_2 + a_3 b_3. \tag{C.6}$$

It should be noted that dot products of unit vectors are as follows:

$$\left.\begin{array}{lll} i \cdot i = 1, & j \cdot j = 1, & k \cdot k = 1 \\ i \cdot j = 0, & i \cdot k = 0, & j \cdot k = 0 \end{array}\right\}. \tag{C.7}$$

In fluid mechanics, a vector and its components usually are functions of position and time. This means that:

$$u(x, y, z, t) = u(x, y, z, t)\,i + v(x, y, z, t)\,j + w(x, y, z, t)\,k. \tag{C.8}$$

Using Eq. (C.8), two other important operations on vector u can be defined as follows:

- the divergence of a vector, u, is a scalar function of vector components:

$$\text{div}u = \frac{\partial u}{\partial x} + \frac{\partial v}{\partial y} + \frac{\partial w}{\partial z}, \tag{C.9}$$

in which $\partial u / \partial x$ is a partial derivative of u with respect to x and so on. Divergence u can also be written in terms of the ∇ operator as:

$$\text{div}u = \nabla \cdot u = \overbrace{\left(\frac{\partial}{\partial x}i + \frac{\partial}{\partial y}j + \frac{\partial}{\partial z}k\right)}^{\nabla} \cdot \overbrace{(ui + vj + wk)}^{u} =$$

$$= \frac{\partial u}{\partial x} + \frac{\partial v}{\partial y} + \frac{\partial w}{\partial z}, \tag{C.10}$$

when properties of unit vectors are used (see Eq. C.7). The symbol ∇, an inverted capital Greek delta, is called 'del' or 'nabla'.

- rotation of a vector u is a vector type quantity:

$$\text{rot}u = \text{curl}\,u = \nabla \times u =$$

$$= \left(\frac{\partial w}{\partial y} - \frac{\partial v}{\partial z}\right)i + \left(\frac{\partial u}{\partial z} - \frac{\partial w}{\partial x}\right)j + \left(\frac{\partial v}{\partial x} - \frac{\partial u}{\partial y}\right)k. \tag{C.11}$$

The expressions in brackets represent a rate of rotation about the x, y and z axes, respectively. The vector rot u is called vorticity. The vorticity is a non-zero quantity when rotation and deformation of fluid elements exist and is superimposed on their translation. Such flow is known as *rotational*. If the motion is purely translational and there is an angular deformation without rotation, the flow is *irrotational*. The physical implications of this definition are explained in Sect. 2.3.3 and a brief summary of the basic properties of an irrotational flow is given in Appendix C.4.

Equation (C.11) suggests that irrotationality of flow requires that rot $\boldsymbol{u} = 0$, or:

$$\frac{\partial w}{\partial y} = \frac{\partial v}{\partial z}, \quad \frac{\partial u}{\partial z} = \frac{\partial w}{\partial x}, \quad \frac{\partial v}{\partial x} = \frac{\partial u}{\partial y}. \tag{C.12}$$

For completeness we define three additional operators, also used in this book:

- gradient of a scalar function $F(x, y, z)$:

$$\mathbf{grad}\, F = \boldsymbol{\nabla} F(x, y, z) = \frac{\partial F}{\partial x}\boldsymbol{i} + \frac{\partial F}{\partial y}\boldsymbol{j} + \frac{\partial F}{\partial z}\boldsymbol{k}. \tag{C.13}$$

Note that the gradient of a scalar function is a vector! Its components are the derivatives in each of the directions, and it has a magnitude and direction giving the maximum rate of change of the scalar quantity.

- Laplacian of scalar function $F(x, y, z)$:

$$\nabla^2 F = \frac{\partial^2 F}{\partial x^2} + \frac{\partial^2 F}{\partial y^2} + \frac{\partial^2 F}{\partial z^2}. \tag{C.14}$$

The Laplacian of scalar function F results in a scalar function. This operator is named after the French mathematician, physicist and astronomer, Pierre Simon Marquis de Laplace.

- Circulation around a contour in a fluid (in the plane Oxz):

$$\Gamma = \oint v_s ds = \left(\frac{\partial u}{\partial z} - \frac{\partial w}{\partial x} \right) dx dz, \tag{C.15}$$

in which $\oint$ denotes the line integral of velocity, v_s, around the closed element in a plane Oxz. The bracket on the right-hand side of Eq. (C.15) is a vorticity component in the plane Oxz. The circulation around a contour is equal to the sum of the vorticity within the area of the contour. By convention, circulation is regarded as positive for anticlockwise direction of integration. The concept of circulation is important in the theory of lift forces on hydrofoils (see Sect. 2.6).

C.2   Derivation of Mass Conservation Equation

Consider a fixed elemental volume of fluid as shown in Fig. C.2. From a physical point of view, it is clear that the change of fluid mass inside the volume $dx\,dy\,dz$ during the small time increment, dt, must be equal to the difference between the rates of influx into and of flux out of the volume. The difference between fluid mass entering through the section ABCD and fluid mass coming out through the section EFGH, during the same interval of time, becomes:

$$\rho_w u\,dy\,dz\,dt - \left[\rho_w u + \frac{\partial(\rho_w u)}{\partial x}dx\right]dy\,dz\,dt = -\frac{\partial(\rho_w u)}{\partial x}dx\,dy\,dz\,dt. \tag{C.16}$$

Similarly, the differences due to fluid motion parallel to the Oy and Oz axes are, respectively: $-[\partial(\rho_w v)/\partial y]dx\,dy\,dz\,dt$ and $-[\partial(\rho_w w)/\partial z]dx\,dy\,dz\,dt$.

Fluid mass after a time dt also changes due to a possible change of density with respect to time. Hence the net change of mass in the element is:

$$\left(\rho_w + \frac{\partial\rho_w}{\partial t}dt\right)dx\,dy\,dz - \rho_w dx\,dy\,dz = \frac{\partial\rho_w}{\partial t}dx\,dy\,dz\,dt. \tag{C.17}$$

Equating the total change of mass contained within the volume under consideration yields:

$$\frac{\partial\rho_w}{\partial t}dx\,dy\,dz\,dt = -\left[\frac{\partial(\rho_w u)}{\partial x} + \frac{\partial(\rho_w v)}{\partial y} + \frac{\partial(\rho_w w)}{\partial z}\right]dx\,dy\,dz\,dt, \tag{C.18}$$

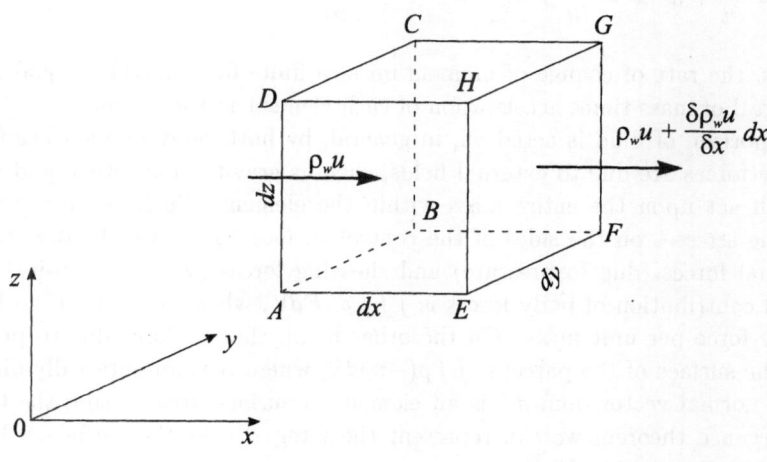

Fig. C.2: Control volume of fluid

and:

$$\frac{\partial}{\partial t}(\rho_w) + \frac{\partial}{\partial x}(\rho_w u) + \frac{\partial}{\partial y}(\rho_w v) + \frac{\partial}{\partial z}(\rho_w w) = 0, \tag{C.19}$$

which constitutes the required conservation of mass – Eq. (2.17). Using notations given in Appendix C.1, the conservation of mass can also be written:

$$\frac{\partial \rho_w}{\partial t} + \nabla \cdot (\rho_w \boldsymbol{u}) = 0. \tag{C.20}$$

C.3   Derivation of the Equations of Fluid Motion

C.3.1   Governing Equations

The equations of motion for fluid provide the relationship between the applied forces on a unit volume of fluid and the change of momentum. The basic equation for derivation of the equation of motion is Newton's second law. Let us consider a finite parcel of fluid. Its momentum is $\int\int\int \rho_w \boldsymbol{u} dV$, where $\boldsymbol{u}$ is velocity, and dV is an element of volume. The change of momentum of that parcel is $d/dt \int\int\int \rho_w \boldsymbol{u} dV$.

Using Eq. (C.20), Batchelor (1967) showed that:

$$\frac{d}{dt}\int\int\int \rho_w \boldsymbol{u} dV = \int\int\int \rho_w \frac{D\boldsymbol{u}}{Dt}, \tag{C.21}$$

where D/Dt is the material derivative:

$$\frac{D}{Dt} = \frac{\partial}{\partial t} + \boldsymbol{u}\cdot\nabla = \frac{\partial}{\partial t} + u\frac{\partial}{\partial x} + v\frac{\partial}{\partial y} + w\frac{\partial}{\partial z}. \tag{C.22}$$

Thus, the rate of change of momentum of a finite fluid parcel is equal to the integral of mass times acceleration of each element in the parcel.

A portion of fluid is acted on, in general, by both body and surface forces. Body forces are due to external fields, such as gravity, magnetism and others which act upon the entire mass within the element. Surface forces are due to the stresses on the sides of the control surface. They can be divided into normal forces (due to pressure) and shearing forces (due to viscosity). The total contribution of body forces is $\int\int\int \rho_w \boldsymbol{F} dV$, when we define $\boldsymbol{F}$ to be the body force per unit mass. On the other hand, the net force due to pressure on the surface of the parcel is $\int\int p(-\boldsymbol{n})dS$, where $\boldsymbol{n}$ is an outwardly-directed unit normal vector, and dS is an element of surface area. Using the Gauss' divergence theorem we can represent the integral over the surface S by the integral over volume V, i.e.:

$$-\int\int p\boldsymbol{n} dS = -\int\int\int \nabla p dV. \tag{C.23}$$

Equating the rate of change of momentum to the forces we obtain:

$$\iiint \rho_w \frac{D\boldsymbol{u}}{Dt} dV = \iiint \rho_w \boldsymbol{F} dV - \iiint \nabla p dV. \tag{C.24}$$

As the parcel is arbitrary, the relationship must hold for all elements, so we have:

$$\frac{D\boldsymbol{u}}{Dt} = \boldsymbol{F} - \frac{1}{\rho_w} \nabla p. \tag{C.25}$$

These equations are called the Euler equations and together with the mass conservation equation (C.19) they form the basis for the study of any flow of an incompressible fluid.

Let us now rewrite material derivative $D\boldsymbol{u}/Dt$ in a slightly different form:

$$\frac{D\boldsymbol{u}}{Dt} = \frac{\partial \boldsymbol{u}}{\partial t} + \boldsymbol{u} \cdot \nabla \boldsymbol{u} = \frac{\partial \boldsymbol{u}}{\partial t} + \nabla \left(\frac{1}{2}|\boldsymbol{u}|^2\right) + \nabla \times \boldsymbol{u} \times \boldsymbol{u}, \tag{C.26}$$

or:

$$\frac{D\boldsymbol{u}}{Dt} = \frac{\partial \boldsymbol{u}}{\partial t} + \nabla \left(\frac{1}{2}|\boldsymbol{u}|^2\right) + (\text{curl}\boldsymbol{u}) \times \boldsymbol{u}, \tag{C.27}$$

where definition (C.11) of rotation has been used. Substituting Eq. (C.27) into Eq. (C.25) provides another representation of the Euler equations:

$$\frac{\partial \boldsymbol{u}}{\partial t} + \nabla \left(\frac{1}{2}|\boldsymbol{u}|^2\right) + (\text{curl}\boldsymbol{u}) \times \boldsymbol{u} = \boldsymbol{F} - \frac{1}{\rho_w} \nabla p. \tag{C.28}$$

This representation clearly demonstrates that the Euler equations also support a rotational flow of incompressible fluid. However, they have no mechanism for vorticity diffusion.

Let us now restrict ourselves to the gravity force only, *i.e.*:

$$\boldsymbol{F} = -g. \tag{C.29}$$

After substituting Eq. (C.29) into Eq. (C.25), we can present the Euler equations for the water motion, induced by gravity, as:

$$\left. \begin{array}{rcl} \rho_w \left(\dfrac{\partial u}{\partial t} + u\dfrac{\partial u}{\partial x} + v\dfrac{\partial u}{\partial y} + w\dfrac{\partial u}{\partial z}\right) &=& -\dfrac{\partial p}{\partial x} \\[2mm] \rho_w \left(\dfrac{\partial v}{\partial t} + u\dfrac{\partial v}{\partial x} + v\dfrac{\partial v}{\partial y} + w\dfrac{\partial v}{\partial z}\right) &=& -\dfrac{\partial p}{\partial y} \\[2mm] \rho_w \left(\dfrac{\partial w}{\partial t} + u\dfrac{\partial w}{\partial x} + v\dfrac{\partial w}{\partial y} + w\dfrac{\partial w}{\partial z}\right) &=& -\dfrac{\partial (p + \rho_w g z)}{\partial z} \end{array} \right\}, \tag{C.30}$$

or in vector notation after using Eq. (C.26):

$$\frac{\partial \mathbf{u}}{\partial t} + \nabla \left(\frac{1}{2} |\mathbf{u}|^2 + gz + \frac{p}{\rho_w} \right) + \nabla \times \mathbf{u} \times \mathbf{u} = 0. \tag{C.31}$$

Many flows in the ocean are rotational due to presence of viscosity friction and the Earth's rotation; they also reveal a lot of diffusion. Extension of the Euler equations to viscous flows is given in the form of the Navier-Stokes equations, named after C.L.M.H. Navier (1785–1836) and Sir George G. Stokes (1819–1903), who are credited with their derivation. The Navier-Stokes equations are usually written as follows:

$$\left. \begin{aligned} \rho \left(\frac{\partial u}{\partial t} + u\frac{\partial u}{\partial x} + v\frac{\partial u}{\partial y} + w\frac{\partial u}{\partial z} \right) &= -\frac{\partial p}{\partial x} + \mu \nabla^2 u \\ \rho \left(\frac{\partial v}{\partial t} + u\frac{\partial v}{\partial x} + v\frac{\partial v}{\partial y} + w\frac{\partial v}{\partial z} \right) &= -\frac{\partial p}{\partial y} + \mu \nabla^2 v \\ \rho \left(\frac{\partial w}{\partial t} + u\frac{\partial w}{\partial x} + v\frac{\partial w}{\partial y} + w\frac{\partial w}{\partial z} \right) &= -\frac{\partial (p + \rho_w gz)}{\partial z} + \mu \nabla^2 w \end{aligned} \right\} \tag{C.32}$$

in which μ is the coefficient of dynamic molecular viscosity, and ∇^2 denotes the Laplacian of velocity u (see Eq. C.14).

If a flow is not uniform, fluid particles undergo deformation. In particular, in a converging flow fluid particles become longer and thinner. This process is called dilatational deformation. When flowing around a bend, fluid particles undergo both deformation and rotation, and motion becomes rotational. In order to emphasize the rotational terms in the Navier-Stokes equation, we rewrite the first equation of (C.32) in a slightly different form:

$$\rho_w \left[\frac{\partial u}{\partial t} + \left(u\frac{\partial u}{\partial x} + v\frac{\partial v}{\partial x} + w\frac{\partial w}{\partial x} \right) + v \left(\frac{\partial u}{\partial y} - \frac{\partial v}{\partial x} \right) + w \left(\frac{\partial u}{\partial z} - \frac{\partial w}{\partial x} \right) \right] =$$
$$-\frac{\partial p}{\partial x} + \mu \nabla^2 u. \tag{C.33}$$

Terms $v \left(\partial u/\partial y - \partial v/\partial x \right)$ and $w \left(\partial u/\partial z - \partial w/\partial x \right)$ are the rotation terms. When these terms vanish, water motion is irrotational. It should be noted that this conclusion is also valid for the Euler equations (C.30). In general, a motion can be assumed irrotational when streamlines converge rapidly, and when the velocity distribution depends on the shape of the boundaries and not on their roughness (Le Méhauté, 1976).

Vector notation allows us to express the Navier-Stokes equation for motion in 3-dimensions in a compact form, i.e.:

$$\frac{\partial \mathbf{u}}{\partial t} + \nabla \left(\frac{1}{2} |\mathbf{u}|^2 + gz + \frac{p}{\rho_w} \right) + \nabla \times \mathbf{u} \times \mathbf{u} + \mu \nabla^2 \mathbf{u} = 0. \tag{C.34}$$

Coming back to the Euler equation (C.31), let us examine its application for the cases of irrotational and rotational flows. For flow to be irrotational, velocity gradients have to satisfy the following equations (see Eq. C.12):

$$\frac{\partial w}{\partial y} - \frac{\partial v}{\partial z} = 0, \quad \frac{\partial u}{\partial z} - \frac{\partial w}{\partial x} = 0, \quad \frac{\partial v}{\partial x} - \frac{\partial u}{\partial y} = 0. \tag{C.35}$$

It can be seen that these conditions are satisfied when velocity components are defined as the derivatives of a function ϕ, such that:

$$u = \frac{\partial \phi}{\partial x}, \quad v = \frac{\partial \phi}{\partial y}, \quad w = \frac{\partial \phi}{\partial z}, \tag{C.36}$$

or in vectorial notation:

$$\boldsymbol{u} = \mathrm{grad}\,\phi = \frac{\partial \phi}{\partial x}\boldsymbol{i} + \frac{\partial \phi}{\partial y}\boldsymbol{j} + \frac{\partial \phi}{\partial z}\boldsymbol{k}. \tag{C.37}$$

The function ϕ is called the **velocity potential function**. As $\boldsymbol{u} = \nabla \phi$ and $\nabla \times \nabla \phi \equiv 0$, Eq. (C.31) becomes:

$$\nabla \left(\frac{\partial \phi}{\partial t} + \frac{1}{2}|\boldsymbol{u}|^2 + gz + \frac{p}{\rho_w} \right) = 0, \tag{C.38}$$

with the result that:

$$\frac{\partial \phi}{\partial t} + \frac{1}{2}|\boldsymbol{u}|^2 + gz + \frac{p}{\rho_w} = f(t), \tag{C.39}$$

or:

$$\frac{\partial \phi}{\partial t} + \frac{1}{2}\left[\left(\frac{\partial \phi}{\partial x}\right)^2 + \left(\frac{\partial \phi}{\partial y}\right)^2 + \left(\frac{\partial \phi}{\partial z}\right)^2 \right] + \frac{p}{\rho_w} + gz = f(t), \tag{C.40}$$

where f is the function of time only. When motion is steady, Eq. (C.39) simplifies as:

$$\frac{1}{2g}|\boldsymbol{u}|^2 + \frac{p}{\rho_w g} + z = \mathrm{Const.} \tag{C.41}$$

This is the well known Bernoulli equation, named after Daniel Bernoulli, who stated it in words in his textbook of 1738, and finally derived in 1755 by Leonard Euler. Equation (C.41) expresses the relationship between velocity, pressure and elevation for steady, frictionless incompressible flow along a streamline.

Let us assume that flow is steady but rotational, then Eq. (C.31) gives:

$$\nabla \left(\frac{1}{2} |\boldsymbol{u}|^2 + gz + \frac{p}{\rho_w} \right) = \boldsymbol{u} \times \nabla \times \boldsymbol{u}, \tag{C.42}$$

which is the Crocco's equation indicating how the Bernoulli constant actually varies across a flow. If we take a scalar product with $\boldsymbol{u}$, then the scalar triple product on the right-hand side of Eq. (C.42) becomes zero, and:

$$\boldsymbol{u} \cdot \nabla \left(\frac{1}{2} |\boldsymbol{u}|^2 + gz + \frac{p}{\rho_w} \right) = 0. \tag{C.43}$$

The left-hand side of this equation represents the dot product of two vectors (see Eq. C.5). Because the dot product vanishes, the gradient of the quantity in brackets is orthogonal to the velocity vector. On the other hand, along the streamline the quantity in brackets is constant. It should be noted that these conclusions are true even for the rotational flow.

There are numerous solutions of the Euler and Bernoulli equations, and many of them are used in this book, but the exact solutions of the Navier-Stokes equations are rather rare. In the next section, an example of such solution for a small sphere moving in the fluid of viscosity μ will be given.

C.3.2 Navier-Stokes Equation Solution for Sphere Moving in Viscous Fluid

To illustrate the applicability of the Navier-Stokes equation we will consider a sphere of diameter D moving in a fluid of viscosity μ. However, it will be convenient to assume the sphere to be stationary. Moreover, the velocity of flow relative to the sphere is assumed to be at infinite distance from the sphere, the velocity of flow and diameter of sphere to be very small, and the Reynolds number to be much smaller than 1. The viscosity effects become predominant and the inertia forces due to transient and convective accelerations can be neglected. Thus, the Navier-Stokes equations (C.32) are simplified as:

$$\mu \nabla^2 u = \frac{\partial p}{\partial x}, \quad \mu \nabla^2 v = \frac{\partial p}{\partial y}, \quad \mu \nabla^2 w = \frac{\partial p}{\partial z}. \tag{C.44}$$

Differentiating the first of Eq. (C.44) with respect to y, and the second equation with respect to x, and subtracting, we obtain:

$$\mu \nabla^2 \left(\frac{\partial v}{\partial x} - \frac{\partial u}{\partial y} \right) = \mu \nabla^2 \zeta = 0. \tag{C.45}$$

In a similar way we obtain:

$$\mu \nabla^2 \left(\frac{\partial w}{\partial y} - \frac{\partial v}{\partial z} \right) = \mu \nabla^2 \xi = 0, \tag{C.46}$$

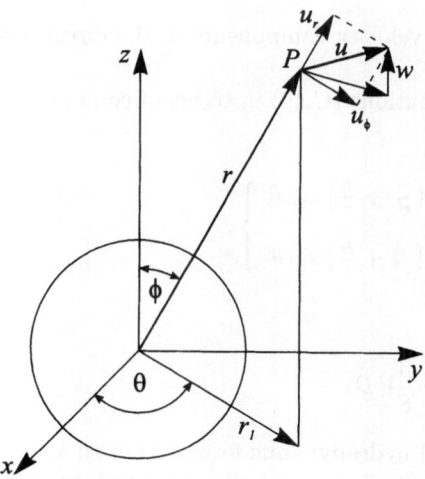

Fig. C.3: Definition of spherical coordinate system

and

$$\mu\nabla^2\left(\frac{\partial u}{\partial z} - \frac{\partial w}{\partial x}\right) = \mu\nabla^2\eta = 0. \tag{C.47}$$

As was shown in Sect. C.2, the quantities ξ, η and ζ are the components of the vorticity vector.

Let us now differentiate Eqs. (C.44) with respect to x, y and z, respectively, and summarize. Thus, after using the equation of mass conservation (C.19), we obtain (Batchelor, 1967):

$$\nabla^2 p = 0. \tag{C.48}$$

Applying the Laplacian operator ∇^2 to Eqs. (C.44) and using Eq. (C.48) gives:

$$\nabla^4 u = 0, \quad \nabla^4 v = 0, \quad \nabla^4 w = 0. \tag{C.49}$$

Because of a sphere geometry, it is convenient to use the cylindrical coordinate system (r, ϕ, θ) – see Fig. C.3. At infinity there is undisturbed flow when:

$$w = V, \quad v_r = V\cos\phi, \quad v_\phi = V\sin\phi. \tag{C.50}$$

On the surface of the sphere $(r = D/2)$ non-slip conditions apply, *i.e.*:

$$v_r = v_\phi = 0, \tag{C.51}$$

where v_r and v_ϕ are velocity components in the directions of increasing r and ϕ, respectively.

The solution of equations (C.49) in terms of components v_r and v_ϕ takes the form (Fung, 1997):

$$\left.\begin{array}{rcl} v_r &=& V\cos\theta - 2\left(\frac{A}{r^3} + \frac{B}{r}\right)\cos\theta \\[2mm] v_\phi &=& -V\sin\theta - \left(\frac{A}{r^3} + \frac{B}{r}\right)\sin\theta \end{array}\right\}, \tag{C.52}$$

in which:

$$A = \frac{1}{32}VD^3, \;\; B = -\frac{3}{8}VD. \tag{C.53}$$

To evaluate the total hydrodynamic force acting on a sphere we must compute all stresses at $r = D/2$. Because of the axisymmetry of the flow, the resultant forces in the x and y directions are zero. The stress in the z direction is (Batchelor, 1967):

$$\tau_z = -p_0 - \frac{3\mu V}{D}, \tag{C.54}$$

in which p_0 is the pressure in the fluid at infinity, and makes no contribution to the total force on the sphere. Integrating the stress τ_z over the sphere we obtain:

$$F = \int\int \tau_z dS = \frac{\pi}{2}D^2 \int_0^\pi \frac{3\mu V}{D}\sin\phi d\phi, \tag{C.55}$$

or:

$$F = 3\pi\rho_w\nu DV. \tag{C.56}$$

This solution is usually known as Stokes' law for the drag on a moving sphere at very small Reynolds number (Stokes, 1851). We use it to determine a terminal velocity in Sect. 2.6.5.

In the Stokes' solution, the inertial terms are entirely neglected. However, for small but finite Reynolds number, some correction due to inertial terms is needed. Oseen (1910) and later Proudman and Pearson (1957) provided expression for the resultant hydrodynamic force as:

$$F = 3\pi\mu DU\left[1 + \frac{3}{8}Re + \frac{9}{40}Re^2\ln(Re)\right]. \tag{C.57}$$

The streamline pattern of the solution based on the Oseen equation shows the development of wake behind a sphere. When the Reynolds number increases, the wake becomes more pronounced and a separation of streamlines from a solid sphere surface is observed at a Reynolds number of about 5.

C.4   Summary of Basic Irrotational Flow Properties

In Sect. C.3 it was shown that the conditions of irrotationality impose an existence of the velocity potential function ϕ, such that velocity components $u(x, y, z, t)$, $v(x, y, z, t)$ and $w(x, y, z, t)$ are the derivatives of function ϕ with respect to x, y and z, correspondingly. The velocity potential function may be used for any kind of irrotational flow, steady or unsteady, and two- or three-dimensional. It is also used in the study of turbulent flow, provided the velocity potential, ϕ, refers to the time-average motion. Note that the dimension of the velocity potential function is $[m^2 s^{-1}]$.

For simplicity, let us assume that fluid density is constant. Then, a substitution of Eq. (C.36) into the equation of mass conservation (C.19) gives:

$$\frac{\partial^2 \phi}{\partial x^2} + \frac{\partial^2 \phi}{\partial y^2} + \frac{\partial^2 \phi}{\partial z^2} = 0 \quad \text{or} \quad \nabla^2 \phi = 0. \tag{C.58}$$

Thus, an irrotational flow of an incompressible fluid satisfies Eq. (C.58), which is called the Laplace equation. This equation appears in many problems in fluid mechanics, electrodynamics and optics, and it possesses a large number of solutions that may not necessarily represent wave motion. For example, it is easy to check that the velocity potential, $\phi(x, z) = Ux$, satisfies Eq. (C.58). In fact, this velocity potential corresponds to a uniform flow in the x-axis direction with constant velocity, U, i.e.:

$$u = \frac{\partial \phi}{\partial x} = U. \tag{C.59}$$

Other examples of the Laplace equation solution are discussed below.

In Sect. 2.3.1, a streamline has been defined as a line tangent everywhere to the local velocity vector. The function which describes a flow pattern in terms of its streamlines is called a **stream function** $\psi(x, y, z)$. This function, along with velocity potential function ϕ, is mainly used to study two-dimensional flow, although theoretically it can also be used for three-dimensional space. However, for simplicity, let us restrict ourselves to two-dimensional flow only. Then, the stream function ψ should satisfy the following relationships:

$$u = \frac{\partial \psi}{\partial y}, \quad v = -\frac{\partial \psi}{\partial x}. \tag{C.60}$$

Substituting Eq. (C.60) into the mass conservation equation gives:

$$\frac{\partial}{\partial x}\left(\frac{\partial \psi}{\partial y}\right) + \frac{\partial}{\partial y}\left(-\frac{\partial \psi}{\partial x}\right) \equiv 0. \tag{C.61}$$

Therefore, stream function ψ satisfies the mass conservation equation automatically. The dimension of a stream function ψ is identical to the velocity

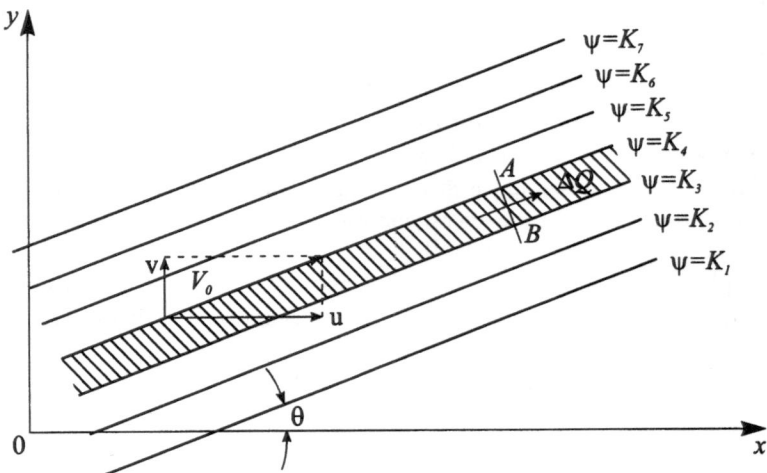

Fig. C.4: Uniform flow in plane Oxy

potential function, namely $[m^2s^{-1}]$. Equation (C.61) also indicates that along a streamline ψ is constant, *i.e.*:

$$\psi(x,y,z) = \text{Const.} \tag{C.62}$$

A noteworthy and useful property of streamlines is that the discharge between two streamlines is given by their difference. Consider, for example, a steady uniform flow in the plane Oxy (Fig. C.4). The velocity of flow, inclined at an angle θ with respect to the x-axis, is V_0. Thus, the velocity components should be:

$$u(x,y) = V_0 \cos\theta, \quad v(x,y) = V_0 \sin\theta. \tag{C.63}$$

Integration of both sides of the first equation in (C.60), with respect to y, yields:

$$\psi(x,y) = \int u\,dy = \int V_0 \cos\theta\,dy + f(x) = V_0 y \cos\theta + f(x), \tag{C.64}$$

and after using the second equation of (C.60) we obtain:

$$-\frac{\partial\psi}{\partial x} = -\frac{\partial f}{\partial x} = V_0 \sin\theta. \tag{C.65}$$

Thus:

$$f(x) = -V_0 x \sin\theta. \tag{C.66}$$

Therefore, the stream function $\psi(x, y)$ finally takes the form:

$$\psi(x, y) = V_0 (y \cos \theta - x \sin \theta), \tag{C.67}$$

and the streamlines are given by equations:

$$V_0 (y \cos \theta - x \sin \theta) = K_n, \tag{C.68}$$

in which K_n is the constant. The subscript n indicates a value of constant K, corresponding to the nth streamline. If we rewrite Eq. (C.68) in another form:

$$y = \tan \theta \, x + \frac{K_n}{V_0 \cos \theta}, \tag{C.69}$$

we find that streamlines are straight lines of slope $\tan \theta$, as should be expected.

Consider now the flow through cross-section A–B in Fig. C.4. The discharge of flow through this cross-section is:

$$\Delta Q = K_4 - K_3. \tag{C.70}$$

Assuming that a difference between two adjacent streamlines is constant and equal to ΔK, the discharge for any two adjacent streamlines becomes:

$$\Delta Q = K_n - K_{n-1} = \Delta K. \tag{C.71}$$

In a similar way it can be found that the velocity potential function, ϕ, corresponding to the uniform flow shown in Fig. C.4, is:

$$\phi(x, y) = V_0(x \cos \theta + y \sin \theta), \tag{C.72}$$

and at equipotential lines where the velocity potential, ϕ, is constant, we have:

$$V_0(x \cos \theta + y \sin \theta) = P_n, \tag{C.73}$$

where P_n is the constant corresponding to the nth equipotential line.

Let us rewrite Eq. (C.73) as:

$$y = -\frac{1}{\tan \theta} x + \frac{P_n}{V_0 \sin \theta}. \tag{C.74}$$

Thus, equipotential lines are also straight lines but of slope $(-1/\tan \theta)$. Comparing this slope to the slope of the streamline (Eq. C.69), we find that equipotential lines are perpendicular to the streamlines as their slopes are negative reciprocals.

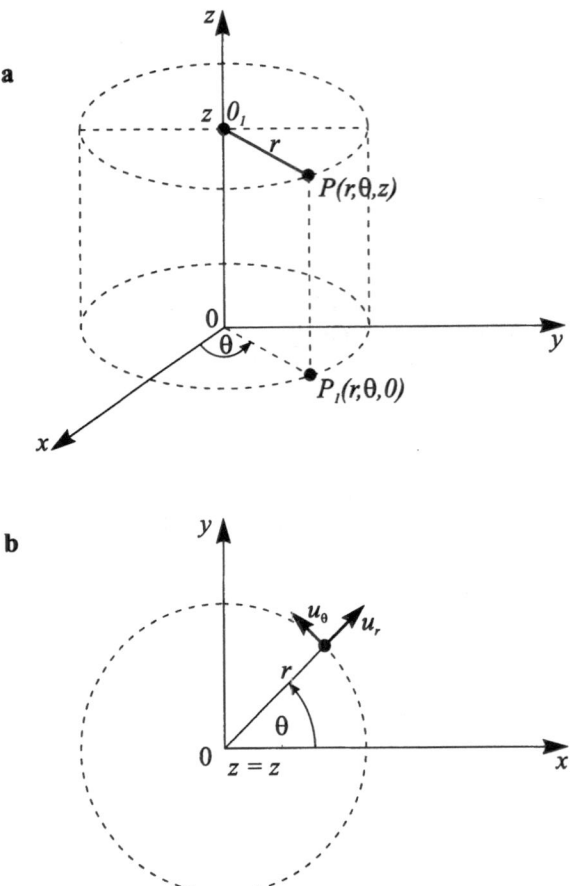

Fig. C.5: Definition of cylindrical coordinate system: **a** cylindrical coordinates of point P, **b** radial and tangential velocities

Some applications of the velocity potential, ϕ, or stream function, ψ, require a formulation of these functions in a coordinate system other than the Cartesian system $Oxyz$, namely the cylindrical coordinate system, $Or\theta$, and the spherical coordinate system, $Or\theta\varphi$. Below, the basic formulas for irrotational flow in the cylindrical coordinate system are listed without proof (see Fig. C.5 for definition of coordinates):

$$x = r\cos\theta, \quad y = r\sin\theta, \quad z = z, \tag{C.75}$$

$$\left.\begin{array}{rcl} u_r &=& u\cos\theta + v\sin\theta \\ u_\theta &=& -u\sin\theta + v\cos\theta \\ u_z &=& w \end{array}\right\}, \tag{C.76}$$

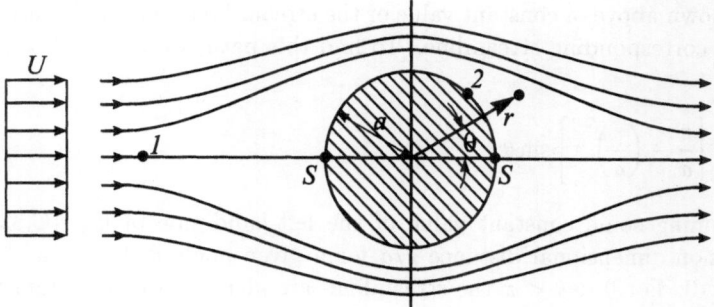

Fig. C.6: Flow around horizontal circular cylinder

and

$$u_r = \frac{\partial \phi}{\partial r}, \quad u_\theta = \frac{1}{r}\frac{\partial \phi}{\partial \theta}, \quad u_z = \frac{\partial \phi}{\partial z}, \tag{C.77}$$

$$\frac{\partial \rho_w}{\partial t} + \frac{1}{r}\frac{\partial}{\partial r}\left(\rho_w r u_r\right) + \frac{1}{r}\frac{\partial}{\partial \theta}\left(\rho_w u_\theta\right) + \frac{\partial}{\partial z}(\rho_w u_z) = 0, \tag{C.78}$$

$$\frac{\partial^2 \phi}{\partial r^2} + \frac{1}{r}\frac{\partial \phi}{\partial r} + \frac{1}{r^2}\frac{\partial^2 \phi}{\partial \phi^2} + \frac{\partial^2 \phi}{\partial z^2} = 0, \tag{C.79}$$

where u_r, u_θ and u_z are the components of velocity along r, θ and z cylindrical coordinates. Equations (C.78) and (C.79) are the mass conservation and Laplace equations, respectively, for the cylindrical coordinate system.

Let us now demonstrate an application of cylindrical coordinates to a flow of perfect fluid around a horizontal, infinitely long, solid cylinder, as shown in Fig. C.6. The stream function, ψ, and velocity potential function, ϕ, for flow around a cylinder are (Lamb, 1932):

$$\psi(r, \theta) = U\left(r - \frac{a^2}{r}\right)\sin \theta, \tag{C.80}$$

and

$$\phi(r, \theta) = U\left(r + \frac{a^2}{r}\right)\cos \theta. \tag{C.81}$$

As was shown above, a constant value of the stream function ψ determines the path of a corresponding streamline. To find this path, we rewrite Eq. (C.80) as:

$$\frac{\psi(r,\theta)}{Ua} = \left[\frac{r}{a} - \left(\frac{r}{a}\right)^{-1}\right]\sin\theta. \tag{C.82}$$

Now assuming some constant value of the left-hand side of Eq. (C.82), we can find non-dimensional distance r/a for a given angle θ, being in a range $-\pi < \theta < 0$. For $0 < \theta < \pi$ the streamlines are identical due to symmetry of the cylinder with respect to the x-axis (see Fig. C.6).

It can be verified that the normal velocity to the cylinder $u_n = \partial\phi/\partial n = \partial\phi/\partial r$ vanishes at each point of the cylinder as:

$$u_n = \frac{\partial\phi}{\partial r}\bigg|_{r=a} = U\left(1 - \frac{a^2}{r^2}\right)\cos\theta\big|_{r=a} \equiv 0, \tag{C.83}$$

but as the fluid is perfect, no velocity gradient is established at the cylinder surface, and the tangent velocity does not vanish, $i.e.$:

$$u_\theta = \frac{1}{r}\frac{\partial\phi}{\partial\theta}\bigg|_{r=a} = -2U\sin\theta. \tag{C.84}$$

A pressure distribution at the cylinder surface follows from the Bernoulli equation. Consider for example, points 1 and 2 at the same streamline (Fig. C.6). Point 1 is assumed to be located very far from the cylinder, thus $u_1 \approx U$ and $p_1 \approx p_0$, where U and p_0 are velocity and pressure at infinity. Point 2 is located at the cylinder circumference where only tangent velocity exists. Therefore, the Bernoulli equation (C.41) gives:

$$\frac{U^2}{2g} + \frac{p_0}{\rho_w g} = \frac{(2U\sin\theta_2)^2}{2g} + \frac{p(\theta_2)}{\rho_w g} = \text{Const}, \tag{C.85}$$

and

$$p(\theta_2) = p_0 + \frac{\rho_w U^2}{2}\left(1 - 4\sin^2\theta_2\right). \tag{C.86}$$

It is interesting to verify the force imposed by the fluid on the cylinder. The integration of pressure along the cylinder circumference yields:

$$F = \int_0^{2\pi} p(\theta)R\cos\theta d\theta \equiv 0. \tag{C.87}$$

This unexpected result can be extended to any submerged body, the total drag force induced by uniform flow of a $perfect$ $fluid$ on a submerged body is nil. (One can have lift, of course). This phenomenon is known as the paradox of D'Alembert.

C.5 Formulas of Small Amplitude Wave Theory

C.5.1 Unknown Quantities and Laplace Equation

In this section a short derivation of various parameters of surface waves is provided, as this section is treated as a supplement to Sect. 4.2 in which only final calculation formulas have been presented.

Consider a regular wave motion as shown in Fig. 4.2. At each point in the water body (say point P), there are five unknown basic wave parameters; horizontal component of orbital velocity $u(x, z, t)$, vertical component of orbital velocity $w(x, z, t)$, wave induced pressure $p(x, z, t)$, surface displacement $\zeta(x, t)$ and wavelength L. To simplify the evaluation of these five unknowns, some specific properties of water and of water motion itself will be used. Firstly, we use the fact that water is an incompressible fluid. Moreover, we assume that waves are uniform in the y-direction (crest lines are perpendicular to the plane $0, x, z$). Therefore, wave motion becomes two-dimensional, satisfying the following equation of mass conservation:

$$\frac{\partial u}{\partial x} + \frac{\partial w}{\partial z} = 0, \tag{C.88}$$

in which u and w are the horizontal and vertical component of velocity, respectively.

Secondly, we take into consideration that wave motion is irrotational. In Sect. C.4 we found that for irrotational flow, there is some function $\phi(x, z, t)$, such that the velocity components $u(x, z)$ and $w(x, z)$ can be expressed as the gradients of the function $\phi(x, z, t)$, *i.e.*:

$$\left.\begin{aligned} u &= \frac{\partial \phi}{\partial x} \\[2mm] w &= \frac{\partial \phi}{\partial z} \end{aligned}\right\}. \tag{C.89}$$

As waves are uniform in the y-direction, dependence of function ϕ on y can be neglected, and all quantities are functions of x, z and t only. The velocity potential ϕ satisfies the Laplace equation:

$$\frac{\partial^2 \phi}{\partial x^2} + \frac{\partial^2 \phi}{\partial z^2} = \nabla^2 \phi = 0. \tag{C.90}$$

In order to find the velocity potential which satisfies Eq. (C.90) and represents wave motion in the water domain as in Fig. 4.2, we must formulate some constraints which have to be satisfied by the velocity potential, $\phi(x, z, t)$. These constraints, known as boundary conditions, control the behaviour of function, ϕ, at the sea surface, sea bottom and at infinite distance.

C.5.2   Boundary Conditions

We start with the condition at the sea bottom (plane $z = -h$ in Fig. 4.2).
If the sea bottom is not permeable, there is no motion through the boundary
($z = -h$), i.e. velocity, normal to the boundary, has to be equal to zero. Hence,
we have:

$$w = \frac{\partial \phi}{\partial z} = 0 \quad \text{at} \quad z = -h. \tag{C.91}$$

This boundary condition is consistent with the particle motion in waves, explained in Sect. 3.4.2 (see also Fig. 3.4). At the sea bottom, the horizontal
velocity component, u, is not necessarily zero while the vertical component vanishes completely as the path of the particle motion deteriorates into horizontal
motion only.

At the sea surface, the boundary conditions are more complicated. First of
all, let us consider a water particle $P_1(x, z, t)$ located at the sea surface (see
Fig. 4.2). Existence of the sea surface means that there is no transfer of matter
across the surface. As this surface moves, the velocity of a point $P_1(x, z, t)$
on the surface, in the direction normal to the surface, is equal to the normal
component of velocity of the fluid particle at the same point on the surface.
This condition, which is known as the **kinematic boundary condition**, can
be expressed in terms of the velocity potential, $\phi(x, z, t)$, as follows (Massel,
1989):

$$\frac{\partial \zeta}{\partial t} - \frac{\partial \phi}{\partial x} \frac{\partial \zeta}{\partial x} = \frac{\partial \phi}{\partial z} \quad \text{at} \quad z = \zeta, \tag{C.92}$$

in which $\zeta(x, t)$ is the surface displacement from the still water level (see
Fig. 4.2). Physically, the term $\partial \zeta / \partial t$ denotes the velocity of vertical movement of the sea surface. The second term on the left-hand side of Eq. (C.92)
expresses the fact that water particles subjected to wave motion move along
the surface. Therefore, a particular point $P(x, z, t)$ not only moves vertically
but also possesses a small back- and forward motion in the horizontal direction.
The summation of the two terms on the left-hand side of Eq. (C.92) provides
the total vertical velocity $\partial \phi / \partial z$ of the water particle. A detailed derivation of
Eq. (C.92) will not be given here, as it can be found in many papers and books
(see, for example, Massel, 1989; Dean and Dalrymple, 1992).

As we assumed that wave steepness was very small, it follows that $\partial \zeta / \partial x$
term is very small and can be neglected. Therefore, Eq. (C.92) simplifies to:

$$\frac{\partial \zeta}{\partial t} = \frac{\partial \phi}{\partial z} \quad \text{at} \quad z = 0. \tag{C.93}$$

Note that the kinematic condition is now applied on the plane $z = 0$, instead
of the surface $z = \zeta$, due to the assumption of low wave steepness (see also
Fig. 4.1).

The existence of the sea surface results also from a dynamic balance between air and water masses. In particular, air pressure at the point $P_1(x, z, t)$ has to be equal to the water pressure at this point. This balance is expressed using the Bernoulli equation. For wave motion with potential ϕ, the Bernoulli equation becomes (see Eq. C.40):

$$\frac{\partial \phi}{\partial t} + \frac{1}{2}\left[\left(\frac{\partial \phi}{\partial x}\right)^2 + \left(\frac{\partial \phi}{\partial z}\right)^2\right] + \frac{p - p_a}{\rho_w} + gz = 0 \quad \text{at} \quad z = \zeta, \tag{C.94}$$

in which p_a is a reference atmospheric pressure and function $f(t)$ is assumed to be zero. Under the assumption that the pressure, p, at the sea surface is constant and equal to atmospheric pressure, Eq. (C.94) can be rewritten in the form of the **dynamic boundary condition**:

$$\frac{1}{2}\left(\frac{\partial \phi}{\partial x}\right)^2 + \frac{1}{2}\left(\frac{\partial \phi}{\partial z}\right)^2 + \frac{\partial \phi}{\partial t} + gz = 0, \quad \text{on } z = \zeta. \tag{C.95}$$

Again, we invoke the assumption of low wave steepness and small amplitude and apply this condition at the plane $z = 0$. After neglecting squares of velocities, Eq. (C.95) is simplified as follows:

$$\frac{\partial \phi}{\partial t} + g\zeta = 0 \quad \text{for} \quad z = 0, \tag{C.96}$$

or:

$$\frac{\partial^2 \phi}{\partial t^2} + g\frac{\partial \zeta}{\partial t} = 0. \tag{C.97}$$

Substituting (C.97) into (C.93) yields:

$$\frac{\partial^2 \phi}{\partial t^2} + g\frac{\partial \phi}{\partial z} = 0 \quad \text{at} \quad z = 0. \tag{C.98}$$

This is the boundary condition on ϕ which holds on the boundary $z = 0$. Equations (C.90), (C.91), (C.93), (C.96) and (C.98) constitute a complete set of equations for solving wave motion in a constant water depth.

C.5.3   Wave Type Solution of the Laplace Equation

We are now in position to determine the velocity potential $\phi(x, z, t)$ which satisfies the Laplace equation and boundary conditions (C.91) and (C.98). To do this we consider the wave motion propagating along the x-axis, from $-\infty$ to ∞, over a constant water depth, h. Wave height is equal to H and wave period is equal to T. Again we assume that wave motion is uniform in the y-direction. The unknown velocity potential, ϕ, is a function of the independent variables

x, z and time, t. The simplest approach is to represent function ϕ as a product of functions $F_1(x)$, $F_2(z)$ and $F_3(t)$, each depending on variable x, z and t, respectively, $i.e.$

$$\phi(x, z, t) = F_1(x)\, F_2(z)\, F_3(t). \tag{C.99}$$

Substitution of (C.99) into Laplace equation (C.90) gives:

$$\frac{1}{F_1(x)}\frac{d^2 F_1(x)}{dx^2} = -\frac{1}{F_2(z)}\frac{d^2 F_2(z)}{dz^2} = -k^2. \tag{C.100}$$

Representation (C.99) constitutes the so called 'separation of variables' method for solving Laplace equations. It is not surprising that the function $F_3(t)$ disappears from (C.100), as the Laplace equation depends on spatial coordinates, not on time. However, $F_3(t)$ was added to express the periodic character of wave motion in time and $F_3(t)$ will be determined later.

The first term on the left-hand side of Eq. (C.100) is a function of x only, while the second term is a function of z only. As these terms have to be equal, they only can be equal to some constant, say $(-k^2)$. Therefore, equation (C.100) can be split up into two equations:

$$\frac{d^2 F_1(x)}{dx^2} + k^2\, F_1(x) = 0, \tag{C.101}$$

and

$$\frac{d^2 F_2(z)}{dz^2} - k^2\, F_2(z) = 0. \tag{C.102}$$

Taking into account that the velocity potential $\phi(x, z, t)$ has to represent wave motion, periodic in the x-direction, the following solution of Eq. (C.101) should be chosen:

$$F_1(x) = \Re(e^{ikx}), \quad \text{or :} \tag{C.103}$$

$$F_1(x) = \cos kx. \tag{C.104}$$

It should be noted that the constant k in Eqs. (C.103) or (C.104), plays the role of the wave number which, however, is still unknown. The e^{ikx} is a complex quantity and symbol $\Re$ denotes the real part of the complex quantity. The representation (C.103) is more general and more convenient for further calculations, but in fact both representations are equivalent as:

$$\Re(e^{ikx}) = \Re\left[\cos(kx) + i\,\sin(kx)\right] = \cos(kx), \tag{C.105}$$

in which $i = \sqrt{-1}$ is an imaginary unit (see Sect. C.6 for more details).

In similar way, for the function $F_2(z)$ we obtain from Eq. (C.102):

$$F_2(z) = \cosh(kz). \tag{C.106}$$

Using Eqs. (C.103) and (C.106) in (C.99) yields:

$$\phi(x, z, t) = A e^{ikx} \cosh(kz) F_3(t), \tag{C.107}$$

where A is a constant, still unknown. As velocity potential $\phi(x, z, t)$ has to satisfy the boundary condition (C.91) at $z = -h$, it will be convenient to rewrite Eq. (C.107) in the form:

$$\phi(x, z, t) = \Re \left[A e^{ikx} \frac{\cosh k(h + z)}{\cosh(kh)} F_3(t) \right]. \tag{C.108}$$

Assuming that the wave period is equal to T, the function $F_3(t)$ should be a periodic function, such as $F_3(t) = e^{-i\omega t}$, where $\omega = 2\pi/T$. Therefore, the function $\phi(x, z, t)$ finally becomes:

$$\phi(x, z, t) = \Re \left[A \frac{\cosh k(z + h)}{\cosh(kh)} e^{i(kx - \omega t)} \right], \tag{C.109}$$

or:

$$\phi(x, z, t) = \Re \left[A \frac{\cosh k(z + h)}{\cosh(kh)} e^{ik(x - Ct)} \right], \tag{C.110}$$

in which $C = \omega/k$ is a phase velocity.

Equation (C.109) allows for the periodicity of wave motion in space as well as in time. Thus, it is reasonable to assume that the surface displacement $\zeta(x, t)$ will behave in the same manner, *i.e.*:

$$\zeta(x, t) = \Re \left[\frac{H}{2} e^{i(kx - \omega t)} \right]. \tag{C.111}$$

Using Eq. (C.109) and (C.111) in (C.96), we obtain:

$$A = \frac{-igH}{2\omega}. \tag{C.112}$$

Thus, the final form of the velocity potential $\phi(x, z, t)$ is:

$$\phi(x, z, t) = \Re \left[\frac{-igH}{2\omega} \frac{\cosh k(z + h)}{\cosh kh} e^{i(kx - \omega t)} \right]. \tag{C.113}$$

The velocity potential is a real quantity. Therefore, for further applications we have to take only the real part of the complex expression (C.113):

$$\phi(x, z, t) = \frac{gH}{2\omega} \frac{\cosh k(z + h)}{\cosh kh} \sin(kx - \omega t). \tag{C.114}$$

Function (C.114) is the required velocity potential for wave motion over a constant water depth, which satisfies the Laplace equation and all boundary conditions.

C.5.4 Wave Parameters Based on Small Amplitude Wave Theory

Orbital Velocity Components. The velocity components of the water particle subjected to wave motion can be found through the differentiation of the velocity potential (C.114). Thus, we have:

- *horizontal velocity components:*

$$u(x, z, t) = \frac{\partial \phi}{\partial x} = \frac{gkH}{2\omega} \frac{\cosh[k(z+h)]}{\cosh(kh)} \cos(kx - \omega t), \quad (C.115)$$

- *vertical velocity components:*

$$w(x, z, t) = \frac{\partial \phi}{\partial z} = \frac{gkH}{2\omega} \frac{\sinh[k(z+h)]}{\cosh(kh)} \sin(kx - \omega t). \quad (C.116)$$

In deep water, the dependence of the velocity potential on z can be presented in a simpler form. We note that (see Sect. C.6):

$$\cosh(x) = \frac{e^x + e^{-x}}{2}. \quad (C.117)$$

Therefore,

$$\frac{\cosh[k(z+h)]}{\cosh(kh)} = \frac{e^{k(z+h)} + e^{-k(z+h)}}{e^{kh} + e^{-kh}} = \frac{e^{kz} + e^{-2kh}e^{-kz}}{1 + e^{-2kh}}. \quad (C.118)$$

For very deep water, $h \to \infty$, and relation (C.118) simplifies as follows:

$$\frac{\cosh[k(z+h)]}{\cosh(kh)}\bigg|_{\lim h \to \infty} = \frac{e^{kz} + e^{-2kh}e^{-kz}}{1 + e^{-2kh}}\bigg|_{\lim h \to \infty} = e^{kz}. \quad (C.119)$$

After substituting Eq. (C.119) into (C.115), the velocity u in deep water becomes:

$$u(z) = \frac{gkH}{2\omega} e^{kz} \cos(kx - \omega t) = \frac{gTH}{2L} e^{(2\pi z/L)} \cos(kx - \omega t). \quad (C.120)$$

The rate of attenuation of velocity, u, with submergence is now controlled by the term $e^{(2\pi z/L)}$. In particular, if submergence, z, is equal to half the wavelength L, *i.e.* $z = -L/2$, Eq. (C.120) yields that velocity, u, at level z is only about 4% of its value at the sea surface. Therefore, the submergence $z = -L/2$ is usually considered as a limiting depth, below which the influence of surface waves can be neglected.

Let us now apply formula (C.115) for very shallow water. We first rewrite this formula as follows::

$$u = \frac{gTH}{2L} \frac{\cosh\left[kh\left(1+\frac{z}{h}\right)\right]}{\cosh(kh)} \cos(kx - \omega t). \tag{C.121}$$

As in shallow water $kh = (2\pi h/L) \to 0$, $\cosh(kh) \to 1$ and $\cosh\left[kh(1 + z/h)\right] \to 1$; thus:

$$u \approx \frac{gTH}{2L} \left\{ 1 + \frac{1}{2}\left[\left(1 + \frac{z}{h}\right)^2 - 1\right](kh)^2 + \cdots \right\}. \tag{C.122}$$

Horizontal velocity in very shallow water attenuates very slowly with depth, and its value is almost constant and equal to its surface value, $gTH/2L$.

Orbital velocity components at the wave crest ($z > 0$) can be determined only approximately within the constraints of linear wave theory, as we assume that the wave displacement is very small and the boundary conditions have been applied at $z = 0$, not at $z = \zeta$. However, for practical applications, the accuracy offered by the linear theory is sufficient.

Pressure in Fluid. In Sect. 2.2 it has been shown that for water at rest, the pressure at any submerged point is equal to the weight of unit water column above the point. The presence of surface waves induces additional dynamic pressure. This pressure can be calculated from Eq. (C.94) after neglecting the quadratic terms of the velocities $(\partial\phi/\partial x)^2$ and $(\partial\phi/\partial z)^2$, which is consistent with the assumption of small amplitude. Thus, after omitting the reference atmospheric pressure, we have:

$$p = -\rho_w g z - \rho_w \frac{\partial\phi}{\partial t}. \tag{C.123}$$

Using Eq. (C.114) in Eq. (C.123) yields:

$$p(x, z, t) = -\rho_w g z + \rho_w g \frac{H}{2} \frac{\cosh\left[k(z + h)\right]}{\cosh(kh)} \cos(kx - \omega t), \tag{C.124}$$

or:

$$p(x, z, t) = -\rho_w g z + \rho_w g \, K_p(z)\, \zeta, \tag{C.125}$$

where:

$$K_p(z) = \frac{\cosh\left[k(z + h)\right]}{\cosh(kh)}. \tag{C.126}$$

The first term on the right-hand side of Eq. (C.125) represents the hydrostatic pressure which exists without the presence of waves. Note that in our notation, the vertical axis z is directed upwards and the term $\rho_w g z$ provides positive pressure. The second term describes the dynamic pressure. To clarify the nature of the dynamic pressure, we rewrite Eq. (C.125) in the form:

$$p(x, z, t) = \rho_w g(-z + \zeta) - \rho_w g \left[1 - K_p(z)\right] \zeta. \tag{C.127}$$

Now, the first term on the right-hand side again has a hydrostatic character and represents the total height of the water column which also includes the contribution associated with the surface displacement. If $K_p(z)$ is equal to 1.0, the pressure is purely hydrostatic:

$$p(x, z, t) = \rho g \left[-z + \zeta(x, t)\right]. \tag{C.128}$$

This is usually the case for shallow water waves when $kh \to 0$ and $K_p(z) \to 1$. Also, at still water level ($z = 0$), $K_p(0) = 1$ for deep as well as for shallow waters.

The second term on the right-hand side of Eq. (C.127) may be rewritten as follows:

$$-\rho_w g \left[1 - K_p(z)\right] \zeta = \frac{-\rho_w g}{gk} \left[1 - K_p(z)\right] \frac{\cosh(kh)}{\sinh[k(z + h)]} \times$$

$$\times \frac{gkH}{2} \frac{\sinh[k(z + h)]}{\cosh(kh)} \cos(kx - \omega t). \tag{C.129}$$

Using Eq. (4.29) in Eq. (C.129) yields:

$$-\rho_w g \left[1 - K_p(z)\right] \zeta = \frac{\rho_w g}{gk} \left[1 - K_p(z)\right] \frac{\cosh(kh)}{\sinh[k(z + h)]} a_z(x, z, t), \tag{C.130}$$

in which a_z is the vertical component of acceleration. Substituting Eq. (C.130) into Eq. (C.127), gives:

$$p(x, z, t) = \rho_w g(-z + \zeta) + \frac{\rho_w g}{gk} \left[1 - K_p(z)\right] \frac{\cosh(kh)}{\sinh[k(z + h)]} a_z(x, z, t). \tag{C.131}$$

Thus, the hydrostatic pressure due to the water column ($-z + \zeta$) is modified by vertical acceleration, which is 180° out of phase with the free surface displacement (see Eq. 4.30). Therefore, the final pressure under the wave crest will be smaller than that resulting from water column weight only.

It should be noted that at mean water level ($z = 0$), pressure (C.131) reduces to the hydrostatic pressure, because $K_p(z = 0) = 1$, $i.e.$:

$$p(x, 0, t) = \rho_w g \zeta, \tag{C.132}$$

while at the sea bottom, the total pressure is:

$$p(x, -h, t) = \rho_w g(h + \zeta) - \rho_w g \left[1 - \frac{1}{\cosh(kh)} \right] \zeta. \tag{C.133}$$

As the surface displacement of a wave trough is negative, the total pressure is greater than that resulting from the water column only.

The pressure distribution above the mean water level is usually assumed to be hydrostatic down to $z = 0$, $i.e.$:

$$p(x, z, t) = \rho_w g(-z + \zeta) \quad \text{for} \quad 0 < z < \zeta. \tag{C.134}$$

Thus, at the sea surface, when $z = \zeta$, $p(x, \zeta, t) = 0$, and at mean water level $(z = 0)$, $p(x, 0, t) = \rho_w g \zeta$, which is in agreement with Eq. (C.132). Examples of the vertical profile of pressure under a wave crest and trough are illustrated in Fig. 4.9. A computer program for calculating wave induced pressure at any water level is given in Appendix D (Program D.44).

Wave Energy. Consider a column of water with mass dm. The potential energy of this mass of the water relative to the bottom is (see Fig. C.7):

$$dE_p = g \frac{h}{2} dm, \tag{C.135}$$

as the centre of gravity is located $h/2$ above the sea bottom. Because $dm = \rho_w h dx$, the potential energy per unit area becomes:

$$E_p = \rho_w g \frac{h^2}{2}. \tag{C.136}$$

When waves are present, the potential energy of the same water column changes as follows (see Fig. C.7b):

$$dE_p = \rho_w g \frac{(h + \zeta)^2}{2} dx. \tag{C.137}$$

Thus, the averaged (over one wavelength) potential energy takes the form:

$$E_p = \frac{1}{L} \int_x^{x+L} dE_p = \frac{1}{L} \int_x^{x+L} \rho_w g \frac{(h + \zeta)^2}{2} dx. \tag{C.138}$$

After substitution of ζ from Eq. (C.111) into Eq. (C.138), we obtain:

$$E_p = \rho_w g \frac{h^2}{2} + \rho_w g \frac{H^2}{16}. \tag{C.139}$$

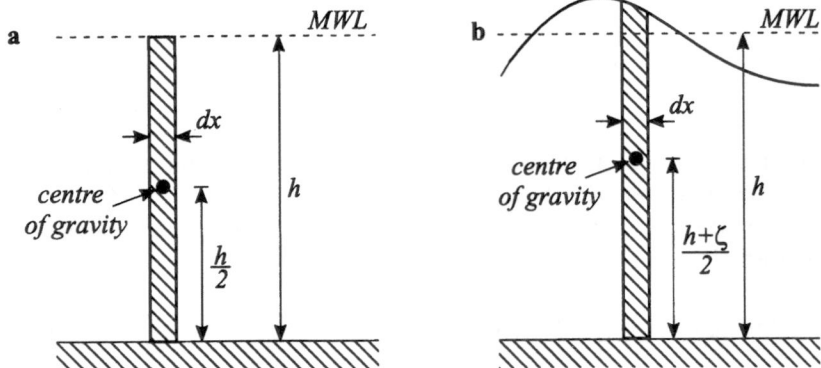

Fig. C.7: Potential energy without and with waves

The first term on the right-hand side of Eq. (C.139) is the potential energy (C.136) existing without the presence of waves. Only the second term is related to the potential energy due to waves, and thus, wave induced potential energy becomes:

$$E_p = \rho_w g \frac{H^2}{16}. \tag{C.140}$$

As water particles move along circular or elliptical paths, their velocity is not zero and they possess some kinetic energy. To find this energy, consider a small element of the water mass $(\rho_w dx dz)$. The kinetic energy associated with this element is:

$$dE_k = \frac{1}{2}\left(u^2 + w^2\right) dm = \frac{1}{2}\rho_w \left(u^2 + w^2\right) dx dz. \tag{C.141}$$

To find the total kinetic energy, we have to integrate dE_k over depth and average the result over one wavelength. After these operations we obtain (Dean and Dalrymple, 1992):

$$E_k = \rho_w g \frac{H^2}{16}. \tag{C.142}$$

This energy is exactly equal to the potential energy. Therefore, small amplitude wave theory predicts an equal partition of the potential and kinetic wave energy and total wave energy, per unit area, is the sum of both energies, *i.e.*:

$$E = E_p + E_k = \rho_w g \frac{H^2}{8}. \tag{C.143}$$

C.6 Circular, Hyperbolic and Complex Functions

A variety of circular and hyperbolic functions are extensively used in this book. For convenience, the definitions of these functions and their graphical representations are summarized in this section. For an acute angle θ, the three most important circular functions are defined, *i.e.* $\sin\theta$, $\cos\theta$ and $\tan\theta$ (see Fig. C.8). Two circular functions $\sin(x)$ and $\cos(x)$ are periodic functions with period 2π; the third circular function, $\tan(x)$, has period π. The most important relationships between $\sin(x)$, $\cos(x)$ and $\tan(x)$ are:

$$\tan(x) = \frac{\sin(x)}{\cos(x)}, \tag{C.144}$$

$$\sin^2(x) + \cos^2(x) = 1, \tag{C.145}$$

$$\sin(-x) = -\sin(x), \quad \cos(-x) = \cos(x), \quad \tan(-x) = -\tan(x), \tag{C.146}$$

$$\sin(2x) = 2\sin(x)\cos(x), \quad \cos(2x) = 2\cos^2(x) - 1, \tag{C.147}$$

$$\left.\begin{array}{ccc} \sin(x) & \to & x \\ \cos(x) & \to & 1 \\ \tan(x) & \to & x \end{array}\right\} \text{ for small } x. \tag{C.148}$$

If $y = \sin x$ then:

$$x = \sin^{-1}(y) = \arcsin(y). \tag{C.149}$$

Function $\sin^{-1}(y)$, or $\arcsin(y)$, is called the inverse sine function. In this book the notation $\arcsin(y)$ is used to avoid confusion with the reciprocal of $\sin(y)$, *i.e.* $1/\sin(y)$. Function $\arcsin(y)$ is valid only when $-1 \leq y \leq 1$.

The inverse cosine function $\arccos(y)$ is handled in a similar way with $-1 \leq y \leq 1$, *i.e.*:

$$y = \cos(x) \quad \text{and} \quad x = \arccos(y). \tag{C.150}$$

When we restrict the tangent function to the interval $(-\pi/2, \pi/2)$, we can define the inverse tangent function $\arctan(y)$ with $-\infty < y < \infty$:

$$\left.\begin{array}{ccll} y & = & \tan(x), & \frac{-\pi}{2} \leq x \leq \frac{\pi}{2} \\ x & = & \arctan(y), & -\infty < y < \infty \end{array}\right\}. \tag{C.151}$$

In some cases, for better clarity of explanation we use *complex numbers* and *complex functions*. We encounter complex numbers, for example, when trying

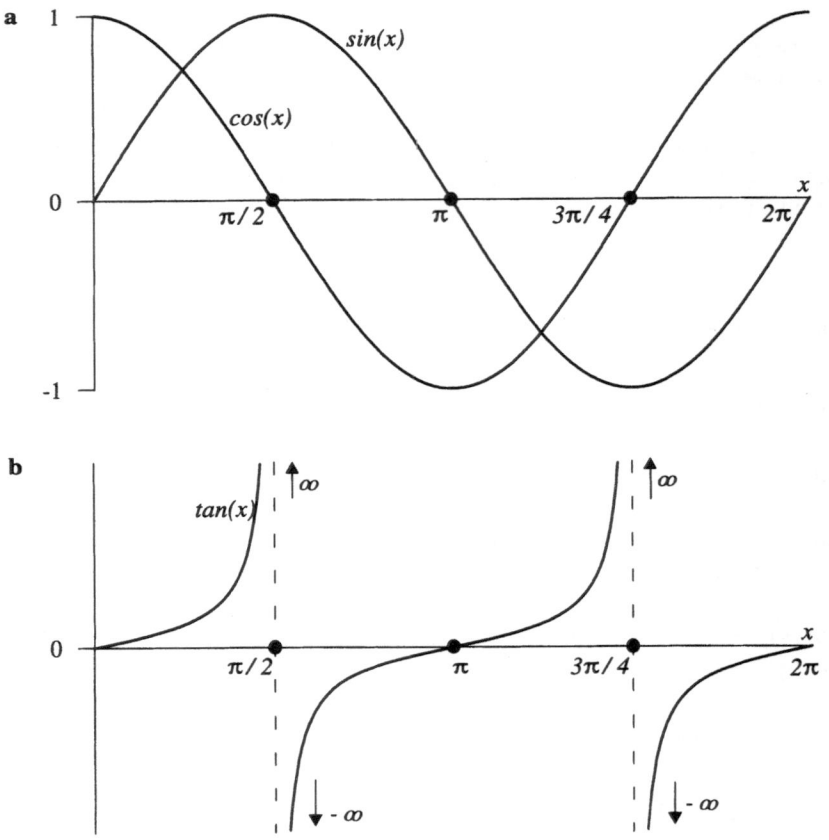

Fig. C.8: Circular functions: **a** $\sin(x)$ and $\cos(x)$, **b** $\tan(x)$

to solve simple equations such as $z^2 = -5$, which has no real-number solution. To find this solution, we define a complex number z as:

$$z = x + iy,$$ (C.152)

where x is the real part of z, y is the imaginary part of z, and i is a number such that $i^2 = -1$. In this book, the following notation for real and imaginary parts of a complex number z is used:

$$\left. \begin{array}{rcl} x &=& Re(z) = \Re(z) \\ y &=& Im(z) = \Im(z) \end{array} \right\}.$$ (C.153)

Using this definition of a complex number, we can write a solution to the equation $z^2 = -5$ as $z = 0 + i\sqrt{5} = i\sqrt{5}$. Indeed, now we have $z^2 = 5i^2 = -5$.

Recall from calculus that the polar form of the point (x, y) is (r, θ), where x, y, r, and θ are related by $x = r \cos \theta$ and $y = r \sin \theta$. Thus, a complex number $z = x + iy$ in polar form can be written as:

$$z = r(\cos \theta + i \sin \theta). \tag{C.154}$$

Complex numbers are also used in combination with exponential functions. However, we need first to give a meaning to the expression e^z when $z = x + iy$ is a complex number, *i.e.*:

$$e^z = e^{x+iy} = e^x e^{iy}. \tag{C.155}$$

Let us denote $iy = u$. The exponential function e^u has the following representation in terms of the Taylor series:

$$e^u = \sum_{n=0}^{\infty} \frac{u^n}{n!} = 1 + u + \frac{u^2}{2!} + \frac{u^3}{3!} + \dots \tag{C.156}$$

If we use the fact that:

$$i^2 = -1, \quad i^3 = -i, \quad i^4 = 1, \quad i^5 = i, \dots, \tag{C.157}$$

we obtain:

$$e^u = e^{iy} = \left(1 - \frac{y^2}{2!} + \frac{y^4}{4!} - \frac{y^6}{6!} + \dots \right) + i \left(y - \frac{y^3}{3!} + \frac{y^5}{5!} - \dots \right). \tag{C.158}$$

The expression within the first parentheses simply is the Taylor series representation of $\cos y$, while the expression within the second parantheses represents the $\sin y$. Thus, we have:

$$e^{iy} = \cos y + i \sin y. \tag{C.159}$$

Substituting the expression (C.159) into Eq. (C.155) we obtain a famous formula, called the Euler's formula:

$$e^z = e^{x+iy} = e^x e^{iy} = e^x (\cos y + i \sin y), \tag{C.160}$$

in which $z = x + iy$ is a complex number.

Hyperbolic functions $\sinh(x)$, $\cosh(x)$, and $\tanh(x)$ are not as familiar as circular functions. The hyperbolic sine is defined as follows (both notations for exponential function are used in this book):

$$\sinh(x) = \frac{1}{2} \left(e^x - e^{-x} \right) = \frac{1}{2} \left[\exp(x) - \exp(-x) \right]. \tag{C.161}$$

For $x = 0$, $\sinh(x) = 0$, as both e^x and e^{-x} equal 1. When x increases, $\sinh(x)$ increases also and for large x (say, greater than 2) $\sinh(x)$ it is approximately equal to $e^x/2$. For negative values of x, $\sinh(x)$ becomes negative.

The complementary function for $\sinh(x)$ is the hyperbolic cosine, $\cosh(x)$:

$$\cosh(x) = \frac{1}{2}\left(e^x + e^{-x}\right) = \frac{1}{2}\left[\exp(x) + \exp(-x)\right]. \qquad (C.162)$$

When $x = 0$, $\cosh(x) = 1$. The $\cosh(x)$ increases with x and for large x it approaches $e^x/2$. $\sinh(x)$ and $\cosh(x)$ are illustrated in Fig. C.9.

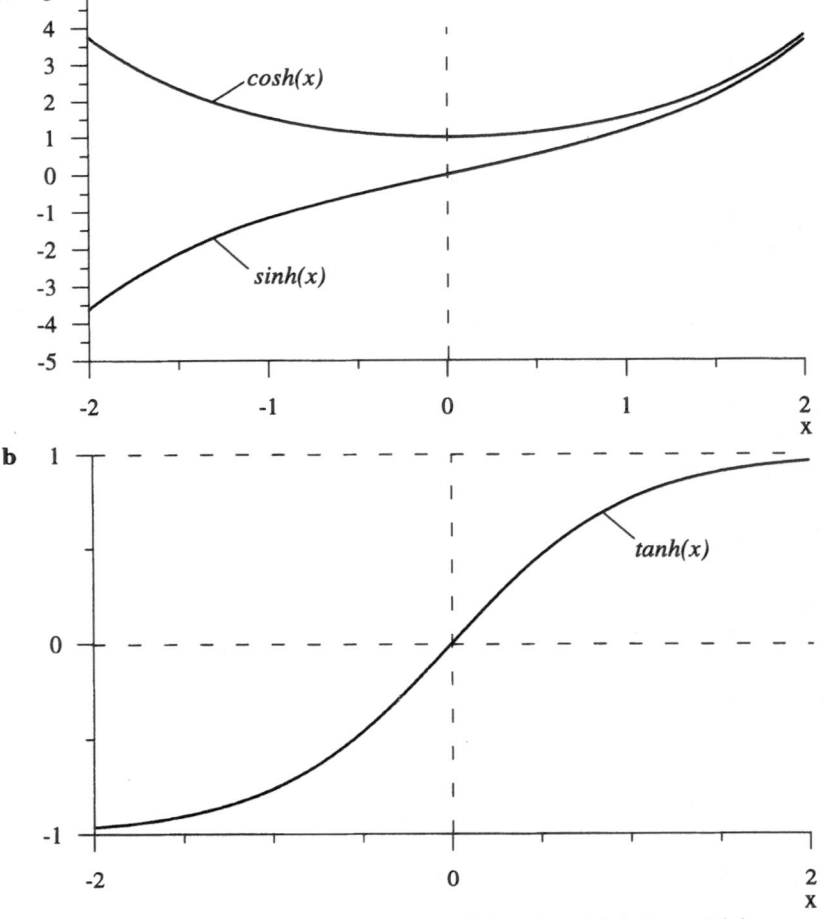

Fig. C.9: Hyperbolic function: **a** $\sinh(x)$ and $\cosh(x)$, **b** $\tanh(x)$

Tanh(x) is defined as a ratio of functions (C.161) and (C.162), *i.e.*:

$$\tanh(x) = \frac{\sinh(x)}{\cosh(x)}, \quad \text{and} \quad \coth(x) = \frac{\cosh(x)}{\sinh(x)}. \tag{C.163}$$

When $x \to \infty$, $\tanh(x) \to 1$, and at negative values of x, $\tanh(x)$ becomes negative. When x is very small, $\sinh(x) \to x$, $\cosh(x) \to 1$, and $\tanh(x) \to x$. Function (C.163) is also shown in Fig. C.9.

A few important relationships related to hyperbolic functions are given below:

$$\cosh^2(x) - \sinh^2(x) = 1, \tag{C.164}$$

$$\sinh(-x) = -\sinh(x), \quad \cosh(-x) = \cosh(x), \quad \tanh(x) = -\tanh(x), \tag{C.165}$$

$$\sinh(2x) = 2\sinh(x)\cosh(x), \quad \cosh(2x) = \cosh^2(x) - 1. \tag{C.166}$$

A very comprehensive list of relationships and tables for circular and hyperbolic functions is given in the handbook by Abramowitz and Stegun (1975).

C.7 Evaluation of Integral (8.60)

A concentration $\bar{c}(x,t)$ in a moving front of contaminants is given by Eq. (8.60), *i.e.*:

$$\bar{c}(x,t) = \frac{Q_i}{\sqrt{4\pi K_x}} \int_0^t \frac{1}{\sqrt{t-\tau}} \exp\left\{-\frac{[x - \bar{u}(t-\tau)]^2}{4K_x(t-\tau)}\right\} d\tau. \tag{C.167}$$

After substitution $z = t - \tau$, and saving the same integration variable Eq. (C.167) becomes:

$$\bar{c}(x,t) = \frac{Q_i}{\sqrt{4\pi K_x}} \int_0^t \frac{1}{\sqrt{\tau}} \exp\left\{-\frac{(x - \bar{u}\tau)^2}{4K_x\tau}\right\} d\tau. \tag{C.168}$$

Let us now denote:

$$a = \frac{\bar{u}x}{4K_x}; \quad b = \frac{\bar{u}}{2}\sqrt{\frac{\tau}{K_x}}; \quad c = \frac{\bar{u}}{2}\sqrt{\frac{t}{K_x}}. \tag{C.169}$$

Thus:

$$\tau = \frac{4K_x}{\bar{u}^2}b^2 \quad \text{and} \quad d\tau = \frac{8K_x}{\bar{u}^2}b\,db. \tag{C.170}$$

Using Eq. (C.169) and (C.170) in Eq. (C.168) gives:

$$\bar{c}(x,t) = \frac{2Q_i}{\sqrt{\pi}\bar{u}} \exp(2a) \int_0^c \exp\left[-\left(\frac{a}{b}\right)^2 - b^2\right] db. \tag{C.171}$$

The integral in above expression can be evaluated as (Prudnikov *et al.*, 1981):

$$\int_0^c \exp\left[-\left(\frac{a}{b}\right)^2 - b^2\right] db = \frac{\sqrt{\pi}}{4}\left\{\exp(2a)\left[\text{erf}\left(c + \frac{a}{c}\right) - 1\right] + \right.$$

$$\left. + \exp(-2a)\left[\text{erf}\left(c - \frac{a}{c}\right) + 1\right]\right\}, \qquad (\text{C.172})$$

where error function $\text{erf}_x(x)$ is given by:

$$\text{erf}(x) = \frac{2}{\sqrt{\pi}}\int_0^x e^{-t^2} dt. \qquad (\text{C.173})$$

Tables for error function values may be found in any standard mathematical tables (for example, Abramowitz and Stegun, 1975).

For practical calculation, the following approximation, suggested by Abramowitz and Stegun (1975), provides sufficiently accurate results:

$$\text{erf}(x) = 1 - \left(a_1 t + a_2 t^2 + a_3 t^3\right)\exp\left(-x^2\right), \qquad (\text{C.174})$$

in which:

$$\left.\begin{array}{l} t = \frac{1}{1+px}; \quad p = 0.4707 \\[2mm] a_1 = 0.34802; \quad a_2 = -0.09588; \quad a_3 = 0.74785 \end{array}\right\}. \qquad (\text{C.175})$$

D Computer Programs Set ApD

D.1 Programs Installation

On a distribution compact disc (CD) attached to the book, series of ready to use computer programs are given. The programs have been written using Microsoft Visual Basic, Version 4, with a friendly menu. References to particular programs are noted accordingly in the text.

The programs ApD run under the Microsoft Windows operating environment and Visual Basic is not needed. All necessary information is provided on the CD. The installation itself comprises the following steps:

1. Start Windows.

2. Insert CD into the CD drive.

3. From Windows Program Manager, select Run from the Start menu.

4. If you are using drive D type d:\setup in the command line and click OK. If you are using a different drive, use that drive letter instead.

5. The installation program copies files to your hard disk and sets up Windows so that you can begin running the program under the Windows operating system.

6. To use the ApD programs click ApD in the Programs folder.

D.2 Brief Program Descriptions

D.2.1 Program D.11: Calculation of Sea Water Density According to UNESCO Formula

- Theoretical background: see Section 1.2.1.
- Input data:

    - temperature T [°C] in the range $0 < T < 40$°C,
    - salinity S (ppm) in the range $0 < S < 42$ ppm.

- Output: value of salt water density ρ_w [kg/m^3].

D.2.2   Program D31: Surface Displacement for Fixed Location or at Given Time Instant

- Theoretical background: see Sects. 3.4.1 and 4.2.2.
- Input data:

    - wave height H [m],
    - wave period T [s],
    - water depth h [m].

- Output:

    - *Representation 1*: wave surface changing in time and observations at a fixed point.
    - *Representation 2*: snapshots of wave surface at selected time instants: $t = 0$, $t = T/4$, and $t = 5T/8$. Comparison of three curves demonstrates the phase shifts between particular time series.

D.2.3   Program D.41: Calculation of Wavelength (or Wave Period) and Phase and Group Velocities

- Theoretical background: see Section 4.2.2.
- Input data:

    - *Option 1*:
        * wave period T [s],
        * wave depth h [m].
    - *Option 2*:
        * wavelength L [m],
        * wave depth h [m].

- Output data:

    - *Option 1*:
        * values of wave frequency ω [rad/s],
        * wavelength L [m],
        * wave number k [1/m],
        * phase velocity C [m/s],
        * group velocity C_g [m/s].
    - *Option 2*:
        * wave period T [s],
        * wave frequency ω [rad/s],
        * wave number k [1/m],
        * phase velocity C [m/s],
        * group velocity C_g [m/s].

D.2.4 Program D.42: Surface Wave Profiles Resulting from Various Wave Theories

- Theoretical background: see Section 4.2.2 and 4.2.3.
- Input data:

    - wave height H [m],
    - wave period T [s],
    - wave depth h [m].

- Output: profiles of wave surface according to small amplitude and Stokes' wave theories, or according to solitary wave theory. In each case, a value of the Ursell number, U, is given. Small amplitude, or Stokes' wave theories, are appropriate for $U < 40$. For $U >> 40$, solitary wave theory should be used.

D.2.5 Program D.43: Vertical Profiles of Horizontal Orbital Velocity Under Wave Crest and Wave Trough

- Theoretical background: see Section 4.2.2.
- Input data:

    - wave height H [m],
    - wave period T [s],
    - water depth h [m].

- Output: vertical profiles of the horizontal velocities under wave crest and wave trough. The maximum values of the horizontal velocities are indicated.

D.2.6 Program D.44: Vertical Profiles of Wave-Induced Pressure Under Wave Crest and Trough

- Theoretical background: see Section 4.2.2.
- Input data:

    - wave height H [m],
    - wave period T [s],
    - water depth h [m].

- Output: vertical profiles of pressure under wave crest and wave trough. The hydrostatic pressure component and total pressure are distinguished and the maximum values of pressure at the sea bottom are indicated.

D.2.7   Program D.45: Wave Shoaling and Refraction (Parallel Bottom Contours)

- Theoretical background: see Section 4.2.4.
- Input data:

  - offshore water depth h [m],

  - incident wave height H [m],

  - wave period T [s],

  - incident wave angle $\theta°$ [°],

  - width of refraction zone l [m].

- Output: graph of wave ray turning within the shallow water zone. At offshore and inshore boundaries of the zone, the values of wave height and wave angle are given.

D.2.8   Program D.46: Harmonic Analysis of a Given Time Series $f(t)$

The program has been carried out assuming that $0 < t < T$.

- Theoretical background: see Section 4.3.3.
- Input data: the name and path of the file containing the values of function $f(t)$ in the format:

| t | $f(t)$ |

For example, your file should be in the format:

0	$f(0)$
Δt	$f(\Delta t)$
$2\Delta t$	$f(2\Delta t)$
$3\Delta t$	$f(3\Delta t)$
...	...
...	...
$N\Delta t = T$	$f(T)$

In the areas provided, enter the value of length of time series T, number of points N, and number of assumed harmonics N_h. As an example, file D46.dat is given (CD drive\data\d46.dat) with $T = 30$ s and $N = 500$.

- Output: graph of initial function $f(t)$ (blue line) and approximation by the Fourier Series (red line). Increasing the number N_h results in a better approximation of function $f(t)$. On the vertical axis, values of both functions are given. The horizontal axis is normalized against T.

D.2.9   Program D.47: Spectral Analysis of Time Series (FFT Method)

- Theoretical background: see Section 4.3.3.

- Input data: the name and path of the file containing the values of time series in the format:

  t $\qquad\qquad\qquad\qquad$ $f(t)$

  For example, your file should be in the format:

Δt	$f(\Delta t)$
$2\Delta t$	$f(2\Delta t)$
	$f(3\Delta t)$
...	...
...	...
$N\Delta t$	$f(N\Delta t)$

  As the procedure with overlapping segments is used, the number of points, N, has to be divisible by 128. In the areas provided, enter the sampling interval Δt and segment length NDB. The NDB value should be equal to 128 or 256. As an example, file D47.dat is given (CD drive\data\d47.dat) with $N = 3072$ and $\Delta t = 0.3906$ s.

- Output: graph of frequency spectrum. On the vertical axis, values of normalized spectrum $S(\omega)/S(\omega_{max})$ are given, where ω_{max} is the peak frequency at which the spectrum reaches its maximum. The peak frequency, as well as the variance (calculated as an integral over the spectrum), are also provided.

D.2.10   Program D.81: Instantaneous Release of Substance in Uniform Flow (One-Dimensional Case)

- Theoretical background: see Section 8.3.4.
- Input data:

  - mass m of released substance [kg/m], in the form of a Gaussian curve,

  - variance of initial distribution σ [m],

  - coefficient of horizontal diffusion K_x [m^2/s],

  - current velocity U [m/s],

  - time t [s] after which a final distribution is required.

- Output: graphs of initial (blue line) and final (red line) concentrations c(x,t). On the vertical axis, concentrations are normalized against initial concentration at $x = 0$. The value of this concentration $c(0,0)$ is indicated. The horizontal distance, x, from the centre of release is normalized by distance $x_{max} = 3\sigma$.

D.2.11   Program D.82:   Steady Release of Substance in Uniform Flow

- Theoretical background: see Section 8.3.4.
- Input data:

    - rate of mass release [kg/s],
    - coefficients of vertical and horizontal diffusion K_z [m^2/s],
    - current velocity U [m/s],
    - maximum water depth for vertical cross-sections h [m],
    - distances of cross-sections (4 points) from point of release and concentration isolines (up to 4 levels).

- Output: results are presented on three graphs. On the first graph, plume cross-sections (normal to flow directions) are given. On the vertical axis, the concentration normalized to the value of concentration at the centre line at distance $x = x_1$ is given. The horizontal axis is normalized with respect to the distance $y_{max} = 3\sigma$.

    The second graph illustrates vertical profiles of concentration at the four distances and at the centre axis of the plume. The concentration values are normalized against concentration at cross-section $x = x_1$, with $y = 0$.

    On the third graph, four isolines of concentration are given. The levels of concentration can be selected as arbitrary, under the condition that they are smaller than the maximum concentration which is shown in the plane view form.

    The cross flow axis is normalized against the maximum width of the c_1 concentration isoline, and the along flow axis shows the distance from the release point.

D.2.12   Program D.83:   Simulation of Substance Diffusion by Monte Carlo Method

- Theoretical background: see Section 8.3.4
- Input data:

    - number of points for simulation N,
    - duration of simulation t [s],
    - time sampling Dt [s] during simulation,
    - coefficient of horizontal diffusion K [m^2/s],
    - current velocity components u and v [m/s].

- Output: graph showing the distribution of N points released at the origin after time t under the influence of advection and diffusion. Dashed line shows the current direction and blue point indicates the centre of 'plume' after time t.

References

Abraham, E.R. (1998): The generation of plankton patchiness by turbulent stirring. Nature **391**, 577–580

Abramowitz, M., Stegun, I.A. (1975): Handbook of Mathematical Functions. Dover Publ., 1045 pp

Alexander, R.McN. (1968): Animal mechanics. Sidgwick and Jackson, London, 346 pp

Alexander, R.McN. (1977): Swimming. In: R.McN. Alexander, G. Goldspink (eds.), Mechanics and Energetics of Animal Locomotion. Chapman and Hall, London, 222–248

Alexander, R.McN. (1982): Locomotion of animals. Blackie, London

Allender, J., Andunson, T., Barstow, S.F., Bjerken, S., Krogstad, H.E., Steinbakke, P., Vartdal, L., Borgman, L.E., Graham, C. (1989): The Wadic project: A comprehensive field evaluation of directional wave instrumentation. Ocean Eng. **16**, 505–536

Alongi, D.M. (1998): Coastal ecosystem processes. CRC Press, Boca Raton, 419 pp

Al-Rabek, A.H., Gunay, N. (1992): On the application of a particle dispersion model. Coastal Eng. **17**, 195–210

Amos, C. (1995): Silicaclastic tidal flats. In: G. Perillo (ed.), Geomorphology and Sedimentology of Estuaries. Developments in Sedimentology. Elsevier, Amsterdam, **53**, 273–306

Anderson, D.L.T., Sarachik, E.S., Webster, P.J., Rothstein, L.M. (1998): The TOGA Decade. Reviewing the progress of El Niño Research and Prediction. Reprint from the Jour. Geoph. Res. vol. 103, Amer. Geoph. Union, Washington, 14169–14510

Antsyferov, S.M., Kosyan, R.D. (1986): Suspended sediments in upper layer of shelf waters. Nauka, Moscow, 223 pp (Russian)

Anwar, H.O. (1983): Turbulence measurements in stratified and well-mixed estuarine flows. Estuarine, Coastal and Shelf Science **17**, 243–260

Arita, M. and Jirka, G.H. (1987): Two-layer model of saline wedge. I. Entrainment and interfacial friction. Jour. Hydr. Eng. **113**, 1229–1243

Arntz, W.E. (1984): El Niño and Peru: Positive aspects. Oceanus **27**, 36–39

Arrigo, K.R., Worthen, D., Schnell, A., Lizotte, M.P. (1998): Primary production in Southern Ocean waters. Jour. Geoph. Res. **102**, 15 587–15 600

Asano, T., Deguchi, H., Kobayashi, N. (1992): Interaction between water waves and vegetation. Proc. 23th Coastal Eng. Conf. **3**, 2710–2723

Atkinson, M., Smith, S.V., Stroup, E.D. (1981): Circulation in Enewetak Atoll lagoon. Limnology and Oceanography **20**, 1074–1083

Au, D., Weihs, D. (1980): At high speeds dolphins save energy by leaping. Nature **284**, 548–550

Barber, N.F. (1969): Water Waves. Wykeham Publ., London, 142 pp

Barber, R.T., Chavez, F.P. (1983): Biological consequences of El Niño. Science **222**, 1203–1210

Barnes, D.J., Lough, J.M. (1993): On the nature and cause of density banding in massive coral skeletons. Jour. Exp. Mar. Biol. Ecol. **167**, 91–108

Barnett, T.P., Graham, N., Cane, M.A., Zebiak, S.E., Dolan, S., O'Brien, J., Legler, D. (1988): On the prediction of the El Niño of 1986–87. Science **245**, 192–196

Barrett, E.C., Curtis, L.F. (1992): Introduction to Environmental Remote Sensing. Chapman & Hall, London, 426 pp

Batchelor, G.K. (1967): An Introduction to Fluid Dynamics. Cambridge University Press, Cambridge, 615 pp

Battjes, J.A., Janssen, J.P.E.M. (1978): Energy loss and set-up due to breaking of random waves. Proc. 16th Coastal Eng. Conf. **1**, 563–587

Battjes, J.A., Stive, M.J.F. (1985): Calibration and verification of a dissipation model for random breaking waves. Jour. Geoph. Res. **90**, 9159–9167

Battjes, J.A., Sobey, R.J., Stive, M.J.F. (1990): Nearshore circulation. In: B. Le Méhauté and D.M. Hanes (eds.). Ocean Engineering Science. A Wiley-Interscience Publ., New York, 467–493

Bearman, G. (ed.) (1997): Waves, Tides and Shallow-Water Processes. Pergamon Press, Oxford, 187 pp

Bell, E.C. (1996): Mechanical design of mussel byssus: Material yield enhances attachment strength. Jour. Exp. Biol. **199**, 1005–1017

Bendat, J.S., Piersol, A.G. (1986): Random Data. Analysis and Measurement Procedures. John Wiley & Sons, New York, 566 pp

Bennett, A.F., Deuman, K.L. (1985): Phytoplankton patchiness: inferences from particle statistics. Jour. Mar. Res. **43**, 307–335

Berg, H.C. (1983): Random Walks in Biology. Princeton Univ. Press, Princeton

Bergamasco, A., Carniel, S., Pastres, R., Pecenik, G. (1998): A unified approach to the modelling of the Venice Lagoon – Adriatic Sea Ecosystem. Estuarine, Coastal and Shelf Science **40**, 483–492

Bernard, E.N. (1998): Program aims to reduce impact of tsunamis on Pacific States. EOS **79** (22), 258 pp

Bernstein, R.L. (1982): Sea surface temperature estimation using the NOAA-6 satellite advances very high resolution radiometer. Jour. Geoph. Res. **87**, 9455–9466

Bjerkness, J. (1969): Atmospheric teleconnections from the equatorial Pacific. Mon. Weather Rev. **97**, 163–172

Blake, J.R. (1972): A model for the microstructure in ciliated organisms. Jour. Fluid Mech. **55**, 1–23

Blake, R.W. (1983): Fish Locomotion. Cambridge Univ. Press, Cambridge, 208 pp

Block, B.A., Booth, D, Carey, F.G. (1992): Direct measurement of swimming speeds and depth of blue marlin. Jour. Exp. Biol. **166**, 267–284

Bockel, M. (1962): Traveaux océanographiques dé 'Origny' a Gibraltar. Cahiers Oéonographique **14**, 325–329

Borgman, L.E. (1963): Risk criteria. Proc. ASCE, Jour. Waterways, Harbors and Coastal Eng. Div. **89** WW3, 1–35.

Bowen, A.J. (1969): Rip currents. 1. Theoretical investigations. Jour. Geoph. Res. **74**, 5467–5478

Bowers, D.G., Simpson, J.H. (1987): Mean position of tidal front in European-shelf seas. Continental Shelf Res. **7**, 35–44

Boje, R., Tomczak, M. (1978): Upwelling Ecosystems. Springer-Verlag, Berlin, 303 pp

Bowman, M.J., Yentsch, C.M., Peterson, W.T. (1986): Tidal Mixing and Plankton Dynamics. Lecture Notes on Coastal and Estuarine Studies. Vol. 17, Springer-Verlag, Berlin, 502 pp

Brennen, C. (1975): Hydrodynamics of propulsion for ciliated microorganisms. In: T.Y. Wu, C.J. Brokaw, C. Brennen (eds.), Swimming and Flying in Nature. Plenum Press, New York **1**, 223–234

Brett, J.R. (1965): The relation of size to rate of oxygen consumption and sustained swimming speed of socheye salmon (*Oncorhynchus nerka*). Jour. Fish. Res. Bd., Canada **22**, 1491–1497

Brink, K.H., Halpern, D., Huyer, A., Smith, R.L. (1983): The physical environment of the Peruvian upwelling system. Prog. Oceanogr. **12**, 285–305

Brinkman, R.M., Massel, S.R., Ridd, P.V., Furukawa, K. (1997): Surface wave attenuation in mangrove forests. Proc. 13th Australasian Coastal and Ocean Engineering Conf. **2**, 941–946.

Bryan, K. (1969): Climate and the ocean circulation: III. The ocean model. Mon. Weather Rev. **97**, 806–827

Bucher, D.J., Harriott, V.J., Roberts, L.G. (1998): Skeletal microdensity, porosity and bulk density of acroporid corals. Jour. Exp. Mar. Biol. Ecol. **228**, 117–136

Bureau of Meteorology (1997): Tropical cyclone 'Justine.' Dept. of Env., Sport and Territories, Brisbane, 15 pp

Burrage, D., Massel, S.R., Steinberg, C., Skirving, W. (1996): Detecting surface and internal wave signatures on the North-West shelf of Australia using the ERS-1 and ERS-2 Active Microwave Instruments (AMI). Proc. First Australian ERS Symposium. CSIRO, Canberra, 11–26

Burton, R. (1977): Antarctica: Rich around the edges. Sea Frontiers **23**, 287–295

Canby, T.Y. (1984): El Niño's ill wind. National Geographic **165**, 144–183

Cane, M.A. (1984): Oceanographic events during El Niño. Science **222**, 1189–1195

Carrington, E. (1990): Drag and dislodgement of an intertidal macroalga: Consequences of morphological variation in *Mastocarpus papillatus* (Kützing). Jour. Exp. Mar. Biol. Ecol. **139**, 185–200

Carter, D.J.T., Challenor, P.G., Srokosz, M.A. (1992): An assessment of GEOSAT wave height and wind speed measurements. Jour. Geoph. Res. **97**, 11383–11392

Cartwright, D.E., Ray, R.D. (1990): Oceanic tides from Geosat altimetry. Jour. Geoph. Res. **95**, 3069–3090

Chamberlain, J.A. (1991): Cephalopod locomotor desing and evolution: the constraints of jet propulsion. In: J.M.V. Rayner, R.J. Wootton (eds.), Biomechanics in Evolution. Cambridge Univ. Press, Cambridge, 57–98

Cheng, F.H. (1985): Statics and Strength of Materials. Macmillan Publ. Comp., New York, 541 pp

Chien, S. (1970): Shear dependence of effective cell volume as a determinant of blood viscosity. Science **168**, 977–978

Childress, S. (1981): Mechanics of Swimming and Flying. Cambridge Univ. Press, Cambridge, 155 pp

Clarke, S., Elliott, A.J. (1998): Modelling suspended sediment concentration in the Fivth of Forth. Estuarine, Coastal and Shelf Science **47**, 235–250

Cloern, J.E. (1987): Turbidity as a control on phytoplankton biomass and productivity in estuaries. Cont. Shelf Res. **7**, 1367–1381

Cooley, J.W., Tukey, J.W. (1965): An algorithm for machine calculation of complex Fourier series. Math. Comput. **19**, 297–301

Cotton, P.D., Carter, D.J.T. (1994): Cross calibration of TOPEX, ERS-1, and Geosat wave heights. Jour. Geoph. Res. **99**, 25025–25033

Crank, J. (1975): The Mathematics of Diffusion. Oxford Univ. Press, Oxford

Csanady, G.T. (1973): Turbulent Diffusion in the Environment. Reidel Publ. Comp., Boston, 248 pp

Csanady, G.T. (1977): Intermittent 'full' upwelling in Lake Ontario. Jour. Geoph. Res. **82**, 397–419

Cushman-Roisin, B. (1994): Introduction to Geophysical Fluid Dynamics. Prentice Hall, Englewood Cliffs, 320 pp

Danabasoglu, G., McWilliams, J.C., Gent, P.R. (1994): The role of mesoscale tracer transport in the global circulation. Science **264**, 1123–1126

Daniel, T., Jordan, C., Grunbaum, D. (1992): Hydrodynamics of swimming. In: R.McN. Alexander (ed.), Advances in Comparative and Environmental Physiology. vol. 11. Mechanics of Animal Locomotion. Springer-Verlag, Heidelberg, 17–49

David, L.T., Kjerfve, B. (1998): Tides and currents in a two-inlet coastal lagoon: Laguna de Términos, México. Continental Shelf Res. **18**, 1057–1079

Davidan, I.N., Lopatukhin, L.I. (1978): Towards the storms. Gidrometeoizdat, Leningrad, 135 pp (Russian)

Davidan, I.N., Lopatukhin, L.I., Rozhkov, W.A. (1985): Wind Waves in the Ocean. Gidrometeoizdat, Leningrad, 256 pp (Russian)

Day, J.W., Hall, C.A., Kemp, W.M., Yanez-Arancila, A. (1989): Estuarine Ecology. John Wiley & Sons, New York, 558 pp

Dean, R.G. (1995): Cross-shore sediment transport processes. In: P.L.F. Liu (ed.), Advances in Coastal and Ocean Engineering **1**, 159– 220

Dean, R.G., Dalrymple, R.A. (1992): Water Wave Mechanics for Engineers and Scientists. World Scientific Publ., Singapore New Jersey London Hong Kong, 353 pp

Defant, A. (1961): Physical Oceanography. Vol. 2. Pergamon Press, Oxford, 598 pp de Jonge, V.N. (1992): Physical Processes and Dynamics of Microphytobenthos in the Ems Estuary (the Netherlands). Directorate-General for Public Works and Water Management, The Hague, 176 pp

De Mont, M.E. (1992): Locomotion of soft bodied animals. In: R.McN. Alexander (ed.), Advances in Comparative and Environmental Physiology, vol. II. Mechanics of Animal Locomotion, Springer-Verlag, Berlin, 167–190

Denny, M.W. (1988): Biology and the Mechanics of the Wave–Swept Environment. Princeton Univ. Press, Princeton, 329 pp

Denny, M.W. (1993): Air and Water. The Biology and Physics of Life's Media. Princeton Univ. Press, Princeton, 341 pp

Denny, M.W., Shibata, M.F. (1989): Consequences of surf-zone turbulence for settlement and external fertilization. The American Naturalist 134, 859–889

Denny, M., Gaylord, B. (1996): Why the urchin lost its spines: Hydrodynamic forces and survivorship in three echinoids. Jour. Exp. Biol. 199, 717–729

Denny, M., Gaylord, B., Helmuth, B., Daniel, T. (1998): The menace of momentum: Dynamic forces on flexible organisms. Limnology and Oceanography 43, 955–968

Denton, E.J. (1974): Buoyancy in marine animals. In: J.J. Head (ed.), Oxford Biology Readers. Oxford Univ. Press, London, 16 pp

Dera, J. (1992): Marine Physics. Elsevier – Polish Scientific Publishers, Amsterdam Warszawa, 516 pp

Dera, J. (1995): Underwater irradiance as a factor affecting primary production. Inst. of Oceanology, Sopot (Poland), Disser. and Monogr. 7, 68 pp

de Swart, H.E., de Jonge, V.N., Vosbeek, M. (1997): Application of the tidal random walk model to calculate water dispersion coefficients in the Ems estuary. Estuarine, Coastal and Shelf Science 45, 123–133

Dingemans, M.W. (1997): Water Wave Propagation over Uneven Bottoms. Part 1 – Linear Wave Propagation. World Scientific Publ., Singapore, 471 pp

Djurfeldt, L. (1994): The influence of physical factors on a subsurface chlorophyll maximum in an upwelling area. Estuarine, Coastal and Shelf Science 39, 389–400

Done, T.J. (1992): Effects of tropical cyclone waves on ecological and geomorphological structures on the Great Barrier Reef. Cont. Shelf Res. 12, 859–872.

Done, T.J., Navin, K.F. (1990): Shallow water benthic communities on coral reefs. In: T.J. Done, K.F. Navin, (eds.), Vanuatu marine resources: report of a biological survey. Australian Institute of Marine Science, Townsville, 10–36

Done, T.J., Potts, D.C. (1992): Influences of habitat and natural disturbances on contributions of massive Porites corals to reef communities. Marine Biology 114, 479–493

Drew, E.A., Ireland, J.F., Muir, C., Robertson, W.A.A., Robinson, J.D. (1982): Photosynthesis, respiration and other factors influencing the growth of Lamina ochroleuca Pyl. below 50 metres in the Strait of Messina. Marine Ecology 3, 335–355

Druet, Cz. (1994): Dynamics of the Stratified Ocean. Wydawn. Naukowe PWN, Warszawa, 225 pp (Polish)

Druet Cz., Zieliński, A. (1994): Modelling the fine structure of the phytoplankton concentration in a stable stratified sea. Oceanologica Acta 17, 79–88

Dubi, A., Tørum, A. (1994): Wave damping by kelp vegetation. Proc. 24th Coastal Eng. Conf., Kobe, **1**, 142–156

Eckman, J.E. (1990): A model of passive settlement by plankton larvae onto bottoms of different roughness. Limnology and Oceanography **39**, 837–901

Ekman, V.W. (1905): On the influence of the earth's rotation on ocean currents. Ark. Math., Astr. Fysik **2**, 1–53

Elder, J.W. (1959): The dispersion of marked fluid in turbulent shear flow. Jour. Fluid Mech. **5**, 544–560

Elliott, A.J., Barr, A.G., Kennan, D. (1997): Diffusion in Irish coastal waters. Estuarine, Coastal and Shelf Science **44** (Supplement A), 15–23

Elwany, M., Hany, S., Flick, R.E. (1996): Relationship between kelp beds and beach width in Southern California. Jour. Waterway, Port, Coastal and Ocean Eng. **122**, 34–37

Emery, W.J., Thomson, R.E. (1997): Data Analysis Method in Physical Oceanography. Pergamon, 634 pp

Fenton, J.D. (1979): A high-order cnoidal wave theory. Jour. Fluid Mech. **94**, 129–161

Fenton, J.D. (1985): A fifth-order Stokes theory for steady waves. Jour. Waterways, Port, Coastal and Ocean Eng. **111**, 216–234

Fenton, J.D. (1990): Nonlinear wave theories. In: B. Le Méhauté, D.M. Hanes, (eds.) Ocean Eng. Science **9**, 3–25

Fiedler, P.C., Philbrick, V., Chavez, F.P. (1991): Oceanic upwelling and productivity in the eastern tropical Pacific. Limnol. and Oceanogr. **36**, 1834–1850

Fischer, H.B. (ed.) (1981): Transport Models for Inland and Coastal Waters. Proc. Symp. on Predictive Ability in Berkeley, 18–20 August, 1980. Academic Press, 542 pp

Fischer, H.B., List, E.J., Koh, R.C.Y., Imberger, J., Brooks, N.H. (1979): Mixing in Inland and Coastal Waters. Academic Press, San Diego, 483 pp

Flohn, H., Fleer, H. (1975): Climatic teleconnections with the equatorial Pacific and the role of ocean/atmosphere coupling. Atmosphere **13**, 96–109

Fonseca, M.S., Cahalan, J.A. (1992): A preliminary evaluation of wave attenuation by four species of seagrass. Estuarine, Coastal and Shelf Science **35**, 565–576

Franks, F. (ed.) (1972): Water. A Comprehensive Treatise. Plenum Press, New York, 596 pp

Fredsøe, J., Deigaard, R. (1992): Mechanics of Coastal Sediment Transport. World Scientific Publ., Singapore, 369 pp

Friedland, M.T., Denny, M.W. (1995): Surviving hydrodynamic forces in a wave-swept environment: Consequences of morphology in the feathers boa kelp *Egregia menziesii* (Turner). Jour. Exp. Mar. Biol. Ecol. **190**, 109–133

Froelich, P.N., Atwood, D.K., Giese, G.S. (1978): Influence of Amazon River discharge on surface salinity and dissolved silicate concentration in the Caribbean Sea. Deep Sea Res. **25**, 735–744

Fu, L.L., Christensen, E.J., Yamarone, C.A., Lefebvre, M., Meuard, Y., Dorrer, M., Escudier, P. (1994): TOPEX/POSEIDON mission overview. Jour. Geoph. Res. **99**, 24369–24381

Fung, Y.C. (1993): Biomechanics: Mechanical Properties of Living Tissues. Springer-Verlag, New York

Fung, Y.C. (1997): Biomechanics: Circulation. Springer Verlag, New York, 571 pp

Furnas, M.J. (1996): Land-sea interactions and oceanographic processes affecting the nutrient dynamics and productivity of Australian marine ecosystems. In: L.P. Zann, P. Kailola (eds.), The State of the Marine Environment Report for Australia. Technical Annex: 1. The Marine Environment. Dept. Environment, Sport and Territories, Canberra, 61-73

Furukawa, K., Wolanski, E. (1996): Sedimentation in mangrove forests. Mangroves and Salt Marshes 1, 3-10

Furukawa, K., Wolanski, E., Mueller, H. (1997): Currents and sediment transport in mangrove forests. Estuarine, Coastal and Shelf Science 44, 301-310

Gallagher, S.M., Davis, C.S., Epstein, A.W., Solow, A.M., Beardsley, R.C. (1996): High-resolution observation of plankton spatial distributions correlated with hydrography in the Great South Channel, Georges Bank. Deep Sea Res. 43, 1627-1663

Gargett, A.F., Hughes, B.A. (1972): On the interaction of surface and internal waves. Jour. Fluid Mech. 52, 179-192

Gargett, A.F., Schmitt, R.W. (1982): Observations of salt fingers in the Central Waters of eastern North Pacific. Jour. Geoph. Res. 87, 8017-8029

Garratt, J.R. (1977): Review of drag coefficients over oceans and continents. Monthly Review 105, 915-929

Garrett, C., Munk, W.H. (1972): Space-time scales of internal waves: a progress report. Jour. Geoph. Res. 3, 225-264

Garrett, C., Munk, W.H. (1975): Space-time scales of internal waves: a progress report. Jour. Geoph. Res. 80, 291-297

Gaylord, B., Blanchette, C., Denny, M.W. (1994): Mechanical consequences of size in wave-swept algae. Ecol. Monogr. 64, 287-313

Gere, J.M., Timoshenko, S.P. (1991): Mechanics of Materials. Chapman & Hall, London, 807 pp

Gibbs, R.J. (1976): Amazon River sediment transport in the Atlantic Ocean. Geology 4, 45-48

Gibbs, R.J., Konwar, L. (1986): Coagulation and settling of Amazon River suspended sediment. Continental Shelf Res. 6, 127-149

Glantz, M.H. (1996): Currents of change: El Niño impact on climate and society. Cambridge University Press, 194 pp

Gnanadesikan, A., Terray, E.A. (1994): A comparison of three wave-measuring buoys. Proc. Symp. Ocean Wave Measurement and Analysis, New Orleans, 1, 287-301

Goda, Y. (1985): Random Sea and Design of Maritime Structures. Univ. Tokyo Press, 323 pp

Godfrey, J.S., Honze, R.A., Johnson, R.H., Lukas, R., Redelsperger, J.L., Sumi, A., Weller, R. (1998): Coupled ocean-atmosphere response experiment (COARE): an interim report. Jour. Geoph. Res. 103, 14395-14450

Godin, G. (1972): The Analysis of Tides. Liverpool Univ. Press, Liverpool, 264 pp

Gonella, J. (1972): A rotary component method for analyzing meteorological and oceanographic vector time series. Deep-Sea Res. 19, 833-846

Gordon, N.D., McMahon, T.A., Finlayson, B.L. (1992): Stream Hydrology. An Introduction for Ecologists. John Wiley & Sons, Inc., New York, 526 pp

Gourlay, M.R. (1988): Coral cays: products of wave action and geological processes in a biogenic environment. Proc. 6th Inter. Coral Reef Symp. **23**, 17–42

Gourlay, M.R. (1994): Wave transformation on a coral reef. Coastal Eng. **2**, 491–496

Gray, J.S., Christiansen, M.E. (eds.) (1985): Marine Biology of Polar Regions and Effects of Stress on Marine Organisms. John Wiley, New York

Gray, J.S., Hancock, G.J. (1955): The propulsion of sea-urchin spermatozoa. Jour. Exp. Biol. **32**, 802–814

Gregg, M.C. (1973): The microstructure of the ocean. Scientific American **228**, 65–77

Groen, P. (1967): On the residual transport of suspended matter by an alternating tidal current. Neth. Jour. Sea Res. **3**, 564–574

Guan, W.B., Wolanski, E., Dong, L.X. (1998): Cohesive sediment transport in the Jiaojiang River Estuary, China. Estuarine, Coastal and Shelf Science **46**, 861–871

Guza, R.T., Inman, D.L. (1975): Edges waves and beach cusps. Jour. Geoph. Res. **80**, 2997–3012

Hallermeier, R.J. (1981): Terminal settling velocity of commonly occurring sand grains. Sedimentology **28**, 859–865

Halpern, D. (1971): Observations of short-period internal waves in Massachusetts Bay. Jour. Mar. Res. **29**, 116–132

Hamilton, A.D., Wilson, R.E. (1980): Nontidal circulation and mixing processes in the Lower Potomac estuary. Estuaries **3**, 11–19

Hamilton, J.M., Aokey, N.S., Kelley, D.E. (1993): Salt finger signatures in microstructure measurements. Jour. Geoph. Res. **98**, 2453–2460

Hardy, T.A., Young, I.R., Nelson, R.C., Gourlay, M.R. (1990): Wave attenuation on an offshore coral reef. Proc. 22nd Coastal Eng. Conf. **1**, 330–344

Harvey, J. (1985): Atmosphere and Ocean: our Fluid Environments. Vision Press Ltd., London, 143 pp

Hasselmann, K., Barnett, T.P., Bouws, E., Carlson, H., Cartwright, D.E., Enke, K., Ewing, J.A., Gienapp, H., Hasselmann, D.E., Kruseman, P., Meerburg, A., Müller, P., Olbers, D.J., Richter, K., Sell, W., Walden, H. (1973): Measurements of wind–wave growth and swell decay during the Joint North Sea Wave Project (JONSWAP). Deutsches Hydr. Zeit **A12**, 1–95

Hedges, T.S. (1995): Regions of validity of analytical wave theories. Proc. Inst. Civ. Eng., Water, Maritime and Energy **112**, 111–114

Hertel, H. (1969): Hydrodynamics of swimming and wave-riding dolphins. In: H.T. Anderson (ed.), The biology of marine mammals. 31–63

Highsmith, R.C. (1981): Coral bioerosion at Enewetak: Agents and Dynamics. Int. Review Geo. Hydrobiol. **66**, 335–375

Hildebrand, F.B. (1965): Methods of applied mathematics. Prentice-Hall, Englewood Cliffs

Holberton, D.V. (1977): Locomotion of *Protozoa* and single cells. In: R.McN. Alexander, G. Goldspink (eds.), Mechanics and energetics of animal locomotion. Chapman and Hall, London, 279–332

Holloway, P.E. (1983): Internal tides on the Australian North-West Shelf: a preliminary investigation. Jour. Phys. Oceanogr. **14**, 1778–1790

Holloway, P.E. (1986): Eddies, waves, circulation and mixing: statistical geofluid mechanics. Ann. Rev. Fluid Mech. **18**, 91–147

Holloway, P.E., Pelinovsky, E., Talipova, T., Barnes, B. (1997): A nonlinear model of internal tide transformation on the Australian North West Shelf. Jour. Phys. Oceanogr. **27**, 871–896

Horne, R.A. (1969): Marine Chemistry. Wiley-Interscience, New York, 568 pp

Hughes, S.A. (1993): Physical Models and Laboratory Techniques in Coastal Engineering. World Scientific Publ., 568 pp

Humborg, C. (1997): Primary productivity regime and nutrient removal in the Danube estuary. Estuarine, Coastal and Shelf Science **45**, 579–589

Hunter, J.R. (1987): The application of Lagrangian particle-tracking techniques to modelling of dispersion in the sea. In: J.B. Noye (ed.). Numerical Modelling: Application to Marine Systems. Elsevier Science Publ., Amsterdam, 257–269

Hurd, C.L., Harrison, P.J., Druehl, L.D. (1996): Effect of seawater velocity on inorganic nitrogen uptake by morphologically distinct forms of *Macrocystis integrifolia* from wave-sheltered and exposed sites. Marine Biology **126**, 205–214

Huthnance, J.M. (1989): Internal tides and waves near the continental shelf edge. Geophys. Fluid Dynamics **48**, 81–106

Huyer, A. (1983): Coastal upwelling in the California Current System. Prog. Oceanogr. **12**, 259–284

Huyer, A., Smith, R.L., Paluszkiewich, T. (1987): Coastal upwelling off Peru during normal and El Niño times, 1981–1984. Jour. Geoph. Res. **92**, 14297–14307

Hwang, P.A., Teague, W.J., Jacobs, G.A. (1998): A statistical comparison of wind speed, wave height and wave period derived from satellite altimeters and ocean buoys in the Gulf of Mexico region. Jour. Geoph. Res. **103**, 10454–10468

Ippen, A.T. (1966): Estuary and Coastline Hydrodynamics. Mc Graw-Hill Book Comp., 744 pp

Jackson, P.S. (1981): On the displacement height in the logarithmic velocity profile. Jour. Fluid Mech. **111**, 15-25

Jackson, G.A., Winant, C.D. (1983): Effect of a kelp forest on coastal currents. Continental Shelf Research **2** (1), 75–80

Jenkins, S.A., Skelly, D.W. (1987): Hydrodynamics of artificial seaweed for shoreline protection. Scripps Inst. Oceano., SIO, **87** (16), 66 pp

Jirka, G.H. (1990): Circulation in the salt wedge estuary. In: R.T. Cheng (ed.), Residual Currents and Long-term Transport. Coastal and Estuarine Studies, Springer-Verlag, New York, **38**, 223–237

Jonsson, I.G. (1966): Wave boundary layers and friction factors. Proc. 10th Coastal Eng. Conf., Tokyo, **1**, 127–148

Jonsson, I.G. (1980): A new approach to oscillatory rough turbulent boundary layers. Ocean Eng. **7**, 109–152

Kamenkovitch, V.M. (1978):Foundations of the theory of large-scale currents). In: V.M. Kamenkovitch, A.S. Monin (eds.), Oceanology. (Ocean Hydrodynamics, vol. 2). Nauka, Moscow, 359–408 (Russian)

Kamphuis, J.W., Readshaw, J.S. (1978): A model study of alongshore sediment transport rate. Proc. 16th Coastal Eng. Conf. **2**, 1656–1674

Kana, T.W., Ward, L.G. (1980): Nearshore suspended sediment load during storm and post-storm conditions. Proc. 17th Coastal Eng. Conf., Sydney, **2**, 1158–1173

Keller, S., Wu, T.Y., Brennen, C. (1975): A traction-layer model for ciliary propulsion. In: T.Y. Wu, C.J. Brokaw, C. Brennen (eds.), Swimming and Flying in Nature. Plenum Press, New York **1**, 253–271

Ketchum, B.H. (1983): Estuaries and Enclosed Seas. Elsevier Scientific Publ. Comp., Amsterdam, 500 pp

King, B.A., Spaniol, S., Wolanski, E. (1997): Modelling of mighty Burdekin River in Flood. Proc. 5th Estuarine and Coastal Modelling Conf., Alexandria (Virginia), 103–115

Kinsman, B. (1965): Wind Waves, their Generation and Propagation on the Ocean Surface. Prentice-Hall, Inc., Englewood Cliffs, 676 pp

Kirk, J.T.O. (1994): Light and Photosynthesis in Aquatic Ecosystems. Cambridge Univ. Press, Cambridge, 509 pp

Kitaigorodskii, S.A. (1962): Application of similarity methods to the analysis of wind waves. Izv. Akad. Nauk, Ser. Geophys. **1**, 105–117 (Russian)

Knauss, J.A. (1961): The structure of the Pacific Equatorial Countercurrent. Jour. Geoph. Res. **66**, 143–155

Komar, P.D. (1976): Beach Processes and Sedimentation. Prentice-Hall, Englewood Cliffs, 428 pp

Komen, G.J., Cavaleri, L., Donelan, M., Hasselmann, K., Hasselmann, S., Janssen, P.A.E.M. (1994): Dynamics and Modelling of Ocean Waves. Cambridge University Press, Cambridge, 532 pp

Koutitas, Ch.G. (1988): Mathematical Models in Coastal Engineering. Pentech Press, London, 156 pp

Kraines, S.B., Yangi, T., Isobe, M., Komiyama, H. (1998): Wind-wave driven circulation on the coral reef at Bora Bay, Miyako Island. Coral Reef **17**, 133–143

Krauss, W. (1972): On the response of a stratified ocean to wind and air pressure. Dtsch. Hydrograph. Zeit. **25**, 49–61

Kriebel, D.L. (1986): Verification study of a dune erosion model. Shore and Beach **54**, 13–20

Kriebel, D.L., Dean, R.G. (1985): Numerical simulation of time-dependent beach and dune erosion. Coastal Eng. **9**, 221–245

Krylov, Y.M. (1986): Wind, Waves and Marine Harbours. Gidrometeoizdat, Leningrad, 264 pp (Russian)

Ku, D.N. (1997): Blood flow in arteries. Ann. Rev. in Fluid Mech. **29**, 399–434

Kuo, C.T., Chiu, Y.F. (1994): Transfer function between wave height and wave pressure for progressive waves. Coastal Eng. **23**, 81–93

LaFond, E.C. (1966): Internal Waves. In: Encyclopedia of Oceanography. Reinhold Publ., New York, 402–408

Laing, A.K. (1996): Variability in the East Auckland current system from satellite altimetry. Proc. of the First Australian ERS Symp., Canberra, 53–63

Lalli, C.M., Parsons, T.R. (1997): Biological Oceanography. An Introduction. Butterworth-Heinemann, Oxford, 314 pp

Lamb, H. (1932): Hydrodynamics. Dover Publ., New York, 738 pp

Lamb, K.G. (1994): Numerical experiments of internal wave generation by strong tidal flow across a finite amplitude bank edge. Jour. Geoph. Res. **99**, 843–864

Lane, A., Prandle, D., Harrison, A.J., Jones, P.D., Jarvis, C.J. (1997): Measuring fluxes in tidal estuaries: Sensitivity to instrumentation and associated data analysis. Estuarine, Coastal and Shelf Science **45**, 433–452

Langmuir, I. (1938): Surface motion of water induced by wind. Science **87**, 119–123

Larson, M. (1995): Model of beach profile change under random waves. Jour. Waterway, Port, Coastal and Ocean Eng. **121**, 172–181

Larson, M., Kraus, N.C. (1989): SBEACH: Numerical Model for Simulating Storm-Induced Beach Change, Report 1: Theory and Model Foundation, U.S. Army, CERC. Tech. Rep. CERC-89-9, 256 pp

Larson, M., Kraus, N.C. (1991): Numerical model of longshore current for bar and trough beaches. Jour. Waterway, Port, Coastal and Ocean Eng. **117**, 326–347

Latif, M., Anderson, D., Barnett, T.P., Cane, M.A., Kleeman, R., Leetmaa, A., O'Brien, J., Rosati, A., Schneider, E. (1998): A review of the predictability and prediction of ENSO. Jour. Geoph. Res. **103**, 14375–14393

Laws, R.M. (1985): The ecology of the Southern Ocean. American Scientist **73**, 26–40

Le Blond, P.H., Mysak, L.A. (1978): Waves in the Ocean. Elsevier Scientific Publ. Comp., Amsterdam, 602 pp

Leetmaa, A., McCreary, J.P., Moore, D.W. (1981): Equatorial currents; observation and theory. In: B.A. Warren, C. Wunsch (eds.). Evolution of Physical Oceanography. MIT Press, Cambridge, 184–196

Le Fevre, J. (1986): Aspects of the biology of frontal systems. Adv. Mar. Biol. **23**, 164–299

Le Méhauté, B. (1976): An Introduction to Hydromechanics and Water Waves. Springer-Verlag, New York Heidelberg Berlin, 315 pp

Le Méhauté, B. (1990): Similitude. In: B. Le Méhauté, D.M. Hanes (eds.), The Sea. (Ocean Engineering Science, vol. 9) A Wiley-Interscience Publ., New York, 955–980

Le Provost, C., Bennett, A.F., Cartwright, D.E. (1995): Ocean tides for and from TOPEX/POSEIDON. Science **267**, 639–642

Le Provost, C., Lyard, F., Genco, M.L., Molines, J.M. (1998a): Global ocean tides through assimilation of oceanographic and altimeter satellite data in a hydrodynamic model. Aviso, Newsletter **6**, 68–71

Le Provost, C., Lyard, F., Molines, J.M., Genco, M.L., Rabilloud, F. (1998b): A hydrodynamic ocean tide model improved by assimilating a satellite altimeter-derived data set. Jour. Geoph. Res. **103**, 5513–5529

Levitus, S. (1982): Climatological Atlas of the World Ocean. NOAA, Prof. Pap., Washington, **13**

Lewis, E.L., Perkin, R.G. (1981): The practical salinity scale 1978: Conversion of existing data. Deep-Sea Res. **28A**, 307–328

Lighthill, M.J. (1969):Hydromechanics of aquatic animal propulsion: a survey. Ann. Rev. of Fluid Mech. **1** 413–446

Lighthill, M.J. (1971): Large-amplitude elongated body theory of fish locomotion. Proc. Roy. Soc., London **50** (179), 125–138

Lighthill, M.J. (1975): Mathematical Biofluiddynamics. Soc. for Ind. and Appl. Math., Philadelphia, 281 pp

Lindsay, P., Balls, P.W., West, J.R. (1996): Influence of tidal range and river discharge of suspended particulate matter fluxes in the Forth Estuary (Scotland). Estuarine, Coastal and Shelf Science 42, 63–82

Llewellyn-Jones, D.T., Lawrence, S.P., Dundas, R.M., Mutlow, C.T. (1996): The observation and analysis of large-scale phenomena using data from ATSR. Proc. of the First Australian ERS Symp., Canberra, 76–84

Loder, J.W., Greenberg, D.A. (1986): Predicted positions of tidal fronts in the Gulf of Maine region. Continental Shelf Res. 6, 397–414

Longuet-Higgins, M.S. (1953): Mass transport in waters. Phil. Trans. Roy. Soc. London, A245, 535–581

Longuet-Higgins, M.S., Stewart, R.W. (1964): Radiation stresses in water waves: a physical discussion with application. Deep Sea Res. 11, 529–562

Lough, J.M., Barnes, D.J. (1992): Comparison of skeletal density variations in Porites from the central Great Barrier Reef. Jour. Exp. Biol. Ecol. 155, 1–25

Lough, J.M., Barnes, D.J., Taylor, R.B. (1996): The potential of massive corals for the study of high-resolution climate variation in the past millennium. In: P.D. Jones, R.S. Bradley, J. Jouzel (eds.), Climate Variations and Forcing Mechanisms of the Last 2000 Years. NATO ASI Series 145, 355–371

Luketina, D. (1998): Simple tidal prism model revisited. Estuarine, Coastal and Shelf Science 46, 77–84

Lutjeharms, J.R.E., van Ballegooyen, R.C. (1988): The retroflection of the Agulhas Current. Jour. Phys. Oceanogr. 18, 1570–1589

Ma, X.C., Shum, C.K., Eanes, R.J., Tapley, B.D. (1994): Determination of ocean tides from the first year of TOPEX/POSEIDON altimeter measurements. Jour. Geoph. Res. 99, 24809–24820

MacIsaac, J.J., Dugdale, R.C., Barber, R.T., Beasco, D., Packard, T.T. (1985): Primary production cycle in an upwelling center. Deep Sea Res. 32, 503–529

Mackas, D.L., Boyd, C.M. (1979): Spectral analysis of zooplankton spatial heterogeneity. Science 204, 62–64

Manabe, S. (1969): Climate and the ocean circulation. (The atmosphere circulation and the effect of heat transfer by ocean currents, vol. II) Mon. Weather Rev. 97, 775–805

Mandelbrot, B.B. (1983): The Fractal Geometry of Nature. Freeman and Comp., New York, 468 pp

Mann, K.H. (1991): Organisms and Ecosystems. In: R.S.K. Barnes, K.H. Mann (eds.), Fundamentals of Aquatic Ecology. Blackwell Scientific Publ., Oxford, 3–26.

Mann, K.H., Lazier, J.R.N. (1996): Dynamics of Marine Ecosystems. Biological-Physical Interactions in the Oceans. Blackwell Science, Cambridge, 394 pp

Massel, B. (1998): Prediction of Sediment Transport and Beach Morphology Changes During Cyclones in the Central Great Barrier Reef Region: A Numerical Approach. Master Degree Thesis. James Cook University, Townsville, Australia. 217 pp

Massel, S.R. (1981): Hydrodynamical Problems of Offshore Structures. Państwowe Wydawnictwo Naukowe, Warszawa, 155 pp (Polish)

Massel, S.R. (1982): Waves in the Atlantic Ocean. In: E. Rühle, J. Zaleski. (eds.), Atlantic Ocean. Państwowe Wydawn. Naukowe, Warszawa, 388–414 (Polish)

Massel, S.R. (1989): Hydrodynamics of Coastal Zone. Elsevier Science Publ., Amsterdam, 336 pp

Massel, S.R. (1992): Wave transformation and dissipation on steep reef slopes. Proc. 11th Australian Fluid Mechanics Conference, Hobart, 319–322

Massel, S.R. (1993a): Extended refraction–diffraction equation for surface waves. Coastal Eng. **19**, 97–126

Massel, S.R. (1993b): Scattering of surface waves by a conical coral. Proc. 11th Australasian Coastal and Ocean Eng. Conf. **2**, 467–471

Massel, S.R. (1994): Impact of surface waves on physical degradation of coral reef. In: J.L. Munro, P.E. Munro (eds.), Management of Coral Reef Resource Systems. ICLARM, Manila, 37–39

Massel, S.R. (1996a): Ocean Surface Waves: their Physics and Prediction. World Scientific Publ., Singapore New Jersey London Hong Kong, 491 pp

Massel, S.R. (1996b): On the largest wave height in water of constant depth. Ocean Eng. **23**, 553–573

Massel, S.R. (1998): The limiting wave height in wind-induced wave trains. Ocean Eng. **25**, 735–752

Massel, S.R., Done, T.J. (1993): Effects of cyclone waves on massive coral assemblages on the Great Barrier Reef: meteorology, hydrodynamics and demography. Coral Reefs **12**, 153–166

Massel, S.R., Brinkman, R.M. (1998a): Measurement and modelling of wave propagation and breaking at steep coral reefs. In: N.C. Saxena (ed.), Recent Advances in Marine Science and Technology, 98, PACON International

Massel, S.R., Brinkman, R.M. (1998b): On the determination of directional wave spectra for practical applications. Applied Ocean Research, **20**, 357–374

Massel, S.R., Furukawa, K., Brinkman, R.M. (1998): Surface wave propagation in mangrove forests. Fluid Dynamics Research, (in press)

Mather, F.J. (1962): Transatlantic migration of two large bluefin tuna. Jour. Cons. Perm. Int. Explor. Mer. **27**, 325–327

Mazda, Y., Magi, M., Kogo, M., Hong, P. N. (1997a): Mangroves as a coastal protection from waves in the Tong King Delta, Vietnam. Mangrove and Salt Marches **1**(2), 127-135

Mazda, Y., Wolanski, E., King, B., Sase, A., Ohtsuka, D. (1997b): Drag force due to vegetation in mangrove swamps. Mangroves and Salt Marches **1**(3), 193–199

McDougal, W.G., Hudspeth, R.T. (1983): Longshore sediment transport on nonplanar beaches. Coastal Eng. **7**, 119–131

McDougal, W.G., Hudspeth, R.T. (1989): Longshore current and sediment transport on composite beach profiles. Coastal Eng. **12**, 315–338

McPhaden, M.J., Busalacchi, A.J., Cheney, R., Donguy, J.R., Gage, K.S., Halpern, D., Ji, M., Julian, P., Mitchum, R.W., Niiler, P.P., Picant, J., Reynolds, R.W., Smith, N., Takeuchi, K. (1998): The tropical Ocean-Global Atmosphere observing system: a decade of progress. Jour. Geoph. Res. **103**, 14169–14240

McWilliams, J.C. (1996): Modelling the oceanic general circulation. Ann. Rev. Fluid Mech. **28**, 215–248

Mei, C.C. (1983): The Applied Dynamics of Ocean Surface Waves. A Wiley - Inter - Science Publication, New York, 734 pp

Mei, C.C., Foda, M.A. (1981): Wave-induced responses in a fluid-filled poro-elastic solid with a free surface – a boundary layer theory. Geophys. Jour. Royal Astr. Soc. **66**, 597–631

Mei, C.C., Liu, P.L.-F. (1977): Effects of topography on the circulation in and near the surf zone-linear theory. Jour. Estuary Coast. Mar. Sci. **5**, 25–37

Mellor, G.L. (1991): User's guide for a three-dimensional, primitive equation, numerical ocean model. Progress in Atmosphere and Ocean Science. Princeton Univ., Princeton, 35 pp

Miles, J.W. (1957): On the generation of surface waves by shear flows. Jour. Fluid Mech. **3**, 185–204

Miles, J.W. (1962): On the generation of surface waves by shear flows. Part 4. Jour. Fluid Mech. **13**, 433–448

Miropolskiy, J.Z., Monin, A.S. (1978): Internal waves. In: V.A. Kamenkovitch, A.S. Monin (eds.) Oceanology. (Ocean hydrodynamics, vol. 2) Nauka, Moscow, 182–228

Mittelstaedt, E. (1983): The upwelling area off Northwest Africa – a description of phenomena related to coastal upwelling. Prog. Oceanogr. **12**, 307–331

Mizuno, K., White, W.B. (1983): Annual and interannual variability in the Kuroshio current system. Jour. Phys. Oceanogr. **13**, 1847–1867

Monin, A.S., Ozmidov, R.V. (1985): Turbulence in the Ocean. Riedel Publ. Comp., Boston, 247 pp

Monin, A.S., Yaglom, A.M. (1971): Statistical Fluid Mechanics (vol. 1). MIT Press, Cambridge, 769 pp

Monin, A.S., Krasitskii, V.P. (1985): Phenomena on Ocean Surface. Gidrometeoizdat, Leningrad, 373 pp (Russian)

Mooers, C.N.K. (1973): A technique for the cross spectrum analysis of pairs of complex-valued time series, with emphasis on properties of polarized components and rotational invariants. Deep-Sea Res. **20**, 1129–1141

Morison, J.R., O'Brien, M.P., Johnson, J.W., Schaaf, S.A. (1950): The force exerted by surface waves on piles. Petroleum Trans., AIME **189**, 149–154

Morris, A.W., Allen, J.I., Howland, R.J.M., Wood, R.G. (1995): The estuary plume zone: source or sink for land-derived nutrient discharges? Estuarine, Coastal and Shelf Science **40**, 387–402

Muller, F.L.L., Tranter, M., Balls, P.W. (1994): Distribution and transport of chemical constituents in the Clyde Estuary. Estuarine, Coastal and Shelf Science **39**, 105–126

Munk, W., Wunsch, C. (1998): Abyssal receipts II: energetics of tidal and wind mixing. Deep Sea Res. **45**, 1977–2010

Munk, W.H., Ewing, C.C., Revelle, R.R. (1949): Diffusion in Bikini Lagoon. Trans. Amer. Geoph. Union **30**, 59–66

Neelin, J.D., Latif, M., Jin, F.F. (1994): Dynamics of coupled ocean-atmosphere models: the tropical problem. Ann. Rev. Fluid Mech. **26**, 617–659

Neelin, J.D., Battisti, D.S., Hirst, A.C., Jin, F.F., Wakata, Y., Yamagata, T., Zebiak, S.E. (1998): ENSO theory. Jour. Geoph. Res. **103**, 14261–14290

Nelson, R.C. (1994): Depth limited design wave heights in very flat regions. Coastal Eng. **23**, 43–59

Nelson, G., Hutchings, L. (1983): The Benguela upwelling area. Prog. Oceanogr. **12**, 333–356

Nepf, H.M., Mugnier, C.G., Zavistoski, R.A. (1997a): The effects of vegetation on longitudinal dispersion. Estuarine, Coastal and Shelf Science **44**, 675–684

Nepf, H.M., Sullivan, J.A., Zavistoski, R.A. (1997b): A model for diffusion within emergent vegetation. Limnology and Oceanography **42**, 1735–1745

Netto, S.A., Lana, P.C. (1997): Influence of *Spartina alterniflora* on superficial sediment characteristics of tidal flats in Paranaguá Bay (South-eastern Brazil). Estuarine, Coastal and Shelf Science **44**, 641–648

Newman, J.N. (1977): Marine Hydrodynamics. The MIT Press, Cambridge, 367 pp (Russian transl.)

Nicholls, N. (1986): A method for predicting Murray Valley encephalitis in Southeast Australia using the Southern Oscillation. Austr. Jour. Exper. Biological and Medical Science **64**, 587–594

Nicholls, N. (1991): The El Niño – Southern Oscillation: Recent Australian research. Bureau of Meteor. Research Centre, Melbourne

Nielsen, P. (1992): Coastal Bottom Boundary Layers and Sediment Transport. World Scientific Publ., Singapore, 324 pp

Noye, J.B. (1987): Numerical Modelling: Applications to Marine Systems. Elsevier Science Publ., Amsterdam, 229 pp

Ochi, M.K. (1990): Applied Probability and Stochastic Processes in Engineering and Physical Sciences. John Wiley & Sons, New York, 499 pp

Ochi, M.K., Hubble, E.N. (1976): On six-parameter wave spectra. Proc. 15th Coastal Eng. Conf. **1**, 301–328

O'Dor, R.K. (1988): The forces acting on swimming squid. Jour. Exp. Biol. **137**, 421–442

O'Dor, R.K., Wells, J., Wells, M.J. (1990): Speed jet pressure and oxygen consumption relationships in free-swimming *Nautilus*. Jour. Exp. Biol. **154**, 386–396

Officer, C.B. (1976): Physical Oceanography of Estuaries (and Associated Coastal Waters). John Wiley & Sons, New York, 465 pp

Okubo, A. (1967): The effect of shear in an oscillatory current on horizontal diffusion from an instantaneous source. Limnology and Oceanography **1**, 194–204

Okubo, A. (1980): Diffusion and Ecological Problems: Mathematical Models. Springer-Verlag, Berlin, 254 pp

Oliver, J.K., Willis, B.L. (1987): Coral-spawn slicks in the Great Barrier Reef: preliminary observations. Marine Biology **94**, 521–529

Oliver, J.K., King, B.A., Willis, B.L., Babcock, R.C., Wolanski, E. (1992): Dispersal of coral larvae from a lagoonal reef. II. Comparisons between model predictions and observed concentrations. Cont. Shelf Res. **12**, 873–889

Osborn, T.R., Cox, C.S. (1972): Oceanic fine structure. Geophys. Fluid Dyn. **3**, 321–345

Osborne, A.R., Burch, T.L. (1980): Internal solitons in the Andaman Sea. Science **208**, 451–460

Otnes, R.K., Enochson, L. (1972): Digital Time Series Analysis. John Wiley & Sons, New York, 388 pp

Oseen, C.W. (1910): Über die Stokessche Formel und über die verwandte Aufgabe in der Hydrodynamic. Arkiv. Mat. Astron. Fysik **6**

Owen, M.W., Thorn, M.F.C. (1978): Effect of waves on sand transport by currents. Proc. 16th Coastal Eng. Conf., Hamburg, **2**, 1675–1687

Ozmidov, R.V. (1986): Diffusion of Contaminants in the Ocean. Gidrometeoizdat, Leningrad, 279 pp (Russian)

Partheniades, E. (1992): Stratified flows salinity intrusion, and transport processes. In: J.B. Herbich (ed.), Handbook of Coastal and Ocean Engineering. Vol. **3**, 881–984

Pedley, T.J. (1980): The Fluid Mechanics of Large Blood Vessels. Cambridge Univ. Press, Cambridge, 446 pp

Pedlosky, J. (1979): Geophysical Fluid Dynamics. Springer-Verlag, New York Heidelberg Berlin, 624 pp

Pennington, J.T. (1985): The ecology of fertilization of echinoid eggs: the consequences of sperm dilution, adult aggregation, and synchronous spawning. Biol. Bull. **169**, 417–430

Pennycuick, C.J. (1992): Newton Rules Biology: A Physical Approach to Biological Problems. Oxford Univ. Press, Oxford, 111pp

Peregrine, D.H. (1990): Theory versus measurements. In: A. Torum, O.T. Gudmestad (eds.), Water Wave Kinematics. Kluwer Academic Publ., 89–101

Perry, R.B., Schimke, G.R. (1965): Large amplitude internal waves observed off the north-west coast of Sumatra. Jour. Geoph. Res. **70**, 2319–2324

Perry, A.H., Walker, J.M. (1977): The ocean-atmosphere system. Longman Group, London, 160 pp

Peters, H., Gregg, M.C., Toole, J.M. (1988): On the parameterization of equatorial turbulence. Jour. Geoph. Res. **93**, 1199–1218

Philander, S.G. (1990): El Niño, La Niña, and the Southern Oscillation. Academic Press Inc., San Diego, 293 pp

Phillips, O.M. (1957): On the generation of waves by turbulent wind. Jour. Fluid Mech. **2**, 417–445

Phillips, O.M. (1958): The equilibrium range in the spectrum of wind-generated waves. Jour. Fluid Mech. **4**, 426–434

Phillips, O.M. (1977): The Dynamics of the Upper Ocean. Second Edition. Cambridge Univ. Press, 336 pp

Pickard, G.L., Emery, W.J. (1982): Descriptive Physical Oceanography. Pergamon Press, Oxford, 249 pp

Pierson, W.J., Moskowitz, L. (1964): A proposed spectral form for fully developed wind seas based on the similarity theory of S.A. Kitaigorodskii. Jour. Geoph. Res. **69**, 5181–5190

Pinet, P.R. (1992): Oceanography: An Introduction to the Planet Oceans. West Publ. Comp., New York, 571 pp

Pond, S., Pickard, G.L. (1983): Introductory Dynamical Oceanography. Pergamon Press, Oxford, 329 pp

Posma, H. (1988): Tidal flat areas. In: Bengt-Owe Jansson (ed.), Coastal-Offshore Ecosystems Interactions. Lecture Notes on Coastal and Estuarine Studies **22**. Springer-Verlag, Heidelberg, 102–121

Prager, E.J. (1991): Numerical simulation of circulation in a Caribbean-type back-reef lagoon. Coral Reefs **10**, 177–182

Primavera, J.H. (1998): Mangroves as nurseries: shrimp populations in mangrove and non-mangrove habitats. Estuarine, Coastal and Shelf Science **46**, 457–464

Proudman, I., Pearson, J.R.A. (1957): Expansions at small Reynolds number for the flow past a sphere and a circular cylinder. Jour. Fluid Mech. **2**, 237–262

Prudnikov, A.P., Britchkov, J.A., Mritchev, O.I. (1981): Integrals and Series. Nauka, Moscow, 797 pp (Russian)

Pugh, D.T. (1987): Tides, Surges and Mean Sea-Level. John Wiley & Sons. Chichester New York Brisbane Toronto Singapore, 472 pp

Quinn, T.P. (1988): Estimated swimming speeds of migrating sockeye salmon. Can. Jour. Zool. **66**, 2160–2163

Qureshi, T.M. (1990): Experimental plantation of rehabilitation of mangrove forests in Pakistan. Mangrove Ecosystems Occasional Papers, UNESCO **4**, 37 pp

Ragueneau, O., Queguiner, B., Treguer, P. (1996): Contrast in biological responses to tidally-induced vertical mixing for two macrotidal ecosystems of Western Europe. Estuarine, Coastal and Shelf Science **42**, 645–665

Rasmusson, E.M., Wallace, J.M. (1984): Meteorological aspects of the El Niño/Southern Oscillation. Science **222**, 1195–1202

RD Instruments (1989): Acoustic Doppler Current Profilers. Principles of operation: a practical primer. San Diego, 36 pp

Redding, R., Yenne, B. (1983): Boeing Planemaker to the World. Arms and Armour Press, London, 256 pp

Richards, F.A. (1981): Coastal Upwelling. American Geophysical Union, Washington, 529 pp

Ridd, P.V., Wolanski, E., Mazda, Y. (1990): Longitudinal diffusion in mangrove-fringed tidal creeks. Estuarine, Coastal and Shelf Science **31**, 541–554

Riedel, H.P., Byrne, A.P. (1986): Random breaking waves – horizontal seabed. Proc. 12th Coastal Eng. Conf. **1**, 903–908

Roberts, J. (1975): Internal Gravity Waves in the Ocean. Marcel Dekker Inc., New York, 274 pp

Robertson, A.I., Alongi, D.M. (1992): Tropical mangrove ecosystems. American Geophysical Union Washington, 329 pp

Rome, L.C., Swank, D., Corda, D. 1993: How fish power swimming. Science **261**, 340–343

Ruddick, B., Walsh, D. (1997): Variations in apparent mixing efficiency in the North Atlantic Central Waters. Jour. Phys. Oceanogr. **27**, 2589–2605

Sargent, F.E., Jirka, G.H. (1987): Experiments on saline wedge. Jour. Hydr. Eng. **113**, 1307–1324

Sarpkaya, T., Isaacson, M.St.Q. (1981): Mechanics of Wave Forces on Offshore Structures. Van Nostrand Reinold Comp., New York, 651 pp

Sasekumar, A., Chong, V.C., Lim, K.H. (1994): Status of mangrove finfish resources in ASEAN. In: C., Wilkinson, S., Sudara, C.L. Ming, (eds.), Proc. of the Third ASEAN-Australia Symposium on Living Coastal Resources, Bangkok, Thailand. Australian Institute of Marine Science, Townsville, Australia, 139–144

Schlichting, H. (1960): Boundary Layer Theory. McGraw Hill Book Comp., New York, 647 pp

Schluessel, P., Shin, H.Y., Emery, W.J., Grassl, H. (1987): Comparison of satellite derived seas surface temperature with in situ skin measurements. Jour. Geoph. Res. **92**, 2859–2874

Schmidt-Nielsen, K. (1989): Scaling: Why is animal size so important. Cambridge Univ. Press, Cambridge, 241 pp

Schoonees, J.S., Theron, A.K. (1994): Accuracy and applicability of the SPM long-shore transport formula. Proc. 24th Coastal Eng. Conf. **3**, 2595–2609

Schott, F. (1983): Monsoon response of the Somali Current and associated upwelling. Prog. Oceanogr. **12**, 357–381

Seymour, R.J. (1989): Nearshore Sediment Transport. Plenum Press, New York, 418 pp

Seymour, R.J., Tegner, M.J., Dayton, P.K., Parnell, P.E. (1989): Storm wave induced mortality of giant kelp *Macrocystis perifera* in Southern California. Estuarine, Coastal and Shelf Science **28**, 277–292

Shaffer, G.R. (1982): On the upwelling circulation over the wide shelf off Peru. I. Circulation. Jour. Mar. Res. **40**, 293–314

Shapiro, D.Y., Hensley, D.A., Appeldoorm, R.S. (1988): Pelagic spawning and egg transport in coral-reef fishes: a skeptical review. Environmental Biology and Fishes, **22**, 3-14

Siddiqi, N.A., Khan, M.A.S. (1990): Two papers on mangrove plantations in Bangladesh. Mangrove Ecosystems Occasional Papers, UNESCO **8**, 19 pp

Simpson, J.H., Hunter, J.R. (1974): Fronts in the Irish Sea. Nature **250**, 404–406

Simpson, J.H., Allen, C.M., Morris, N.C.G. (1978): Fronts on the continental shelf. Jour. Geoph. Res. **83**, 4607–4614

Sleath, J.F.A. (1987): Turbulent oscillatory flow over rough beds. Jour. Fluid Mech. **182**, 369–409

Smith, F.G. (1973): The Seas in Motion. Thomas Y. Crowell Comp., New York, 248 pp

Smith, R.L. (1981): A comparison of the structure and variability of the flow field in three coastal upwelling regions: Oregon, Northwest Africa, and Peru. In: F.A. Richards (ed.), Coastal Upwelling. American Geophysical Union, Washington, 107-1 18

Smith, C.L., Richards, K.J., Fashman, M.J.R. (1996): The impact of mesoscale eddies on plankton dynamics in the upper ocean. Deep Sea Res. **43**, 1807–1832

Sobey, R.J. (1993): Quantifying coastal and ocean processes. Proc. 11th Australasian Conference on Coastal and Ocean Engineering, Townsville **1**, 1–10

SPM (Shore Protection Manual) (1984): Shore Protection Manual (vol. 1–3). Coastal Engineering Research Center, Department of the Army, Washington

Steele, K.E., Teng, C.C., Wang, D.W.C. (1992): Wave direction measurements using pitch-roll buoys. Ocean Eng. **19**, 349–375

Steven, G.A. (1950): Swimming of dolphins. Science Progr. **38**, 524–525

Stevens, E.D., Lightfoot, E.N. (1986): Hydrodynamics of water flow in front of and through the gills of skipjack tuna. Comp. Biochem. Physiol. **83A**, 255–259

Stokes, G.G. (1847): On the theory of oscillatory waves. Transactions of the Cambridge Philosophical Society 8, 441–455

Stokes, G.G. (1851): On the effect of the internal friction of fluids on the motion of pendulums. Trans. of the Cambridge Phil. Soc. 1, 1–141

Stommel, H. (1965): The Gulf Stream. Univ. of California Press, Berkeley, 248 pp

Stommel, H., Yoshida, K. (1972): Kuroshio, Physical Aspects of the Japan Current. Univ. of Washington Press, Seatle, 517 pp

Strahler, A.H., Strahler, A.N. (1992): Modern Physical Geography. John Wiley & Sons, Inc. New York. 638 pp

Strekalov, S.S., Massel, S.R. (1971): On the spectral analysis of wind waves. Arch. Hydrot. 18, 457-485 (Polish)

Sulaiman, D.M., Tsutsui, S., Yoshioka, H., Yamashita, T., Oshiro, S., Tsuchiya, Y. (1994): Prediction of the maximum wave on the coral flat. Proc. 24th Coastal Eng. Conf., Kobe, 1, 609–623

Svendsen, I.A. (1987): Analysis of surf zone turbulence. Jour. Geoph. Res. 92, 5115–5124

Sverdrup, H.U. (1945): Oceanography for Meteorologists. Allen & Unwin, 246 pp

Sverdrup, H.U. (1947): Wind driven currents in a baroclinic ocean with application to the equatorial currents of the Eastern Pacific. Proc. Nat. Acad. Sci. U.S. 33, 318–326

Sverdrup, H.U., Munk, W.H. (1947): Wind, sea and swell; theory of relations for forecasting. U.S. Navy Hydrographic Office, H.O. Publ. No. 601

Tapley, B.D., Chambers, D.P., Shum, C.K., Eanes, R.J., Ries, J.C. (1994): Accuracy assessment of the large-scale dynamic ocean topography from TOPEX/ POSEIDON altimetry. Jour. Geoph. Res. 99, 24605–24617

Taylor, G.I. (1953): Dispersion of soluble matter in solvent moving slowly through a tube. Proc. Roy. Soc., London, Ser. A, 219, 186–203

Taylor, J., Bucens, P. (1989): Laboratory experiments on the structure of salt fingers. Deep-Sea Res. 36. 1675–1704

Thorn, M.F.C. (1979): The effects od waves on the tidal transport of sand. Hydr. Res. Station, Wallingford, Notes 21, 4–5

Thornton, E.B., Guza, R.T. (1983): Transformation of wave height distribution. Jour. Geoph. Res. 88, 5925–5938

Thornton, E.B., Guza, R.T. (1986): Surf zone longshore currents and random waves: field data and models. Jour. Phys. Oceanogr. 16, 1165–1178

Tide Tables (1998): Austr. Nat. Tide Tables, Dept. of Defense, Canberra, 372 pp

Tomczak, M., Godfrey, J.S. (1994): Regional Oceanography: An Introduction. Pergamon, Oxford, 422 pp

Tørum, A. (1996): Kelp–wave damping – beach and dune erosion. Progress report. SINTET, Trondheim, 50 pp

Townsend, A.A. (1976): The Structure of Turbulent Shear Flow. Cambridge University Press, Cambridge, 429 pp

Townsend, M., Fenton, J.D. (1995): Numerical comparison of wave analysis methods. Proc. First Australasian Coastal and Ocean Eng. Conf. 1, 169–173

Tremblay, J-E., Legendre, L., Therriault, J-C. (1997): Size-differential effects on vertical stability on the biomass and production of phytoplankton in a large estuarine system. Estuarine, Coastal and Shelf Science 45: 415–431

Trenberth, K.E., Branstator, G.W., Karoly, D., Kumar, A., Lau, N.C., Ropelewski, Ch. (1998): Progress during TOGA in understanding and modelling global teleconnections associated with tropical sea surface temperatures. Jour. Geoph. Res. **103** , 14291–14324

Trueman, E.R. (1975): The locomotion of soft-bodied animals. Arnolds, London.

Tucker, M.J., Car, A.P., Pitt, E.G. (1983): The effect of an off-shore bank in attenuating waves. Coastal Eng. **7**, 133–144

Tucker, M.J. (1989): Interpreting directional data from large pitch-roll-heave buoys. Ocean Eng. **16**, 173–192

Tunnicliffe, V. (1979): The effects of wave-induced flow on a reef coral. Jour. Exp. Mar. Biol. Ecol. **64**, 1–10

Turner, J.S. (1967): Salt fingers across a density interface. Deep-Sea Res. **14**, 599–611

Umar, M.J., McCook, L.J., Price, I.R. (1998): Effects of sediment deposition on the seaweed *Sargassum* on a fringing coral reef. Coral Reefs. **17**, 169–177

UNESCO (1987): International oceanographic tables. UNESCO Technical Papers in Marine Science (vol. 40). UNESCO, Paris

Utter, B.D., Denny, M.W. (1996): Wave induced forces on the giant kelp *Macrocystis pyrifera* (Agardh): Field test of a computational model. Jour. Exp. Biol. **199**, 2645–2654

Valiela, I. (1995): Marine Ecological Processes. Springer-Verlag, New York, 686 pp

Van Woesik, R., Ayling, A.M., Mapstone, B. (1991): Impact of tropical cyclone 'Ivor' on the Great Barrier Reef, Australia. Jour. Coastal Res. **7**, 551–558

Veron, J.E.N. (1986): Corals of Australia and Indo-Pacific. Univ. of Hawaii Press, Honolulu, 644 pp

Videler, J.J. (1993): Fish Swimming. Chapman and Hall, London, 260 pp

Videler, J.J., Weihs, D. (1982): Energetic advantages of burst-and-coast swimming of fish at high speeds. Jour. Exp. Biol. **97**, 169–178

Vincent, J.F.V. (1982): Structural Biomaterials. John Wiley & Sons, New York, 206 pp

Vlymen, W.J. (1974): Swimming energetics of the larval anchovy, *Engraulis mordax*. Fishery Bulletin **72**, 885–899

Vogel, S. (1994): Life in Moving Fluids. The Physical Biology in Flow. Princeton Univ. Press, Princeton, 467 pp

von Arx, W. (1977): An Introduction to Physical Oceanography. Addison-Wesley Publ. Comp., Reading, 422 pp

Wainwright, S.A., Biggs, W.D., Currey, J.D., Gosline, J.M. (1976): Mechanical Design in Organisms. Edward Arnold, London, 423 pp

Wardle, C.S., Videler, J.J., Arimoto, T., Franco, J.M., He, P. (1989): The muscle twitch and the maximum speed of giant bluefin tuna, *Thunnus thynnus*. Jour. Fish Biol. **35**, 129–137

Washington, W.M., Semtner, A.J., Meehl, G.D., Kinght, D.J., Mayer, T.A. (1980): A general circulation experiment with a coupled atmosphere, ocean and sea ice model. Jour. Phys. Oceanogr. **10**, 1887–1908

Webb, P.W. (1984): Body form, locomotion and foraging in aquatic vertebrates. Amer. Zool., **24**, 107–120

Webb, P.W., Sims, D., Schultz, W.W. (1991): The effects of an air/water surface on the fast-start performance of rainbow trout *Oncorhynchus mykiss*. Jour. Exp. Biol. **155**, 219–226

Weihs, D. (1973): Optimal fish cruising speed. Nature **245**, 48–50

Weihs, D. (1974): Energetic advantages of burst swimming of fish. Jour. Theor. Biol. **48**, 215–229

Weir, D.J., McManus, J. (1987): The role of wind in generating turbidity maxima in the Tay Estuary. Cont. Shelf Res. **7**, 1315–1318

Wells, J.T. (1983): Dynamics of coastal fluid muds in low-, moderate- and high-tide-range environments. Can. Jour. Fish. Aquat. Sci. **40** (suppl.), 130–142

White, F.M. (1994): Fluid Mechanics. Mc Graw-Hill, Inc., New York, 736 pp

Whitehead, J.A. (1995): Thermohaline ocean processes and models. Ann. Rev. Fluid Mech. **27**, 89–113

Willis, B.L., Oliver, J.K. (1990): Direct tracking of coral larvae: implications for dispersal studies of planktonic larvae in topographically complex environments. Ophelia **32**, 145–162

Wilson, B.W. (1972): Seiches. Adv. in Hydroscience **8**, 1–94

Wilson, B.W., Reid, R.O. (1963): Wave force coefficients for offshore pipelines. Jour. Waterways and Harbour Div., ASCE, **89**,

Wolanski, E. (1994): Physical oceanographic processes of the Great Barrier Reef. CRC Press, Boca Raton, 194 pp

Wolanski, E. (1995): Transport of sediment in mangrove swamps. Hydrobiol. **295**, 31–42

Wolanski, E., van Senden, D. (1983): Mixing of Burdekin River flood waters in the Great Barrier Reef. Austr. Jour. Mar. and Freshwater Res. **34**, 49–63

Wolanski, E., Hamner, W.M. (1988): Topographically controlled fronts in the ocean and their biological influence. Science **241**, 177–181

Wolanski, E., Drew, E.A., Abel, K.M., O'Brien, J. (1988): Tidal jets, nutrient upwelling and their influence on the productivity of the alga *Halimeda* in the Ribbon Reefs, Great Barrier Reef. Estuarine, Coastal and Shelf Science **26**, 169–201

Wolanski, E., Burrage, D., King, B.A. (1989): Trapping and dispersion of coral eggs around Bowden Reef, Great Barrier Reef following mass coral spawning. Continental Shelf Res. **9**, 479–496

Wolanski, E., Eagle, M. (1991): Oceanography and fine sediment transport, Fly River Estuary and Gulf of Papua. Proc. 10th Australian Coastal and Ocean Eng. Conf., 453–457

Wolanski, E., Mazda, Y., Riedel, P. (1992): Mangrove hydrodynamics. In: A.I. Robertson, D.M. Alongi (eds.), Tropical Mangrove Ecosystems. Amer. Geoph. Union, Washington, 43–62

Wolanski, E., Delesalle, B., Dufour, V., Aubanel, A. (1993): Modelling the fate of pollutants in the Tiahura Lagoon, Moorea, French Polynesia. Proc. 11th Australasian Coastal and Ocean Eng. Conf. **2**, 583–588

Wolanski, E., King, B., Galloway, D. (1995): Dynamics of the turbidity maximum in the Fly River Estuary, Papua New Guinea. Estuarine, Coastal and Shelf Scien. **40**, 321–337

Wolanski, E., Gibbs, R.J. (1995): Flocculation of suspended sediment in the Fly River Estuary, Papua New Guinea. Jour. Coastal Res. **11**, 754-762

Wu, T.Y. (1971): Hydromechanics of swimming fishes and cetaceans. Adv. in Appl. Math. **11**, 1-63

Yang, S.L. (1998): The role of *Scirpus* marh in attenuation of hydrodynamics and retention of fine sediment in the Yangtze Estuary. Estuarine, Coastal and Shelf Science **47**, 227-233

Yeh, H., Titov, V., Gusiakov, V., Pelinovsky, E., Khramushin, V., Kaistrenko, V. (1995): The 1994 Shikotau Earthquake Tsunamis. Pure and Applied Geophysics **144**: 855-874

Yin, K., Harrison, P.J., Ponds, S., Beamish, R.J. (1995a): Entrainment of nitrate in the Fraser River estuary and its biological implications. (Effects of the salt wedge, Part I) Estuarine, Coastal and Shelf Science **40**, 505-528

Yin, K., Harrison, P.J., Ponds, S., Beamish, R.J. (1995b): Entrainment of nitrate in the Fraser River estuary and its biological implications. (Effects of spring vs. neap tides and river discharge, Part II) Estuarine, Coastal and Shelf Science **40**, 529-544

Yin, K., Harrison, P.J., Ponds, S., Beamish, R.J. (1995c): Entrainment of nitrate in the Fraser River estuary and its biological implications. (Effects of winds, Part III) Estuarine, Coastal and Shelf Science **40**, 545-558

Young, I.R. (1994): Global ocean wave statistics obtained from satellite observations. Applied Ocean Res. **16**, 235-248

Young, I.R. (1998): An intercomparison of GEOSAT, TOPEX and ERSI measurements of wind speed and wave height. Ocean Eng. **26**, 67-81

Young, I.R., Holland, G.J. (1996): Atlas of the Oceans: Winds and Wave Climate. Pergamon, Oxford, 241 pp

Zabusky, M.J., Kruskal, M.D. (1965): Interaction of 'solitons' in a collisionless plasma and the recurrence of initial states. Phys. Rev. Letters **15**, 240-243

Zebiak, S.E., Cane, M.A. (1987): A model El Niño – Southern Oscillation. Monthly Weather Review **115**, 2262-2278

Zhung, W.Y., Shebel, J. (1991): Influences of seagrass on tidal flat sedimentation. Acta Oceanographica Sinica **13**, 230-239

Zwarts, C.M.G. (1974): Transmission line wave height transducer. Proc. Inter. Symp. Ocean Wave Measurement and Analysis, New Orleans, **1**, 605-620

Author Index

Subject Index

Colour Plates

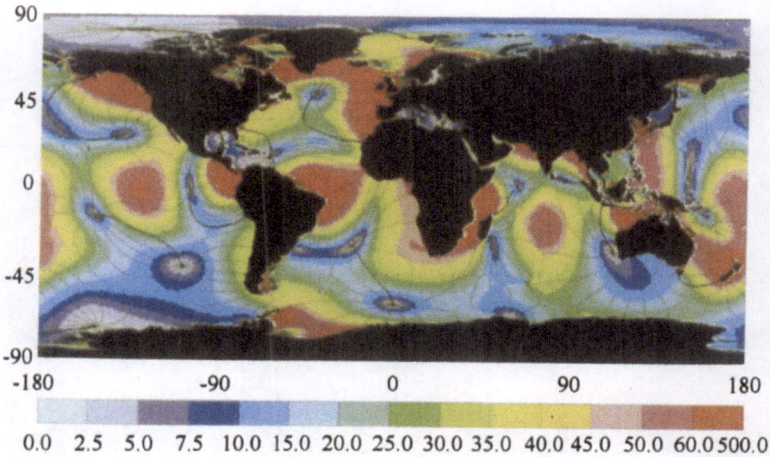

Fig. 5.13: The amplitudes (cm) and co-tidal phases (degrees) of the lunar tide constituent, M_2, resulting from the FES 95.2 model; the colour code denotes the range of amplitudes and co-tidal lines are drawn with a 30° interval; 0° phase is marked with the thicker line; amplitudes and phases in the 0.5° x 0.5° grid; in the model, 26 main tidal constituents have been included, with some of them corrected by assimilation (adapted from Le Provost *et al.*, 1998a)

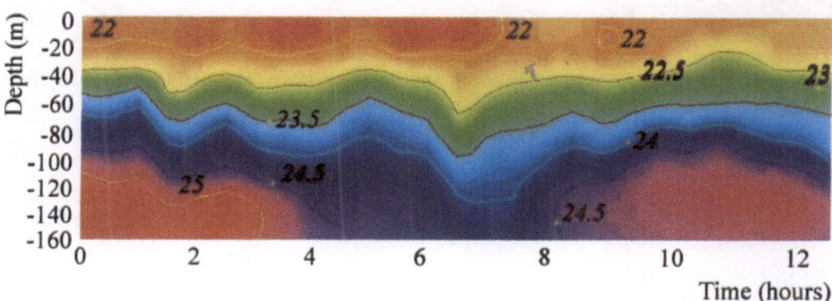

Fig. 6.1: Time series of density in σ_t units at shelf break on the Australian north-west shelf (adapted from Burrage *et al.*, 1996)

Fig. 7.5: Tropical cyclone Gladys photographed from Apollo 7 spacecraft on October 13, 1968 (photo courtesy of Steinberg, 1998)

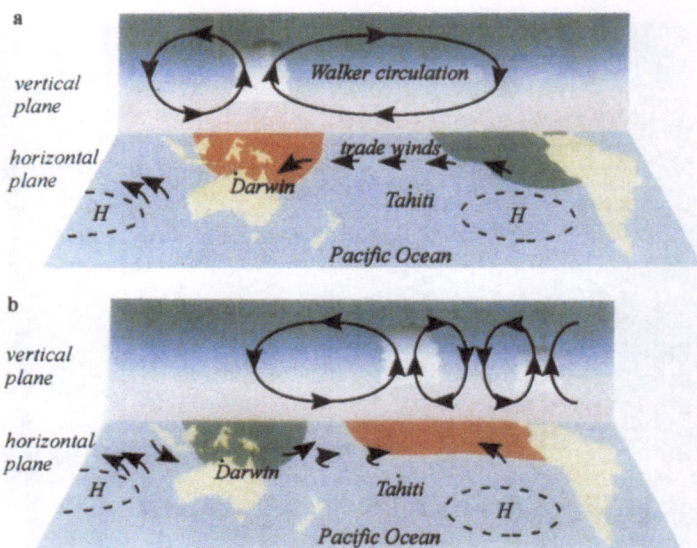

Fig. 7.10: Features of the Walker Circulation across the Pacific Ocean: **a** typical circular pattern, **b** circulation during an El Nino; *H* denotes typical southern summer position of high pressure systems, red colour denotes warmer seas and green denotes cooler seas (adapted from Young and Holland, 1996)

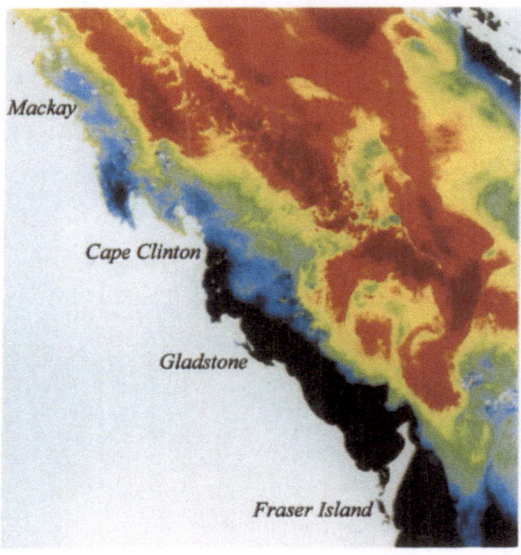

Fig. 7.13: Mesoscale eddies of the East Australian Current seen from the NOAA9 satellite on September 19, 1988 (image courtesy of Skirving, 1998)

Fig. 14.1: Mangrove forest at Nadara River on Iriomote Island, Japan (photo courtesy of Furukawa, 1998)